# Springer Optimization and Its Applications

## Volume 139



*Aims and Scope*
Optimization has been expanding in all directions at an astonishing rate during the last few decades. New algorithmic and theoretical techniques have been developed, the diffusion into other disciplines has proceeded at a rapid pace, and our knowledge of all aspects of the field has grown even more profound. At the same time, one of the most striking trends in optimization is the constantly increasing emphasis on the interdisciplinary nature of the field. Optimization has been a basic tool in all areas of applied mathematics, engineering, medicine, economics and other sciences.

The series *Springer Optimization and Its Applications* publishes undergraduate and graduate textbooks, monographs and state-of-the-art expository works that focus on algorithms for solving optimization problems and also study applications involving such problems. Some of the topics covered include nonlinear optimization (convex and nonconvex), network flow problems, stochastic optimization, optimal control, discrete optimization, multi-objective programming, description of software packages, approximation techniques and heuristic approaches.

More information about this series at http://www.springer.com/series/7393

# Springer Optimization and Its Applications

Boris Goldengorin
Editor

# Optimization Problems in Graph Theory

## In Honor of Gregory Z. Gutin's 60th Birthday

*Editor*
Boris Goldengorin
Department of Information Systems
and Decision Science
Merrick School of Business
University of Baltimore
Baltimore, MD, USA

ISSN 1931-6828 ISSN 1931-6836 (electronic)
Springer Optimization and Its Applications
ISBN 978-3-030-06921-6 ISBN 978-3-319-94830-0 (eBook)
https://doi.org/10.1007/978-3-319-94830-0



This Springer imprint is published by the registered company Springer Nature Switzerland AG
The registered company address is: Gewerbestrasse 11, 6330 Cham, Switzerland

*This book is dedicated to Professor Gregory Z. Gutin on the occasion of his 60th birthday.*

# Preface

This book is a collection of papers related to the conference on 7 and 8 January 2017 organised by Simon Blackburn, Jason Crampton and Stefanie Gerke at Royal Holloway University of London on the occasion of Professor Gregory Z. Gutin's 60th birthday. The invited speakers Noga Alon, Jørgen Bang-Jensen, Fedor Fomin, Mark Jones, Daniel Karapetyan, Eunjung Kim, Michael Krivelevich, Igor Razgon, Saket Saurabh, Benny Sudakov, Stefan Szeider and Anders Yeo have contributed to the success of this conference with the following:

**Programme**

**Saturday 7th January 2017, Horton Lecture Theatre 1**

10.20–10.50 Coffee/tea break (McCrea 219)
10.50–11.00 Welcome
11.00–11.30 Noga Alon, Tel Aviv University, Israel. Universal tournaments.

**Abstract** Tournaments have been the subject of much of the work of Gutin. In the spirit of his work, I will discuss the problem, raised by Moon in the 1960s, of estimating the minimum possible number of vertices in a tournament that contains every $k$-vertex tournament as an induced subgraph. The main result is that this number is $(1 + o(1))2^{(k-1)/2}$, improving earlier estimates of several researchers. The proof combines combinatorial and probabilistic techniques with group theoretic tools.

11.35–12.05 Jørgen Bang-Jensen, University of Southern Denmark, Denmark. 2-partitions of digraphs.

**Abstract** We report on recent results with various co-authors on the complexity of deciding whether a given digraph $D$ has a vertex partition into two digraphs $D'$ and $D''$ such that these have prespecified properties. These could be minimum out-degree at least $k$, being strongly connected, acylic and lots of others. Among the results on which we report is a complete classification of the complexity for 120 natural such problems. This part is joint work with Havet and Cohen.

12.10–12.40 Anders Yeo, Singapore University of Technology and Design. Perfect forests in digraphs

**Abstract** A spanning subgraph F of a graph G is called perfect if F is a forest, the degree of each vertex in F is odd, and each tree of F is an induced subgraph of G. Alex Scott (Graphs and Combinatorics, 2001) proved that every connected graph G contains a perfect forest if and only if G has an even number of vertices. We consider four generalisations to directed graphs of the concept of a perfect forest. While the problem of existence of the most straightforward one is NP-hard, for the three others, this problem is polynomial time solvable. Moreover, every digraph with only one strong component contains a directed forest of each of these three generalisation types. One of our results extends Scott's theorem to digraphs in a nontrivial way.

12.40–14.30 Lunch break (McCrea 219)

14.30–15.00 Michael Krivelevich, Tel Aviv University, Israel. Long cycles in locally expanding graphs, with applications

**Abstract** We provide sufficient conditions for the existence of long cycles in locally expanding graphs. Time permitting, we will present applications of our conditions and techniques to Ramsey theory, random graphs and positional games.

15.05–15.35 Eunjung Kim, Paris Dauphine University. A polynomial kernel for distance-hereditary vertex deletion

**Abstract** A graph is distance-hereditary if, for any pair of vertices, their distance in every connected induced subgraph containing both vertices is the same as their distance in the original graph. Distance hereditary graphs are exactly the graphs with rank-width at most 1. The Distance-Hereditary Vertex Deletion problem asks, given a graph $G$ on $n$ vertices and an integer $k$, whether there is a set $S$ of at most $k$ vertices in $G$ such that $G^S$ is distance-hereditary. It was shown by Eiben, Ganian and Kwon (MFCS'16) that Distance-Hereditary Vertex Deletion can be solved in time $2^{O(k)}n^{O(1)}$, and they asked whether the problem admits a polynomial kernelisation. We show that this problem admits a polynomial kernel, answering this question positively. For this, we use a similar idea for obtaining an approximate solution for Chordal Vertex Deletion due to Jansen and Pilipczuk (SODA'17) to obtain an approximate solution with $O(k^4)$ vertices when the problem is a Yes-instance and use Mader's $S$-path theorem to hit all obstructions containing exactly one vertex of the approximate solution. Then, we exploit the structure of split decompositions of distance-hereditary graphs to reduce the total size. Using Mader's $S$-path theorem in the context of kernelisation might be of independent interest.

15.35–16.10 Coffee/tea break (McCrea 219)

16.10–16.40 Benny Sudakov. Swiss Federal Institute of Technology, Switzerland. Rainbow cycles and trees in properly edge-colored complete graphs

**Abstract** A rainbow subgraph of a properly edge-colored complete graph is a subgraph all of whose edges have different colors. One reason to study such subgraphs arises from the canonical version of Ramsey's theorem, proved by Erdõs and Rado. Another motivation comes from problems in design theory. In this talk, we discuss several old conjectures about finding spanning rainbow cycles and trees in properly edge-colored complete graphs and present some recent progress on these problems. Joint work with A. Pokrovskiy and in part with N. Alon.

16.45–17.15 Mark Jones, Royal Holloway, University of London. Enforcing information flow policies through chain and tree-based enforcement schemes

**Abstract** In an information flow policy, users have different access permissions based on their position in a hierarchy or partial order. In most enforcement schemes that use symmetric cryptographic primitives, each user is assigned a single secret and derives decryption keys using this secret and publicly available information. Recent work has challenged this approach by developing schemes, based on chain or tree partitions of the information flow policy, that do not require public information for key derivation, the trade-off being that a user may need to be assigned more than one secret. In this talk, we show how to construct chain and tree-based cryptographic enforcement schemes and give polynomial-time algorithms to find such enforcement schemes using the minimum number of secrets.

18.30 Dinner (Large Board Room, Founders) Only if you have reserved a seat!

**Sunday 8th January 2018, McCrea 219**

9.30–10.00 Fedor Fomin, University of Bergen, Norway. Finding detours is fixed-parameter tractable

**Abstract** We consider the following natural "above guarantee" parameterisation of the classical longest path problem: For given vertices $s$ and $t$ of a graph $G$ and an integer $k$, the problem longest detour asks for an $(s, t)$-path in $G$ that is at least $k$ longer than a shortest $(s, t)$-path. Using insights into structural graph theory, we prove that Longest Detour is fixed-parameter tractable on undirected graphs and actually even admits a single-exponential algorithm, that is, one of running time $\exp(O(k)) poly(n)$. This matches (up to the base of the exponential) the best algorithms for finding a path of length at least $k$. Joint work with Ivona Bezáková, Radu Curticapean and Holger Dell.

10.05–10.35 Stefan Szeider, Vienna University of Technology, Austria. Backdoors for constraint satisfaction

**Abstract** We will review some recent parameterised complexity results for the constraint satisfaction problem (CSP), considering parameters that arise from strong backdoor sets. A strong backdoor set of a CSP instance is a set of variables with the property that any instantiation of these variables moves the instance into a polynomial-time tractable class. We will focus on tractable classes defined by restricting the involved constraint relations. Joint work with R. Ganian, S. Gaspers, N. Misra, S. Ordyniak, M.S. Ramanujan and S. Živný.

10.35–11.00 Coffee/tea break (McCrea 237)

11.00–11.30 Saket Saurabh, University of Bergen, Norway. Gregory: The "tree" of knowledge

**Abstract** In this talk, I will document my association with Gregory via algorithms for finding trees with certain properties. This will include some old and some more modern developments in the area.

11.35–12.05 Igor Razgon, Birkbeck University of London, United Kingdom. Well quasi-orderability vs. clique-width

**Abstract** Well quasi-orderability is an important topic of structural graph theory. The famous result of Robertson and Seymour showing that the class of all graphs is well-quasi-ordered (WQO) by the graph minors relation inspired researchers to consider other order relations on graphs. One such relation is "induced subgraph". This relation is easy to show to be non-WQO; however, many hereditary graph classes are WQO. Up to some moment, all known WQO classes were of bounded clique-width, and this led researchers to a question whether this situation is true in general (i.e. whether a class of graphs that is WQO under the induced subgraph relation is of bounded clique-width). A wide belief was that it is indeed the case. V. Lozin, V. Zamaraev, and myself have demonstrated the first counterexample: a hereditary class of unbounded clique-width which is WQO by the induced subgraph relation. A preliminary version of our result appeared in WG15. In this talk, I will overview the result and state several interesting open problems.

12.10–12.40 Daniel Karapetyan, University of Essex, United Kingdom. Practically efficient algorithms for the workflow satisfiability problem and its optimisation version.

**Abstract** We consider an interesting satisfiability problem finding applications in access control. The problem is known to be fixed-parameter tractable (FPT), but existing algorithms are relatively inefficient in practice. Our new algorithm more than doubled the value of the parameter that could be practically tackled. The key idea of this new approach is also incorporated into a pseudo-Boolean formulation, with promising results. In the second part of the talk, we discuss single- and bi-objective optimisation extensions of the problem. While providing much greater modelling power, these extensions are still FPT, and our algorithms need only small modification to be used for them. Some conclusions of this research may be applicable to other FPT problems.

This book's focus is on the recent research in modern optimisation problems on graphs and their computational complexities. Researchers, students and engineers will benefit from the original contributions and overviews included in this book. The book is of great interest to researchers in algorithmical graph theory and its applications to max-clique and stable set problems, computing the line index of balance in general graphs, branching in digraphs with many and few leaves, dominance certificates for combinatorial optimisation problems, improved upper bounds for 12 computationally difficult KG instances for the simple plant location problem, an algorithmic answer to the Ore-type version of Dirac's question on disjoint cycles formulated in 1963, efficient heuristics for solving optimal patrol problem on graph against random and strategic attackers, branch-and-cut-type algorithm for the network design problem with cut constraints, heuristic algorithm for the sequencing problem in distributed manufacturing planning process as well as sharp Nordhaus-Gaddum-type lower bounds for proper connection numbers of graphs.

The book presents open problems in graph theory including applied optimisation problems on graphs and networks which have many applications in markets and data analysis, design of efficient algorithms and software for solving optimisation problems in industrial and systems engineering. Undergraduate, graduate and PhD students as well as theoreticians in computer science, big data analysis, applied mathematics, operations research, design of algorithms, artificial intelligence and software engineering will benefit from the state-of-the-art results in modern graph theory and its applications presented in this book.

Baltimore, MD, USA
October 2018

Boris Goldengorin

# Acknowledgements

I would like to acknowledge Simon Blackburn, Jason Crampton and Stefanie Gerke who have organised a great conference and supported my idea to publish this book. I am thankful to the reviewers for their comprehensive feedback on every submitted paper and their timely replies. They fundamentally improved the quality of submitted contributions and hence of this volume.

The project of this book was supported by Panos M. Pardalos. His careful editing contributed enormously to the production of this book.

Technical assistance with reformatting and compilations of several versions of this book by Arkopaul Sarkar (PhD candidate in the Industrial and Systems Engineering Department, Ohio University, Athens, OH, USA) is greatly appreciated. I would like to thank all my colleagues from the Department of Industrial and Sys-tems Engineering, the Russ College of Engineering and Technology, and Ohio University, Athens, OH, USA, especially the chair and Russ professor, Robert Judd for providing me with unlimited freedom in research activities and creative atmosphere to work within C. Paul Stocker visiting professor position.

The research and travel grant supported by the Chair of Department of Information Systems and Decision Science, Danielle Fowler, and granted by the Dean of Merrick School of Business, Murray Dalziel, University of Baltimore Maryland, USA is greatly appreciated.

The support of my wife, Ljana, and children, Mark, Vitaliy, Nicolai and Polina, stimulated to complete this book with the highest quality.

# ACKNOWLEDGMENTS

# Contents

# Dr. Gregory Gutin – Short Bio

Gregory Gutin received his MSc in mathematics in 1979 from Gomel State University, Belarus. He worked in high school and research institutes of Belarus from 1979 to 1990. He studied for PhD under Professor Noga Alon at the School of Mathematics, Tel Aviv University, Israel, and received his PhD (with distinction) in 1993. Between 1993 and 1996, he held visiting positions in the Department of Mathematics and Computer Science, Odense University, Denmark, and then became a lecturer in the Department of Mathematics, Brunel University, United Kingdom. Since 2000, Gregory has been professor of computer science, Department of Computer Science, Royal Holloway, University of London.

Gregory is recepient of the following prestigious scientific awards. In 1992, he received the Wolf Prize for PhD students scholarship for excellence. This is the most prestigious Israeli prize awarded to PhD students. The Wolf Foundation also grants the international Wolf Prize to senior researchers. The Wolf Foundation Prize Committee elects the candidates according to high-standard excellence criteria, regardless to the institution where the research is conducted.

In 1996, he received the Kirkman Medal of International Institute of Combinatorics and Applications to recognise outstanding achievements by members who are within 4 years past their PhD.

In 2014, he received Royal Society Wolfson Research Merit Award and the Best Paper Awards at SACMAT 2015 and 2016. In 2017, he became a member of Academia Europaea.

Currently, Gregory Gutin's h-index is 38 with 8400 citations including 3542 citations since 2013.

Professor Gutin's main research interests include graphs and combinatorics (theory, algorithms and applications), parameterised algorithmics and combinatorial optimisation. Dr. Gutin has more than 200 papers published or accepted for publication in refereed journals and conference proceedings. He published nine chapters/sections in books and two editions of the following monograph: J. Bang-Jensen and G. Gutin, *Digraphs: Theory, Algorithms and Applications*,

Springer-Verlag 2000 (1st Ed.) and 2009 (2nd Ed.). The 1st Edition was published in Chinese in 2009. He co-edited (with A.P. Punnen) the book *Traveling Salesman Problem and Its Variations*, Springer 2002. He was or is on the editorial board of the following journals: *Discrete Optimization, Order, Algorithmic Operations Research, Memetic Computing*.

# Gregory Gutin and Graph Optimization Problems

**Noga Alon**

This is a very brief contribution to the book "Optimization Problems In Graph Theory" devoted to the 60th birthday of Gregory (Zvi) Gutin, whom I first met when he came as a graduate student to Tel Aviv University nearly 30 years ago. I always knew that I learned far more from my graduate students than they have learned from me, and this has certainly been the case with Gregory.

When he came to Tel Aviv University he already had several beautiful results, most notably his 1986 paper showing that in any multipartite tournament containing at most one vertex of indegree 0, there is a vertex $v$ so that for every $u$ there is a directed path of length at most 4 from $v$ to $u$.

His work in Combinatorial Optimization and his results in Parameterized Complexity are of top quality, contributing to the topics discussed in the present book which deals with optimization problems on graphs and their computational complexity. The topics studied in this book represent well the interests of Gutin. The interesting papers here include a discussion of the algorithmic aspects of problems dealing with disjoint cycles in graphs, with the maximum independent set problem in some special classes of graphs, and with the existence of branchings with many and few leaves in directed graphs. There is also a paper providing a general technique for proving that a heuristic algorithm for an optimization problem provides a solution which is not worse than many others, an investigation of a network design problem with cut constraints, and a discussion of algorithms for computing the frustration index of a signed graph. Other results provide a heuristic approach for studying a patrol problem on a graph and an investigation of the

---

N. Alon (✉)
Sackler School of Mathematics and Blavatnik School of Computer Science, Tel Aviv University, Tel Aviv, Israel
e-mail: nogaa@tau.ac.il



B. Goldengorin (ed.), *Optimization Problems in Graph Theory*,
Springer Optimization and Its Applications 139,
https://doi.org/10.1007/978-3-319-94830-0_1

minimum possible sum and product of the proper connection number of a graph of a given order and that of its complement.

A subject considered often by Gutin is Hamilton cycles and the Traveling Salesman Problem. I thus chose to include here a simple yet intriguing result on this topic. This is a solution of a problem I heard from Amin Rezaie that compares two heuristics to the (path version of) the Traveling Salesman Problem. The first seems natural, the second seems terrible, yet it requires some thought to show that the first indeed always gives a result which is at least as good as that provided by the second. I believe that the reasoning in the proof fits this book and the way of thinking of Gutin.

The brief technical part follows.

Let $K_n$ be a complete, undirected graph with edges of non-negative weights. Starting from a vertex $v$ construct a greedy Hamilton path $P$ by always selecting the cheapest edge leading to a yet uncovered vertex. Starting from a vertex $u$ (which may be equal to $v$ or not) construct an anti-greedy Hamilton path $Q$ by always selecting the most expensive edge leading to a yet uncovered vertex. Is it necessarily true that the total weight of $P$ is at most the total weight of $Q$?

We prove that this is indeed the case. Without loss of generality we may assume that all weights are positive reals. For a set of edges $F$ and a real $x$ let $f(x, F)$ denote the number of edges in $F$ whose weight is at least $x$. It is easy to see that the total weight of the edges in $F$ is exactly $\int_{x=0}^{\infty} f(x, F)dx$. Thus, in order to prove the desired result it suffices to show that for any fixed $x$, $f(x, P) \leq f(x, Q)$. We proceed with a proof of that. Let $P(x)$ denote the set of all edges of $P$ whose weight is at least $x$. Then $P(x)$ is a union of connected components, and each of them is a path. Let $C = c_1c_2 \ldots c_k$ be such a path that forms a component with at least one edge, thus $|C| = k \geq 2$. Assume that $C$ was selected greedily in this order, from $c_1$ to $c_k$. Note that the weight of every edge $c_ic_j$ with $1 \leq i < j \leq k$ is at least $x$. Indeed, this weight is at least the weight of $c_ic_{i+1}$, as $P$ was chosen greedily. Consider now the path $Q$ chosen anti-greedily. As $Q$ is a Hamilton path all the vertices of $C$ appear in $Q$ in some order. We claim that for every $i$, $1 \leq i < k$, the edge of $Q$ emanating from the vertex of $C$ that appears in place number $i$ in this order is of weight at least $x$. Indeed, when we chose this edge anti-greedily, there was still at least one yet uncovered vertex of $C$ and by the argument above the weight of the edge leading to it is at least $x$. Thus the weight of the edge chosen was indeed at least $x$. This means that among the edges emanating from vertices of $C$ at least $k - 1$ have weight at least $x$, and as $C$ has $k - 1$ edges, summing over all components implies that indeed $f(x, Q) \geq f(x, P)$, completing the proof. □

**Note Added in Proof** After the completion of this note I have heard from Gutin that the problem discussed above has been solved much earlier by V.M. Kirzhner and V.I. Rublinetskii in their paper: On the procedure "go to the nearest" of the Traveling Salesman Problem, Vychislitel'naya Matematika i Vychislitel'naya Technika, no. 4 (1973), 40–41 (in Russian). Their proof is different than the one presented here. This is yet another demonstration of the fact mentioned in the first paragraph of this note that I learned (and am still learning) more from Gutin than he has learned from me!

# On Graphs Whose Maximal Cliques and Stable Sets Intersect

**Diogo V. Andrade, Endre Boros, and Vladimir Gurvich**

## 1 Introduction

### *1.1 CIS-Graphs and Simplicial Vertices*

Given a graph $G$, we say that it has the *CIS-property*, or equivalently that $G$ is a *CIS-graph*, if every maximal clique $C$ and every maximal stable set $S$ in $G$ intersects. Obviously, they may have at most one common vertex and hence $|C \cap S| = 1$. It is convenient to represent a CIS-graph $G$ as a *2-dimensional box partition*, that is, a matrix whose rows and columns are labeled, respectively, by the maximal cliques and stable sets of $G$ and whose entries are the (unique) vertices of the corresponding intersections. For example, Figure 1 shows four CIS-graphs and their intersection matrices. More examples are given in Figures 6, 7, and 10.

The CIS-property appears in the survey [8] (under the name *clique-kernel intersection property*) but no related results are mentioned. Indeed, natural problems of efficient characterization and recognition of the CIS-graphs look difficult and remain open. Perhaps, one of the reasons is that the CIS-property is not hereditary. Indeed, if $C \cap S = \{v\}$ then $C \setminus \{v\}$, and $S \setminus \{v\}$ may become disjoint maximal clique and stable set after $v$ is deleted.

D. V. Andrade
Google Inc NYC, New York, NY, USA

E. Boros
RUTCOR, Rutgers University, Piscataway, NJ, USA
e-mail: endre.boros@rutgers.edu

V. Gurvich (✉)
National Research University Higher School of Economics (HSE), Moscow, Russia
e-mail: vgurvich@hse.ru


B. Goldengorin (ed.), *Optimization Problems in Graph Theory*,
Springer Optimization and Its Applications 139,
https://doi.org/10.1007/978-3-319-94830-0_2

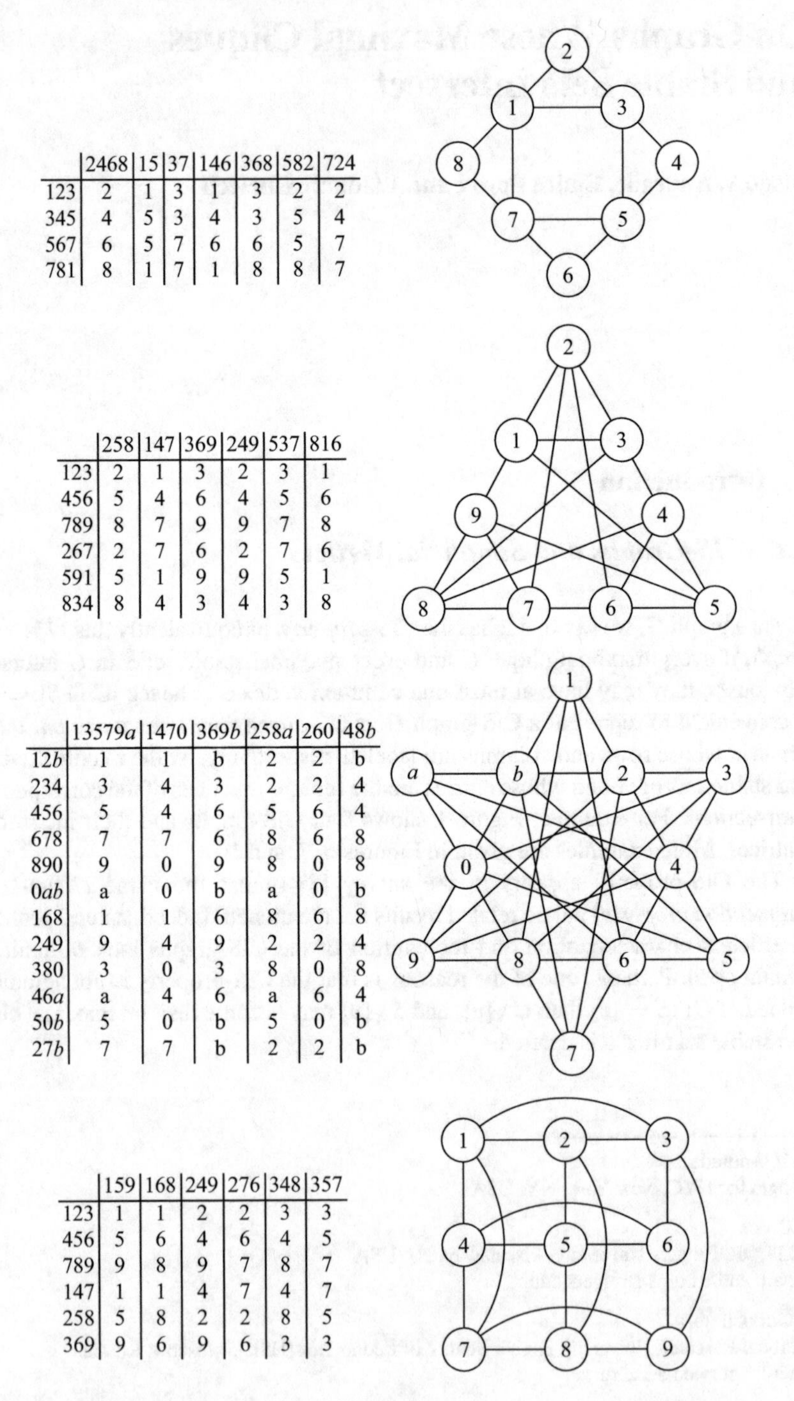

|  | 2468 | 15 | 37 | 146 | 368 | 582 | 724 |
|---|---|---|---|---|---|---|---|
| 123 | 2 | 1 | 3 | 1 | 3 | 2 | 2 |
| 345 | 4 | 5 | 3 | 4 | 3 | 5 | 4 |
| 567 | 6 | 5 | 7 | 6 | 6 | 5 | 7 |
| 781 | 8 | 1 | 7 | 1 | 8 | 8 | 7 |

|  | 258 | 147 | 369 | 249 | 537 | 816 |
|---|---|---|---|---|---|---|
| 123 | 2 | 1 | 3 | 2 | 3 | 1 |
| 456 | 5 | 4 | 6 | 4 | 5 | 6 |
| 789 | 8 | 7 | 9 | 9 | 7 | 8 |
| 267 | 2 | 7 | 6 | 2 | 7 | 6 |
| 591 | 5 | 1 | 9 | 9 | 5 | 1 |
| 834 | 8 | 4 | 3 | 4 | 3 | 8 |

|  | 13579*a* | 1470 | 369*b* | 258*a* | 260 | 48*b* |
|---|---|---|---|---|---|---|
| 12*b* | 1 | 1 | b | 2 | 2 | b |
| 234 | 3 | 4 | 3 | 2 | 2 | 4 |
| 456 | 5 | 4 | 6 | 5 | 6 | 4 |
| 678 | 7 | 7 | 6 | 8 | 6 | 8 |
| 890 | 9 | 0 | 9 | 8 | 0 | 8 |
| 0*ab* | a | 0 | b | a | 0 | b |
| 168 | 1 | 1 | 6 | 8 | 6 | 8 |
| 249 | 9 | 4 | 9 | 2 | 2 | 4 |
| 380 | 3 | 0 | 3 | 8 | 0 | 8 |
| 46*a* | a | 4 | 6 | a | 6 | 4 |
| 50*b* | 5 | 0 | b | 5 | 0 | b |
| 27*b* | 7 | 7 | b | 2 | 2 | b |

|  | 159 | 168 | 249 | 276 | 348 | 357 |
|---|---|---|---|---|---|---|
| 123 | 1 | 1 | 2 | 2 | 3 | 3 |
| 456 | 5 | 6 | 4 | 6 | 4 | 5 |
| 789 | 9 | 8 | 9 | 7 | 8 | 7 |
| 147 | 1 | 1 | 4 | 7 | 4 | 7 |
| 258 | 5 | 8 | 2 | 2 | 8 | 5 |
| 369 | 9 | 6 | 9 | 6 | 3 | 3 |

**Fig. 1** Four CIS-graphs and their intersection matrices

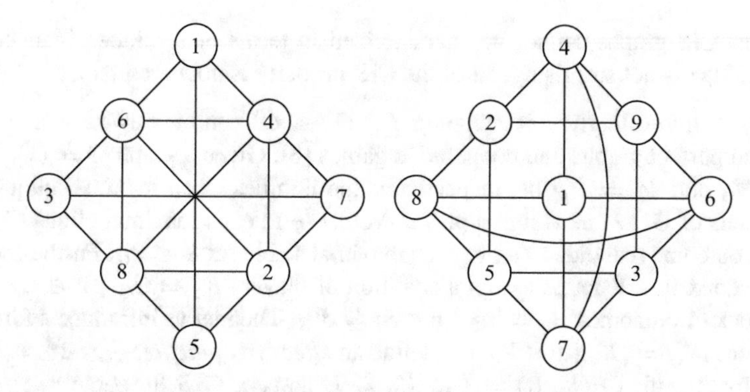


**Fig. 2** Complements to the first two graphs in the previous figure (Obviously, for every graph $G$ the intersection matrix of $\overline{G}$ is the transpose of the intersection matrix of $G$.)

On the positive side, by definition, the CIS-property is self-complementary, that is, $G$ is a CIS-graph if and only if the complementary graph $\overline{G}$ is a CIS-graph.

We start with a simple sufficient condition. Given a graph $G = (V, E)$, a vertex $v \in V$ is called *simplicial* if its neighborhood $N[v]$ is a clique.

Clearly, if a maximal clique $C$ of $G$ contains a simplicial vertex $v$, then it is a *private* vertex of $C$, that is, $v$ cannot belong to any other maximal clique, except $C$. Vice versa, every private vertex $v$ of a maximal clique $C$ is simplicial, since in this case $N[v] = C$.

Moreover, in this case $C \cap S \neq \emptyset$ for every maximal stable set $S$ in $G$. Indeed, if $S \cap (C \setminus \{v\}) = \emptyset$, then $v \in S$, since $S$ is maximal. Thus, we obtain the following statement.

**Proposition 1** *Graph $G$ is a CIS-graph whenever*
*(s) every maximal clique of $G$ has a simplicial vertex.* □

Let us remark that condition **(s)**: is only sufficient but not necessary. For example, **(s)** holds for the first graph in Figure 1 but not for the other three graphs. Let us also remark that **(s)** does not hold for both graphs in Figure 2. Furthermore, **(s)** holds for the graphs of Figures 6, 7, and 10 and **(s)** does not hold for the graphs of Figures 4, 5, and 9, because they are not CIS-graphs.

By Proposition 1, given an arbitrary graph $G$, we can get a CIS-graph $G^s$ just adding a simplicial (private) vertex $v_C$ to each maximal clique $C$ of $G$ that does not have one.

Let us remark that we have to add such a vertex to $C$ even when $C \cap S \neq \emptyset$ for each maximal stable set $S$ in $G$, since otherwise $C$ may become disjoint from a new maximal stable set of $G^s$; consider, for example, $G = \overline{C_6}$.

Obviously, the size of $G^s$ may be exponential in the size of $G$.

**Corollary 1** *Any graph $G$ is an induced subgraph of a CIS-graph.*

*Proof* Indeed, for any graph $G$ the CIS-graph $G^s$ contains $G$ as an induced subgraph. □

Thus, CIS-graphs cannot be characterized in terms of forbidden induced subgraphs. This is not surprising, since the CIS-property is not hereditary.

*Remark 1* Interestingly, this mapping $f : G \to G^s$ can be viewed as a "bridge" between perfect graphs and cooperative games [3]. Given a graph $G = (V, E)$, let $\mathscr{C} = \mathscr{C}_G$ and $\mathscr{S} = \mathscr{S}_G$ be, respectively, the families of all maximal cliques and stable sets of $G$. Let us assign a player (voter) $i_C$ to each maximal clique $C \in \mathscr{C}_G$ and an outcome (candidate) $a_S$ to each maximal stable set $S \in \mathscr{S}_G$. Furthermore, to every vertex $v \in V$ let us assign a coalition of players $K_v = \{i_C \mid v \in C\} \subseteq \mathscr{C}_G$ and block of outcomes $B_v = \{a_S \mid v \in S\} \subseteq \mathscr{S}_G$. Then let us introduce a family of coalitions $\mathscr{K}_G = \{K_v \mid v \in V\}$ and define an *effectivity function* $\mathscr{E}_G : 2^{\mathscr{C}} \times 2^{\mathscr{S}} \to \{0, 1\}$ by formula $\mathscr{E}_G(K, B) = 1$ iff $K_v \subseteq K$ and $B_v \subseteq B$ for some $v \in V$. It is proved in [3–5] that the following claims are equivalent:

(i) Graph $G$ is perfect;
(ii) Effectivity function $\mathscr{E}_G$ is stable;
(iii) Family of coalitions $\mathscr{K}_G$ is stable;
(iv) Family of coalitions $\mathscr{K}_{G^s}$ is partitionable.

A family of sets is called *partitionable* if every of its minimal balanced subfamily is a partition. A family of coalitions or an effectivity function is called *stable* if the corresponding *core* is not empty for any *utility function*. We refer to [3–5] for accurate definitions.

## 1.2 Almost CIS-Graphs and Split Graphs

We will call a graph $G = (V, E)$ an *almost CIS-graph* if every (maximal) clique $C$ and stable set $S$ in $G$ intersects, except for a unique pair $(C_0, S_0)$.

By definition, almost CIS-graphs are closed under complementation. However, unlike CIS-graphs, they are not closed under substitution.

Notice that, by definition, the families of CIS- and almost CIS-graphs are disjoint.

Let us recall that $G = (V, E)$ is a *split* graph if $V = C_0 \cup S_0$, where $C_0$ and $S_0$ are (maximal) clique and stable set, respectively. Foldes and Hammer [19] showed that split graphs are exactly $(C_4, \overline{C_4}, C_5)$-free graphs.

It is not difficult to see that every split graph is either a CIS-graph or an almost CIS-graph. More precisely, the following claim holds.

**Proposition 2** *Let $G = (V, E)$ be a split graph in which $C_0$ and $S_0$ are maximal and $V = C_0 \cup S_0$. If $C_0 \cap S_0 \neq \emptyset$, then $G$ is a CIS-graph, otherwise, if $C_0 \cap S_0 = \emptyset$, then $G$ is an almost CIS-graph in which $(C_0, S_0)$ is the only disjoint pair.*

*Proof* Obviously, for each maximal clique $C$ and stable set $S$ in $G$ we have: $C_0 \cap S \neq \emptyset$ unless $S = S_0$ and $C \cap S_0 \neq \emptyset$ unless $C = C_0$. Let us assume that both intersections are non-empty (then, clearly, each of them consists of a single vertex) and denote $C_0 \cap S$ by $v_S$ and $C \cap S_0$ by $v_C$. If $v_C = v_S$, then $C \cap S = \{v_S\} = \{v_S\}$.

Otherwise, if $(v_C, v_S) \in E$, then $C \cap S = \{v_S\}$; if $(v_C, v_S) \notin E$, then $C \cap S = \{v_C\}$. In any case $C \cap S \neq \emptyset$.

Thus, if $C \cap S = \emptyset$, then $C = C_0$, $S = S_0$, and $C_0 \cap S_0 = \emptyset$. □

**Theorem 1** *Every almost CIS-graph is a split graph.*

By [19], it is sufficient to show that almost CIS graphs are $(C_4, \overline{C_4}, C_5)$-free.

This statement of Theorem 1 was conjectured in [7], where some partial results were obtained. Then, the proof was given in [43].

## 1.3 $P_4$-Free CIS-Graphs

We proceed with the following simple observation: every $P_4$-free graph is a CIS-graph; see, e.g., [16, 17, 20, 22, 23, 25, 28, 30, 42]. In fact, a stronger claim holds. We say that a set $T \subseteq V$ is a transversal of the hypergraphs $\mathscr{H} \subseteq 2^V$ if $T \cap H \neq \emptyset$ for all hyperedges $H \in \mathscr{H}$. The family of minimal transversals of $\mathscr{H}$ is denoted by $\mathscr{H}^d$ and is called the *dual* of $\mathscr{H}$. Given a graph $G = (V, E)$ we assign to it two hypergraphs, $\mathscr{C} = \mathscr{C}_G$ the collection of all maximal cliques of $G$, and $\mathscr{S} = \mathscr{S}_G$ the collections of all its maximal stable sets.

**Proposition 3 ([22, 25, 28, 30])** *A graph G has no induced $P_4$ if and only if the hypergraphs $\mathscr{C}$ and $\mathscr{S}$ of all maximal cliques and stable sets of G are dual hypergraphs.* □

Furthermore, $P_4$-free graphs are closely related to *read-once* Boolean functions and 2-person positional games, see for definitions, e.g., [21, 24, 25, 30].

*Remark 2* Read-once Boolean functions can be efficiently characterized, since their co-occurrence graphs are $P_4$-free [6, 16, 17, 22, 23, 25–27, 30]. Moreover, the normal forms of positional 2-person games with perfect information can be characterized by Proposition 3 [23–25]. Such a normal form is exactly the intersection matrix of the maximal cliques and stable sets of the corresponding graph, where the final positions (outcomes) of the game are in one-to-one correspondence with the vertices of this graph. See an example in Figure 3, where the monotone Boolean functions $F_S = 13 \vee 24$ and $F_C = (1 \vee 3)(2 \vee 4)$ corresponding to the hypergraphs $\mathscr{S} = \{(1, 3), (2, 4)\}$ and $\mathscr{C} = \{(1, 2), (2, 3), (3, 4), (4, 1)\}$ are read-once.

However, the absence of induced $P_4$s is only sufficient but not necessary for the CIS-property to hold. Let a graph $G$ contain an induced $P_4$ defined by $(v_1, v_1'), (v_2, v_2'), (v_1, v_2)$. The clique $\{v_1, v_2\}$ and stable set $\{v_1', v_2'\}$ are disjoint. Hence, they cannot be maximal in $G$ if it is a CIS-graph. In other words, $G$ must contain a fifth vertex $v_0$ such that $(v_0, v_1), (v_0, v_2)$ are edges, while $(v_0, v_1'), (v_0, v_2')$ are not. In this case we will say that $P_4$ *is settled* by $v_0$, cf. [2, 37]. Let us note that the graph induced by $\{v_0, v_1, v_2, v_1', v_2'\}$ is a CIS-graph, see Figure 6.

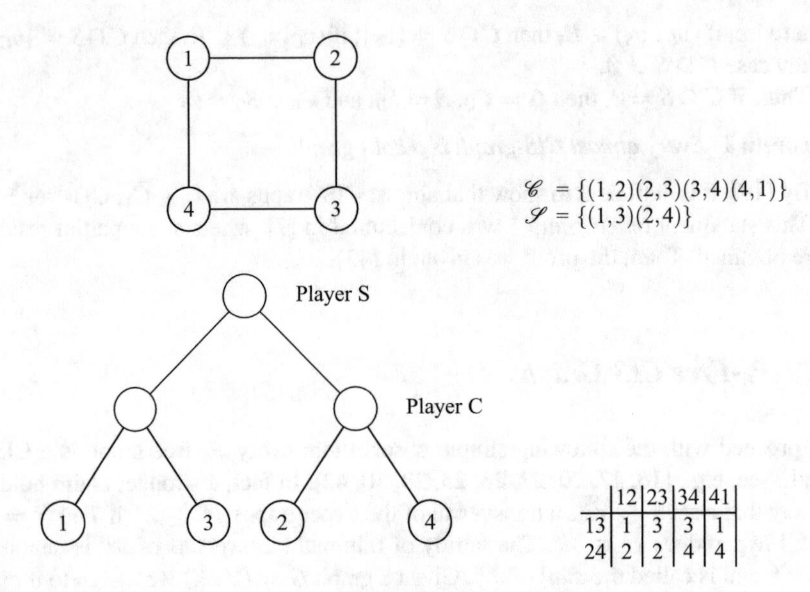


**Fig. 3** A $P_4$-free graph and the corresponding positional and normal game forms

Thus, every induced $P_4$ in a CIS-graph must be settled. This condition is necessary, as we argued above, yet, it is not sufficient, according to the following examples.

## 1.4 Combs and Anti-combs

Given an integer $k \geq 2$, a *comb* (or $k$-comb) $S_k$ is defined as a graph with $2k$ vertices $k$ of which form a clique $C = \{v_1, \ldots, v_k\}$, while the remaining $k$ form a stable set $S = \{v'_1, \ldots, v'_k\}$. In addition, $S_k$ contains the perfect matching $(v_i, v'_i)$ for $i = 1, \ldots, k$, and there are no more edges in $S_k$. Let us note that graphs $S_2$ and $P_4$ are isomorphic. Furthermore, $S_3$ contains 3 induced $S_2$ and all 3 are settled. More generally, $S_k$ contains $k$ induced $S_{k-1}$ and they all are settled. Figure 4 shows $S_k$, for $k = 2, 3$, and 4.

The complementary graph $\overline{S_k}$ is called an *anti-comb* (or $k$-anti-comb). Figure 5 shows $\overline{S_k}$ for $k = 2, 3$, and 4.

Clearly, the sets $S$ and $C$ are switched in $S_k$ and $\overline{S_k}$. It is also clear that combs and anti-combs are not CIS-graphs, since they contain a maximal clique $C$ and stable set $S$ that are disjoint. Hence, if a CIS-graph $G$ contains an induced comb $S_k$ (respectively, anti-comb $\overline{S_k}$), then it must be *settled*, that is, $G$ must contain a vertex $v_0$ adjacent to each vertex of $C$ and to no vertex of $S$. Thus, the following condition is necessary for the CIS-property to hold.

**(COMB)** Every induced comb and anti-comb must be settled in $G$.

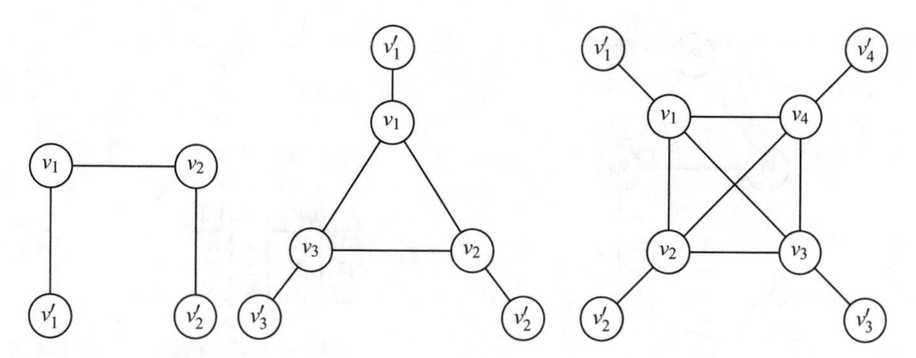


**Fig. 4** Combs $S_k$, for $k = 2$, 3 and 4

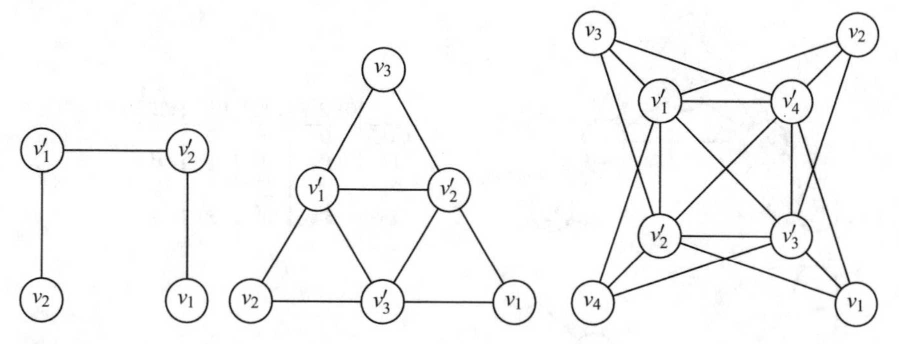


**Fig. 5** Anti-combs $\overline{S_k}$, for $k = 2$, 3 and 4

For $k = 2$ this observation was made by Berge [2] in 1985.

Figures 6 and 7 show settled combs and anti-combs. It is easy to verify that they are complementary CIS-graphs. Hence, the corresponding intersection matrices are mutually transposed.

The following obvious properties of combs and anti-combs are worth summarizing:

- The 2-comb $S_2$ and 2-anti-comb $\overline{S_2}$ are isomorphic, while the $k$-comb $S_k$ and $k$-anti-comb $\overline{S_k}$ are not isomorphic for $k > 2$.
- The $k$-comb $S_k$ contains $\binom{k}{m}$ induced $m$-combs $S_m$ that are all settled in $S_k$, yet, it contains no induced $m$-anti-combs $\overline{S_m}$ for $m > 2$, respectively, the $k$-anti-comb $\overline{S_k}$ contains $\binom{k}{m}$ induced $m$-anti-combs $\overline{S_m}$ that are all settled in $\overline{S_k}$, yet, it contains no induced $m$-combs $S_m$ for $m > 2$.
- The settled $k$-comb and anti-comb are complementary CIS-graphs.

Obviously, **COMB** is a necessary condition for the CIS-property to hold. Yet, it is not sufficient, as we will see in Section 1.5. Let us introduce the following stronger condition.

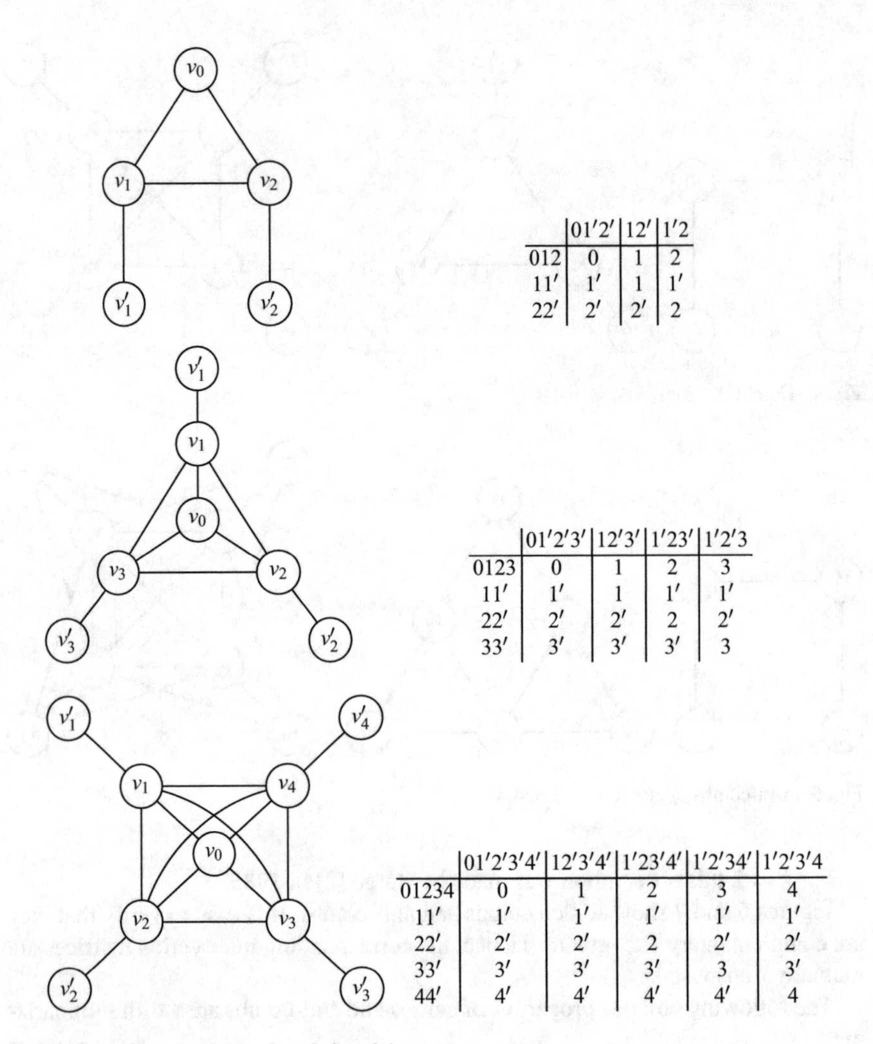


| | 01′2′ | 12′ | 1′2 |
|---|---|---|---|
| 012 | 0 | 1 | 2 |
| 11′ | 1′ | 1 | 1′ |
| 22′ | 2′ | 2′ | 2 |

| | 01′2′3′ | 12′3′ | 1′23′ | 1′2′3 |
|---|---|---|---|---|
| 0123 | 0 | 1 | 2 | 3 |
| 11′ | 1′ | 1 | 1′ | 1′ |
| 22′ | 2′ | 2′ | 2 | 2′ |
| 33′ | 3′ | 3′ | 3′ | 3 |

| | 01′2′3′4′ | 12′3′4′ | 1′23′4′ | 1′2′34′ | 1′2′3′4 |
|---|---|---|---|---|---|
| 01234 | 0 | 1 | 2 | 3 | 4 |
| 11′ | 1′ | 1′ | 1′ | 1′ | 1′ |
| 22′ | 2′ | 2′ | 2 | 2′ | 2′ |
| 33′ | 3′ | 3′ | 3′ | 3 | 3′ |
| 44′ | 4′ | 4′ | 4′ | 4′ | 4 |

**Fig. 6** Settled combs $S_k$, for $k = 2, 3$ and 4

**COMB**(3, 3) There is no induced 3-comb or 3-anti-comb, and every induced 2-comb is settled in $G$. This stronger condition already implies the CIS-property.

**Theorem 2** *A graph G is a CIS-graph whenever it satisfies* ***COMB****(3, 3).*

This was conjectured by Chvatal in early 90s. The first partial results were published by Zang [44]. Then, the statement was proven in [14, 15]. In Section 2 we give an alternative proof, which is of independent interest. It still contains a complicated case analysis in which one of the cases results in a remarkable graph that is "almost" a counterexample to Theorem 2. This graph $2\mathscr{P}$ (see Figure 8) consists of two identical copies of the Petersen graph induced by the vertices $v_0, \ldots, v_9$ and $v'_0, \ldots, v'_9$, respectively. Furthermore, $(v'_i, v_j)$ is an edge if and only

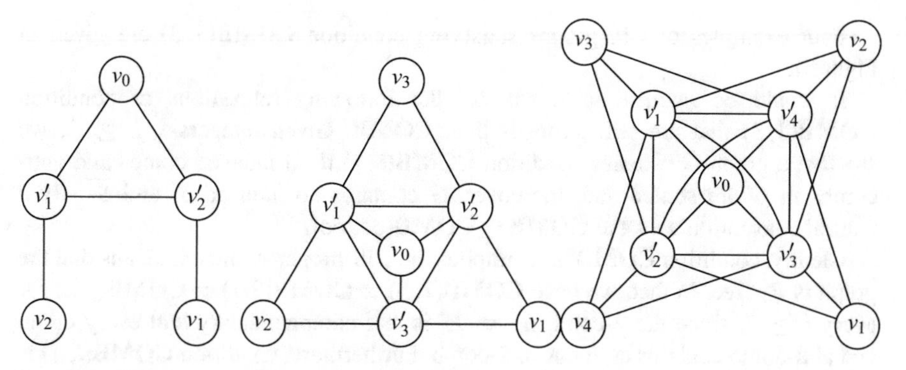


**Fig. 7** Settled anti-combs $\overline{S_k}$, for $k = 2$, 3 and 4

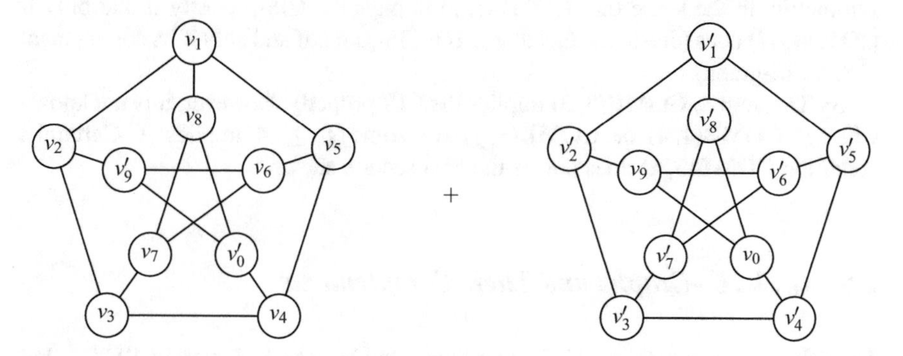


**Fig. 8** Graph $2\mathscr{P}$

if $(v_i, v_j)$ is not, for all $i \neq j$. Ten remaining pairs, $(v_i, v'_i)$, $i = 0, \ldots, 9$, are uncertain, that is, configuration $2\mathscr{P}$ represents in fact $2^{10}$ possible graphs rather than one graph. The following properties of $2\mathscr{P}$ are easy to see.

(a) $2\mathscr{P}$ is isomorphic to its complement.
(b) $2\mathscr{P}$ is regular of "degree 9.5," that is, each vertex is incident to 9 edges and belongs to one uncertain pair.
(c) For every two vertices $u$, $v$ there is an automorphism $\alpha$ of $2\mathscr{P}$ such that $\alpha(u) = v$.
(d) None of the $2^{10}$ graphs of $2\mathscr{P}$ contains an induced 3-comb or 3-anti-comb.
(e) Every induced 2-comb in all $2^{10}$ graphs of $2\mathscr{P}$ involves a pair $v_i, v'_i$ for some $i = 0, \ldots, 9$.

In fact, 36 induced 2-combs appear, whenever we substitute a pair $v_i, v'_i$ by an edge (or by a non-edge). It is easy to see that none of these 2-combs can be settled by a vertex of $2\mathscr{P}$, and if it is settled by a new vertex then an unsettled 3-comb or 3-anti-comb always appears. Thus, the case under consideration does not lead to a counterexample, and a complete case analysis yields the proof of Theorem 2, see Section 2.

Four examples of CIS-graphs satisfying condition **COMB**(3, 3) are given in Figure 1.

It would be interesting to analyze the following relaxations of condition **COMB**(3, 3) that are still stronger than **COMB**. Given integers $i, j \geq 2$, we say that a graph $G$ satisfies condition **COMB**($i, j$) if all induced combs and anti-combs in $G$ are settled and, moreover, $G$ contains no induced $S_i$ and $\overline{S_j}$. By a natural convention we have **COMB** = **COMB**($\infty, \infty$).

Clearly, condition **COMB**(2, 2) implies the CIS-property, since it means that the graph is $P_4$-free. In fact, we have **COMB**(2, 2) $\equiv$ **COMB**(2, $i$) $\equiv$ **COMB**($i$, 2) for every $i \geq 2$, since the 2-comb $S_2 \equiv P_4$ is self-complementary and every comb and anti-comb contains an induced 2-comb. Furthermore, condition **COMB**($i, j$) is monotone in the sense that it implies **COMB**($i', j'$) for all $i \leq i'$ and $j \leq j'$, and symmetric, in the sense that **COMB**($i, j$) implies the CIS-property if and only if **COMB**($j, i$) does (due to the fact that $G$ is a CIS-graph if and only if its complement $\overline{G}$ is a CIS-graph).

By Theorem 2, **COMB**(3, 3) implies the CIS-property. However, it is not known whether **COMB**(4, 4) or **COMB**(3, $j$) for some $j \geq 4$ implies it. Certainly, condition **COMB**(5, 4) does not, as the next section shows.

## 1.5 $(n, k, \ell)$-Graphs and Their Complements

The following graph $G = (V, E)$ was suggested by Ron Holzman in 1994. It has $\binom{5}{1} + \binom{5}{2} = 5 + 10 = 15$ vertices, where subsets $S = \{v_1, \ldots, v_5\}$ and $C = \{v_{12}, \ldots, v_{45}\}$ induce a stable set and clique, respectively; $V = C \cup S$ (hence, $G$ is a split graph); furthermore, every pair $(v_i, v_{ij})$, where $i, j = 1, \ldots, 5$ and $i \neq j$, is an edge, and there are no more edges. Let us denote this graph by $G(5, 1, 2)$, see Figure 9.

It is easy to verify that $G(5, 1, 2)$ contains no induced 5-combs and 4-anti-combs. In Section 3 we will show that all induced combs and anti-combs in $G(5, 1, 2)$ are settled. For example, the 4-comb induced by vertices $(v_{12}, v_{13}, v_{14}, v_{15}, v_2, v_3, v_4, v_5)$ is settled by $v_1$ and the 3-anti-comb induced by $(v_{12}, v_{13}, v_{23}, v_1, v_2, v_3)$ is settled by $v_{45}$, etc. Thus, the graph $G(5, 1, 2)$ satisfies

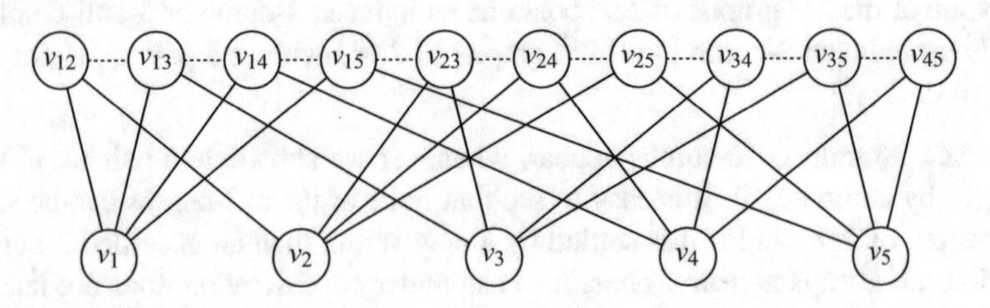


**Fig. 9** Graph $G(5, 1, 2)$ was constructed by Ron Holzman in 1994

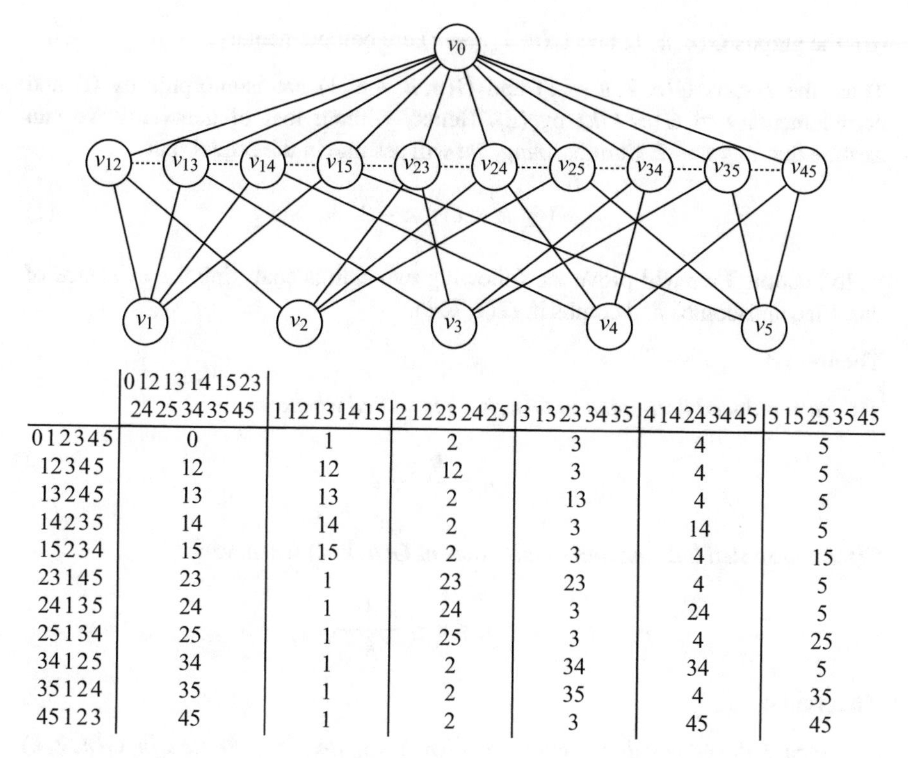


| | 0 12 13 14 15 23 24 25 34 35 45 | 1 12 13 14 15 | 2 12 23 24 25 | 3 13 23 34 35 | 4 14 24 34 45 | 5 15 25 35 45 |
|---|---|---|---|---|---|---|
| 0 1 2 3 4 5 | 0 | 1 | 2 | 3 | 4 | 5 |
| 12 3 4 5 | 12 | 12 | 12 | 3 | 4 | 5 |
| 13 2 4 5 | 13 | 13 | 2 | 13 | 4 | 5 |
| 14 2 3 5 | 14 | 14 | 2 | 3 | 14 | 5 |
| 15 2 3 4 | 15 | 15 | 2 | 3 | 4 | 15 |
| 23 1 4 5 | 23 | 1 | 23 | 23 | 4 | 5 |
| 24 1 3 5 | 24 | 1 | 24 | 3 | 24 | 5 |
| 25 1 3 4 | 25 | 1 | 25 | 3 | 4 | 25 |
| 34 1 2 5 | 34 | 1 | 2 | 34 | 34 | 5 |
| 35 1 2 4 | 35 | 1 | 2 | 35 | 4 | 35 |
| 45 1 2 3 | 45 | 1 | 2 | 3 | 45 | 45 |

**Fig. 10** Settled $G(5, 1, 2)$

condition **COMB**(5, 4), however, it is not a CIS-graph, since $C \cap S = \emptyset$. Let us notice that the settled extension of $G(5, 1, 2)$ is a CIS-graph, see Figure 10.

We generalize the above example as follows. Given integers $n, k, \ell$ such that $n > k \geq 1$ and $n > \ell \geq 1$, consider a set $S$ (respectively, $C$) consisting of $\binom{n}{k}$ (respectively, $\binom{n}{\ell}$) vertices labeled by $k$-subsets (respectively, by $\ell$-subsets) of a ground $n$-set. Let us introduce the graph $G(n, k, \ell)$ on the vertex-set $C \cup S$ such that $S$ is a stable set, $C$ is a clique, and a vertex of $S$ is adjacent to a vertex of $C$ if and only if the corresponding $k$-set is either a subset or a superset of the corresponding $\ell$-set. Obviously, $G(n, k, \ell)$ is not a CIS-graph, since $C \cap S = \emptyset$. However, some of these graphs satisfy the condition **COMB**, for example, $G(5, 1, 2)$. Moreover, $G(5, 1, 2)$ satisfies the stronger condition **COMB**(5, 4).

By definition, $G(n, 1, 1) = S_n$ is an $n$-comb and $G(n, n - 1, 1) = \overline{S_n}$ is an $n$-anti-comb. Furthermore, it is easy to see that

(i) the graphs $G(n, k, \ell)$ and $G(n, n - k, n - \ell)$ are isomorphic.

Hence, without loss of generality we can assume that $k \leq \ell$ and even that $k < \ell$, since $G(n, k, k)$ is just a comb $S_{\binom{n}{k}}$. Then, from the simple fact that a set contains an element if and only if the complementary set does not contain it, we derive

(ii) the graphs $G(n, k, 1)$ and $G(n, 1, n-k)$ are complementary.

Thus, the graphs $G(n, k, n-1)$ and $G(n, n-k, 1)$ are isomorphic by (i) and complementary to $G(n, 1, k)$ by (ii). Hence, without loss of generality we can assume that $\ell \leq n-2$. Summarizing, we will assume in the sequel that

$$1 \leq k < \ell \leq n-2. \tag{1}$$

In Section 3 we will prove the following two claims analyzing the existence of unsettled anti-combs and combs in $G(n, k, \ell)$.

**Theorem 3**

*(i) Each induced anti-comb in $G(n, k, \ell)$ is settled whenever*

$$n > \frac{k+1}{k}\ell.$$

*(ii) An unsettled induced anti-comb exists in $G(n, k, \ell)$ whenever*

$$k+\ell \leq n \leq \frac{k+1}{k}\ell.$$

**Theorem 4**

*(a) Each induced comb is settled in $G(n, 1, \ell)$, and it is settled in $G(n, 2, \ell)$ whenever*

$$n < 2\ell - 3.$$

*(b) An unsettled induced comb exists in $G(n, k, \ell)$ for $k \geq 2$ whenever*

$$n \geq \frac{k}{k-1}\ell - \frac{r}{k-1} \text{ or } n = \frac{k}{k-1}\ell - \frac{r}{k-1} - 1 \text{ and } \ell > r + k^2 - k,$$

*where $r \equiv \ell \pmod{k-1}$ and $r \in \{2, 3, \ldots, k\}$.*

Let us denote by $\mathbf{G}$ the subfamily of graphs $G(n, k, \ell)$ whose induced combs and anti-combs are all settled and $n, k, \ell$ satisfy (1).

**Corollary 2** *For $k = 1$ and $k = 2$ the membership in $\mathbf{G}$ is characterized as follows:*

$$G(n, 1, \ell) \in \mathbf{G} \text{ iff } n > 2\ell$$

$$G(n, 2, \ell) \in \mathbf{G} \text{ iff } 2\ell - 3 > n > (3/2)\ell.$$

*Proof* By (1) we have $n \geq \ell + 2 \geq \ell + k$, whenever $k \leq 2$, and thus, by Theorem 3, all induced anti-combs are settled in $G(n, k, \ell)$ for $k \leq 2$ if and only if $n > \frac{k+1}{k}\ell$. This and (a) of Theorem 4 then imply the claim for $k = 1$.

If $k = 2$, then $G(n, 2, \ell)$ has an unsettled comb, by (b) of Theorem 4, if $n \geq 2\ell - 2$ or if $n = 2\ell - 3$ and $\ell > 4$, since $r = 2$ in this case. However, if $n = 2\ell - 3$ then $\ell \geq 5$ by (1). Hence, the second condition holds automatically, and therefore by (a) and (b) of Theorem 4, we can conclude that $G(n, 2, \ell)$ has an unsettled comb if and only if $n \geq 2\ell - 3$. □

Thus, for $k = 1$ we get $\{G(5, 1, 2), G(6, 1, 2), G(7, 1, 2), G(7, 1, 3), \dots\} \subseteq \mathbf{G}$ and for $k = 2$ we get $\{G(14, 2, 9), G(16, 2, 10), G(17, 2, 11), G(18, 2, 11), G(19, 2, 12), G(20, 2, 13), \dots\} \subseteq \mathbf{G}$.

*Remark 3* Notice that conditions (i) and (ii) of Theorem 3 provide an almost complete characterization of the existence of unsettled anti-combs in $G(n, k, \ell)$. However, it is not clear if condition $n \geq k + \ell$ in part (ii) is necessary. Note that if $k \leq 2$, then this condition holds automatically by (1). For instance, we do not know if $G(8, 3, 6)$ has an unsettled anti-comb. Computer experiments show that there are no unsettled $m$-anti-combs for $m \leq 10$. In any case, $G(8, 3, 6)$ has an unsettled 6-comb, by Theorem 4.

Let us also note that we know much less about combs. For instance, we could only treat the case of $k \leq 2$ in (a) of Theorem 4, though we conjecture that a similar claims can hold for all $k$. For example, $G(10, 3, 8)$ is the smallest graph for which we do not know if it contains an unsettled comb or anti-comb.

Based on the proofs of the above theorems and on several numerical examples we conjecture that membership in $\mathbf{G}$ can be characterized by inequalities of the approximate form

$$\frac{k}{k-1}\ell + O(k) \geq n \geq \frac{k+1}{k}\ell - O(k).$$

This is certainly the case for $k \leq 2$, by Corollary 2.

By definition, in a graph $G = G(n, k, \ell) \in \mathbf{G}$, as well as in its complement $\overline{G}$, all induced combs and anti-combs are settled, that is, both $G$ and $\overline{G}$ satisfy the condition **COMB**. Let us notice, however, that $\overline{G}$ is not an $(n, k, \ell)$-graph unless $k = 1$. (Recall that $G(n, 1, \ell)$ and $G(n, n - \ell, 1)$ are complementary.)

It seems that every non-CIS-graph satisfying **COMB** contains either an induced $G(n, k, \ell) \in \mathbf{G}$ or its complement. At least, we have no counterexample for this claim.

Let us add that, unlike the case of combs and anti-combs, one graph from $\mathbf{G}$ may contain another as an unsettled induced subgraph. For example, $G(6, 1, 2)$ contains an unsettled induced $G(5, 1, 2)$, while in $G(7, 1, 2)$ all induced $G(5, 1, 2)$ are settled. Yet, in $G(7, 1, 2)$ there is an unsettled induced $G(6, 1, 2)$. Vice versa, in $G(7, 1, 3)$ each induced $G(6, 1, 2)$ is settled but there are unsettled induced $G(5, 1, 2)$. Further, in $G(8, 1, 3)$, all induced $G(5, 1, 2)$ and $G(7, 1, 2)$ are settled but there are unsettled induced $G(6, 1, 2)$ and $G(7, 1, 3)$. Due to this "non-transitivity," in order to enforce the CIS-property for a graph $G$, it seems easier to assume that all induced subgraphs from $\mathbf{G}$ as well as their complements are settled in $G$. Of course, it is even simpler to assume that $G$ does not contain such subgraphs at all.

*Conjecture 1* If $G$ contains no induced $G(5, 1, 2)$ nor its complement $G(5, 3, 1)$ and all induced combs and anti-combs are settled in $G$, then $G$ is a CIS-graph.

We remark here that $G(n, k, l)$ contains an induced $G(n', k', l')$ whenever $n' \leq n$, $k' \leq k$, and $l' \leq l$.

*Remark 4* Let us note that CIS-graphs and perfect graphs look somewhat similar. Both classes are closed with respect to complementation and substitution. Odd holes and anti-holes are similar to combs and anti-combs. The following two tests look similar too: whether $G$ contains an induced odd hole or anti-hole and whether $G$ contains an induced unsettled comb or anti-comb. It seems that CIS-graphs, like perfect graphs, may allow a simple characterization and a polynomial recognition algorithm (that may be very difficult to obtain, though).

However, there are dissimilarities, too. The property of perfectness is hereditary, unlike the CIS-property. Also, there are non-CIS-graphs in which all induced combs and anti-combs are settled. (By Conjecture 1, every such graph contains an induced $G(5, 1, 2)$ or its complement $G(5, 3, 1)$.)

*Remark 5* CIS-graphs were recently mentioned (under the name of *stable* graphs) in [45], where it is shown that recognition of stable graphs is a special case of a difficult problem (strongly bipartite bihypergraph recognition problem) introduced in this paper. Based on this observation, the authors conjecture that recognition of stable graphs is co-NP-complete. However, we conjecture that this problem is polynomial.

The following relaxation of the CIS-property was considered in [31] and [41].

**Triangle Condition** For every maximal stable set $S$ and every edge $(u, v)$ such that $u, v \notin S$ there exists a vertex $w \in S$ such that vertices $u, v, w$ induce a clique.

Obviously, each CIS-graph has this property.

## 1.6 Gallai's and CIS-d-Graphs

Let us generalize the concept of a CIS-graph as follows. For a given integer $d \geq 2$, a complete graph whose edges are colored by $d$ colors $\mathcal{G} = (V; E_1, \ldots, E_d)$ is called a *d-graph*. To a given $d$-graph $\mathcal{G}$ let us assign a family of $d$ hypergraphs $\mathcal{C} = \mathcal{C}(G) = \{\mathcal{C}_i \mid i = 1, \ldots, d\}$ on the common vertex-set $V$, where the hyperedges of $\mathcal{C}_i$ are all inclusion maximal subsets of $V$ containing no edges of color $i$. We say that $\mathcal{G}$ is a *CIS-d-graph* (has the *CIS-d-property*) if $\bigcap_{i=1}^{d} C_i \neq \emptyset$ for all selections $C_i \in \mathcal{C}_i$ for $i = 1, \ldots, d$. Obviously, such an intersection can contain at most one vertex. If $d = 2$, then we obtain the original concept of $CIS$-graphs. (More accurately, CIS-2-graph is a pair of two complementary CIS-graphs.) Similarly to CIS-graphs, CIS-$d$-graphs also satisfy a natural requirement that can be considered as a generalization of settling. Assume that $X_i$ is a clique in the subgraph $G_i = (V, \cup_{j \neq i} E_j)$ for $i = 1, \ldots, d$, and that $\cap_{i=1}^{d} X_i = \emptyset$. Then, these cliques cannot all be maximal and, hence, there must be a vertex $x \in V$ such that for every $i = 1, \ldots, d$ and $y \in X_i$ we have $(x, y) \notin E_i$. We will say in this case that $\{X_1, \ldots, X_d\}$ are *settled* by $x$.

Given a CIS-$d$-graph $\mathscr{G}$, let us assign to it a $d$-dimensional table $g = g(\mathscr{G})$, that is, a mapping $g : \mathscr{C}_1 \times \cdots \times \mathscr{C}_d \to V$ defined by the rule: $g(C_1, \ldots, C_d) = v$ whenever $\{v\} = \cap_{i=1}^{d} C_i$. Let us observe that this $d$-dimensional array is partitioned by the elements of $V$ into $n = |V|$ sub-arrays called *boxes*, since the following implication holds:

if $g(C_1', \ldots, C_d') = g(C_1'', \ldots, C_d'') = v$, then $v$ belongs to all these $2d$ sets, and hence,

$g(C_1, \ldots, C_d) = v$ for all $2^d$ choices $C_i \in \{C_i', C_i''\}, i = 1, \ldots, d$.

Let us further introduce two special edge colored graphs. Let $\Pi$ denote the 2-colored graph whose both chromatic components form a $P_4$, that is, $V = \{v_1, v_2, v_3, v_4\}$; $E_1 = \{(v_1, v_2), (v_2, v_3), (v_3, v_4)\}$, and $E_2 = \{(v_2, v_4), (v_4, v_1), (v_1, v_3)\}$. Furthermore, let $\Delta$ denote the 3-colored triangle, for which $V = \{v_1, v_2, v_3\}$, $E_1 = \{(v_1, v_2)\}$, $E_2 = \{(v_2, v_3)\}$, and $E_3 = \{(v_3, v_1)\}$. Figure 11 illustrates these graphs.

**Proposition 4 ([23, 25])** *Every $\Pi$- and $\Delta$-free d-graph is a CIS-d-graph.* □

In fact, a stronger claim holds.

**Proposition 5 ([23–25])** *A d-graph $\mathscr{G}$ is $\Pi$- and $\Delta$-free if and only if the corresponding mapping $g(\mathscr{G})$ defines the normal form of a positional d-person game with perfect information whose final positions (outcomes of the game) are in one-to-one correspondence with the vertices of $\mathscr{G}$.* □

For example, let us consider the $\Pi$- and $\Delta$-free 3-graph $\mathscr{G}$ given in Figure 12. For this graph we have $\mathscr{C}_1 = \{(1, 3), (2, 4)\}$, $\mathscr{C}_2 = \{(1, 2, 4), (2, 3, 4)\}$, and $\mathscr{C}_3 = \{(1, 2, 3), (1, 3, 4)\}$. The mapping $g(\mathscr{G})$ and the corresponding positional game are shown in Figure 12.

Another example of a $\Pi$- and $\Delta$-free 3-graph is given in Figure 13. In this case $\mathscr{C}_1 = \{(1), (2, 3, 4)\}$, $\mathscr{C}_2 = \{(1, 3), (1, 2, 4)\}$, and $\mathscr{C}_3 = \{(1, 2, 3), (1, 3, 4)\}$. The mapping $g(\mathscr{G})$ and the corresponding positional game are shown in Figure 13.

Of course, the condition that a $d$-graph $\mathscr{G}$ must be $\Pi$- and $\Delta$-free is only sufficient but not necessary for the CIS-$d$-property to hold. On the other hand, the

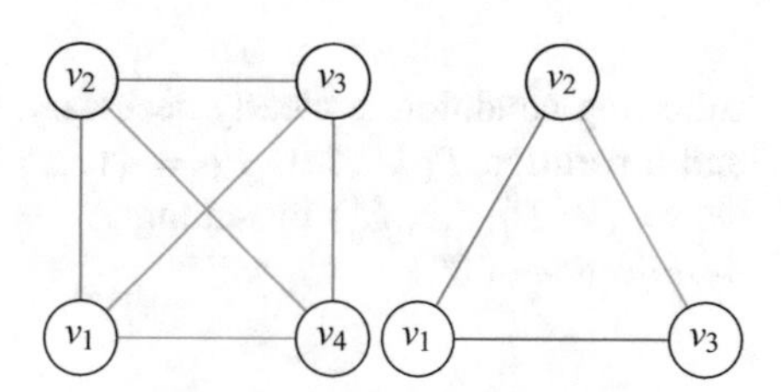


**Fig. 11** Colored $\Pi$ and $\Delta$

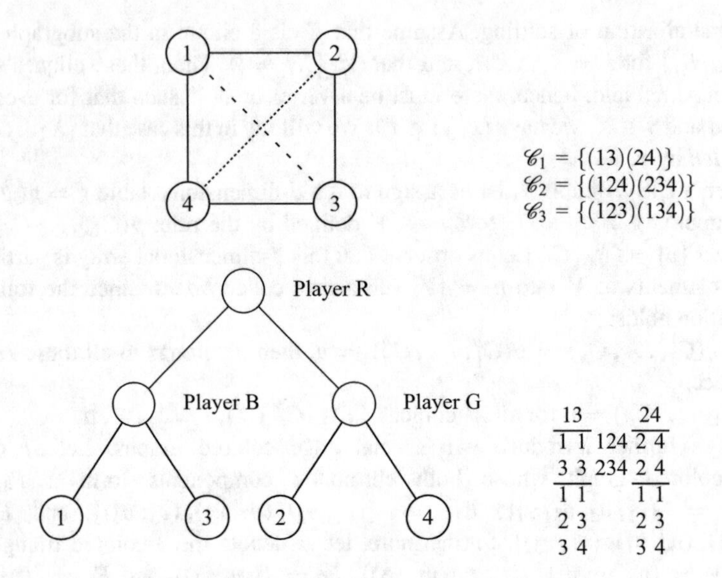


**Fig. 12** A $\Pi$- and $\Delta$-free 3-graph and the corresponding positional and normal game forms

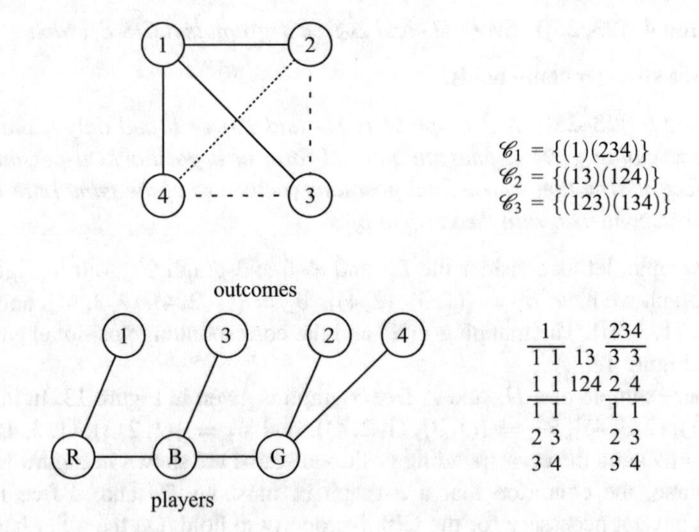


**Fig. 13** A $\Pi$- and $\Delta$-free 3-graph and the corresponding positional and normal game forms

following condition is clearly necessary. Given a $d$-graph $\mathscr{G} = (V; E_1, \dots, E_d)$ and a partition $P_1 \cup \dots \cup P_\delta = \{1, \dots, d\}$ of its colors, let us define a $\delta$-graph $\mathscr{G}' = (V; E'_1, \dots, E'_\delta)$ by setting $E'_i = \cup_{j \in P_i} E_j$, $i = 1, \dots, \delta$ and call $\mathscr{G}'$ the *$\delta$-projection* of $\mathscr{G}$.

**Proposition 6** *Let $\mathscr{G}$ be a CIS-d-graph whose set of colors $\{1, \ldots, d\}$ is partitioned into $\delta$ non-empty subsets ($2 \leq \delta \leq d$) then the corresponding $\delta$-graph $\mathscr{G}'$ is a CIS-$\delta$-graph.*

In particular, in case $\delta = 2$ we must get two complementary CIS-graphs.
The following conjecture is open since 1978.

*Conjecture 2 ([23])* Every CIS-$d$-graph is $\Delta$-free.

By Proposition 6, it would suffice to prove this conjecture for $d = 3$. In this case, it was verified up to $n = 12$ vertices by a computer code written by Steven Jaslar in 2003. We will consider this conjecture in Section 4 and show that, similarly to combs and anti-combs, all $\Delta$s in a CIS-$d$-graph must be settled, and it takes two vertices to settle a $\Delta$ (see Section 4.2). Although there are $d$-graphs in which all $\Delta$s are settled, yet, it seems impossible to have settled simultaneously all combs and anti-combs in all 2-projections of these $d$-graphs, a condition that is necessary by Proposition 6.

In the literature $\Delta$-free $d$-graphs are called *Gallai's graphs*, since they were introduced by Gallai in [20]. We will call them Gallai's $d$-graphs which is more accurate. They are well studied [1, 10–12, 18, 29, 32, 33]. Conjecture 2 means that CIS-$d$-graphs form a subfamily of Gallai's $d$-graphs. Next, we will characterize Gallai's CIS-$d$-graphs in terms of CIS-graphs. Hence, to characterize CIS-$d$-graphs it would suffice to do it for $d = 2$ and prove Conjecture 2.

First, let us note that both Gallai's and CIS-$d$-graphs are closed under substitution. (For Gallai's $d$-graphs this is well known [10, 29].) Moreover, the inverse claims hold too.

**Proposition 7** *Let us substitute a d-graph $\mathscr{G}''$ for a vertex $v$ of a d-graph $\mathscr{G}'$ and denote the obtained d-graph by $\mathscr{G} = \mathscr{G}(\mathscr{G}', v, \mathscr{G}'')$. Then $\mathscr{G}$ is a Gallai (respectively, CIS-) d-graph if and only if both $\mathscr{G}'$ and $\mathscr{G}''$ are Gallai's (respectively, CIS-) d-graphs.*

In case $d = 2$ this proposition implies the similar property for CIS-graphs.

**Proposition 8** *Let us substitute a graph $G''$ for a vertex $v$ of a graph $G'$ and denote the obtained graph by $G = G(G', v, G'')$. Then $G$ is a CIS-graph if and only if both $G'$ and $G''$ are CIS-graphs.* □

Let us recall, however, that CIS-$d$-property is not hereditary, that is, an induced subgraph of a CIS-$d$-graph may have no CIS-$d$-property. In particular, for $d = 2$, this means that an induced subgraph of a CIS-graph may have no CIS-property.

Here and in the sequel we assume that the set of colors $[d] = \{1, \ldots, d\}$ is the same for all considered $d$-graphs, while some chromatic components may be trivial (edge-empty). For example, by a 2-graph we mean a $d$-graph with at most 2 non-trivial chromatic components.

It is known that each Gallai $d$-graph can be obtained from 2-graphs by substitutions. More precisely, the following claim holds.

**Proposition 9 (Cameron and Edmonds [10]; Gyárfás and Simonyi [29])** *For each Gallai d-graph $\mathscr{G}$ there exist a 2-graph $\mathscr{G}_0$ with n vertices and n Gallai d-graphs $\mathscr{G}_1, \ldots, \mathscr{G}_n$ such that $\mathscr{G}$ is obtained by substituting $\mathscr{G}_1, \ldots \mathscr{G}_n$ for n vertices of $\mathscr{G}_0$.*

In [29], this claim is derived from the following Lemma.

**Lemma 1 ([10, 20], and [29])** *Every Gallai d-graph $\mathscr{G} = (V; E_1, \ldots, E_d)$ with $d \geq 3$ has a color $i \in [d]$ that does not span V, or in other words, the graph $G_i = (V, E_i)$ is not connected.*

*Remark 6* It is interesting to compare Lemma 1 with the following Lemma from [23, 25]. If a $d$-graph $\mathscr{G}$ is $\Pi$- and $\Delta$-free, then there exists a unique color $i \in [d]$ such that *the complement* of the $i$-th chromatic component, $\overline{G_i}$, is disconnected.

Gyárfás and Simonyi remark that Lemma 1 "is essentially a content of Lemma (3.2.3) in [20]" and they derive Proposition 9 from it as follows. If $d \leq 2$, we are done. Otherwise, we have a color $i \in [d]$ such that graph $G_i = (V, E_i)$ is disconnected. It is not difficult to show that for each two of its connected components all edges between them are of the same color $j$ (clearly, $j \neq i$), since otherwise a $\Delta$ appears.

Collapsing these components into vertices we get a smaller $(d-1)$-graph which is still $\Delta$-free, by Proposition 7. By induction, $\mathscr{G}_1, \ldots, \mathscr{G}_n$ and $\mathscr{G}_0$ can be constructed as required. □

Moreover, applying the above decomposition recursively, we can represent an arbitrary Gallai $d$-graph $\mathscr{G} = (V; E_1, \ldots, E_d)$ by a substitution-tree $T(\mathscr{G})$ whose leaves are associated to 2-graphs. If $d \leq 2$, then $\mathscr{G}$ itself is a 2-graph and $T(\mathscr{G})$ is reduced to one vertex. If $d \geq 3$ then, by Lemma 1, there is a color $i \in [d]$ such that the $i$-th component $G_i = (V, E_i)$ does not span $V$, or in other words, it is disconnected. Let $W \subset V$ be a connected component of $G_i$. Furthermore, let $G'' = G[W]$ be the subgraph of $\mathscr{G}$ induced by $W$, while $G'$ be obtained from $\mathscr{G}$ by contracting $W$ to a single new vertex $v$. Then, as it was shown above, substituting $\mathscr{G}''$ for $v$ in $\mathscr{G}'$ we get $\mathscr{G} = \mathscr{G}(\mathscr{G}', v, \mathscr{G}'')$; see Figure 14. If $\mathscr{G}'$ (or $\mathscr{G}''$) is a 2-graph, then it becomes a leaf of $T(\mathscr{G})$. Otherwise, if $\mathscr{G}'$ (or $\mathscr{G}''$) has more than 2 non-trivial chromatic components, we decompose it further in the same way until only 2-graphs remain. They are the leaves of the obtained decomposition tree $T(\mathscr{G})$, as required.

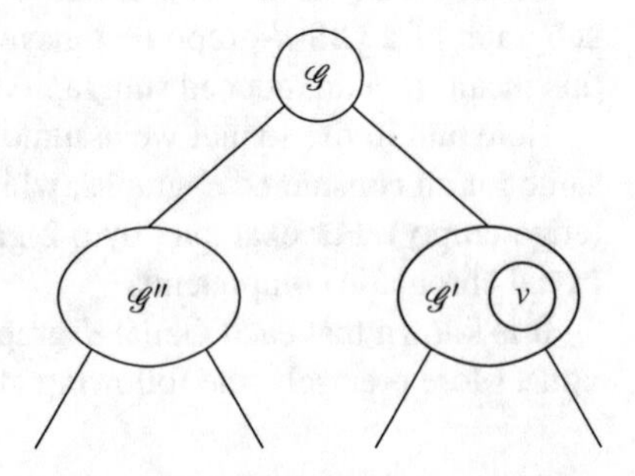


**Fig. 14** Decomposing $\mathscr{G}$ by the tree $T(\mathscr{G})$; substituting $\mathscr{G}''$ for $v$ in $\mathscr{G}'$ to get $\mathscr{G}$

It is wellknown that decomposing a given graph into connected components can be executed in linear time. Hence, given $\mathscr{G}$, its decomposition tree $T(\mathscr{G})$ can be constructed in linear time, too.

*Remark 7* As defined above, tree $T(\mathscr{G})$ is not unique, since several chromatic components of $\mathscr{G}$ may be disconnected and any connected component of any chromatic component can be chosen as $W$ for the decomposition. Let us note, however, that the corresponding vertex sets are *nested*. More precisely, if $E_i^a, E_j^b$ are connected components of colors $i, j \in [d]$, then the corresponding vertex-sets $V_i^a, V_j^b \subseteq V$ are either disjoint or one of them is a subset of the other. Yet, the latter case cannot take place when $i = j$.

Let us also note that in general $T(\mathscr{G})$ can be extended further, since some 2-graphs also can be decomposed by substitution. Obviously, the decomposition of a 2-graph $\mathscr{G} = (V; E_1, E_2)$ is reduced to a decomposition of a graph, namely, of a chromatic component, $G_1 = G(V, E_1)$ or $G_2 = G(V, E_2)$.

In general, decomposing graphs (as well as $d$-graphs, digraphs, Boolean functions, etc.) by substitution is known as their *modular* decomposition. A *module* is a set $X \subseteq V$ such that no member of $V \setminus X$ distinguishes members of $X$. A set family $\mathscr{F}$ is called *decomposable* if $X \cap Y$, $X \cup Y$, $X \setminus Y$, $Y \setminus X$, and $X \Delta Y = (X \setminus Y) \cup (Y \setminus X)$ are in $\mathscr{F}$ whenever $X, Y \in \mathscr{F}$ and $X \cap Y \neq \emptyset$. Möring [39] proved that the family of modules is decomposable and hence, there is a unique canonical modular decomposition tree.

In general, modular decomposition is more complicated than decomposition of Gallai's $d$-graphs. There have been a number of $O(n^4)$, $O(n^3)$, $O(mn)$, $O(n^2)$, $O(n + m \log n)$ algorithms. Finally, $O(m + n)$ algorithms were given by Cournier and Habib [13] and McConnell and Spinrad [38]. Some linear time algorithms work for graphs, $d$-graphs, digraphs, and Boolean functions. See [8, 9, 38–40] for a survey on modular decomposition.

We make use of the decomposition tree $T(\mathscr{G})$ to recognize whether $\mathscr{G}$ is a CIS-$d$-graph. Obviously, by Proposition 7, we can extend Proposition 9 as follows.

**Proposition 10** *A Gallai d-graph $\mathscr{G}$ has the CIS-d-property if and only if all $n + 1$ d-graphs $\mathscr{G}_1, \ldots, \mathscr{G}_n$ and $\mathscr{G}_0$ from Proposition 9 have this property.* □

Thus, every Gallai's $CIS$-$d$-graph can be obtained from $CIS$-2-graphs by recursive substitutions, and hence, a characterization or polynomial recognition algorithm of CIS-graphs would provide one for the Gallai $CIS$-$d$-graphs too.

From Propositions 6, 9, and 10 we will derive the following two claims.

**Proposition 11** *A Gallai d-graph $\mathscr{G}$ is a CIS-d-graph if and only if all d of its chromatic components are CIS-graphs.*

The "only if" part follows from Proposition 6 and "if" part can be strengthened as follows.

**Proposition 12** *Given a Gallai d-graph $\mathscr{G}$ such that at least $d-1$ of its d chromatic components are CIS-graphs, then $\mathscr{G}$ is a CIS-d-graph.*

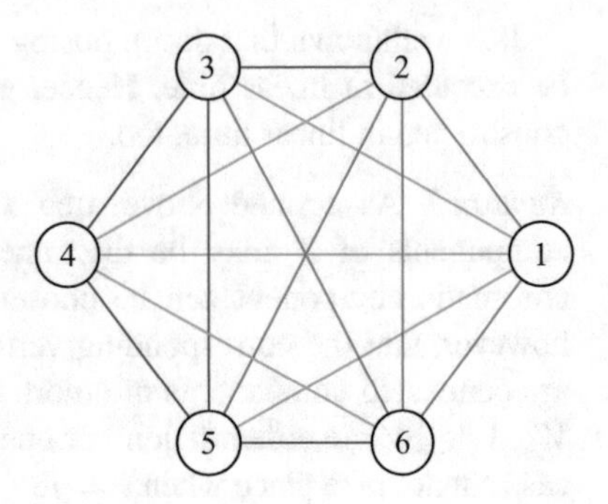


**Fig. 15** A non-Gallai 3-graph in which $G_1$ and $G_2$ are CIS-graphs, while $G_3$ is not

In particular, the remaining chromatic component of $\mathscr{G}$ must be a CIS-graph.

In the next subsection we generalize the last claim by showing that it holds not only for CIS-graphs but also for perfect graphs and, in fact, for every family of graphs satisfying some simple requirements.

Yet, of course, it is essential that $\mathscr{G}$ is a Gallai $d$-graph. For example, let us consider a 3-graph $\mathscr{G}$ in Figure 15. Graphs $G_1$ and $G_2$ are isomorphic, each of them is a settled 2-comb with one isolated vertex. Hence, they are CIS-graphs. Yet, $G_3$ is not, since the stable set $S = \{2, 3, 5, 6\}$ and clique $C = \{1, 4\}$ are disjoint. However, $\mathscr{G}$ is not Gallai's 3-graph, for example, $\{5, 6, 1\}$ as well as $\{1, 2, 3\}$ is a $\Delta$.

## 1.7 Extending Cameron-Edmonds-Lovász' Theorem

Cameron, Edmonds, and Lovász [11] proved the statement similar to Proposition 12 for perfect graphs: given a Gallai $d$-graph, if at least $d - 1$ of its chromatic components are perfect graphs, then the remaining component is a perfect graph, too. Later, Cameron and Edmonds [10] showed that, in fact, the statement holds for any family of graphs that is closed under: (i) substitution, (ii) complementation, and (iii) taking induced subgraphs. For example, it holds for $P_4$-free graphs, or in other words, for the components of $\Pi$- and $\Delta$-free $d$-graphs [23].

However, CIS-graphs satisfy only (i) and (ii) but not (iii). Nevertheless, the statement holds for them too; see Proposition 12.

In general, one can substitute the following property for (iii).

Let us say that a family of graphs (or $d$-graphs) $\mathscr{F}$ is *exactly closed under substitution* $G = G(G', v, G'')$ whenever $G \in \mathscr{F}$ if and only if both $G'$ and $G''$ belong to $\mathscr{F}$.

For example, CIS-graphs are exactly closed under substitution, by Propositions 8, and both, Gallai's and CIS-$d$-graphs, by Propositions 7.

**Proposition 13** *If $\mathscr{F}$ is closed under substitution and taking induced subgraphs, then $\mathscr{F}$ is exactly closed under substitution.*

*Proof* Indeed, if $G = G(G', v, G'')$, then both $G'$ and $G''$ are induced subgraphs of $\mathcal{G}$. □

We say that the family of graphs $\mathcal{F}$ has the *CES-property* and call it a *CES-family* if $\mathcal{F}$ is closed under complementation and exactly closed under substitution.

For example, the families of perfect graphs and CIS-graphs have the CES-property.

We strengthen Cameron-Edmonds' theorem as follows.

**Theorem 5** *Let $\mathcal{F}$ be a CES-family of graphs and $\mathcal{G} = (V; E_1, \dots, E_d)$ be a Gallai d-graph such that at least $d - 1$ of its chromatic components, say, $G_i = (V, E_i)$ for $i = 1, \dots, d - 1$, belong to $\mathcal{F}$. Then*

*(a) the last component $G_d = (V, E_d)$ is in $\mathcal{F}$ too, and moreover,*
*(b) all $2^d$ projections of $\mathcal{G}$ belong to $\mathcal{F}$, that is, for each subset $I \subseteq [d] = \{1, \dots, d\}$ the graph $G_I = (V, \cup_{i \in I} E_i)$ is in $\mathcal{F}$.*

We will prove this Theorem in Section 4.1. By Proposition 13, part (a) implies Cameron-Edmonds' theorem. Since CIS-graphs form a CES-family, we obtain the following claim.

**Corollary 3** *Let $\mathcal{G} = (V; E_1, \dots, E_d)$ be a Gallai d-graph such that at least $d - 1$ of its chromatic components are CIS-graphs. Then the remaining chromatic component of $\mathcal{G}$ is a CIS-graph too; hence, $\mathcal{G}$ is a CIS-d-graph and all its $2^d$ projections are CIS-graphs.* □

## 1.8 On families of Graphs Closed with Respect to Substitution

To get more examples of CES-families let us, first, consider hereditary classes. Each such class is a family of graphs $\mathcal{F}$ defined by a family, finite or infinite, of forbidden subgraphs $\mathcal{F}'$. By definition, $G \in \mathcal{F}$ if and only if $G$ contains no induced subgraph isomorphic to a $G' \in \mathcal{F}'$.

Let us call a graph (or $d$-graph) $G$ *substitution-prime* (or just, prime, for brevity) if it is not decomposable by substitution, or more precisely, if $G = G(G', v, G'')$ for no $G'$, $G''$ and $v$, except for two trivial cases: ($G = G'$ and $V(G'') = \{v\}$) or ($G = G''$ and $V(G') = \{v\}$).

Suppose that $G$ is decomposable, $G = G(G', v, G'')$. Then, as we already mentioned, both $G'$ and $G''$ are induced subgraphs of $G$. Hence, if $G'$ or $G''$ contains an induced subgraph $G_0$, then $G$ also contains it. However, $G$ may contain $G_0$ even if $G'$ and $G''$ do not. Yet, clearly, in this case $G_0$ is not substitution-prime. Thus, for both, graphs and $d$-graphs, we obtain the following statement.

**Proposition 14** *Family $\mathcal{F}$ is exactly closed under substitution if all $d$-graphs in $\mathcal{F}'$ are substitution-prime.* □

The inverse holds too if we assume (by the way, without any loss of generality) that no ($d$-)graph of $\mathscr{F}'$ contains another one as an induced subgraph. Thus, $\mathscr{F}$ is a CES-family (and, hence, it satisfies all conditions of Theorem 5) whenever $\mathscr{F}'$ is closed under complementation ($G \in \mathscr{F}'$ if and only if $\overline{G} \in \mathscr{F}'$) and $\mathscr{F}'$ contains only substitution-prime graphs.

For example, these two properties hold for the odd holes and anti-holes. In this case, $\mathscr{F}$ is the family of Berge graphs. Thus, Theorem 5 and the Strong Perfect Graph Theorem imply the Cameron-Edmonds-Lovász Theorem [11]. Of course, it can be proved simpler: first, show that perfect graphs are exactly closed under substitution, [35], and then apply Lovász' perfect graph theorem [34, 35], instead of the strong one.

Another example is provided by the family $\mathscr{F}$ of $P_4$-free graphs. In this case $\mathscr{F}' = \{P_4\}$ and all conditions of Theorem 5 hold, since $P_4$ is self-complementary and prime.

*Remark 8* Moreover, in this case, it is easy to verify directly claims (a) and (b) of Theorem 5, see [23]. The following implication is instrumental: if a $d$-graph $\mathscr{G} = (V; E_1, \dots, E_d)$ is $\Pi$- and $\Delta$-free, then every its 2-projection $\mathscr{G}' = (V; E_1', E_2')$ is $\Pi$-free too.

A similar example is given by the family $\mathscr{F}$ of $A$-free graphs. In this case $\mathscr{F}' = \{A\}$, where $A$ is the settled $P_4$ (or in other words, settled 2-comb, or bull-graph). Like $P_4$, it is also self-complementary and substitution-prime.

However, if $\mathscr{F}'$ contains a decomposable graph, e.g., $C_4$, then $\mathscr{F}$ may be not closed under substitution. For example, let $\mathscr{F}' = \{C_4, \overline{C_4}\}$ and consider the Gallai 3-graph in Figure 12. Two of its chromatic components belong to $\mathscr{F}$, while the third one, $C_4$, does not.

As another example, let us consider $\mathscr{F}' = \{C_4, \overline{C_4}, C_5\}$. By [19], $\mathscr{F}$ is the family of split graphs. This family is self-complementary, yet, it is not closed under substitution. Indeed, substituting, for example, a non-edge for the middle vertex of $P_3$ we get $C_4$.

There are also non-hereditary families of graphs (respectively, $d$-graphs) closed under substitution; for example, CIS-graphs (respectively, CIS-$d$-graphs). It is not difficult to give more examples of such families and even to characterize them. Given a family $\mathscr{F}'$, finite or infinite, of ($d$-)graphs, let us denote by $cl(\mathscr{F}')$ its closure with respect to substitution.

**Proposition 15** *A family $\mathscr{F}$ of (d-)graphs is exactly closed under substitution if and only if $\mathscr{F} = cl(\mathscr{F}')$, where $\mathscr{F}'$ is a family, finite or infinite, of substitution-prime (d-)graphs. Furthermore, $\mathscr{F}$ is closed under complementation whenever $\mathscr{F}'$ is.*

*Proof* The second claim makes sense only for graphs and it is obvious. The first one follows from uniqueness of the canonical modular decomposition [39]. □

The obtained family $\mathscr{F} = cl(\mathscr{F}')$ is not hereditary if and only if there are substitution-prime ($d$)-graphs $G \in \mathscr{F}'$ and $G' \notin \mathscr{F}'$ such that $G'$ is an induced subgraph of $G$. For example, let $\mathscr{F}' = \{A\}$ contain only the bull-graph $A$ then $\mathscr{F} = cl(\mathscr{F}')$ contains no 2-comb.

However, the characterization of the CES-families by Proposition 15 is not constructive. For example, the substitution-prime perfect or CIS-graphs form infinite families that are difficult to describe explicitly.

## 1.9 *Almost CIS-d-Graphs*

A $d$-graph $\mathscr{G} = (V; E_1, \ldots, E_d)$ will be called an *almost CIS-d-graph* if $\bigcap_{i=1}^{d} C_i = \emptyset$ for exactly one $d$-tuple $C_1, \ldots, C_d$, where $C_i \subseteq V$ is an inclusion maximal vertex-set containing no edges of color $i$, that is, for each $i \in [d] = \{1, \ldots, d\}$, we have $(v, v') \in E_i$ for no $v, v' \in C_i$,

Notice that, by definition, the families of CIS- and almost CIS-$d$-graphs are disjoint.

For $d = 2$ we return to the definition of almost CIS-graphs. More precisely, an almost CIS-2-graph is a pair of two complementary almost CIS-graphs.

By Proposition 2, any split graph is either a CIS- or almost CIS-graph. Moreover, by Theorem 1, except split graphs, there are no other almost CIS-graphs. The latter are in a natural one-to-one correspondence with the split almost CIS-2-graphs. Let us recall that we may have $d \geq 2$ for a 2-graph. In particular, for an arbitrary $d \geq 2$ and almost CIS graph $G = (V, E)$ let us define a $d$-graph $\mathscr{G} = (V; E_1, \ldots, E_d)$ by setting $E_1 = E$, $E_2 = \overline{E}$, and $E_i = \emptyset$ for each $i > 2$. It is easy to see that $\mathscr{G}$ is an almost CIS-$d$-graph.

Let us note that already the 3-graph $\Delta$ is not almost CIS, since it has two distinct triplets $C_1 = \{v_2, v_3\}$, $C_2 = \{v_3, v_1\}$, $C_3 = \{v_1, v_2\}$ and $C_1' = \{v_3, v_1\}$, $C_2' = \{v_1, v_2\}$, $C_3' = \{v_2, v_3\}$ such that $C_1 \cap C_2 \cap C_3 = \emptyset$ and $C_1' \cap C_2' \cap C_3' = \emptyset$; see Figure 16 and also Section 4.2 for more details.

However, it is not difficult to extend $\Delta$ to an almost CIS-3-graph. Indeed, let us add to $\Delta$ a new vertex $v_4$ such that $(v_1, v_4) \in E_1$, $(v_2, v_4) \in E_2$, $(v_3, v_4) \in E_3$, and denote the obtained 3-graph by $\Delta'$. In other words, $\Delta' = (V; E_1, E_2, E_3)$, where

$$V = \{v_1, v_2, v_3, v_4\};$$
$$E_1 = \{(v_1, v_2), (v_1, v_4)\},\ E_2 = \{(v_2, v_3), (v_2, v_4)\},\ E_3 = \{(v_3, v_1), (v_3, v_4)\}.$$

It is easy to see that in $\Delta'$ vertices $v_1, v_2, v_3$ induce $\Delta$ and that

$\mathscr{C}_1 = \{(v_3, v_1), (v_2, v_3, v_4)\}$, $\mathscr{C}_2 = \{(v_1, v_2), (v_3, v_1, v_4)\}$,

$\mathscr{C}_3 = \{(v_2, v_3), (v_1, v_2, v_4)\}$.

Thus, $\Delta'$ is an almost CIS-3-graph, since only one of its eight triplets has the empty intersection: $\{v_3, v_1\} \cap \{v_1, v_2\} \cap \{v_2, v_3\} = \emptyset$.

*Remark 9* We can say that vertex $v_4$ settles one of the above two triplets of $\Delta$, namely, $(C_1, C_2, C_3)$. However, if we introduce one more vertex $v_5$ to settle $(C_1', C_2', C_3')$, too, then we have to choose a color for $(v_4, v_5)$. It is easy to verify

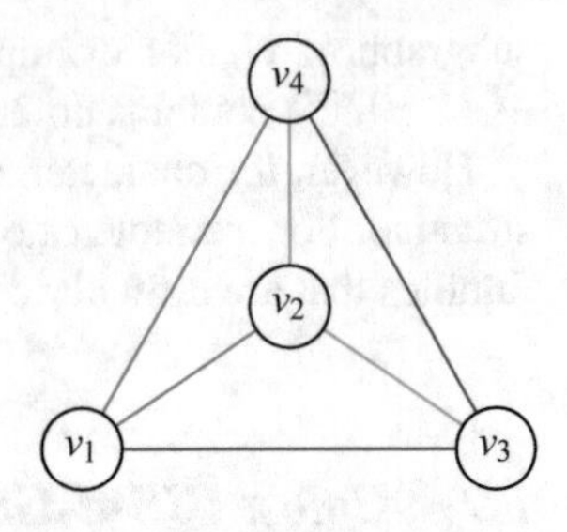


**Fig. 16** Almost CIS-3-graph $\Delta'$

that for each coloring of it a new $\Delta$ appears that should be, in its turn, settled, etc.; see Section 4.2 for more details.

Furthermore, it is not difficult to verify that $\Delta'$ is a unique almost CIS-3-graph with four vertices. Let us recall that there are also two CIS-3-graphs given in Figures 12 and 13.

Standardly, for any $d \geq 3$ we obtain an almost CIS-3-graph $\mathscr{G} = (V; E_1, \ldots, E_d)$ setting $E_1$, $E_2$, $E_3$, and $V$ by (1.9) and $E_i = \emptyset$ for each $i > 3$.

Let us also remark that, unlike CIS-$d$-graphs, almost CIS-$d$-graphs (and, in particular, almost CIS-graphs) are not closed under substitution. Nevertheless, we get an almost CIS-3-graph substituting $\Delta'$ for a vertex of a 1-graph $\mathscr{G}$. More precisely, $\mathscr{G}$ is a monochromatic clique whose all edges are colored by one of the three colors of $\Delta'$.

However, if all edges of $\mathscr{G}$ are colored by a new, fourth, color, then the obtained 4-graph is not almost CIS. Similarly, we won't get an almost CIS-3- or 4-graph by substituting $\Delta'$ for more than one vertex of $\mathscr{G}$, nor, viceversa, by substituting $\mathscr{G}$ for a vertex of $\Delta'$.

Finally, let us mention that we are not aware of any other almost CIS-$d$-graphs.

## 2 Proof of Theorem 2

In this section we prove Theorem 2 which claims that graphs satisfying condition COMB(3, 3) are CIS-graphs. First we describe the structure of our proof and a few main lemmas, then we give the complete proofs which are technical, long, and partially computer assisted.

### *2.1 Plan of the Proof of Theorem 2*

Let us assume by contradiction that there is a graph $G$ such that

(i) it contains no induced 3-combs and 3-anti-combs,
(ii) each induced 2-comb is settled in $G$, and

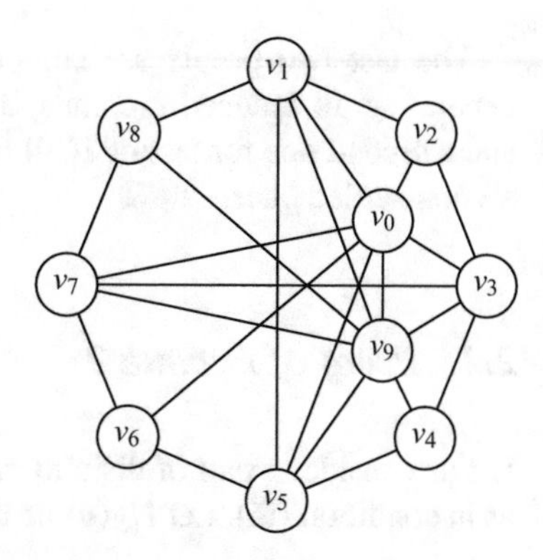

**Fig. 17** Graph $G_{10}$

(iii) there exist a maximal clique $C$ and a maximal stable set $S$ in $G$ such that $S \cap C = \emptyset$.

First, we will prove that $G$ must contain an induced subgraph $G_{10}$, shown in Figure 17.

**Lemma 2** *If $G$ satisfies conditions (i), (ii), and (iii), then $G$ must contain an induced $G_{10}$.*

Graph $G_{10}$ contains no induced 3-combs and 3-anti-combs, yet it contains several unsettled induced 2-combs. To settle them we have to introduce 10 new vertices that, somewhat surprisingly, induce a graph isomorphic to $G_{10}$ itself (since otherwise an induced 3-comb or 3-anti-comb would appear). Moreover, the obtained 20-vertex graph is the sum of two Petersen graphs, that is, the graph $2\mathscr{P}$ described in Section 1.4, Figure 8.

**Lemma 3** *If $G$ contains an induced $G_{10}$ and satisfies conditions (i) and (ii), then $G$ must contain an induced $2\mathscr{P}$.*

Let us recall that $2\mathscr{P}$ contains 10 uncertain pairs of vertices each of which can be either an edge or non-edge. Hence in fact, $2\mathscr{P}$ represent $2^{10} = 1024$ graphs. We will show that all these 1024 graphs contain no induced 3-combs and 3-anti-combs and, moreover, each induced 2-comb in $2\mathscr{P}$ (that contains no uncertain pair) is settled. However, 36 induced 2-combs appear in $2\mathscr{P}$ whenever we fix any uncertain pair either as an edge or as a non-edge. It is easy to see that none of these 2-combs are settled in $2\mathscr{P}$. We will show that they cannot be settled in $G$ either, because if a vertex of $G$ were settling one of them then an induced 3-comb or 3-anti-comb would exist in $G$. We can reformulate this result as follows.

**Lemma 4** *If $G$ satisfies conditions (i) and (ii), then it cannot contain an induced $2\mathscr{P}$.*

Obviously, the above 3 lemmas prove Theorem 2 by contradiction. We will prove Lemmas 2, 3, and 4 below in Sections 2.2, 2.3, and 2.4, respectively.

The last two proofs are computer assisted. We use two procedures, one for generating all induced 2-combs, 3-combs, and 3-anti-combs of a given graph $G$, and a second one for testing if all induced 2-combs are settled in $G$, and outputting all non-settled ones.

## 2.2 *Proof of Lemma 2*

Let us consider a pair of disjoint maximal clique $C$ and maximal stable set $S$ of $G$, as in condition (iii). Let $N_S(v)$ be the set of neighbors of $v$ in $S$. Notice that

$$\bigcap_{v \in C} N_S(v) = \emptyset, \tag{2}$$

because $C$ is maximal. Moreover,

$$N_S(v) \neq \emptyset \text{ for all } v \in C, \tag{3}$$

because $S$ is maximal.

We assume that $G$ satisfies conditions (i), (ii), and (iii). The following series of claims will imply the lemma.

*Claim 1* Given a maximal clique $C$ and a (not necessarily maximal) stable set $S$ in $G$ such that $C \cap S = \emptyset$, there exists vertices $u, v \in C$ such that $N_S(u) \cap N_S(v) = \emptyset$.

*Proof* Assume by contradiction that for all pairs of vertices $u, v \in C$, we have $N_S(u) \cap N_S(v) \neq \emptyset$. By this assumption, $|C| \geq 3$, otherwise $C$ would not be maximal.

So let $I = \{v_1, v_2, \ldots, v_k\}$ be a minimal subset of $C$ such that $\bigcap_{v \in I} N_S(v) = \emptyset$. Such a minimal subset of $C$ exists according to (2). Furthermore, by our assumption $|I| \geq 3$.

Now, define $u_i \in \bigcap_{j \neq i}^{k} N_S(v_j)$ for $i = 1, \ldots, k$. Note that $u_i \neq u_j$, due to the minimality of $I$. Thus, any 3 vertices $v_1, v_2, v_3 \in I$ with the corresponding $u_1, u_2, u_3$ form an $\overline{S_3}$ (see Figure 18), contradicting condition (i).

Note that for this claim we only need that $G$ is $S_3$-free.

From Claim 1, it follows that there are some pairs of vertices $u, v \in C$ such that $N_S(u) \cap N_S(v) = \emptyset$. Hence, there exist $x \in N_S(u)$ and $y \in (N_S(v))$ such that $x, u, v, y$ form an $S_2$ not settled by any vertex of $S$. The following claim states a useful property of any vertex $w \in V(G)$ settling such an $S_2$.

*Claim 2* We have $N_S(w) \subseteq N_S(u) \cup N_S(v)$.

*Proof* First notice that $x, y \notin N_S(w)$ because $w$ is a settling vertex. Then, assume by contradiction that there is a vertex $z \in N_S(w) \setminus (N_S(u) \cup N_S(v))$. Then, vertices $u, v, w, x, y, z$ form an $S_3$ (see Figure 19), contradicting condition (i).

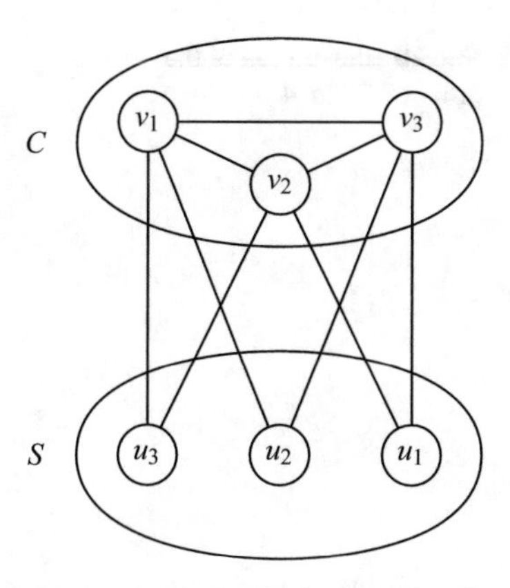


**Fig. 18** Illustration of the proof of Claim 1

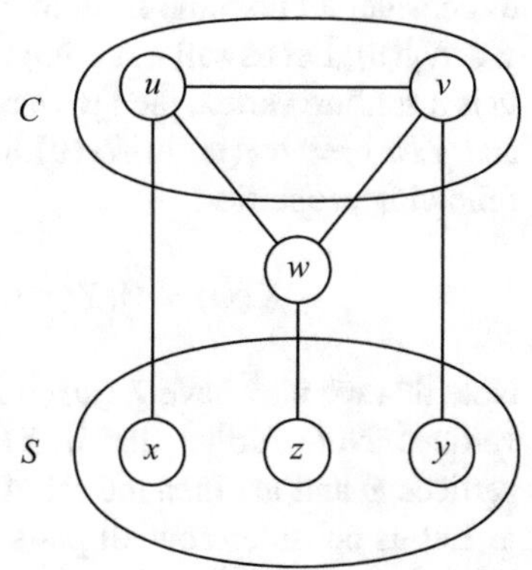


**Fig. 19** Illustration of the proof of Claim 2

For the remainder of the proof we fix a maximal clique $C$, a maximal stable set $S$, and vertices $u, v \in C$ such that

(iv) $C \cap S = \emptyset$, $N_S(u) \cap N_S(v) = \emptyset$, and $N_S(u) \cup N_S(v)$ is minimal,

among all possible choices of such sets $C$, $S$ and vertices $u, v \in C$ satisfying the conditions of (iv). Let us note that by (2) and (3), we have such a selection of $C$, $S$, $u$, and $v$ for which $N_S(u) \neq \emptyset$, $N_S(v) \neq \emptyset$, and hence $u \neq v$.

*Claim 3* Let $x \in N_S(u)$, $y \in N_S(v)$, and $w$ be a vertex of $V(G)$ that settles $S_2 = \{x, u, v, y\}$. Then, $N_S(w) \cap N_S(u) \neq \emptyset$ and $N_S(w) \cap N_S(v) \neq \emptyset$.

*Proof* From Claim 2, we know that $N_S(w) \subseteq N_S(u) \cup N_S(v)$. Assume by contradiction that e.g., $N_S(w) \cap N_S(u) = \emptyset$. This implies that $N_S(w) \subseteq N_S(v) \setminus \{y\}$ (since $w$ is settling $S_2$).

Then, consider a maximal clique $C' \supseteq \{u, w\}$. Notice that $C' \cap S = \emptyset$ because $N_S(w) \cap N_S(u) = \emptyset$. But $N_S(u) \cup N_S(w) \subsetneq N_S(u) \cup N_S(v)$, since $y \notin N_S(u) \cup N_S(w)$, contradicting property (iv), that is, the minimality of $N_S(u) \cup N_S(v)$.

We define next a minimal collection of settling vertices $\mathscr{W}$. Given a maximal clique $C$, a maximal stable set $S$, and vertices $u, v \in C$ satisfying property (iv), let

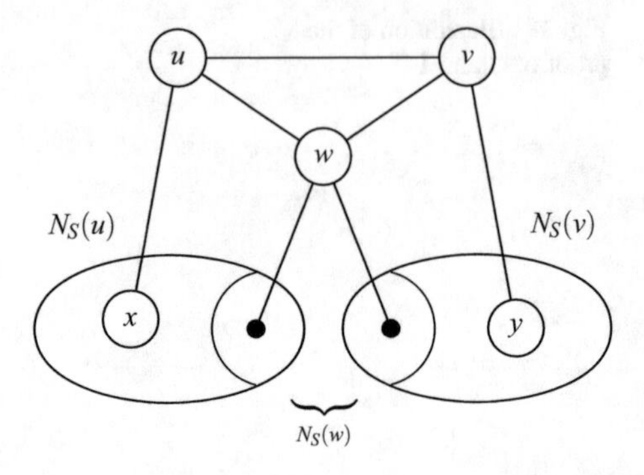


**Fig. 20** Illustration of the proof of Claim 4

us consider all possible 2-combs induced by $\{x, u, v, y\}$ in $G$, where $x \in N_S(u)$ and $y \in N_S(v)$. Let us call a *settling vertex* a vertex $w$ of $G$ that settles such a 2-comb. If $w$ is a settling vertex, then we have by Claims 2 and 3 that $X(w) = N_S(w) \cap N_S(u)$ and $Y(w) = N_S(w) \cap N_S(v)$ are subsets, uniquely defined by $w$, satisfying the following properties:

$$X(w) \neq \emptyset, Y(w) \neq \emptyset, \text{ and } N_S(w) = X(w) \cup Y(w). \tag{4}$$

Note that we may have $X(w) = X(w')$ and $Y(w) = Y(w')$ for two distinct settling vertices. Note further that if $X(w) \subseteq X(w')$ and $Y(w) \subseteq Y(w')$ hold for two vertices $w$ and $w'$, then the set of $S_2$ subgraphs settled by $w'$ are also settled by $w$.

Let us consider now all pairs of subsets $(X, Y)$ such that $X = X(w)$ and $Y = Y(w)$ for some settling vertex $w$. Let us call $(X, Y)$ *minimal*, if for there is no settling vertex $w'$ such that $X(w') \subseteq X$, $Y(w') \subseteq Y$ and $X(w') \cup Y(w') \subsetneq X \cup Y$, and let $\mathscr{X}\mathscr{Y}$ denote the collection of all such minimal pairs. For each pair $(X, Y) \in \mathscr{X}\mathscr{Y}$ let us choose one settling vertex $w = w_{XY}$ for which $X = X(w)$ and $Y = Y(w)$, and denote by $\mathscr{W} = \{w_{XY} | (X, Y) \in \mathscr{X}\mathscr{Y}\}$ the collection of these vertices; see Figure 20.

*Claim 4* There are at least two distinct vertices in $\mathscr{W}$.

*Proof* The statement follows from the definition of $\mathscr{W}$ and (4). Indeed, if $w_{XY} \in \mathscr{W}$, then by (4) there are vertices $x \in X$ and $y \in Y$, and hence the 2-comb $S_2$ induced by $\{x, u, v, y\}$ is not settled by $w_{XY}$. Let $w$ be a vertex settling this 2-comb. By the minimality of $(X, Y)$ the pair $(X(w), Y(w))$ is not comparable to $(X, Y)$, and hence we must have a pair $(X', Y') \in \mathscr{X}\mathscr{Y}$ such that $X' \subseteq X$ and $Y' \subseteq Y$. Consequently, $w_{X'Y'} \in \mathscr{W}$ and $w_{XY} \neq w_{X'Y'}$.

In the sequel we consider pairs of vertices from $\mathscr{W}$ and derive some containment relations for the corresponding sets. First we consider pairs which are edges of $G$.

*Claim 5* If $(w_{XY}, w_{X'Y'}) \in E(G)$ and $X \cap X' \neq \emptyset$, then $Y \subseteq Y'$ or $Y' \subseteq Y$.

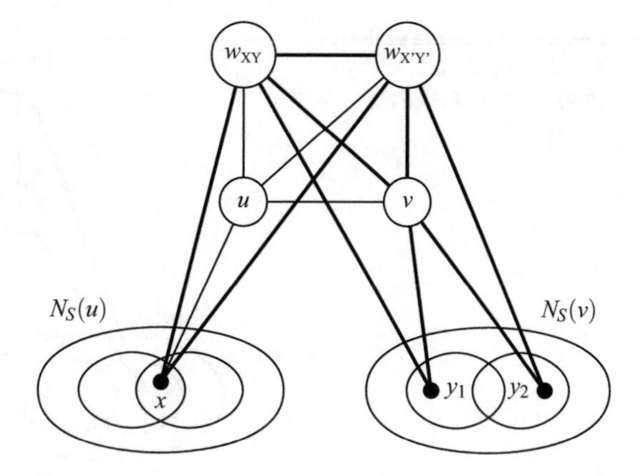


**Fig. 21** Illustration of the proof of Claim 5

*Proof* Assume by contradiction that there is a vertex $x \in X \cap X'$, but $Y \not\subseteq Y'$ and $Y' \not\subseteq Y$, that is, there are vertices $y_1 \in Y \setminus Y'$ and $y_2 \in Y' \setminus Y$. Then, an $\overline{S_3}$ is formed by $w_{XY}, w_{X'Y'}, v, x, y_1, y_2$ (see Figure 21), in contradiction to (i).

We next show a stronger version of the above claim, by proving proper containments.

*Claim 6* If $(w_{XY}, w_{X'Y'}) \in E(G)$ and $X \cap X' \neq \emptyset$, then either $Y \subsetneq Y'$ or $Y' \subsetneq Y$.

*Proof* Assume by contradiction that $X \cap X' \neq \emptyset$ and $Y = Y'$. By this assumption $Y \cap Y' \neq \emptyset$. Hence, we can apply Claim 5 (with the roles of $X$ and $Y$ exchanged), and conclude that $X \subseteq X'$ or $X' \subseteq X$.

Say, e.g., that $X \subseteq X'$. Then, $X \cup Y \subseteq X' \cup Y'$, and consequently we would not have both $w_{X,Y}$ and $w_{X',Y'}$ in $\mathscr{W}$, by its definitions.

*Claim 7* If $(w_{XY}, w_{X'Y'}) \in E(G)$, then exactly one of the following holds:

(a) $X \cap X' = Y \cap Y' = \emptyset$,
(b) $(X \subsetneq X'$ and $Y' \subsetneq Y)$,
(c) $(X' \subsetneq X$ and $Y \subsetneq Y')$.

*Proof* This follows from Claim 6 by applying it twice: once directly and once exchanging the roles of $X$ and $Y$. Since $X$, $Y$, $X'$ and $Y'$ are nonempty sets by (4), cases (a), (b), and (c) are pairwise exclusive.

Next we consider pairs of settling vertices that are not edges of $G$.

*Claim 8* If $(w_{XY}, w_{X'Y'}) \notin E(G)$, then either $X \subseteq X'$ or $Y \subseteq Y'$.

*Proof* If not, then there are vertices $x \in X \setminus X'$ and $y \in Y \setminus Y'$ such that $\{w_{XY}, u, v, x, y, w_{X'Y'}\}$ form a 3-anti-comb $\overline{S_3}$ (see Figure 22), in contradiction to condition (i).

Note that we cannot have both containments in the claim, because of the minimality of pairs in $\mathscr{X}\mathscr{Y}$.

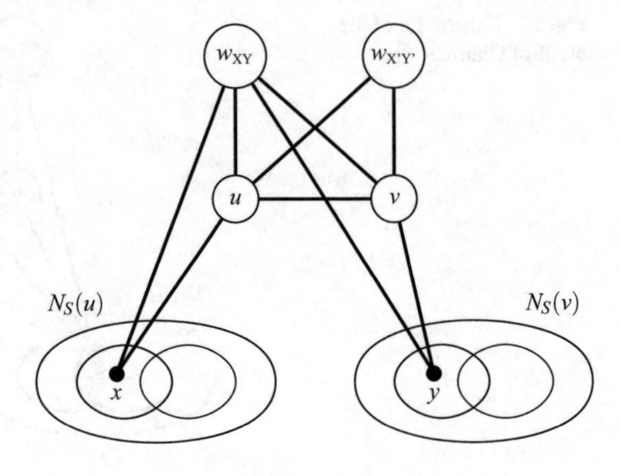


**Fig. 22** Illustration of the 3-anti-comb $\overline{S_3}$ induced by $\{w_{XY}, u, v, x, y, w_{X'Y'}\}$

*Claim 9* If $(w_{XY}, w_{X'Y'}) \notin E(G)$, then exactly one of the following must hold:

(a) $X \subsetneq X'$ and $Y' \subsetneq Y$,
(b) $X' \subsetneq X$ and $Y \subsetneq Y'$,
(c) $X = X'$,
(d) $Y = Y'$.

*Proof* Since the roles of $(X, Y)$ and $(X', Y')$ are symmetric, it follows directly by Claim 8 that one of (a), (b), (c), or (d) holds. To see that exactly one of them holds, it is enough to note that (c) and (d) together would contradict the minimality of the pairs $(X, Y) \in \mathscr{X}\mathscr{Y}$.

We are going to show next that if (c) or (d) holds in the previous claim for some vertices $w_{XY}, w_{X'Y'} \in \mathscr{W}$, then $G$ contains an induced $G_10$, as claimed in Lemma 2. For this end, let us first observe that if e.g., (d) holds, then we cannot have $X \subseteq X'$ or $X' \subseteq X$, by the minimality and uniqueness of pairs in $\mathscr{X}\mathscr{Y}$. Consequently, we can choose vertices $x \in X \setminus X'$, and $x' \in X' \setminus X$. Let us also choose an arbitrary vertex $y \in Y = Y'$ (which exists by (4)), and consider first the 2-comb $S_2$ induced by $\{x, u, v, y\}$. This 2-comb is settled by neither $w_{XY}$ nor $w_{X'Y'}$, and therefore there must be a vertex $w_{AB} \in \mathscr{W}$ settling it, since all 2-combs, containing $(u, v)$ as their middle edge, are settled by some vertices in $\mathscr{W}$.

*Claim 10* If $Y = Y'$, then $(w_{AB}, w_{XY}) \in E(G)$.

*Proof* Since $x \notin A$ and $y \notin B$ we have

$$X \nsubseteq A \text{ and } Y \nsubseteq B \tag{5}$$

implied. Assume indirectly that $(w_{AB}, w_{XY}) \notin E(G)$, then the previous observation implies that in Claim 9 applied to $w_{XY}$ and $w_{AB}$ none of (a), (b), (c) or (d) could hold. This contradiction proves the claim.

*Claim 11* If $Y = Y'$, then $A \cap X = B \cap Y = \emptyset$, $A \cup X = N_S(u)$ and $B \cup Y = N_S(v)$.

*Proof* Due to (5) only (a) of Claim 7 is possible, that is $A \cap X = B \cap Y = \emptyset$ is implied. Therefore the neighborhoods of $w_{AB}$ and $w_{XY}$ within $S$ are disjoint, and since they are subsets of the neighborhoods of $u$ and $v$, they cannot be proper subsets by property (iv), implying the statement.

*Claim 12* If $Y = Y'$, then $(w_{AB}, w_{X'Y'}) \notin E(G)$.

*Proof* Since $y \in Y' \setminus B$ and $x \in X \setminus A$ (since $w_{AB}$ is settling $\{x, u, v, y\}$), cases (b) and (c) of Claim 7 cannot hold for the pair $w_{AB}$ and $w_{X'Y'}$. Thus, if $(w_{AB}, w_{X'Y'}) \in E(G)$, then $A \cap X' = B \cap Y' = \emptyset$ would follow by Claim 7. Therefore, the neighborhoods of $w_{AB}$ and $w_{X'Y'}$ in $S$ are disjoint, and their union is a proper subset of $N_S(u) \cup N_S(v)$, in contradiction with property (iv). This contradiction proves the claim.

*Claim 13* If $Y = Y'$, then $A = X' = N_S(u) \setminus X$ and $Y = Y' = N_S(v) \setminus B$.

*Proof* Claims 11 and 9 applied to $w_{AB}$ and $w_{X'Y'}$ implies that only (c) of Claim 9 can hold. Thus, the statement implied by Claim 11 and (c) of Claim 9.

Let us still assume $Y = Y'$ and consider next the 2-comb induced by $\{x', u, v, y\}$ (where $x' \in X' \setminus X$). None of the vertices $w_{XY}$, $w_{X'Y'}$ and $w_{AB}$ settle this 2-comb, hence, there is a vertex $w_{A'B'} \in \mathscr{W}$ that settles it. By exchanging the roles of $w_{XY}$ and $w_{X'Y'}$ in Claims 10–13, we can conclude that

$$(w_{A'B'}, w_{XY}) \notin E(G), (w_{A'B'}, w_{X'Y'}) \in E(G), A' = X' \text{ and } B = B'. \tag{6}$$

*Claim 14* If $Y = Y'$ or $X = X'$, then $G$ contains an induced $G_{10}$.

*Proof* Note that the roles of conditions (c) and (d) in Claim 9 are perfectly symmetric, thus we could arrive to the same conclusions from both assumptions. Starting with $Y = Y'$ we arrived to the equalities of Claim 13 and (6). Choosing one vertex from each of the sets $X$, $Y$, $A$, and $B$, these four vertices together with $u$, $v$, $w_{XY}$, $w_{X'Y'}$, $w_{AB}$, and $w_{A'B'}$ form an induced $G_{10}$ by the above claims and definitions; see Figure 23.

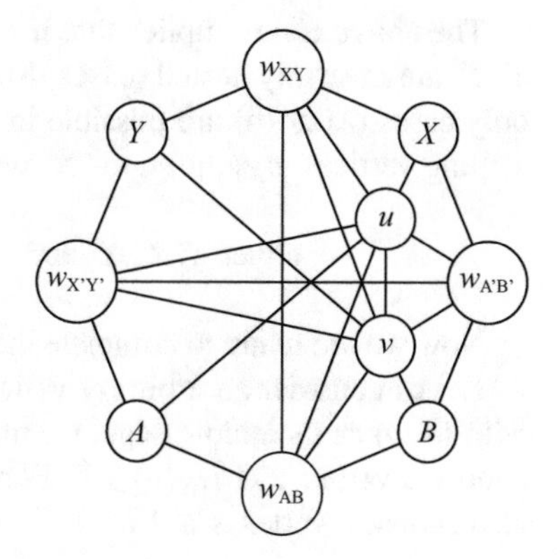


**Fig. 23** Illustration of the induced $G_{10}$ that appears by adding the settling vertices $w_{XY}, w_{X'Y'}, w_{AB}, w_{A'B'}$

For the rest of the proof, we assume that (a) or (b) of Claim 9 holds for every non-edge $(w_{XY}, w_{X'Y'}) \notin E(G)$. We are going to derive a contradiction from this assumption, completing the proof of Lemma 2.

First, we show that under the above assumption, case (a) of Claim 7 never holds.

*Claim 15* If $(w_{XY}, w_{X'Y'}) \in E(G)$, then either $X \cap X' \neq \emptyset$ or $Y \cap Y' \neq \emptyset$.

*Proof* Assume by contradiction that (a) of Claim 9 holds, that is that $X \cap X' = Y \cap Y' = \emptyset$. Then, by the minimality of $N_S(u) \cup N_S(v)$ as stated in property (iv), and by Claim 2, we know that $N_S(u) = X \cup X'$ and $N_S(v) = Y \cup Y'$.

Let us consider vertices $x \in X$ and $y \in Y'$ such that the set $\{x, u, v, y\}$ forms a 2-comb. This 2-comb is settled neither by $w_{XY}$ nor by $w_{X'Y'}$. Since every 2-comb with $(u, v)$ as a middle edge is settled by a vertex of $\mathscr{W}$, this 2-comb is also settled by one, say by a vertex $w_{AB} \in \mathscr{W}$. Let us now check the connections of this vertex to $w_{XY}$ and $w_{X'Y'}$. We consider two cases:

Case 1. If $(w_{AB}, w_{XY}) \notin E(G)$, then by Claim 9 we must have $A \subset X$ and $Y \subset B$, because $x \notin A$, and because we assumed that only cases (a) or (b) are possible in Claim 9.

If $(w_{AB}, w_{X'Y'}) \notin E(G)$, then by similar reasoning based on by Claim 9 and the fact that $y \notin B$ we can conclude that $X' \subset A$ and $B \subset Y'$. This, however, leads to a contradiction, since $A \subseteq X$ and $X \cap X' = \emptyset$.

Hence, we must have $(w_{AB}, w_{X'Y'}) \in E(G)$ in this case. Then by Claim 7 either $X' \cap A = Y' \cap B = \emptyset$ or $A, X'$ and $B, Y'$ are inversely nested. However, the latter is not possible, since $A \subset X$ and $X \cap X' = \emptyset$. In this case the neighborhoods of $w_{AB}$ and $w_{X'Y'}$ are disjoint in $S$, and their union is a proper subset of $N_S(u) \cup N_S(v)$ (since $x \notin A$), in contradiction with property (iv).

Case 2. If $(w_{AB}, w_{XY}) \in E(G)$, then (b) of Claim 7 is not possible, since $x \in X \setminus A$. If (a) holds, that is if $X \cap A = Y \cap B = \emptyset$, then the neighborhoods of $w_{AB}$ and $w_{XY}$ are disjoint in $S$, and their union is a proper subset of $N_S(u) \cup N_S(v)$ (since $y \in Y' \setminus B$), contradicting to property (iv). Consequently, case (c) holds, that is $A \subset X$ and $Y \subset B$, and consequently we can proceed as in Case 1.

In both cases we arrived to a contradiction, completing the proof of the claim.

The above claim implies that if $(w_{XY}, w_{X'Y'}) \in E(G)$, then the sets $X, X'$ and $Y, Y'$ are inversely nested (cases (b) or (c) in Claim 7). Since we also assumed that only cases (a) or (b) are possible in Claim 9, we can conclude that for all pairs of settling vertices $w_{XY}, w_{X'Y'} \in \mathscr{W}$ we have

$$\text{either } X \subset X' \text{ and } Y' \subset X \text{ or } X' \subset X \text{ and } Y \subset Y'. \tag{7}$$

Now we are ready to complete the proof of the lemma; see Figure 24.

Let us consider an arbitrary vertex $w_{XY} \in \mathscr{W}$. Since $w_{XY}$ is settling a 2-comb with $(u, v)$ as its middle edge, we must have $Y \neq N_S(v)$, and consequently we can choose a vertex $y \in N_S(v) \setminus Y$. Furthermore, we have $X \neq \emptyset$ by (4), thus we can also choose a vertex $x \in X$.

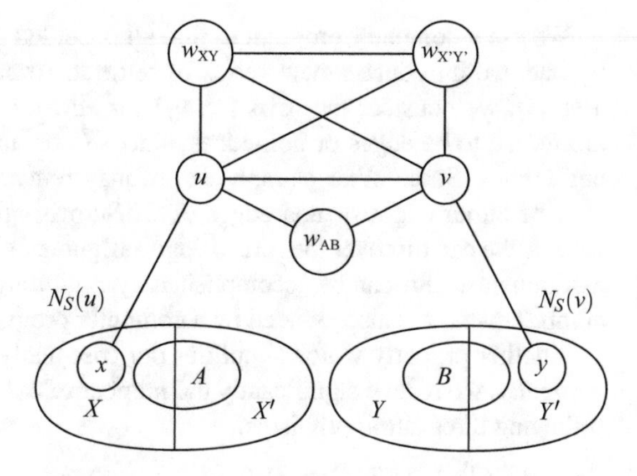


**Fig. 24** Illustration to the end of the proof of Lemma 2

Then, the 2-comb $S_2$ induced by $\{x, u, v, y\}$ is not settled by $w_{XY}$, and therefore there is a vertex $w_{X'Y'} \in \mathscr{W}$ settling this 2-comb. Then, by (7) we must have $X' \subseteq X \setminus \{x\}$ and $Y \subset Y'$, since $x \notin X'$.

Then, $X' \neq \emptyset$ by (4), so we can choose a vertex $x' \in X' \subsetneq X$. The 2-comb induced by $\{x', u, v, y\}$ is not settled by either $w_{XY}$ or $w_{X'Y'}$, and therefore there is a vertex $w_{X''Y''} \in \mathscr{W}$ settling this 2-comb.

Clearly, we can repeat the same arguments, and choose a vertex $x'' \in X'' \subsetneq X' \subsetneq X$, etc., resulting in an infinite chain $X \supsetneq X' \supsetneq X'' \supsetneq \cdots$ of strictly nested nonempty subsets, contradicting the finiteness of $G$. This concludes the proof of the lemma. □

## 2.3 *Proof of Lemma 3*

In this section we present the proof of Lemma 3, claiming that if $G$ contains $G_{10}$ as an induced subgraph and satisfies conditions (i) and (ii) of Section 2.1, then it must have an induced $2\mathscr{P}$ configuration (see Figures 8 and 17).

The proof is a case analysis that was assisted by a computer program. We assume by contradiction that there is a graph that has an induced $G_{10}$, has all 2-combs settled, and does not contain 3-combs and 3-anti-combs. The graph $G_{10}$ itself contains neither 3-combs nor 3-anti-combs, but it has several 2-combs that are not settled in it. For instance, such 2-combs are induced by $\{v_2, v_1, v_5, v_4\}$, $\{v_6, v_7, v_3, v_4\}$, $\{v_1, v_2, v_3, v_7\}$, etc. Therefore, some other vertices of $G$ must settle these 2-combs.

We show that in order to settle all 2-combs of $G_{10}$, the graph $G$ must contain a disjoint copy of $G_{10}$ such that the 20 vertices of these two $G_{10}$ subgraphs form an induced $2\mathscr{P}$ configuration. Since we do not know $G$, we try to extend $G_{10}$, and we show that this can be done essentially in a unique way.

We use a computer program to find all unsettled 2-combs of $G_{10}$. For each, one by one, we introduce a new vertex to settle it. After adding a settling vertex $v' \notin V(G_{10})$, we consider the pairs $(v', v_j)$ for all $v_j \in V(G_{10})$. Some of these pairs are forced to be edges or non-edges, since $G$ contains no induced 3-combs and 3-anti-combs. Some other pairs, however, may remain *uncertain*, that is those pairs may be either edges or non-edges of $G$. Surprisingly, all but one of the pairs are forced. We can discover the forced edge assignments by excluding all other possible assignments. This can be accomplished by exhibiting an induced 3-comb or 3-anti-comb. This task is also assisted by a computer program.

Another property which simplifies our case analysis is the symmetry of $G_{10}$. In particular, we reduce significantly the number of cases in our proof by means of the following three automorphisms:

$A_1$: $(3)(7)(1, 5)(2, 4)(6, 8)(0, 9)$
$A_2$: $(1)(5)(2, 8)(3, 7)(4, 6)(0, 9)$
$A_3$: $(7, 5, 3, 1)(8, 6, 4, 2)(0, 9)$

They are given in the *cycle* notation, that is $(i_1, i_2, \ldots, i_n)$ means the cyclic mapping $i_1 \mapsto i_2, i_2 \mapsto i_3, \ldots, i_n \mapsto i_1$. Figure 25 shows the graphs after the application of these automorphisms.

From now on we will choose some of the unsettled 2-combs to be settled, and try to fix as many edges and non-edges as possible. Even though the order that we pick the 2-combs may seem arbitrary, we follow an order that reduces the number of cases to be considered.

Let us choose first the 2-comb induced by $\{v_2, v_3, v_7, v_8\}$, and denote by $v'_1$ the vertex that settles it. The pairs $(v'_1, v_3)$ and $(v'_1, v_7)$ are forced to be edges, while $(v'_1, v_2)$ and $(v'_1, v_8)$ are forced to be non-edges, by the definition of settling. There are six more pairs, connecting $v'_1$ with $v_0, v_1, v_4, v_5, v_6$ and $v_9$, that remain uncertain.

Let us note first that $(v'_1, v_5)$ has to be a non-edge, since otherwise the vertices $\{v_3, v_7, v'_1, v_2, v_8, v_5\}$ form a 3-comb. Unlike $(v'_1, v_5)$, the pairs $(v'_1, v_0)$, $(v'_1, v_4)$, $(v'_1, v_6)$, $(v'_1, v_9)$ cannot be fixed if treated individually. But analyzing them together, we conclude that $(v'_1, v_4)$ and $(v'_1, v_6)$ are edges, while $(v'_1, v_0)$ and $(v'_1, v_9)$ are non-edges. Table 1 shows that in any other case there is an induced 3-comb or 3-anti-comb.

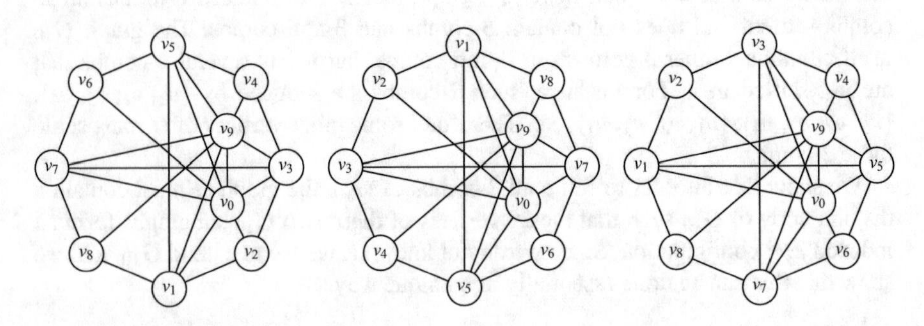


**Fig. 25** Graphs $A_1(G_{10})$, $A_2(G_{10})$, and $A_3(G_{10})$

**Table 1** Case analysis for the pairs $(v_1', v_0)$, $(v_1', v_4)$, $(v_1', v_6)$, $(v_1', v_9)$

| $(v_1', v_4)$ | $(v_1', v_6)$ | $(v_1', v_0)$ | $(v_1', v_9)$ | $S_3$ or $\overline{S_3}$ |
|---|---|---|---|---|
| 0 | 0 | 0 | 0 | $S_3: \{v_3, v_0, v_9, v_1', v_6, v_8\}$ |
| 0 | 0 | 0 | 1 | $\overline{S_3}: \{v_2, v_5, v_1', v_3, v_0, v_9\}$ |
| 0 | 0 | 1 | 0 | $\overline{S_3}: \{v_4, v_8, v_1', v_3, v_7, v_9\}$ |
| 0 | 0 | 1 | 1 | $\overline{S_3}: \{v_4, v_6, v_1', v_5, v_0, v_9\}$ |
| 0 | 1 | 0 | 0 | $S_3: \{v_5, v_6, v_0, v_2, v_4, v_1'\}$ |
| 0 | 1 | 0 | 1 | $S_3: \{v_3, v_9, v_1', v_2, v_6, v_8\}$ |
| 0 | 1 | 1 | 0 | $\overline{S_3}: \{v_4, v_8, v_1', v_3, v_7, v_9\}$ |
| 0 | 1 | 1 | 1 | $S_3: \{v_3, v_9, v_1', v_2, v_6, v_8\}$ |
| 1 | 0 | 0 | 0 | $S_3: \{v_3, v_0, v_9, v_6, v_8, v_1'\}$ |
| 1 | 0 | 0 | 1 | $\overline{S_3}: \{v_2, v_5, v_1', v_3, v_0, v_9\}$ |
| 1 | 0 | 1 | 0 | $S_3: \{v_4, v_5, v_9, v_6, v_8, v_1'\}$ |
| 1 | 0 | 1 | 1 | $S_3: \{v_7, v_0, v_1', v_2, v_4, v_8\}$ |
| 1 | 1 | 0 | 0 | None |
| 1 | 1 | 0 | 1 | $S_3: \{v_3, v_9, v_1', v_2, v_6, v_8\}$ |
| 1 | 1 | 1 | 0 | $S_3: \{v_7, v_0, v_1', v_2, v_4, v_8\}$ |
| 1 | 1 | 1 | 1 | $S_3: \{v_3, v_9, v_1', v_2, v_6, v_8\}$ |

**Table 2** Connections between $v_1'$ and $G_{10}$

| | $v_1$ | $v_2$ | $v_3$ | $v_4$ | $v_5$ | $v_6$ | $v_7$ | $v_8$ | $v_9$ | $v_0$ |
|---|---|---|---|---|---|---|---|---|---|---|
| $v_1'$ | $*$ | 0 | 1 | 1 | 0 | 1 | 1 | 0 | 0 | 0 |

**Table 3** Connections between $v_5'$ and $G_{10}$

| | $v_1$ | $v_2$ | $v_3$ | $v_4$ | $v_5$ | $v_6$ | $v_7$ | $v_8$ | $v_9$ | $v_0$ |
|---|---|---|---|---|---|---|---|---|---|---|
| $v_5'$ | 0 | 1 | 1 | 0 | $*$ | 0 | 1 | 1 | 0 | 0 |

Only one pair $(v_1', v_1)$ remains uncertain, since no induced $S_3$ nor $\overline{S_3}$ appears whether this pair is an edge or not.

Tables 2–16 show the connections between all pairs of vertices. For each pair, entry 1 means an edge, 0 means a non-edge, while $*$ means an uncertain pair.

Table 2 shows the connections between $v_1'$ and the vertices of $G_{10}$.

Next, we use automorphisms to simplify case analysis for the three 2-combs induced by $\{v_4, v_3, v_7, v_6\}$, $\{v_6, v_5, v_1, v_8\}$, and $\{v_2, v_1, v_5, v_4\}$, respectively, and not settled by $v_1'$.

Let us denote by $v_5'$ the vertex that settles $\{v_4, v_3, v_7, v_6\}$. By applying the automorphism $A_1$ to $G_{10}$, the 2-comb $\{v_2, v_3, v_7, v_8\}$ settled by $v_1'$ becomes $\{v_4, v_3, v_7, v_6\}$. Consequently, $v_5'$ should have the same connections as $v_1'$ has after applying $A_1$. Table 3 shows the connections between $v_5'$ and $G_{10}$.

Analogously, let us denote by $v_3'$ the vertex that settles $\{v_2, v_1, v_5, v_4\}$. By applying $A_3$ to $G_{10}$, $\{v_2, v_3, v_7, v_8\}$ becomes $\{v_2, v_1, v_5, v_4\}$. Therefore, $v_3'$ should have the same connections as $v_1'$ after transformation $A_3$. Table 4 shows the connections between $v_3'$ and $G_{10}$.

**Table 4** Connections between $v_3'$ and $G_{10}$

| | $v_1$ | $v_2$ | $v_3$ | $v_4$ | $v_5$ | $v_6$ | $v_7$ | $v_8$ | $v_9$ | $v_0$ |
|---|---|---|---|---|---|---|---|---|---|---|
| $v_3'$ | 1 | 0 | * | 0 | 1 | 1 | 0 | 1 | 0 | 0 |

**Table 5** Connections between $v_7'$ and $G_{10}$

| | $v_1$ | $v_2$ | $v_3$ | $v_4$ | $v_5$ | $v_6$ | $v_7$ | $v_8$ | $v_9$ | $v_0$ |
|---|---|---|---|---|---|---|---|---|---|---|
| $v_7'$ | 1 | 1 | 0 | 1 | 1 | 0 | * | 0 | 0 | 0 |

**Table 6** Case analysis for the pairs $(v_2', v_6)$, $(v_2', v_7)$, $(v_2', v_8)$, $(v_2', v_0)$

| $(v_2', v_6)$ | $(v_2', v_7)$ | $(v_2', v_8)$ | $(v_2', v_0)$ | $S_3$ or $\overline{S_3}$ |
|---|---|---|---|---|
| 0 | 0 | 0 | 0 | $S_3 : \{v_1, v_5, v_0, v_3, v_8, v_2'\}$ |
| 0 | 0 | 0 | 1 | $\overline{S_3} : \{v_6, v_8, v_2', v_7, v_0, v_9\}$ |
| 0 | 0 | 1 | 0 | $S_3 : \{v_4, v_5, v_2', v_3, v_6, v_8\}$ |
| 0 | 0 | 1 | 1 | $S_3 : \{v_4, v_5, v_2', v_3, v_6, v_8\}$ |
| 0 | 1 | 0 | 0 | $S_3 : \{v_1, v_5, v_0, v_3, v_8, v_2'\}$ |
| 0 | 1 | 0 | 1 | $\overline{S_3} : \{v_1, v_4, v_7, v_5, v_0, v_2'\}$ |
| 0 | 1 | 1 | 0 | $S_3 : \{v_4, v_5, v_2', v_3, v_6, v_8\}$ |
| 0 | 1 | 1 | 1 | $S_3 : \{v_4, v_5, v_2', v_3, v_6, v_8\}$ |
| 1 | 0 | 0 | 0 | $S_3 : \{v_1, v_5, v_0, v_3, v_8, v_2'\}$ |
| 1 | 0 | 0 | 1 | $\overline{S_3} : \{v_1, v_4, v_6, v_0, v_9, v_2'\}$ |
| 1 | 0 | 1 | 0 | $\overline{S_3} : \{v_1, v_7, v_2', v_5, v_6, v_0\}$ |
| 1 | 0 | 1 | 1 | $\overline{S_3} : \{v_1, v_4, v_6, v_0, v_9, v_2'\}$ |
| 1 | 1 | 0 | 0 | $S_3 : \{v_1, v_5, v_0, v_3, v_8, v_2'\}$ |
| 1 | 1 | 0 | 1 | $\overline{S_3} : \{v_1, v_4, v_6, v_0, v_9, v_2'\}$ |
| 1 | 1 | 1 | 0 | None |
| 1 | 1 | 1 | 1 | $\overline{S_3} : \{v_1, v_4, v_6, v_0, v_9, v_2'\}$ |

**Table 7** Connections between $v_2'$ and $G_{10}$

| | $v_1$ | $v_2$ | $v_3$ | $v_4$ | $v_5$ | $v_6$ | $v_7$ | $v_8$ | $v_9$ | $v_0$ |
|---|---|---|---|---|---|---|---|---|---|---|
| $v_2'$ | 0 | * | 0 | 1 | 1 | 1 | 1 | 1 | 1 | 0 |

Next, let us denote by $v_7'$ the vertex that settles $\{v_8, v_1, v_5, v_6\}$. By applying $A_3$ then $A_2$ to $G_{10}$, $\{v_2, v_3, v_7, v_8\}$ becomes $\{v_8, v_1, v_5, v_6\}$. Thus, $v_3'$ should have the same connections as $v_1'$ after transformations $A_3$ then $A_2$ (or the same connections as $v_3'$ after $A_2$). Table 5 shows the connections between $v_7'$ and $G_{10}$.

Next, let us consider four 2-combs induced by $\{v_5, v_1, v_2, v_3\}$, $\{v_1, v_5, v_4, v_3\}$, $\{v_7, v_3, v_4, v_5\}$, and $\{v_1, v_2, v_3, v_7\}$. They are not settled by any of the vertices of $G_{10}$, nor by $v_1', v_3', v_5', v_7'$.

Let $v_2'$ denote the vertex settling $\{v_3, v_4, v_5, v_1\}$. By definition of settling, the pairs $(v_2', v_4)$ and $(v_2', v_5)$ are edges, while $(v_2', v_1)$ and $(v_2', v_3)$ are non-edges. The pair $(v_2', v_9)$ must be an edge, since otherwise $\{v_1, v_3, v_2', v_4, v_5, v_9\}$ forms a 3-anti-comb. Table 6 shows the case analysis for the pairs $(v_2', v_6)$, $(v_2', v_7)$, $(v_2', v_8)$, and $(v_2', v_0)$. The only possible configuration is that $(v_2', v_6)$, $(v_2', v_7)$, $(v_2', v_8)$ are edges, and $(v_2', v_0)$ is not. The pair $(v_2', v_2)$ remains uncertain. Table 7 shows the connections between $v_2'$ and the vertices of $G_{10}$.

**Table 8** Connections between $v_4'$ and $G_{10}$

| | $v_1$ | $v_2$ | $v_3$ | $v_4$ | $v_5$ | $v_6$ | $v_7$ | $v_8$ | $v_9$ | $v_0$ |
|---|---|---|---|---|---|---|---|---|---|---|
| $v_4'$ | 1 | 1 | 0 | * | 0 | 1 | 1 | 1 | 0 | 1 |

**Table 9** Connections between $v_6'$ and $G_{10}$

| | $v_1$ | $v_2$ | $v_3$ | $v_4$ | $v_5$ | $v_6$ | $v_7$ | $v_8$ | $v_9$ | $v_0$ |
|---|---|---|---|---|---|---|---|---|---|---|
| $v_6'$ | 1 | 1 | 1 | 1 | 0 | * | 0 | 1 | 1 | 0 |

**Table 10** Connections between $v_8'$ and $G_{10}$

| | $v_1$ | $v_2$ | $v_3$ | $v_4$ | $v_5$ | $v_6$ | $v_7$ | $v_8$ | $v_9$ | $v_0$ |
|---|---|---|---|---|---|---|---|---|---|---|
| $v_8'$ | 0 | 1 | 1 | 1 | 1 | 1 | 0 | * | 0 | 1 |

Let $v_4'$ denote the vertex settling $\{v_5, v_1, v_2, v_3\}$. By applying $A_1$ to $G_{10}$, the subgraph $\{v_1, v_5, v_4, v_3\}$ becomes $\{v_5, v_1, v_2, v_3\}$. Therefore, vertex $v_4'$ must have the same connections as $v_2'$ after transformation $A_1$. Table 8 shows the connections between $v_4'$ and $G_{10}$.

Next, let $v_6'$ denote the vertex settling $\{v_7, v_3, v_4, v_5\}$. By applying transformations, first $A_1$ and then $A_3$, to $G_{10}$, the subgraph $\{v_1, v_5, v_4, v_3\}$ becomes $\{v_7, v_3, v_4, v_5\}$. Thus, $v_6'$ must have the same connections as $v_2'$ after the transformation $A_3 \circ A_1$. Table 9 shows the connections between $v_6'$ and $G_{10}$.

Let us next denote by $v_8'$ the vertex that settles $\{v_1, v_2, v_3, v_7\}$. By applying $A_3^{-1}$, to $G_{10}$, the subgraph $\{v_1, v_5, v_4, v_3\}$ becomes $\{v_1, v_2, v_3, v_7\}$. Therefore, $v_8'$ should have the same connections as $v_2'$ after $A_3^{-1}$. Table 10 shows the connections between $v_8'$ and $G_{10}$.

At this point, all $S_2$ subgraphs of $G_{10}$ are settled by some of the vertices $v_1'$, $v_2'$, ..., $v_8'$. Yet, nothing was said about the connections between those vertices. Nevertheless, all 3-combs and 3-anti-combs that appeared to indicate contradictions were independent from those connections; in other words, each of those subgraphs contains only one vertex $v_i'$ and the remaining five vertices are in $G_{10}$.

Interestingly, the connections between these eight vertices are uniquely implied. Table 11 shows the only possible assignments of edges and non-edges between the vertices $v_i'$ and $v_j'$, for $i, j = 1, \ldots, 8$, $i \neq j$. Each entry of the table contains the assignment, and the corresponding 3-comb or 3-anti-comb that would appear if the entry was reversed.

Let us notice that the pairs $(v_i, v_i')$ still remain uncertain. This means that all $2^8$ possible graphs have no induced 3-combs and 3-anti-combs. Yet, they contain some unsettled induced 2-combs.

Next, we introduce the automorphism $A_4$ of the current configuration, induced by the 18 vertices $V(G_{10}) \cup \{v_1', \ldots, v_8'\}$.

$A_4$: $(1, 3, 5, 7)(2, 4, 6, 8)(0, 9)(1', 3', 5', 7')(2', 4', 6', 8')$.

Let us further consider the unsettled 2-comb induced by $\{v_2, v_1', v_5', v_6\}$, and denote by $v_0'$ the vertex that settles it. By definition, $(v_0', v_1')$ and $(v_0', v_5')$ are edges, while $(v_0', v_2)$ and $(v_0', v_6)$ are non-edges. The pair $(v_0', v_9)$ cannot be an edge, since otherwise $\{v_1', v_5', v_0', v_2, v_6, v_9\}$ forms a 3-comb. Table 16 shows that $(v_0', v_4)$ and $(v_0', v_8)$ must be edges, while $(v_0', v_1)$, $(v_0', v_3)$, $(v_0', v_5)$ and $(v_0', v_7)$ must be non-

**Table 11** Case analysis for the connections between $v'_1, \ldots, v'_8$

| Edge | $S_3$ or $\overline{S_3}$ | Edge | $S_3$ or $\overline{S_3}$ |
|---|---|---|---|
| $(v'_1, v'_2) = 1$ | $S_3 : \{v_4, v_5, v'_2, v_8, v_0, v'_1\}$ | $(v'_1, v'_3) = 0$ | $S_3 : \{v_6, v'_1, v'_3, v_4, v_8, v_0\}$ |
| $(v'_1, v'_4) = 0$ | $S_3 : \{v_6, v'_1, v'_4, v_3, v_5, v_8\}$ | $(v'_1, v'_5) = 1$ | $\overline{S_3} : \{v_2, v_8, v'_1, v_3, v_7, v'_5\}$ |
| $(v'_1, v'_6) = 0$ | $S_3 : \{v_4, v'_1, v'_6, v_2, v_5, v_7\}$ | $(v'_1, v'_7) = 0$ | $S_3 : \{v_4, v'_1, v'_7, v_2, v_6, v_9\}$ |
| $(v'_1, v'_8) = 1$ | $S_3 : \{v_5, v_6, v'_8, v_2, v_9, v'_1\}$ | $(v'_2, v'_3) = 1$ | $S_3 : \{v_7, v_8, v'_2, v_4, v_0, v'_3\}$ |
| $(v'_2, v'_4) = 0$ | $\overline{S_3} : \{v_1, v_3, v'_2, v_7, v_0, v'_4\}$ | $(v'_2, v'_5) = 0$ | $S_3 : \{v_8, v'_5, v'_2, v_1, v_3, v_6\}$ |
| $(v'_2, v'_6) = 0$ | $\overline{S_3} : \{v_1, v_4, v_7, v_8, v'_2, v'_6\}$ | $(v'_2, v'_7) = 0$ | $S_3 : \{v_4, v'_7, v'_2, v_1, v_3, v_6\}$ |
| $(v'_2, v'_8) = 0$ | $\overline{S_3} : \{v_1, v_3, v'_2, v_5, v_0, v'_8\}$ | $(v'_3, v'_4) = 1$ | $S_3 : \{v_6, v_7, v'_4, v_2, v_9, v'_3\}$ |
| $(v'_3, v'_5) = 0$ | $S_3 : \{v_8, v'_5, v'_3, v_2, v_6, v_9\}$ | $(v'_3, v'_6) = 0$ | $S_3 : \{v_8, v'_3, v'_6, v_2, v_5, v_7\}$ |
| $(v'_3, v'_7) = 1$ | $\overline{S_3} : \{v_2, v_4, v'_3, v_1, v_5, v'_7\}$ | $(v'_3, v'_8) = 0$ | $S_3 : \{v_6, v'_3, v'_8, v_1, v_4, v_7\}$ |
| $(v'_4, v'_5) = 1$ | $S_3 : \{v_1, v_2, v'_4, v_6, v_9, v'_5\}$ | $(v'_4, v'_6) = 0$ | $\overline{S_3} : \{v_3, v_5, v'_4, v_1, v_9, v'_6\}$ |
| $(v'_4, v'_7) = 0$ | $S_3 : \{v_2, v'_7, v'_4, v_3, v_5, v_8\}$ | $(v'_4, v'_8) = 0$ | $\overline{S_3} : \{v_1, v_3, v_6, v_2, v'_4, v'_8\}$ |
| $(v'_5, v'_6) = 1$ | $S_3 : \{v_1, v_8, v'_6, v_4, v_0, v'_5\}$ | $(v'_5, v'_7) = 0$ | $S_3 : \{v_2, v'_5, v'_7, v_4, v_8, v_0\}$ |
| $(v'_5, v'_8) = 0$ | $S_3 : \{v_2, v'_5, v'_8, v_1, v_4, v_7\}$ | $(v'_6, v'_7) = 1$ | $S_3 : \{v_3, v_4, v'_6, v_8, v_0, v'_7\}$ |
| $(v'_6, v'_8) = 0$ | $\overline{S_3} : \{v_1, v_7, v'_8, v_3, v_9, v'_6\}$ | $(v'_7, v'_8) = 1$ | $S_3 : \{v_2, v_3, v'_8, v_6, v_9, v'_7\}$ |

**Table 12** Connections between $v'_0$ and $G_{10}$

| | $v_1$ | $v_2$ | $v_3$ | $v_4$ | $v_5$ | $v_6$ | $v_7$ | $v_8$ | $v_9$ | $v_0$ |
|---|---|---|---|---|---|---|---|---|---|---|
| $v'_0$ | 0 | 0 | 0 | 1 | 0 | 0 | 0 | 1 | 0 | * |

**Table 13** Connections between $v'_9$ and $G_{10}$

| | $v_1$ | $v_2$ | $v_3$ | $v_4$ | $v_5$ | $v_6$ | $v_7$ | $v_8$ | $v_9$ | $v_0$ |
|---|---|---|---|---|---|---|---|---|---|---|
| $v'_9$ | 0 | 1 | 0 | 0 | 0 | 1 | 0 | 0 | 0 | * |

edges. Furthermore, the pairs $(v'_0, v'_2)$, $(v'_0, v'_3)$, $(v'_0, v'_6)$, and $(v'_0, v'_7)$ must be edges, since otherwise one of the following 3-combs would appear: $\{v_4, v_5, v'_2, v_1, v_7, v'_0\}$, $\{v_1, v_8, v'_3, v_2, v_6, v'_0\}$, $\{v_1, v_8, v'_6, v_3, v_5, v'_0\}$, or $\{v_1, v_8, v_9, v_3, v'_7, v'_0\}$. The pairs $(v'_0, v'_4)$ and $(v'_0, v'_8)$ cannot be edges, since otherwise the 3-combs induced by $\{v_1, v_2, v'_4, v_3, v_5, v'_0\}$ and $\{v_2, v_3, v'_8, v_1, v_7, v'_0\}$ would appear. Finally, the pair $(v'_0, v_0)$ remains uncertain. Table 12 shows the connections between $v'_0$ and $G_{10}$.

Next, let us consider the 2-comb induced by $\{v'_3, v'_7, v_4, v_8\}$ and denote by $v'_9$ the vertex settling it. Notice that this 2-comb can be obtained from $\{v'_1, v'_5, v_2, v_6\}$ by applying transformation $A_4$. Therefore $v'_9$ must have the same connections as $v'_0$ after applying $A_4$. Table 13 shows the connections between $v'_9$ and $G_{10}$.

We summarize the connections between vertices... in Table 14; between... and... in Table 15, and finally between $v'_0$ and $v_1, v_3, v_4, v_5, v_7, v_8$ in Table 16.

Interestingly, the graph induced by $v'_1, \ldots, v'_9, v'_0$ is an isomorphic copy of $G_{10}$. Moreover, $(v_i, v'_j)$ for $i \neq j$ is an edge if and only if $(v_i, v_j)$ is not an edge, while the pairs $(v_i, v'_i)$, $i = 0, 1, \ldots, 9$ are uncertain. Thus, this configuration is the sum of two copies of $G_{10}$, that is, the graph $2G_{10}$; see Figure 26. Let us recall that to any graph $G$ we can apply the same operation and obtain the sum $G + G = 2G$.

**Table 14** Connections between vertices $v'_1, \ldots, v'_9, v'_0$

| | $v'_1$ | $v'_2$ | $v'_3$ | $v'_4$ | $v'_5$ | $v'_6$ | $v'_7$ | $v'_8$ | $v'_9$ | $v'_0$ |
|---|---|---|---|---|---|---|---|---|---|---|
| $v'_1$ | – | 1 | 0 | 0 | 1 | 0 | 0 | 1 | 1 | 1 |
| $v'_2$ | 1 | – | 1 | 0 | 0 | 0 | 0 | 0 | 0 | 1 |
| $v'_3$ | 0 | 1 | – | 1 | 0 | 0 | 1 | 0 | 1 | 1 |
| $v'_4$ | 0 | 0 | 1 | – | 1 | 0 | 0 | 0 | 1 | 0 |
| $v'_5$ | 1 | 0 | 0 | 1 | – | 1 | 0 | 0 | 1 | 1 |
| $v'_6$ | 0 | 0 | 0 | 0 | 1 | – | 1 | 0 | 0 | 1 |
| $v'_7$ | 0 | 0 | 1 | 0 | 0 | 1 | – | 1 | 1 | 1 |
| $v'_8$ | 1 | 0 | 0 | 0 | 0 | 0 | 1 | – | 1 | 0 |
| $v'_9$ | 1 | 0 | 1 | 1 | 1 | 0 | 1 | 1 | – | 1 |
| $v'_0$ | 1 | 1 | 1 | 0 | 1 | 1 | 1 | 0 | 1 | – |

**Table 15** Connections between vertices $v_1, \ldots, v_9, v_0$ and $v'_1, \ldots, v'_9, v'_0$

| | $v_1$ | $v_2$ | $v_3$ | $v_4$ | $v_5$ | $v_6$ | $v_7$ | $v_8$ | $v_9$ | $v_0$ |
|---|---|---|---|---|---|---|---|---|---|---|
| $v'_1$ | * | 0 | 1 | 1 | 0 | 1 | 1 | 0 | 0 | 0 |
| $v'_2$ | 0 | * | 0 | 1 | 1 | 1 | 1 | 1 | 1 | 0 |
| $v'_3$ | 1 | 0 | * | 0 | 1 | 1 | 0 | 1 | 0 | 0 |
| $v'_4$ | 1 | 1 | 0 | * | 0 | 1 | 1 | 1 | 0 | 1 |
| $v'_5$ | 0 | 1 | 1 | 0 | * | 0 | 1 | 1 | 0 | 0 |
| $v'_6$ | 1 | 1 | 1 | 1 | 0 | * | 0 | 1 | 1 | 0 |
| $v'_7$ | 1 | 1 | 0 | 1 | 1 | 0 | * | 0 | 0 | 0 |
| $v'_8$ | 0 | 1 | 1 | 1 | 1 | 1 | 0 | * | 0 | 1 |
| $v'_9$ | 0 | 1 | 0 | 0 | 0 | 1 | 0 | 0 | * | 0 |
| $v'_0$ | 0 | 0 | 0 | 1 | 0 | 0 | 0 | 1 | 0 | * |

Another remarkable property of the obtained configuration is as follows: if we exchange $v_0$ with $v'_0$ and $v_9$ with $v'_9$ then the resulting graph becomes the sum of two Petersen graphs, that is, $2\mathscr{P} \equiv 2G_{10}$, as shown in Figure 27.

This completes the proof of Lemma 3. □

## 2.4 *Proof of Lemma 4*

We prove that if a graph $G$ contains an induced $2\mathscr{P}$, then it must have either an unsettled 2-comb or an induced 3-comb or 3-anti-comb.

Let us recall that $2\mathscr{P}$ still has 10 uncertain edges. Hence, it gives us in fact 1024 possible graphs, one of which is an induced subgraph of $G$. Since we do not know which one, we will prove the statement by considering each such possible subgraphs.

Remarkably, none of these 1024 graphs contains an induced 3-comb or 3-anti-comb, as verified by computer.

**Table 16** Case analysis for the pairs $(v_0', v_1)$, $(v_0', v_3)$, $(v_0', v_4)$, $(v_0', v_5)$, $(v_0', v_7)$, and $(v_0', v_8)$

| $(v_0', v_1)$ | $(v_0', v_3)$ | $(v_0', v_4)$ | $(v_0', v_5)$ | $(v_0', v_7)$ | $(v_0', v_8)$ | $S_3$ or $\overline{S_3}$ |
|---|---|---|---|---|---|---|
| 0 | 0 | 0 | 0 | 0 | 0 | $S_3 : \{v_2, v_3, v_5', v_1, v_4, v_0'\}$ |
| 0 | 0 | 0 | 0 | 0 | 1 | $S_3 : \{v_1, v_8, v_9, v_2, v_4, v_0'\}$ |
| 0 | 0 | 0 | 0 | 1 | 0 | $S_3 : \{v_2, v_3, v_5', v_1, v_4, v_0'\}$ |
| 0 | 0 | 0 | 0 | 1 | 1 | $S_3 : \{v_1, v_8, v_9, v_2, v_4, v_0'\}$ |
| 0 | 0 | 0 | 1 | 0 | 0 | $S_3 : \{v_1, v_5, v_9, v_2, v_7, v_0'\}$ |
| 0 | 0 | 0 | 1 | 0 | 1 | $S_3 : \{v_1, v_5, v_9, v_2, v_7, v_0'\}$ |
| 0 | 0 | 0 | 1 | 1 | 0 | $S_3 : \{v_2, v_3, v_5', v_1, v_4, v_0'\}$ |
| 0 | 0 | 0 | 1 | 1 | 1 | $S_3 : \{v_1, v_8, v_9, v_2, v_4, v_0'\}$ |
| 0 | 0 | 1 | 0 | 0 | 0 | $S_3 : \{v_3, v_4, v_9, v_2, v_8, v_0'\}$ |
| 0 | 0 | 1 | 0 | 0 | 1 | None |
| 0 | 0 | 1 | 0 | 1 | 0 | $S_3 : \{v_3, v_4, v_9, v_2, v_8, v_0'\}$ |
| 0 | 0 | 1 | 0 | 1 | 1 | $S_3 : \{v_3, v_7, v_9, v_2, v_5, v_0'\}$ |
| 0 | 0 | 1 | 1 | 0 | 0 | $S_3 : \{v_1, v_5, v_9, v_2, v_7, v_0'\}$ |
| 0 | 0 | 1 | 1 | 0 | 1 | $S_3 : \{v_1, v_5, v_9, v_2, v_7, v_0'\}$ |
| 0 | 0 | 1 | 1 | 1 | 0 | $S_3 : \{v_3, v_4, v_9, v_2, v_8, v_0'\}$ |
| 0 | 0 | 1 | 1 | 1 | 1 | $S_3 : \{v_4, v_5, v_0', v_3, v_6, v_8\}$ |
| 0 | 1 | 0 | 0 | 0 | 0 | $S_3 : \{v_3, v_7, v_9, v_1, v_6, v_0'\}$ |
| 0 | 1 | 0 | 0 | 0 | 1 | $S_3 : \{v_1, v_8, v_9, v_2, v_4, v_0'\}$ |
| 0 | 1 | 0 | 0 | 1 | 0 | $\overline{S_3} : \{v_4, v_8, v_0', v_3, v_7, v_9\}$ |
| 0 | 1 | 0 | 0 | 1 | 1 | $S_3 : \{v_1, v_8, v_9, v_2, v_4, v_0'\}$ |
| 0 | 1 | 0 | 1 | 0 | 0 | $S_3 : \{v_1, v_5, v_9, v_2, v_7, v_0'\}$ |
| 0 | 1 | 0 | 1 | 0 | 1 | $S_3 : \{v_1, v_5, v_9, v_2, v_7, v_0'\}$ |
| 0 | 1 | 0 | 1 | 1 | 0 | $S_3 : \{v_3, v_7, v_0', v_2, v_5, v_8\}$ |
| 0 | 1 | 0 | 1 | 1 | 1 | $S_3 : \{v_1, v_8, v_9, v_2, v_4, v_0'\}$ |
| 0 | 1 | 1 | 0 | 0 | 0 | $S_3 : \{v_3, v_7, v_9, v_1, v_6, v_0'\}$ |
| 0 | 1 | 1 | 0 | 0 | 1 | $S_3 : \{v_3, v_4, v_0', v_2, v_5, v_8\}$ |
| 0 | 1 | 1 | 0 | 1 | 0 | $S_3 : \{v_4, v_5, v_9, v_6, v_8, v_0'\}$ |
| 0 | 1 | 1 | 0 | 1 | 1 | $S_3 : \{v_3, v_4, v_0', v_2, v_5, v_8\}$ |
| 0 | 1 | 1 | 1 | 0 | 0 | $S_3 : \{v_1, v_5, v_9, v_2, v_7, v_0'\}$ |
| 0 | 1 | 1 | 1 | 0 | 1 | $S_3 : \{v_1, v_5, v_9, v_2, v_7, v_0'\}$ |
| 0 | 1 | 1 | 1 | 1 | 0 | $S_3 : \{v_3, v_7, v_0', v_2, v_5, v_8\}$ |
| 0 | 1 | 1 | 1 | 1 | 1 | $S_3 : \{v_3, v_1', v_0', v_2, v_6, v_8\}$ |
| 1 | 0 | 0 | 0 | 0 | 0 | $S_3 : \{v_1, v_5, v_9, v_3, v_6, v_0'\}$ |
| 1 | 0 | 0 | 0 | 0 | 1 | $S_3 : \{v_1, v_5, v_9, v_3, v_6, v_0'\}$ |
| 1 | 0 | 0 | 0 | 1 | 0 | $S_3 : \{v_1, v_5, v_9, v_3, v_6, v_0'\}$ |
| 1 | 0 | 0 | 0 | 1 | 1 | $S_3 : \{v_1, v_5, v_9, v_3, v_6, v_0'\}$ |
| 1 | 0 | 0 | 1 | 0 | 0 | $S_3 : \{v_3, v_7, v_1', v_2, v_8, v_0'\}$ |
| 1 | 0 | 0 | 1 | 0 | 1 | $S_3 : \{v_3, v_7, v_5', v_4, v_6, v_0'\}$ |
| 1 | 0 | 0 | 1 | 1 | 0 | $S_3 : \{v_1, v_5, v_0', v_2, v_4, v_7\}$ |

(continued)

**Table 16** (continued)

| 1 | 0 | 0 | 1 | 1 | 1 | $S_3 : \{v_1, v_5, v_0', v_2, v_4, v_7\}$ |
|---|---|---|---|---|---|---|
| 1 | 0 | 1 | 0 | 0 | 0 | $S_3 : \{v_1, v_5, v_9, v_3, v_6, v_0'\}$ |
| 1 | 0 | 1 | 0 | 0 | 1 | $S_3 : \{v_1, v_5, v_9, v_3, v_6, v_0'\}$ |
| 1 | 0 | 1 | 0 | 1 | 0 | $S_3 : \{v_1, v_5, v_9, v_3, v_6, v_0'\}$ |
| 1 | 0 | 1 | 0 | 1 | 1 | $S_3 : \{v_1, v_5, v_9, v_3, v_6, v_0'\}$ |
| 1 | 0 | 1 | 1 | 0 | 0 | $S_3 : \{v_3, v_4, v_9, v_2, v_8, v_0'\}$ |
| 1 | 0 | 1 | 1 | 0 | 1 | $S_3 : \{v_1, v_8, v_0', v_2, v_4, v_7\}$ |
| 1 | 0 | 1 | 1 | 1 | 0 | $S_3 : \{v_3, v_4, v_9, v_2, v_8, v_0'\}$ |
| 1 | 0 | 1 | 1 | 1 | 1 | $S_3 : \{v_4, v_5, v_0', v_3, v_6, v_8\}$ |
| 1 | 1 | 0 | 0 | 0 | 0 | $S_3 : \{v_6, v_7, v_1', v_5, v_8, v_0'\}$ |
| 1 | 1 | 0 | 0 | 0 | 1 | $S_3 : \{v_3, v_1', v_0', v_2, v_6, v_8\}$ |
| 1 | 1 | 0 | 0 | 1 | 0 | $S_3 : \{v_3, v_7, v_0', v_1, v_4, v_6\}$ |
| 1 | 1 | 0 | 0 | 1 | 1 | $S_3 : \{v_3, v_7, v_0', v_1, v_4, v_6\}$ |
| 1 | 1 | 0 | 1 | 0 | 0 | $S_3 : \{v_1, v_5, v_0', v_3, v_6, v_8\}$ |
| 1 | 1 | 0 | 1 | 0 | 1 | $S_3 : \{v_3, v_1', v_0', v_2, v_6, v_8\}$ |
| 1 | 1 | 0 | 1 | 1 | 0 | $S_3 : \{v_1, v_5, v_0', v_2, v_4, v_7\}$ |
| 1 | 1 | 0 | 1 | 1 | 1 | $S_3 : \{v_1, v_5, v_0', v_2, v_4, v_7\}$ |
| 1 | 1 | 1 | 0 | 0 | 0 | $S_3 : \{v_4, v_5, v_9, v_6, v_8, v_0'\}$ |
| 1 | 1 | 1 | 0 | 0 | 1 | $S_3 : \{v_1, v_8, v_0', v_2, v_4, v_7\}$ |
| 1 | 1 | 1 | 0 | 1 | 0 | $S_3 : \{v_4, v_5, v_9, v_6, v_8, v_0'\}$ |
| 1 | 1 | 1 | 0 | 1 | 1 | $S_3 : \{v_3, v_4, v_0', v_2, v_5, v_8\}$ |
| 1 | 1 | 1 | 1 | 0 | 0 | $S_3 : \{v_1, v_5, v_0', v_3, v_6, v_8\}$ |
| 1 | 1 | 1 | 1 | 0 | 1 | $S_3 : \{v_1, v_8, v_0', v_2, v_4, v_7\}$ |
| 1 | 1 | 1 | 1 | 1 | 0 | $S_3 : \{v_1, v_5, v_0', v_3, v_6, v_8\}$ |
| 1 | 1 | 1 | 1 | 1 | 1 | $S_3 : \{v_3, v_1', v_0', v_2, v_6, v_8\}$ |

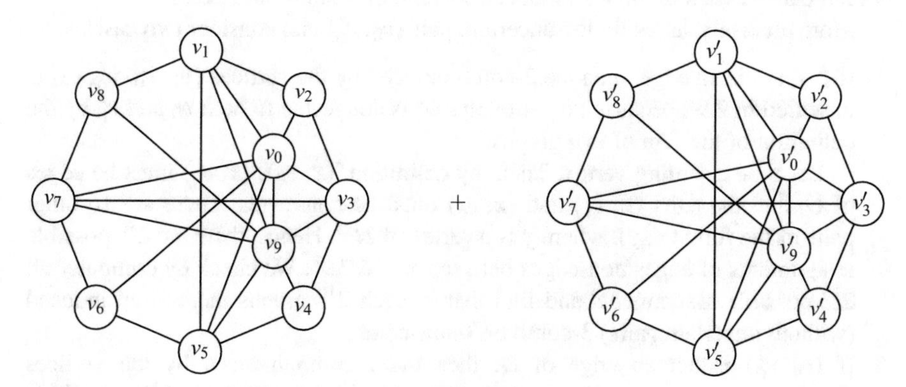


**Fig. 26** The sum of two graphs $G_{10}$ or $2G_{10}$

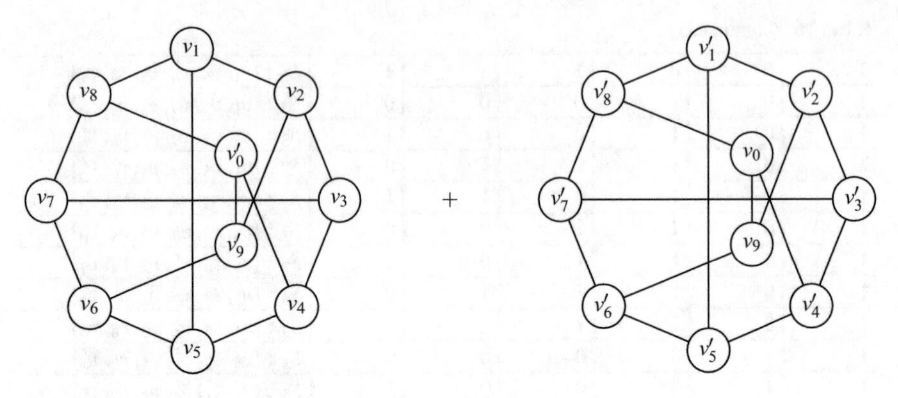


**Fig. 27** The graph $2\mathscr{P}$, isomorphic to $2G_{10}$ by exchanging $v_0$, $v'_0$ and $v_9$, $v'_9$

Furthermore, $2\mathscr{P}$ itself contains no induced 2-combs either. (Since $2\mathscr{P}$ contains uncertain pairs, we call a subgraph of $2\mathscr{P}$ an induced one only if it does not involve any uncertain pair.) However, each of the 1024 graphs obtained from $2\mathscr{P}$ contains many 2-combs each of which involves exactly one pair of vertices $v_i$ and $v'_i$ for some index $i$.

Now we will fix one of the uncertain pairs (once as an edge and once as a non-edge), while keeping all others uncertain. Several (36) unsettled induced 2-combs appear that contain the fixed uncertain pair. Each of these 2-combs must be settled in $G$ by our assumption (i), thus there exists a vertex $x$ settling it. There are 16 pairs $(x, y)$, where $y$ is a vertex of $2\mathscr{P}$, not belonging to the unsettled 2-comb. We check all $2^{16}$ possible edge/non-edge assignments to these 16 pairs, and find by computer search that for each of them an induced 3-comb or 3-anti-comb exists.

More precisely, let us fix the uncertain pair $(v_0, v'_0)$ and consider two cases:

1. If $(v_0, v'_0)$ is an edge, then the 2-comb induced by the vertices $\{v_1, v_0, v'_0, v_4\}$ is unsettled in $2\mathscr{P}$, because no vertex in $2\mathscr{P}$ is connected to both $v_0$ and $v'_0$ by the definition of the sum of two graphs.

   Let $x$ be a settling vertex. Then, by definition, $(x, v_0)$, $(x, v'_0)$ must be edges of $G$, and the pairs $(x, v_1)$ and $(x, v_4)$ must be non-edges. There are 16 other pairs of the form $(x, y)$, where $y$ is a vertex of $2\mathscr{P}$. Hence, there are $2^{16}$ possible assignments of edges/non-edges between $x$ and $2\mathscr{P}$. We check by computer all $2^{16}$ possible assignments and find that in each $2^{16}$ graphs there is an induced (without uncertain pairs) 3-comb or 3-anti-comb.
2. If $(v_0, v'_0)$ is not an edge of $G$, then the 2-comb induced by the vertices $\{v_0, v_1, v_8, v'_0\}$ is not settled in $2\mathscr{P}$. Since it must be settled in $G$ by condition (i), there is a vertex $x$ of $G$ that settles it. Similarly to the previous case, we again consider all $2^{16}$ graphs, and find by computer search that all of them contain an induced 3-comb or 3-anti-comb.

This concludes the proof of Lemma 4. □

## 3 Proof of Theorems 3 and 4

*Proof of Theorem 3* Recall by (1) that we can reduce the case analysis by assuming that $1 \le k < \ell \le n-2$.

**We Start By Proving (i)** Assume by contradiction that there exists an unsettled $\overline{S_m} = \{B_1, \ldots, B_m, A_1, \ldots, A_m\}$, $|B_i| = k$, $|A_i| = \ell$. Then, by assumption we must have

$$A_i \supset B_j \text{ for all } j \neq i \text{ and } A_i \not\supset B_i. \tag{8}$$

Let us recall that $\overline{S_m}$ is settled by a $k$-set $K$ iff $K \subseteq \bigcap_{j=1}^m A_j$, and it is settled by an $\ell$-set $L$ iff $L \not\supset B_i$ for $i = 1, \ldots, m$.

Let $\mathscr{B} = \{B_1, \ldots, B_m\}$, and let $X \subseteq [n]$ be the set that contains the elements that are in more than one of the $B_i$'s, i.e. $X = \{x \in [n] \mid \deg_{\mathscr{B}}(x) > 1\}$. Notice that $X \subseteq \bigcap_{j=1}^m A_j$ because by (8) we have that every vertex belonging to two or more of the sets from $\mathscr{B}$ must belong to all sets $A_i$, $i = 1, \ldots, m$. Clearly $|X| < k$, otherwise $\overline{S_m}$ would be settled by a $k$-set in $X$.

In the following steps of the proof, we will derive some inequalities, to arrive to a contradiction. First, we need some more definitions.

Let $a_p$, $p = 0, 1, \ldots q \le |X| < k$, be the number of sets $B_i \in \mathscr{B}$ for which $|B_i \cap X| = p$, and let $\mathscr{H} = \{B_i \cap X | i = 1, \ldots, m\}$. Let us observe first that $\tau(\mathscr{B}) \le \tau(\mathscr{H}) + a_0$, where $\tau$ denotes the size of a minimum vertex cover. To see this inequality, let us first cover the intersecting hyperedges of $\mathscr{B}$ optimally by $\tau(\mathscr{H})$ vertices, and then cover the rest by choosing one vertex from each remaining set outside of $X$ (i.e., by at most $a_0$ additional vertices). Moreover, we have $\tau(\mathscr{B}) > n - \ell$, since otherwise there exists an $\ell$-set settling $\overline{S_m}$. Thus, we can conclude that

$$\tau(\mathscr{H}) + a_0 \ge n - \ell + 1 \tag{9}$$

Assume w.l.o.g. that $|B_1 \cap X| \le |B_2 \cap X| \le \ldots \le |B_m \cap X|$. Since we know by (8) that $\bigcup_{j=1}^{m-1} B_j \subseteq A_m$, we have:

$$|\bigcup_{i=1}^{m-1} B_j| = |X| + \sum_{p=0}^{q} (k-p)a_p - (k-q) \le \ell \tag{10}$$

Let us now take away $k$ times Equation (9) from (10) and obtain

$$|X| + \sum_{p=0}^{q} (k-p)a_p - (k-q) - k(\tau(\mathscr{H}) + a_0) \le \ell - k(n - \ell + 1)$$

which can be simplified to

$$|X| + \sum_{p=1}^{q}(k-p)a_p + q - k\tau(\mathscr{H}) \le (k+1)\ell - kn \tag{11}$$

Notice that the right-hand side of (11) is negative by our initial assumption of $kn > (k+1)\ell$. Thus, to arrive to a contradiction, it is enough to prove that

$$k\tau(\mathscr{H}) \le |X| + \sum_{p=1}^{q}(k-p)a_p + q. \tag{12}$$

Let us observe next that $\sum_{p=1}^{q}(k \ a_p) = k|\mathscr{H}|$, and that $\sum_p(p \ a_p) = \sum_{H\in\mathscr{H}}|H|$. Thus, we can equivalently rewrite inequality (12) as:

$$k(|\mathscr{H}| - \tau(\mathscr{H})) \ge \sum_{H\in\mathscr{H}}|H| - |X| - q \tag{13}$$

To show (13), let us construct a cover $C$ of $\mathscr{H}$ as follows. First we choose into $C$ a vertex of the highest degree in $\mathscr{H}$. This vertex covers at least $\frac{\sum_{H\in\mathscr{H}}|H|}{|X|}$ hyperedges of $\mathscr{H}$. We cover the remaining edges by choosing one vertex from each. This simple procedure shows that

$$\tau(\mathscr{H}) \le |C| \le |\mathscr{H}| - \frac{\sum_{H\in\mathscr{H}}|H|}{|X|} + 1. \tag{14}$$

From this simple inequality we can derive the following:

$$\begin{aligned} k(|\mathscr{H}| - \tau(\mathscr{H})) &\ge \tfrac{k}{|X|}\textstyle\sum_{H\in\mathscr{H}}|H| - k \\ &= \textstyle\sum_{H\in\mathscr{H}}|H| + \tfrac{k-|X|}{|X|}\textstyle\sum_{H\in\mathscr{H}}|H| - k \\ &\ge \textstyle\sum_{H\in\mathscr{H}}|H| - |X| \end{aligned}$$

where the second inequality follows from $|X| \le \sum_{H\in\mathscr{H}}|H|$, which is true, since every vertex of $X$ has degree at least 2 in $\mathscr{B}$. The above inequalities then prove (13), since $q \ge 0$, which then yields the desired contradiction, completing the proof of (i). □

**We Prove Next (ii)** We will show, by a construction that an unsettled $\overline{S_m}$ exists in $G(n, k, \ell)$, whenever $kn \le (k+1)\ell$ and $n \ge k + \ell$.

For this let us set $r \equiv \ell \pmod{k}$, $0 \le r < k$, $m = \frac{\ell+k-r}{k}$, and let $B_1, \ldots, B_m$, and $R$ be pairwise disjoint subsets of $[n] = \{1, 2, \ldots, n\}$, such that $|R| = r$ and $|B_i| = k$ for $i = 1, \ldots, m$. Notice that

$$|R \cup B_1 \cup \cdots \cup B_m| = km + r = \ell + k. \tag{15}$$

Thus, it is possible to choose such pairwise disjoint subsets, since $k + \ell \leq n$ by our assumption. Let us further define

$$A_i = R \cup \left( \bigcup_{j \neq i} B_j \right) \text{ for } i = 1, \ldots, m.$$

With these definitions, we have $|A_i| = r + k(m-1) = r + (\ell - r) = \ell$ for all $i = 1, \ldots, m$. Furthermore, $A_i \supseteq B_j$ if and only if $i \neq j$. Thus, the sets $A_1, \ldots, A_m$, and $B_1, \ldots, B_m$ are vertices of $G(m, k, \ell)$ forming an $\overline{S_m}$.

We show that this $\overline{S_m}$ is unsettled in $G(n, k, \ell)$. For this, observe first that $|\bigcap_{i=1}^m A_i| = |R| = r < k$, and consequently, no $k$-set can settle $\overline{S_m}$.

Next, let us assume indirectly that there is an $\ell$-set $L$ which settles $\overline{S_m}$. Hence, $L$ cannot be connected in $G(n, k, \ell)$ to any of the $B_i$'s. In other words, $L \not\supseteq B_i$ for $i = 1, \ldots, m$. It follows that $|L \cap B_i| \leq k - 1$ for all $i = 1, \ldots, m$, implying

$$|L| \leq m(k-1) + r + (n - k - \ell). \tag{16}$$

That is, we can take at most $k-1$ elements from each of the $k$-sets, and the remaining $r + n - k - \ell$ elements of $[n]$, as implied by (15). It is now enough to show that $|L| < \ell$, because this contradicts the assumption that $L$ is an $\ell$-set. To do this, let us rewrite (16) as

$$|L| \leq m(k-1) + r + (n - k - \ell) = \frac{\ell + k - r}{k}(k-1) + r - n - k - \ell,$$

which implies

$$\begin{aligned} k|L| + \ell &\leq (\ell + k - r)(k-1) + k(r - n - k - \ell) + \ell \\ &= k\ell - \ell + k^2 - k - kr + r + kr + kn - k^2 - k\ell + \ell \\ &= kn - (k - r) \ < \ kn \ \leq \ (k+1)\ell \end{aligned}$$

where the last two inequalities follow by $k > r$ and our assumption that $kn \leq (k+1)\ell$. Thus, $|L| < \ell$ follows, completing the proof of (ii). □

This completes the proof of Theorem 3. □

*Proof of Theorem 4*

**We Prove First (a)** Even though this claim is only for $k \leq 2$, let us first disregard this restriction. Assume by contradiction that there exists an unsettled $S_m$ in $G(m, k, \ell)$ defined by the sets $\{B_1, \ldots, B_m, A_1, \ldots, A_m\}$, where $|B_i| = k$, $|A_i| = \ell$, for $i = 1, \ldots, m$, and $B_j \subseteq A_i$, iff $i = j$. Set $\mathscr{B} = \{B_1, \ldots, B_m\}$ and $\mathscr{A} = \{A_1, \ldots, A_m\}$.

By definitions, an $\ell$-set $L$ can settle $S_m$ only if $[n] \setminus L$ is a vertex cover of the hypergraph $\mathscr{B}$. Furthermore, a $k$-set $K$ can settle $S_m$, only if $K \subseteq A_i$ for all

$i = 1, \ldots, m$. Since $S_m$ is assumed to be unsettled in $G(n, k, \ell)$, we must have the following properties.

(i) $\tau(\mathscr{B}) \geq n - \ell + 1$, since otherwise the complement of a minimum vertex cover of $\mathscr{B}$ would contain a settling $\ell$-set.

(ii) $|\bigcap_{i=1}^{m} A_i| < k$, since otherwise the intersection of the sets of $\mathscr{A}$ would contain a settling $k$-set.

Let us also observe that $B_j \subseteq A_i$ if and only if $i = j$ implies that $\overline{A_i} = [n] \setminus A_i$ is a vertex cover for $\mathscr{B} \setminus B_i$, implying $|\overline{A_i}| = n - \ell \geq \tau(\mathscr{B} \setminus \{B_i\}) \geq \tau(\mathscr{B}) - 1$. This, together with (i), implies that

$$n - \ell \;=\; \tau(\mathscr{B}) - 1 \;=\; \tau(\mathscr{B} \setminus \{B_i\}) \tag{17}$$

for all $i = 1, \ldots, m$.

Let us now consider the subset

$$X \;=\; [n] \setminus \bigcup_{i=1}^{m} B_i.$$

Equation (17) imply that $X \subseteq A_i$ for all $i = 1, \ldots, m$. Thus, by property (ii) we must have

$$|X| \;\leq\; k - 1 \tag{18}$$

Another consequence of (17) is that the hypergraph $\mathscr{B}$ is $\tau$-critical, i.e., the minimum vertex cover size strictly decreases whenever we remove a hyperedge from $\mathscr{B}$. This also implies that $\mathscr{B}$ is $\alpha$-critical, where $\alpha(\mathscr{B})$ is the size of the largest *independent set* of $\mathscr{B}$, i.e., the largest set not containing a hyperedge of $\mathscr{B}$. This is because $\alpha(\mathscr{B}) + \tau(\mathscr{B}) = n$ for all hypergraphs $\mathscr{B}$.

Let us now consider the case of $k = 1$. In this case we have $|\mathscr{B}| = \tau(\mathscr{B})$ and by (18) $X = \emptyset$, implying that $|\mathscr{B}| = n$, which together with the previous equality and (17) implies

$$n = |\mathscr{B}| = \tau(\mathscr{B}) = n - \ell + 1$$

from which $\ell = 1$ follows, contradicting (1).

Let us next consider the case of $k = 2$. In this case $\mathscr{B}$ is an $\alpha$-critical graph $G$ on vertex-set $V = [n] \setminus X$, with $\alpha(G) = \alpha(\mathscr{B}) - |X| = \ell - 1 - |X|$.

We apply a result attributed to Erdős and Gallai (see Exercise 8.20 in [36]; see also the proof of Exercise 8.10 by Hajnal), stating that in an $\alpha$-critical graph $G$ with no isolated vertices we have $|V| \geq 2\alpha(G)$. This implies for our case that $n - |X| \geq 2(\ell - 1 - |X|)$, from which

$$n \geq 2\ell - 2 - |X|$$

follows. Since by (18) we have $|X| \leq k - 1 = 1$, the above inequality implies

$$n \geq 2\ell - 3$$

contradicting (a) of Theorem 4, according to which we have $n < 2\ell - 3$. □

*Remark 10* We could extend the above line of arguments for $k \geq 3$, if the inequality $n \geq \frac{k}{k-1}\alpha(\mathscr{B})$ were valid for $\alpha$-critical $k$-uniform hypergraphs, in general. However, this is not the case, as the following example shows: let $n = 10$, $k = 3$ and $\mathscr{B} = \{\{1, 2, 3\}, \{3, 4, 5\}, \{5, 6, 7\}, \{7, 8, 9\}, \{9, 10, 1\}\}$. In this case we have $\alpha(\mathscr{B}) = 7$, and $10 \not\geq (3/2)7 = 21/2$.

**Now We Prove (b)** We will now provide a construction for an unsettled $S_m$. Let $L = \{2, 3, \ldots, k\}$, and choose $r \in L$, such that $r \equiv \ell \pmod{k-1}$ (for instance, if $k = 2$ then we have $r = 2$).

Let us next partition $[n]$ as

$$[n] = X \cup \bigcup_{j=1}^{p} Q_j,$$

where $|X| = r - 1$, $p = \frac{\ell - r}{k-1}$, and where the sets $Q_1, \ldots, Q_p$ are almost equal, i.e., $|Q_i| \sim \frac{n-r+1}{p}$.

Then, we construct an unsettled $S_m = \{B_1, \ldots, B_m, A_1, \ldots, A_m\}$ as follows. We define $\binom{m=\sum_{j=1}^{p}|Q_i|}{k}$, and the sets $B_i, i = 1, \ldots, m$ are the$k$-subsets of the $Q_j$-s, i.e.,

$$\binom{\{B_1, .., B_m\} = \bigcup_{i=1}^{p} Q_i}{k.}$$

Finally, we set for $i = 1, \ldots, m$

$$A_i = X \cup B_i \cup \bigcup_{\substack{1 \leq j \leq p \\ j \neq j^*}} R_{ij},$$

where $B_i \subseteq Q_{j^*}$ and $R_{ij} \subseteq Q_j$, $|R_{ij}| = k - 1$ for all $j \neq j^*$. In other words, each $A_i$ contains $X$, the corresponding set $B_i$, and $k - 1$ points from each set $Q_j$ not containing $B_i$.

It is easy to see that $|A_i| = \ell$. Indeed,

$$\begin{aligned} |A_i| &= k + r - 1 + (p - 1)(k - 1) \\ &= k + r - 1 + \left(\frac{\ell - r}{k - 1} - 1\right)(k - 1) \\ &= r + \ell - r = \ell \end{aligned}$$

Let us observe first that by the above calculations no $\ell$-set can settle $S_m$. This is because all $\ell$-sets must intersect at least one of the $Q_j$'s in $k$ or more points, therefore any $\ell$-set contains at least one of the $B_i$'s.

Furthermore, we can show that $|Q_j| \geq k$, for $j = 1, \ldots, p$. By our assumption we have $n(k-1) \geq \ell k - r - k + 1$ from which we can derive the following chain of inequalities:

$$\begin{aligned} n &\geq \ell\frac{k}{k-1} - \frac{k+r-1}{k-1} \\ n(k-1) &\geq k\ell - k - r + 1 \\ n(k-1) - kr + k + r - 1 &\geq k\ell - kr \\ (n-r+1)(k-1) &\geq k\ell - kr \\ (n-r+1) &\geq k\frac{\ell - r}{k-1} = kp \\ \frac{n-r+1}{p} &\geq k, \end{aligned}$$

which implies that $|Q_j| \geq \lfloor\frac{n-r+1}{p}\rfloor \geq k$.

Finally we have to prove that no $k$-set can settle $S_m$. For this, as we remarked earlier, it is enough to show that $|\bigcap_{i=1}^m A_i| < k$, which will follow from

$$\left(\bigcap_{i=1}^m A_i\right) \cap Q_j = \emptyset \tag{19}$$

for $j = 1, \ldots, p$, since then $(\bigcap_{i=1}^m A_i) \subseteq X$ is implied, and we have $|X| = k - 1$.

To see (19) let us consider the following cases:

*Case 1* If $|Q_j| > k$, then for all $v \in Q_j$, there is an index $i$ such that $B_i \subset Q_j \setminus \{v\}$, implying by the definitions that $v \notin A_i$. Hence, (19) follows.

*Case 2* If $|Q_j| = k$ and $m \geq k + 1$, then we have $Q_j = B_{i^*}$ for exactly one index $i^* \in \{1, \ldots, m\}$. For all other indices $i$ we have $Q_j \cap A_i = R_{ij}$ of size $k - 1$. Thus, since $m \geq k + 1$, we can choose for each $v \in Q_j$ an index $i \neq i^*$ such that $v \notin A_i$, implying (19).

*Case 3* If $m \leq k$, then we must have $|Q_1| = |Q_2| = \ldots = |Q_p| = k$, $m = p \leq k$, since we already know that $|Q_j| \geq k$ for all $j = 1, \ldots, p$, and if $|Q_j| > k$ for at least one index $j$, then $m \geq k + 1$ would be implied. Thus, we have

$$\begin{aligned} n &= |X| + \sum_{i=1}^{p} |Q_i| \\ &= r - 1 + pk \\ &= r - 1 + k\frac{\ell - r}{k-1} \\ &= \ell\frac{k}{k-1} - \frac{r}{k-1} - 1, \end{aligned}$$

and hence, by our assumption, we must have $\ell \geq r+k^2-k+1$. However, $p \leq k$ implies that $p = \frac{\ell-r}{k-1} \leq k$ from which $\ell \leq r + k^2 - k$ follows. □

This completes the proof of Theorem 4. □

# 4 More About CIS-$d$-Graphs

## 4.1 Proofs of Propositions 6, 7, 11 and Theorem 5

*Proof of Proposition 6* Obviously, every partition of colors can be realized by successive identification of two colors. Hence, the following Lemma implies Proposition 6.

Given a $(d+1)$-graph $\mathscr{G} = (V; E_1, \ldots, E_d, E_{d+1})$, let us identify the last two colors $d$ and $d+1$ and consider the $d$-graph $\mathscr{G}' = (V; E_1, \ldots, E_{d-1}, E_{\mathbf{d}})$, where $E_{\mathbf{d}} = E_d \cup E_{d+1}$.

**Lemma 5** *If $\mathscr{G}$ is a CIS-$(d+1)$-graph, then $\mathscr{G}'$ is a CIS-$d$-graph.*

*Proof* Suppose that $\mathscr{G}'$ does not have the CIS-$d$-property, that is, there are $d$ vertex-sets $C_1, \ldots, C_{d-1}, C_{\mathbf{d}} \subseteq V$ such that they have no vertex in common, where $C_i$ is a maximal subset of $V$ avoiding color $i$ for $i = 1, \ldots, d-1$, and $C_{\mathbf{d}}$ is a maximal subset of $V$ avoiding both colors $d$ and $d+1$. Clearly, there exist maximal vertex-sets $C_d$ and $C_{d+1}$ avoiding colors $d$ and $d+1$, respectively, and such that $C_d \cap C_{d+1} = C_{\mathbf{d}}$. Then $C_1, \ldots, C_{d-1}, C_d, C_{d+1} \subseteq V$ are maximal vertex-sets avoiding colors $1, \ldots, d-1, d, d+1$, respectively, and with no vertex in common. Hence, the $(d+1)$-graph $\mathscr{G}'$ does not have the CIS-$(d+1)$-property, either. □

A little later we will need the following similar claim.

**Lemma 6** *If $\mathscr{G}$ is a Gallai $(d+1)$-graph then $\mathscr{G}'$ is a Gallai $d$-graph.*

*Proof* It is obvious. If $\mathscr{G}'$ contains a $\Delta$, then the same three vertices form a $\Delta$ in $\mathscr{G}$ too. □

*Proof of Proposition 7* It follows by a routine case analysis from the definitions.

First, let us consider Gallai's property. Suppose that $\mathscr{G}$ has a $\Delta$. Clearly, it cannot contain exactly one edge in $\mathscr{G}''$, since then two remaining edges are of the same color. If this $\Delta$ contains 2 edges in $\mathscr{G}''$, then the third one is there, too, and hence $G''$ contains a $\Delta$. If all 3 edges are in $\mathscr{G}'$, then $G'$ contains a $\Delta$. Conversely, if $\mathscr{G}'$ or $\mathscr{G}''$ contains a $\Delta$ then, clearly, $\mathscr{G}$ contains it too, since both $\mathscr{G}'$ and $\mathscr{G}''$ are induced subgraphs of $\mathscr{G}$.

Now let us consider the CIS-property. To simplify the notation we restrict ourselves by the case $d = 2$, though exactly the same arguments work in general. Obviously, all maximal cliques (respectively, stable sets) of $\mathscr{G}'$ which do not contain $v$ remain unchanged in $\mathscr{G}$, while a maximal clique $C'$ (respectively, a maximal stable set $S'$) of $\mathscr{G}'$ which contains $v$ and for every maximal clique $C''$ (respectively, every maximal stable set $S''$) of $\mathscr{G}''$ the set $C = C' \cup C'' \setminus \{v\}$ (respectively, $S = S' \cup S'' \setminus \{v\}$ is a maximal clique (respectively, a maximal stable set) of $\mathscr{G}$ and moreover, there are no other maximal cliques (respectively, maximal stable sets) in $\mathscr{G}$.

It is not difficult to verify that every maximal clique $C = C' \cup C'' \setminus \{v\}$ and every maximal stable set $S = S' \cup S'' \setminus \{v\}$ in $\mathscr{G}$ intersects if and only if every maximal clique $C'$ intersects every maximal stable set $S'$ of $G'$ and every maximal clique $C''$ intersects every maximal stable set $S''$ of $G''$. Indeed, if $C' \cap S' = \{v'\} \neq \{v\}$, then $C \cap S = \{v'\}$ for any $C''$ and $S''$. If $C' \cap S' = \{v\}$, then $C \cap S = C'' \cap S''$ and hence $C \cap S \neq \emptyset$ if and only if $C'' \cap S'' \neq \emptyset$. If $C \cap S \neq \emptyset$, then both $C' \cap S'$ and $C'' \cap S''$ must be non-empty. □

*Proof of Theorem 5*

**Part (a)** By Proposition 7, $\mathscr{G}$ is exactly closed under substitution. By Proposition 9, $\mathscr{G}$ can be obtained from 2-graphs by substitutions. Such a decomposition of $\mathscr{G}$ is given by a tree $T(\mathscr{G})$ whose leaves correspond to 2-graphs. It is easy to see that by construction each chromatic component of $\mathscr{G}$ is decomposed by the same tree $T(\mathscr{G})$. Hence, all we have to prove is that both chromatic components of every 2-graph belong to $\mathscr{F}$. For colors $1, \ldots, d-1$ this holds, since $\mathscr{F}$ is exactly closed under substitution, and for the color $d$ it holds, too, since $\mathscr{F}$ is also closed under complementation.

**Part (b)** It follows easily from part (a). As in Lemma 5, given a $(d+1)$-graph $\mathscr{G} = (V; E_1, \ldots, E_d, E_{d+1})$, let us identify the last two colors $d$ and $d+1$ and consider the $d$-graph $\mathscr{G}' = (V; E_1, \ldots, E_{d-1}, E_{\mathbf{d}})$, where $E_{\mathbf{d}} = E_d \cup E_{d+1}$. We assume that $\mathscr{G}$ is $\Delta$-free and that $G_i = (V, E_i) \in \mathscr{F}$ for $i = 1, \ldots, d-1$. Then, by Lemma 6, $\mathscr{G}'$ is $\Delta$-free, too, and it follows from part (a) that $G_{\mathbf{d}} = (V, E_{\mathbf{d}})$ is also in $\mathscr{F}$. Hence, the union of any two colors is in $\mathscr{F}$. From this by induction we derive that the union of any set of colors is in $\mathscr{F}$. □

*Proof of Proposition 11* Given $\mathscr{G}$, let us again consider the decomposition tree $T(\mathscr{G})$, fix an arbitrary its leaf $v$, and consider the corresponding 2-graph $\mathscr{G}_v$. Both its chromatic components are CIS-graphs, by Proposition 8. Hence, $\mathscr{G}_v$ is a CIS-$d$-graph. Thus, $\mathscr{G}$ is a CIS-$d$-graph, too, by Proposition 7. □

## 4.2 Settling $\Delta$

Let $V = \{v_1, v_2, v_3\}$ and assume that $E_1 = \{(v_1, v_2)\}$, $E_2 = \{(v_2, v_3)\}$, and $E_3 = \{(v_3, v_1)\}$ form a $\Delta$, see Figure 11. Obviously, $\Delta$ is not a CIS-3-graph. Indeed, let us consider $C_1 = \{v_2, v_3\}$, $C_2 = \{v_3, v_1\}$, and $C_3 = \{v_1, v_2\}$. There is no edge from $E_i$ in $C_i$ for $i = 1, 2, 3$ and $C_1 \cap C_2 \cap C_3 = \emptyset$. Hence, if a CIS-3-graph $\mathscr{G} = (V; E_1, E_2, E_3)$ contains a $\Delta$, then it must contain a vertex $v_4$ such that the sets $C_1' = \{v_2, v_3, v_4\}$, $C_2' = \{v_3, v_1, v_4\}$, and $C_3' = \{v_1, v_2, v_4\}$ contain no edges from $E_1$, $E_2$, and $E_3$, respectively.

Fig. 28 Settling $\Delta$

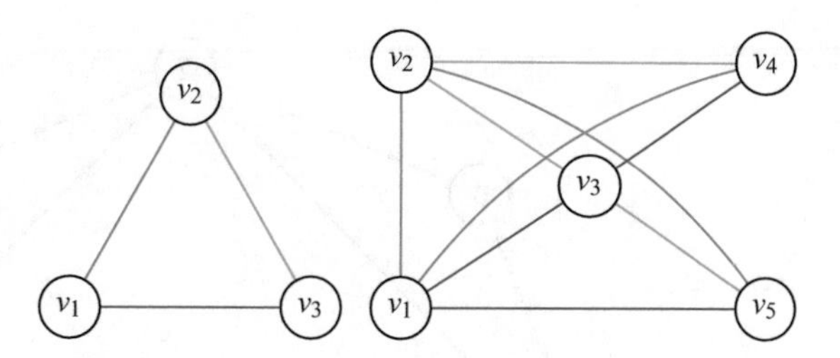


Similarly, let us consider the sets $C_1 = \{v_3, v_1\}$, $C_2 = \{v_1, v_2\}$, and $C_3 = \{v_2, v_3\}$. Again, there is no edge from $E_i$ in $C_i$ for $i = 1, 2, 3$ and $C_1 \cap C_2 \cap C_3 = \emptyset$. Hence, if a CIS-3-graph $\mathcal{G} = (V; E_1, E_2, E_3)$ contains a $\Delta$, then it must contain a vertex $v_5$ such that $C_1' = \{v_3, v_1, v_5\}$, $C_2' = \{v_1, v_2, v_5\}$, and $C_3' = \{v_2, v_3, v_5\}$ contain no edges from $E_1$, $E_2$, and $E_3$, respectively.

It is easy to check that $v_4 \neq v_5$ and that we must have $(v_4, v_1), (v_1, v_2), (v_2, v_5) \in E_1$, $(v_4, v_2), (v_2, v_3), (v_3, v_5) \in E_2$, $(v_4, v_3), (v_3, v_1), (v_1, v_5) \in E_3$, see Figure 11. This leaves only one pair $(v_4, v_5)$ whose color is not implied. Yet, let us note that for any coloring of $(v_4, v_5)$ a new $\Delta$ appears. For example, if $(v_4, v_5) \in E_1$, then vertices $(v_3, v_4, v_5)$ form a $\Delta'$ given in Figure 28.

## 4.3 A Stronger Conjecture

We say that two vertices $v_4$ and $v_5$ settle $\Delta$. Note however that $v_1$ and $v_2$ do not settle $\Delta'$. So we need more vertices to settle it. Nevertheless, there are $d$-graphs whose all $\Delta$s are settled. The first such example was given by Andrey Gol'berg in 1984, see Figure 30.

We call this construction a 4-cycle. It has 4 $\Delta$s and they are all settled. Yet, if we partition its three colors into two sets, we will get 44 2-combs none of which is settled. Hence, by Proposition 6, the 4-cycle is not a CIS-3-graph.

Moreover, in the next section we give examples of 3-graphs whose all $\Delta$s and 2-combs are settled, however, their 2-projections have unsettled induced 3-combs or 3-anti-combs.

*Conjecture 3* Let $\mathcal{G}$ be a non-Gallai 3-graph with chromatic components $G_1, G_2, G_3$, then there is an unsettled $\Delta$ in $\mathcal{G}$ or $G_i$ has an unsettled induced comb or anti-comb for some $i = 1, 2, 3$.

Obviously, Proposition 6 and Conjecture 3 imply Conjecture 2.

*Remark 11* It is not difficult to show that for every fixed $k, d \in \mathbb{Z}_+$ and $\varepsilon > 0$ there is $n = n(k, d, \varepsilon)$ such that in a random $d$-graph $\mathcal{G}$ with a fixed $|V(G)| \geq k$ all $\Delta$s as well as all induced $m$-combs and $m$-anti-combs for $m \leq k$ in all projections of $\mathcal{G}$ are settled with probability greater than $1 - \varepsilon$.

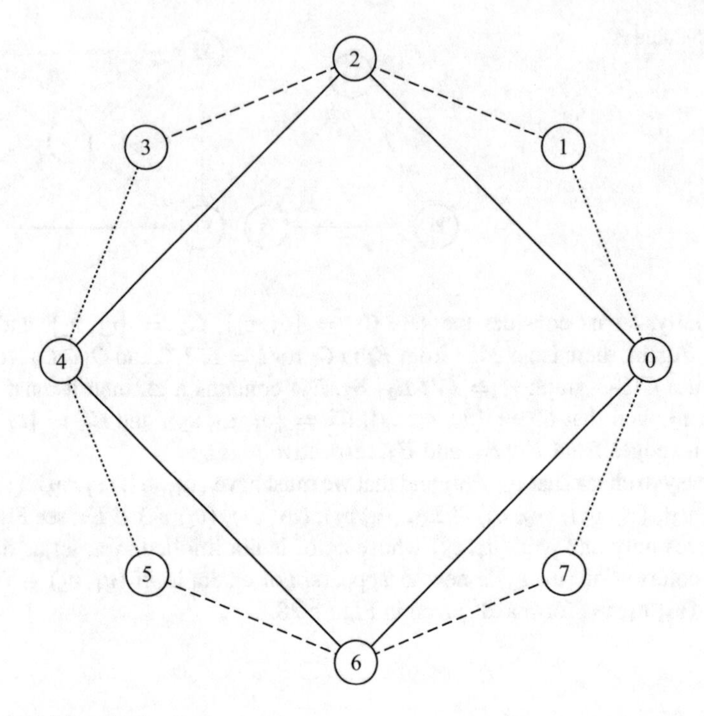


**Fig. 29** Initial 4-cycle structure

Yet, for $m > k$, unsettled induced $m$-combs and $m$-anti-combs exist with high probability.

## 4.4 *Even Cycles and Flowers*

In this section we describe some interesting 3-graphs in support of Conjecture 3. They have all $\Delta$s settled, and sometimes even all 2-combs are settled in their 2-projections. However, then unsettled 3-combs, or 3-anti-combs, or 4-combs appear.

Let us consider four $\Delta$s in Figure 29. They form a cycle.

This construction can be extended (uniquely) to a 3-graph, shown in Figure 30, in which all four $\Delta$s are settled "counterclockwise" (i.e., $\Delta$s induced by the triplets $\{0, 1, 2\}$, $\{2, 3, 4\}$, $\{4, 5, 6\}$, and $\{6, 7, 0\}$ are settled by the pairs $\{3, 4\}$, $\{5, 6\}$, $\{7, 1\}$, and $\{1, 2\}$, respectively), and no new $\Delta$ appears. However, 2-projections of this 3-graph contain 44 unsettled 2-combs (induced by the quadruples $\{0, 5, 1, 4\}$, $\{3, 2, 6, 7\}$, $\{4, 1, 2, 3\}$, $\{0, 5, 6, 7\}$, etc.) as shown in Figure 30.

Level 1: GBBGGBBG
Level 2: RGRBRGRB
Level 3: RBRGRBRG
Level 4: GBBRGBBR

4 settled $\Delta$s
44 $S_2$: 0 settled

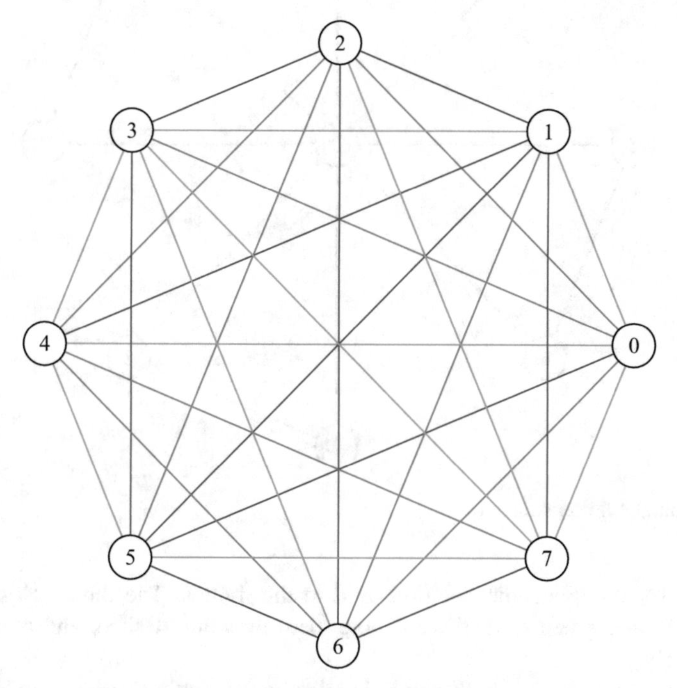


**Fig. 30** 4-cycle with all $\Delta$s settled. This 3-graph was constructed by Andrey Gol'berg in 1984

Now, let us consider four $\Delta$s with one common vertex as shown in Figure 31. This construction we call a 4-*flower*. It can be extended to a 3-graph, as shown in Figure 31, in which all four $\Delta$s are settled "counterclockwise" (i.e., $\Delta$s induced by the triplets $\{0, 1, 2\}$, $\{0, 3, 4\}$, $\{0, 5, 6\}$, and $\{0, 7, 8\}$ are settled by the pairs $\{3, 4\}$, $\{5, 6\}$, $\{7, 8\}$, and $\{1, 2\}$, respectively). Although four more $\Delta$s (induced by the triplets $\{0, 1, 6\}$, $\{0, 2, 5\}$, $\{0, 4, 7\}$, and $\{0, 3, 8)\}$ appear in this extension), yet they are settled too. Moreover, 2-projections of this 3-graph contain twenty induced 2-combs that are all settled. However, there exist also eight induced 3-combs that are not settled.

Using a computer, we analyzed also some larger flowers (namely, $2j$-flowers for $j = 3, 4, 5$, and 6) shown below. In all these examples all $\Delta$s are settled. However, in agreement with Conjecture 3, for each of these 3-graphs always there is a 2-projection that contains an unsettled comb or anti-comb.

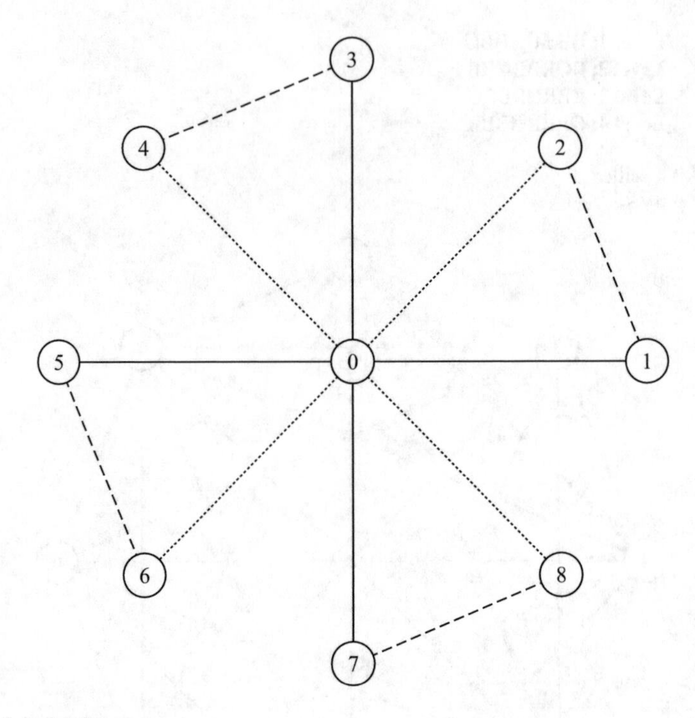


**Fig. 31** Initial 4-flower structure

We have to explain the notation used in the figures. The three colors are red $R$, blue $B$, and green $G$, and we denote them by solid, dashed, and dotted lines, respectively.

In a $2j$-flower we denote the central vertex by 0 and other vertices are labeled by $1, 2, \ldots, 2j-1, 2j$. Due to the symmetry, we can describe this 3-graph in terms of a list of colors $L$ present in *level* $i$, where level $i$ contains all edges $(a, b)$ such that $a - b = \pm i \pmod{n}$. Clearly, we only need to provide the color lists from level 1 to $j$, since level $i$ gives the same assignment as level $2j - i$. Finally Level 0 shows the coloring of the radial edges. For example, the 4-flower in Figure 31 is colored as follows:

Level 0: the edges (0, 1), (0, 2), (0, 3)(0, 4), (0, 5), (0, 6), (0, 7), (0, 8) are colored by $RGRGRGRG$.

Level 1: the edges (1, 2), (2, 3), (3, 4), (4, 5), (5, 6), (6, 7), (7, 8), (8, 1) are colored by $BGBGBGBG$;

Level 2: the edges (1, 3), (2, 4), (3, 5), (4, 6), (5, 7), (6, 8), (7, 1), (8, 2) are all colored by $BBBBBBBB$;

Level 0: RGRGRGRG
Level 1: BGBGBGBG
Level 2: BBBBBBBB
Level 3: RBRBRBRB
Level 4: RGRGRGRG

8 $\Delta$s: 8 settled
20 $S_2$: 20 settled
8 $S_3$: 0 settled

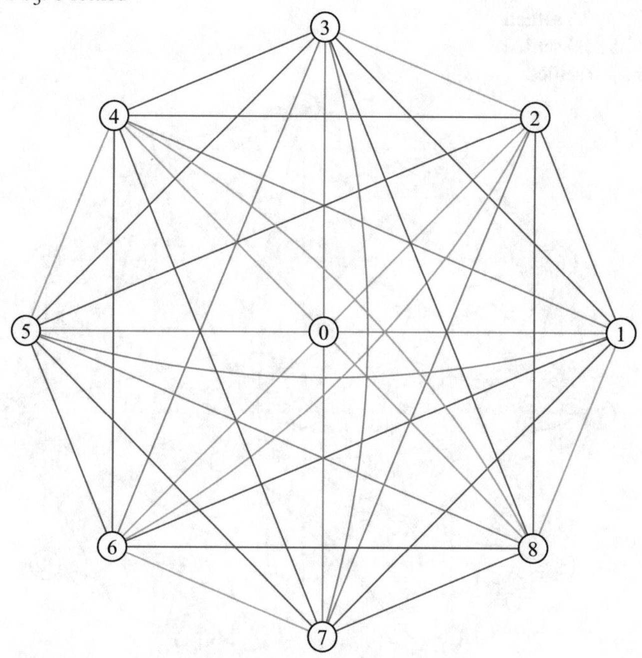


**Fig. 32** 4-flower example

Level 3: the edges $(1, 4)$, $(2, 5)$, $(3, 6)$, $(4, 7)$, $(5, 8)$, $(6, 1)$, $(7, 2)$, $(8, 3)$ are colored by $RBRBRBRB$;

Level 4: the edges $(1, 5)$, $(2, 6)$, $(3, 7)$, $(4, 8)$, $((5, 1)$, $(6, 2)$, $(7, 3)$, $(8, 4))$ are colored by $RGRG(RGRG)$ (Figures 32, 33, 34, 35, and 36).

Level 0: RGRGRGRGRGRG
Level 1: BGBGBGBGBGBG
Level 2: BBBBBBBBBBBB
Level 3: RBRBRBRBRBRB
Level 4: RGRGRGRGRGRG
Level 5: BRBRBRBRBRBR
Level 6: BBBBBBBBBBBB

18 $\Delta$s: 18 settled
66 $S_2$: 66 settled
38 $S_3$: 20 settled
6 $S_4$: 0 settled

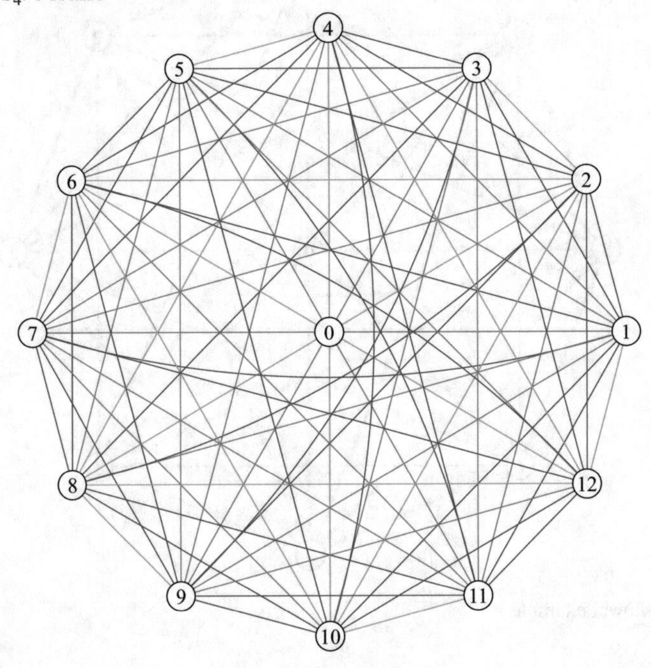


**Fig. 33** 6-flower example

Level 0: RGRGRGRGRGRGRGRG
Level 1: BGBGBGBGBGBGBGBG
Level 2: BBBBBBBBBBBBBBBB
Level 3: RBRBRBRBRBRBRBRB
Level 4: RGRGRGRGRGRGRGRG
Level 5: BRBRBRBRBRBRBRBR
Level 6: BBBBBBBBBBBBBBBB
Level 7: GBGBGBGBGBGBGBGB
Level 8: RRRRRRRRRRRRRRRR

32 $\Delta$s: 32 settled
192 $S_2$: 192 settled
256 $S_3$: 0 settled

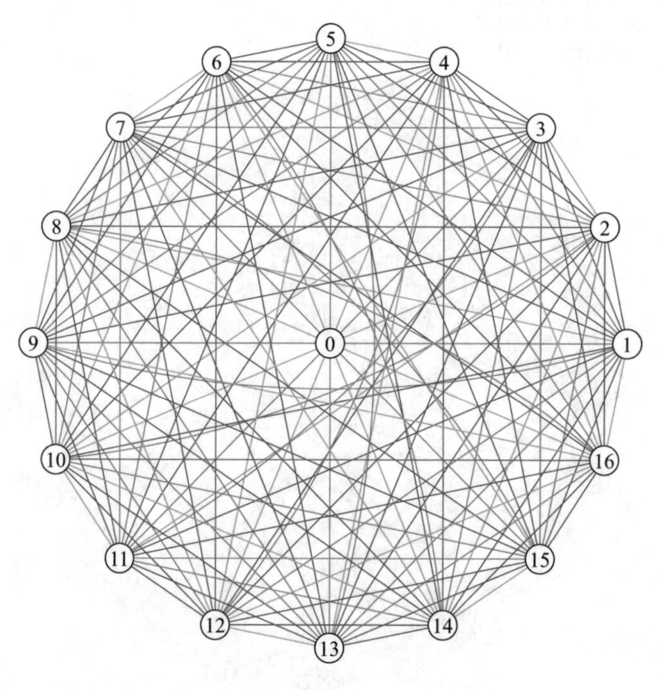


**Fig. 34** 8-flower example

Level 0: RGRGRGRGRGRGRGRGRGRG
Level 1: BGBGBGBGBGBGBGBGBGBG
Level 2: BBBBBBBBBBBBBBBBBBBB
Level 3: RBRBRBRBRBRBRBRBRBRB
Level 4: RGRGRGRGRGRGRGRGRGRG
Level 5: BRBRBRBRBRBRBRBRBRBR
Level 6: BBBBBBBBBBBBBBBBBBBB
Level 7: RBRBRBRBRBRBRBRBRBRB
Level 8: RGRGRGRGRGRGRGRGRGRG
Level 9: BRBRBRBRBRBRBRBRBRBR
Level 10: BBBBBBBBBBBBBBBBBBBB

50 $\Delta$s: 50 settled
290 $S_2$: 290 settled
220 $S_3$: 120 settled
110 $S_4$: 0 settled

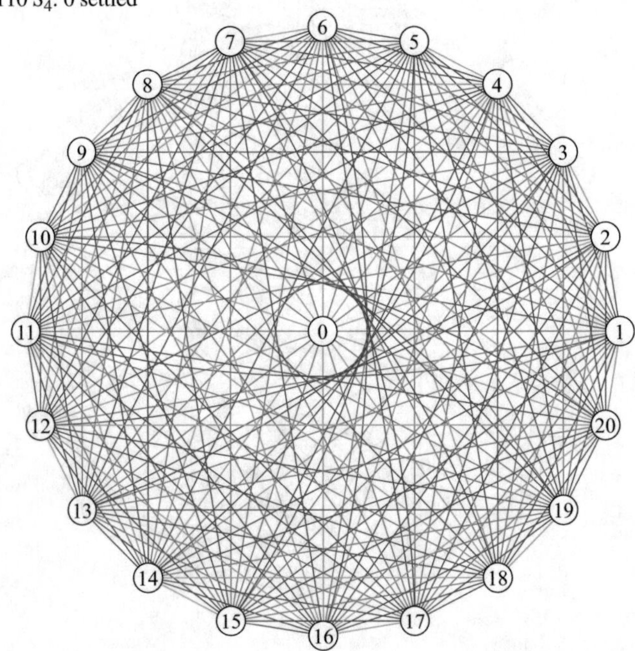


**Fig. 35** 10-flower example

Level 0: RGRGRGRGRGRGRGRGRGRGRGRG
Level 1: BGBGBGBGBGBGBGBGBGBGBGBG
Level 2: BBBBBBBBBBBBBBBBBBBBBBBB
Level 3: RBRBRBRBRBRBRBRBRBRBRBRB
Level 4: RRRRRRRRRRRRRRRRRRRRRRRR
Level 5: BRBRBRBRBRBRBRBRBRBRBRBR
Level 6: BBBBBBBBBBBBBBBBBBBBBBBB
Level 7: RBRBRBRBRBRBRBRBRBRBRBRB
Level 8: RGRGRGRGRGRGRGRGRGRGRGRG
Level 9: BGBGBGBGBGBGBGBGBGBGBGBG
Level 10: BBBBBBBBBBBBBBBBBBBBBBBB
Level 11: RBRBRBRBRBRBRBRBRBRBRBRB
Level 12: RRRRRRRRRRRRRRRRRRRRRRRR

72 settled $\Delta$s
600 $S_2$: 600 settled
184 $S_3$: 76 settled
24 $S_4$: 0 settled

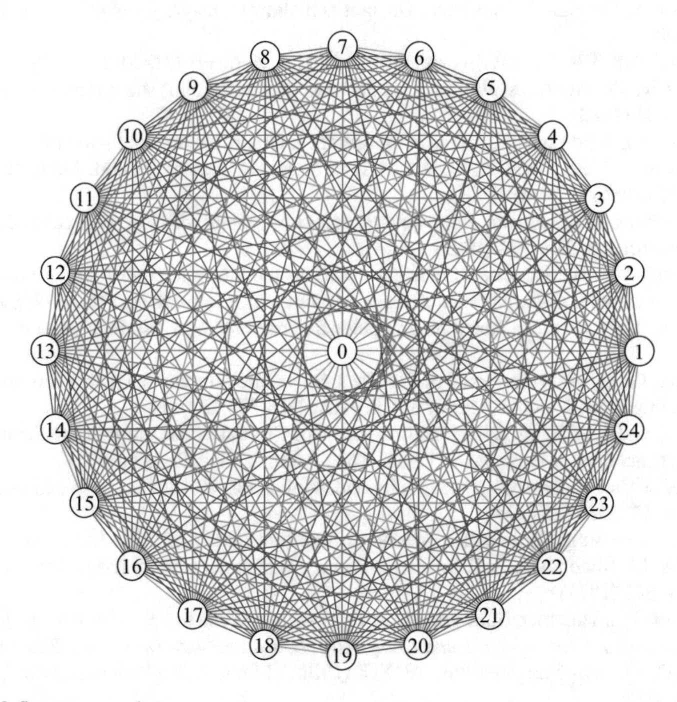


**Fig. 36** 12-flower example

**Acknowledgements** The third author was partially funded by the Russian Academic Excellence Project '5-100'.

## References

1. R.N. Ball, A. Pultr, P. Vojtěchovský, Colored graphs without colorful cycles. Combinatorica **27**(4), 407–427 (2007)
2. C. Berge, Problems 9.11 and 9.12, in *Graphs and Order*, ed. by I. Rival (Reidel, Dordrecht, 1985), pp. 583–584
3. E. Boros, V. Gurvich, Perfect graphs are kernel-solvable. Discret. Math. **159**, 35–55 (1996)
4. E. Boros, V. Gurvich, Stable effectivity functions and perfect graphs. Math. Soc. Sci. **39**, 175–194 (2000)
5. E. Boros, V. Gurvich, A. Vasin, Stable families of coalitions and normal hypergraphs. Math. Soc. Sci. **34**, 107–123 (1997). RUTCOR Research Report, RRR-22-1995, Rutgers University
6. E. Boros, V. Gurvich, P.L. Hammer, Dual subimplicants of positive Boolean functions. Optim. Methods Softw. **10**, 147–156 (1998)
7. E. Boros, V. Gurvich, I. Zverovich, On split and almost CIS-graphs. Aust. J. Comb. **43**, 163–180 (2009)
8. A. Brandstädt, V.B. Le, J.P. Spinrad, *Graph Classes: A Survey* (SIAM, Philadelphia, 1999)
9. H. Buer, R. Möring, A fast algorithm for decomposition of graphs and posets. Math. Oper. Res. **3**, 170–184 (1983)
10. K. Cameron, J. Edmonds, Lambda composition. J. Graph Theory **26**, 9–16 (1997)
11. K. Cameron, J. Edmonds, L. Lovász, A note on perfect graphs. Period. Math. Hung. **17**(3), 441–447 (1986)
12. F.R.K. Chung, R.L. Graham, Edge-colored complete graphs with precisely colored subgraphs. Combinatorica **3**, 315–324 (1983)
13. A. Cournier, M. Habib, A new linear algorithm for modular decomposition, in *Proceedings of 19th International Colloquium on Trees in Algebra and Programming (CAAP-94), Edinburgh*, ed. by S. Tison. Lecture Notes in Computer Science, vol. 787 (Springer, Berlin, 1994), pp. 68–82
14. X. Deng, G. Li, W. Zang, Proof of Chvatal's conjecture on maximal stable sets and maximal cliques in graphs. J. Comb. Theory Ser. B **91**(2), 301–325 (2004)
15. X. Deng, G. Li, W. Zang, Corrigendum to proof of Chvatal's conjecture on maximal stable sets and maximal cliques in graphs. J. Comb. Theory Ser. B **94**, 352–353 (2005)
16. T. Eiter, Exact transversal hypergraphs and application to Boolean $\mu$-functions. J. Symb. Comput. **17**, 215–225 (1994)
17. T. Eiter, Generating Boolean $\mu$-expressions. Acta Informatica **32**, 171–187 (1995)
18. P. Erdős, M. Simonovits, V.T. Sos, Anti-Ramsey theorems. Colloq. Math. Soc. Janos Bolyai **10**, 633–643 (1973)
19. S. Foldes, P.L. Hammer, Split graphs, in *Proceedings of the 8th Southeastern Conference on Combinatorics, Graph Theory, and Computing (Louisiana State University, Baton Rouge, LA, 1977)*. Congressus Numerantium, vol. XIX (Utilitas Mathematica Publisher, Winnipeg, 1977), pp. 311–315
20. T. Gallai, Transitiv orientierbare graphen. Acta Math. Acad. Sci. Hungar. **18**(1–2), 25–66 (1967). English translation by F. Maffray, M. Preissmann, Chapter 3: Perfect graphs, ed. by J.L.R. Alfonsin, B.A. Reed (Wiley, Hoboken, 2001)
21. M.C. Golumbic, V. Gurvich, Read-once Boolean functions, in *Boolean Functions: Theory, Algorithms, and Applications*, ed. by Y. Crama, P.L. Hammer (Cambridge University Press, Cambridge, 2011), pp. 448–486
22. V. Gurvich, On repetition-free Boolean functions. Russ. Math. Surv. **32**(1), 183–184 (1977) (in Russian)

23. V. Gurvich, Applications of Boolean functions and contact schemes in game theory, section 5, Repetition-free Boolean functions and normal forms of positional games, Ph.D. thesis, Moscow Institute of Physics and Technology, Moscow, USSR (in Russian), 1978
24. V. Gurvich, On the normal form of positional games. Soviet Math. Dokl. **25**(3), 572–575 (1982)
25. V. Gurvich, Some properties and applications of complete edge-chromatic graphs and hypergraphs. Soviet Math. Dokl. **30**(3), 803–807 (1984)
26. V. Gurvich, Criteria for repetition-freeness of functions in the Algebra of Logic. Russ. Acad. Sci. Dokl. Math. **43**(3), 721–726 (1991)
27. V. Gurvich, Positional game forms and edge-chromatic graphs. Russ. Acad. Sci. Dokl. Math. **45**(1), 168–172 (1992)
28. V. Gurvich, Decomposing complete edge-chromatic graphs and hypergraphs, revisited. Discret. Appl. Math. **157**, 3069–3085 (2009)
29. A. Gyárfás, G. Simonyi, Edge coloring of complete graphs without tricolored triangles. J. Graph Theory **46**, 211–216 (2004)
30. M. Karchmer, N. Linial, L. Newman, M. Saks, A. Wigderson, Combinatorial characterization of read-once formulae. Discret. Math. **114**, 275–282 (1993)
31. T. Kloks, C.-M. Lee, J. Liu, H. Müller, On the recognition of general partition graphs, in *Graph-Theoretic Concepts of Computer Science*. Lecture Notes in Computer Science, vol. 2880 (Springer, Berlin, 2003), pp. 273–283
32. J. Körner, G. Simonyi, Graph pairs and their entropies: modularity problems. Combinatorica **20**, 227–240 (2000)
33. J. Körner, G. Simonyi, Zs. Tuza, Perfect couples of graphs. Combinatorica **12**, 179–192 (1992)
34. L. Lovász, Normal hypergraphs and the weak perfect graph conjecture. Discret. Math. **2**(3), 253–267 (1972)
35. L. Lovász, A characterization of perfect graphs. J. Comb. Theory Ser. B **13**(2), 95–98 (1972)
36. L. Lovász, *Combinatorial Problems and Exercises* (North-Holland Publishing, Amsterdam, 1979)
37. K. McAvaney, J. Robertson, D. DeTemple, A characterization and hereditary properties for partition graphs. Discret. Math. **113**(1–3), 131–142 (1993)
38. R.M. McCollel, J.P. Spinrad, Modular decomposition and transitive orientation. Discret. Math. **201**, 189–241 (1999)
39. R. Möring, Algorithmic aspects of the substitution decomposition in optimization over relations, set systems, and Boolean functions. Ann. Oper. Res. **4**, 195–225 (1985/1986)
40. J. Muller, J. Spinrad, Incremental modular decomposition. J. ACM **36**(1), 1–19 (1989)
41. Yu.L. Orlovich, I.E. Zverovich, Independent domination and the triangle condition. Electron. Notes Discrete Math. **28**, 341–348 (2007)
42. G. Ravindra, Strongly perfect line graphs and total graphs, in *Finite and Infinite Sets, 6-th Hungarian Combinatorial Colloquium, vol. 2, Eger, 1981*. Colloquia Mathematica Societatis Janos Bolyai, vol. 37 (North Holland, Amsterdam, 1984) 621–633.
43. Y. Wu, W. Zang, C.-Q. Zhang, A characterization of almost CIS graphs. SIAM J. Discret. Math. **23**(2), 749–753 (2009)
44. W. Zang, Generalizations of Grillet's theorem on maximal stable sets and maximal cliques in graphs. Discret. Math. **143**, 259–268 (1995)
45. I. Zverovich, I. Zverovich, Bipartite hypergraphs: a survey and new results. Discret. Math. **306**, 801–811 (2006)

# Computing the Line Index of Balance Using Integer Programming Optimisation

**Samin Aref, Andrew J. Mason, and Mark C. Wilson**

## 1 Introduction

Graphs with positive and negative edges are referred to as *signed graphs* [67] which are very useful in modelling the dual nature of interactions in various contexts. Graph-theoretic conditions [11, 30] of the structural balance theory [30, 36] define the notion of *balance* in signed graphs. If the vertex set of a signed graph can be partitioned into $k \leq 2$ subsets such that each negative edge joins vertices belonging to different subsets, then the signed graph is balanced [11]. For graphs that are not balanced, a distance from balance (a measure of partial balance [7]) can be computed.

Among various measures is the *frustration index* that indicates the minimum number of edges whose removal results in balance [1, 32, 65]. This number was originally proposed in oblique form and referred to as *complexity* by Abelson et al. [1]. One year later, Harary proposed the same idea much more clearly with the name *line index of balance* [32]. More than two decades later, Toulouse used the term *frustration* to discuss the minimum energy of an Ising spin glass model [63]. Zaslavsky has made a connection between the line index of balance and spin glass concepts and introduced the name frustration index [65]. We use both names, line index of balance and frustration index, interchangeably in this chapter.

S. Aref (✉) · M. C. Wilson
Department of Computer Science, University of Auckland, Auckland, New Zealand
e-mail: sare618@aucklanduni.ac.nz

A. J. Mason
Department of Engineering Science, University of Auckland, Auckland, New Zealand



B. Goldengorin (ed.), *Optimization Problems in Graph Theory*,
Springer Optimization and Its Applications 139,
https://doi.org/10.1007/978-3-319-94830-0_3

## 2 Literature Review

Except for a normalised version of the frustration index [7], measures of balance used in the literature [11, 18, 42, 53, 61] do not satisfy key axiomatic properties [7]. Using cycles [11, 53], triangles [42, 61], Laplacian matrix eigenvalues [43], and closed walks [18] to evaluate distance from balance has led to conflicting observations [18, 19, 45].

Besides applications as a measure of balance, the frustration index is a key to frequently stated problems in several fields of research [4]. In biology, optimal decomposition of biological networks into monotone subsystems is made possible by calculating the line index of balance [38]. In finance, portfolios whose underlying signed graph has negative edges and a frustration index of zero have a relatively low risk [34]. In physics, the line index of balance provides the minimum energy state of atomic magnets [8, 39, 60]. In international relations, alliance and antagonism between countries can be analysed using the line index of balance [15]. In chemistry, bipartite edge frustration indicates the stability of fullerene, a carbon allotrope [16]. For a discussion on applications of the frustration index, one may refer to [4].

Detecting whether a graph is balanced can be solved in polynomial time [29, 33, 64]. However, calculating the line index of balance in general graphs is an NP-hard problem equivalent to the ground state calculation of an unstructured Ising model [50]. Computation of the line index of balance can be reduced from the graph maximum cut (MAXCUT) problem, in the case of all negative edges, which is known to be NP-hard [37].

Similar to MAXCUT for planar graphs [27], the line index of balance can be computed in polynomial time for planar graphs [40]. Other special cases of related problems can be found among the works of Hartmann and collaborators who have suggested efficient algorithms for computing ground state in 3-dimensional spin glass models [47] improving their previous contributions in 1-, 2-, and 3-dimensional [14, 35, 49] spin glass models. Recently, they have used a method for solving 0/1 optimisation models to compute the ground state of 3-dimensional models containing up to $268^3$ nodes [24].

A review of the literature shows 5 algorithms suggested for computing the line index of balance between 1963 and 2002. The first algorithm [21, pp. 98–107] is developed specifically for complete graphs. It is a naive algorithm that requires explicit enumeration of all possible combinations of sign changes that may or may not lead to balance. With a run time exponential in the number of edges, this is clearly not practical for graphs with more than 8 nodes that require billions of cases to be checked. The second algorithm is an optimisation method suggested by Hammer [28]. This method is based on solving an unconstrained binary quadratic model. We will discuss a model of this type in Section 4.2 and other more efficient models later in this chapter. The third computation method is an iterative algorithm suggested in [29, algorithm 3, p. 217]. The iterative algorithm is based on removing edges to eliminate negative cycles of the graph and only provides an upper bound on the line index of balance. A fourth method suggested by Harary and Kabell

[33, p. 136] is based on extending a balance detection algorithm. This method is inefficient according to Bramsen [10] who in turn suggests an iterative algorithm with a run time that is exponential in the number of nodes. Using Bramsen's suggested method for a graph with 40 nodes requires checking trillions of cases to compute the line index of balance which is clearly impractical. Doreian and Mrvar have recently attempted computing the line index of balance using a polynomial time algorithm [15]. However, our computations on their data show that their solutions are not optimal and thus do not give the line index of balance.

This review of literature shows that computing the line index of balance in general graphs lacks extensive and systematic investigation.

We provide an efficient method for computing the line index of balance in general graphs of the sizes found in many application areas. Starting with a quadratic programming model based on signed graph switching equivalents, we suggest several optimisation models. We use powerful mathematical programming solvers like Gurobi [25] to solve the optimisation models.

This chapter begins with the preliminaries in Section 3. Three mathematical programming models are developed in Section 4. The results on synthetic data are provided in Section 5. Numerical results on real social and biological networks are provided in Section 6 including graphs with up to 3215 edges. Section 7 summarises the key highlights of the research.

## 3 Preliminaries

### 3.1 Basic Notation

We consider undirected signed networks $G = (V, E, \sigma)$. The set of nodes is denoted by $V$, with $|V| = n$. The set $E$ of edges is partitioned into the set of positive edges $E^+$ and the set of negative edges $E^-$ with $|E^-| = m^-$, $|E^+| = m^+$, and $|E| = m = m^- + m^+$. For clarity, we sometimes use $m^-_{(G)}$ to refer to the number of negative edges in $G$. The sign function, denoted by $\sigma$, is a mapping of edges to signs $\sigma : E \rightarrow \{-1, +1\}$. We represent the $m$ undirected edges in $G$ as ordered pairs of vertices $E = \{e_1, e_2, \ldots, e_m\} \subseteq \{(i, j) \mid i, j \in V, i < j\}$, where a single edge $e_k$ between nodes $i$ and $j$, $i < j$, is denoted by $e_k = (i, j), i < j$. We denote the graph density by $\rho = 2m/(n(n-1))$. The entries of the adjacency matrix $\mathbf{A} = (a_{ij})$ are defined in (1).

$$a_{ij} = \begin{cases} \sigma_{(i,j)} & \text{if } (i,j) \in E \\ \sigma_{(j,i)} & \text{if } (j,i) \in E \\ 0 & \text{otherwise} \end{cases} \tag{1}$$

The number of positive (negative) edges incident on the node $i \in V$ is the *positive (negative) degree* of the node and is denoted by $d^+(i)$ $(d^-(i))$. The *net degree* of a node is defined by $d^+(i) - d^-(i)$. The *degree* of node $i$ is represented by $d(i) = d^+(i) + d^-(i)$ and equals the total number of edges incident on node $i$.

A *walk* of length $k$ in $G$ is a sequence of nodes $v_0, v_1, \ldots, v_{k-1}, v_k$ such that for each $i = 1, 2, \ldots, k$ there is an edge between $v_{i-1}$ and $v_i$. If $v_0 = v_k$, the sequence is a *closed walk* of length $k$. If the nodes in a closed walk are distinct except for the endpoints, it is a *cycle* of length $k$. The *sign* of a cycle is the product of the signs of its edges. A balanced graph is one with no negative cycles [11].

## 3.2 Node Colouring and Frustration Count

For each signed graph $G = (V, E, \sigma)$, we can partition $V$ into two sets, denoted $X \subseteq V$ and $\bar{X} = V \backslash X$. We think of $X$ as specifying a colouring of the nodes, where each node $i \in X$ is coloured black, and each node $i \in \bar{X}$ is coloured white.

We let $x_i$ denote the colour of node $i \in V$ under $X$, where $x_i = 1$ if $i \in X$ and $x_i = 0$ otherwise. We say that an edge $(i, j) \in E$ is *frustrated* under $X$ if either edge $(i, j)$ is a positive edge ( $(i, j) \in E^+$) but nodes $i$ and $j$ have different colours ($x_i \neq x_j$), or edge $(i, j)$ is a negative edge ( $(i, j) \in E^-$) but nodes $i$ and $j$ share the same colour ($x_i = x_j$). We define the *frustration count* $f_G(X)$ as the number of frustrated edges in $G$ under $X$:

$$f_G(X) = \sum_{(i,j)\in E} f_{ij}(X)$$

where for $(i, j) \in E$:

$$f_{ij}(X) = \begin{cases} 0, & \text{if } x_i = x_j \text{ and } (i, j) \in E^+ \\ 1, & \text{if } x_i = x_j \text{ and } (i, j) \in E^- \\ 0, & \text{if } x_i \neq x_j \text{ and } (i, j) \in E^- \\ 1, & \text{if } x_i \neq x_j \text{ and } (i, j) \in E^+. \end{cases} \tag{2}$$

The frustration index $L(G)$ of a graph $G$ can be obtained by finding a subset $X^* \subseteq V$ of $G$ that minimises the frustration count $f_G(X)$, i.e. solving Equation (3).

$$L(G) = \min_{X \subseteq V} f_G(X) \tag{3}$$

## 3.3 Minimum Deletion Set and Switching Function

For each signed graph, there are sets of edges, called *deletion sets*, whose deletion results in a balanced graph. A minimum deletion set $E^* \subseteq E$ is a deletion set with the minimum size. The frustration index $L(G)$ equals the size of a minimum deletion set: $L(G) = |E^*|$.

We define the *switching function* $g(X)$ operating over a set of vertices, called the *switching set*, $X \subseteq V$ as follows in (4).

$$\sigma_{(i,j)}^{g(X)} = \begin{cases} \sigma_{(i,j)} \text{ if } i, j \in X \text{ or } i, j \notin X \\ -\sigma_{(i,j)} \text{ if } (i \in X \text{ and } j \notin X) \text{ or } (i \notin X \text{ and } j \in X) \end{cases} \quad (4)$$

The graph resulting from applying switching function $g$ to signed graph $G$ is called $G$'s *switching equivalent* and denoted by $G^g$. The switching equivalents of a graph have the same value of the frustration index, i.e. $L(G^g) = L(G) \forall g$ [66]. It is straightforward to prove that the frustration index is equal to the minimum number of negative edges in $G^g$ over all switching functions $g$. An immediate result is that any balanced graph can switch to an equivalent graph where all the edges are positive [66]. Moreover, in a switched graph with the minimum number of negative edges, called a *negative minimal graph* and denoted by $G^{g^*}$, all vertices have a non-negative net degree. In other words, every vertex $i$ in $G^{g^*}$ satisfies $d^-(i) \leq d^+(i)$.

## 3.4 Bounds for the Line Index of Balance

An obvious upper bound for the line index of balance is $L(G) \leq m^-$ which states the result that removing all negative edges gives a balanced graph. Recalling that acyclic signed graphs are balanced, the circuit rank of the graph can also be considered as an upper bound for the frustration index [22, p. 8]. Circuit rank, also known as the cyclomatic number, is the minimum number of edges whose removal results in an acyclic graph.

Petersdorf [56] proves that among all sign functions for complete graphs with $n$ nodes, assigning negative signs to all the edges, i.e. putting $\sigma : E \rightarrow \{-1\}$, gives the maximum value of the frustration index which equals $\lfloor (n-1)^2/4 \rfloor$. Petersdorf's proof confirms a conjecture by Abelson and Rosenberg[1] that is also proved in [62] and further discussed in [2].

Akiyama et al. provide results indicating that the frustration index of signed graphs with $n$ nodes and $m$ edges is bounded by $m/2$ [2]. They also show that the frustration index of signed graphs with $n$ nodes is maximum in all complete graphs with no positive 3-cycles and is bounded by $\lfloor (n-1)^2/4 \rfloor$ [2, Theorem 1]. This group of graphs also contains complete graphs with nodes that can be partitioned into two classes such that all positive edges connect nodes from different classes and all negative edges connect nodes belonging to the same class [62]. Akiyama et al. refer to these graphs as *antibalanced* [2] which is a term coined by Harary in [31] and also discussed in [66].

# 4 Mathematical Programming Models

In this section, we formulate three mathematical programming models in (5), (8), and (11) to calculate the frustration index by optimising an objective function formed using integer variables.

## 4.1 A Quadratically Constrained Quadratic Programming Model

We formulate a mathematical programming model in Equation (5) to maximise $Z_1$, the sum of entries of $\mathbf{A}^g$, the adjacency matrix of the graph switched by $g$, over different switching functions. Bearing in mind that the frustration index is the number of negative edges in a negative minimal graph, $L(G) = m^-(G^{g^*})$, then maximising $Z_1$ will effectively calculate the line index of balance. We use decision variables, $y_i \in \{-1, 1\}$ to define node colours. Then $X = \{i \mid y_i = 1\}$ gives the black-coloured nodes (alternatively nodes in the switching set). The restriction $y_i \in \{-1, 1\}$ for the variables is formulated by $n$ quadratic constraints $y_i^2 = 1$. Note that the switching set $X = \{i \mid y_i = 1\}$ creates a negative minimal graph with the adjacency matrix entries given by $a_{ij} y_i y_j$. The model can be represented as Equation (5) in the form of a continuous quadratically constrained quadratic programming (QCQP) model with $n$ decision variables and $n$ constraints.

$$
\begin{aligned}
\max_{y_i} Z_1 &= \sum_{i \in V} \sum_{j \in V} a_{ij} y_i y_j \\
\text{s.t.} \quad y_i^2 &= 1 \quad \forall i \in V
\end{aligned}
\tag{5}
$$

Maximising $\sum_{i \in V} \sum_{j \in V} a_{ij} y_i y_j$ is equivalent to computing $m^-(G^{g^*}) = |\{(i, j) \in E : a_{ij} y_i y_j = -1\}|$. Note that choosing $y_i, i \in V$ to maximise $\sum_{i \in V} \sum_{j \in V} a_{ij} y_i y_j$ is equivalent to choosing $g$ to minimise $m^-(G^g)$. The optimal value of the objective function, $Z_1^*$, is equal to the sum of entries in the adjacency matrix of a negative minimal graph which can be represented by $Z_1^* = 2m^+(G^{g^*}) - 2m^-(G^{g^*}) = 2m - 4L(G)$. Therefore, the graph frustration index can be calculated by $L(G) = (2m - Z_1^*)/4$.

While the model expressed in (5) is quite similar to the non-linear energy function minimisation model used in [17, 19, 20, 46] and the Hamiltonian of Ising models with $\pm 1$ interactions [60], the feasible region in model (5) is neither convex nor a second order cone. Therefore, the QCQP model in (5) only serves as an easy-to-understand optimisation model clarifying the node colouring (alternatively selecting nodes to switch) and how it relates to the line index of balance.

## 4.2 An Unconstrained Binary Quadratic Programming Model

The optimisation model (5) can be converted into an unconstrained binary quadratic programming (UBQP) model (8) by changing the decision variables into binary variables $y_i = 2x_i - 1$ where $x_i \in \{0, 1\}$. Note that the binary variables, $x_i$, define the black-coloured nodes $X = \{i \mid x_i = 1\}$ (alternatively, nodes in the switching set). The optimal solution represents a subset $X^* \subseteq V$ of $G$ that minimises the resulting frustration count.

Furthermore, by substituting $y_i = 2x_i - 1$ into the objective function in (5) we get (6). The terms in the objective function can be modified as shown in (6)–(7) in order to have an objective function whose optimal value, $Z_2^*$, equals $L(G)$.

$$\begin{aligned} Z_1 &= \sum_{i\in V}\sum_{j\in V}(4a_{ij}x_ix_j - 2x_ia_{ij} - 2x_ja_{ij} + a_{ij}) \\ &= \sum_{i\in V}\sum_{j\in V}(4a_{ij}x_ix_j - 4x_ia_{ij}) + (2m - 4m^-_{(G)}) \end{aligned} \tag{6}$$

$$Z_2 = (2m - Z_1)/4 \tag{7}$$

Note that the binary quadratic model in Equation (8) has $n$ decision variables and no constraints.

$$\begin{aligned} \min_{x_i} Z_2 &= \sum_{i\in V}\sum_{j\in V}(a_{ij}x_i - a_{ij}x_ix_j) + m^-_{(G)} \\ \text{s.t.} \quad & x_i \in \{0,1\}\ i \in V \end{aligned} \tag{8}$$

The optimal value of the objective function in Equation (8) represents the frustration index directly as shown in (9).

$$Z_2^* = (2m - Z_1^*)/4 = (2m - (2m - 4L(G)))/4 = L(G) \tag{9}$$

The objective function in Equation (8) can be interpreted as initially starting with $m^-_{(G)}$ and then adding 1 for each positive frustrated edge (positive edge with different endpoint colours) and $-1$ for each negative edge that is not frustrated (negative edge with different endpoint colours). This adds up to the total number of frustrated edges.

## 4.3 The 0/1 Linear Model

The linearised version of (8) is formulated in Equation (11). The objective function of (8) is first modified as shown in Equation (10) and then its non-linear term $x_ix_j$

is replaced by $|E|$ additional binary variables $x_{ij}$. The new decision variables $x_{ij}$ are defined for each edge $(i, j) \in E$ and take value 1 whenever $x_i = x_j = 1$ and 0 otherwise. Note that $d_i = \sum_{j \in V} a_{ij}$ is a constant that equals the net degree of node $i$.

$$\begin{aligned} Z_2 &= \sum_{i \in V} \sum_{j \in V} a_{ij} x_i - \sum_{i \in V} \sum_{j \in V} a_{ij} x_i x_j + m^-_{(G)} \\ &= \sum_{i \in V} x_i \sum_{j \in V} a_{ij} - \sum_{i \in V} \sum_{j \in V, j > i} 2a_{ij} x_i x_j + m^-_{(G)} \\ &= \sum_{i \in V} x_i d_i - \sum_{i \in V} \sum_{j \in V, j > i} 2a_{ij} x_i x_j + m^-_{(G)} \end{aligned} \tag{10}$$

The dependencies between the $x_{ij}$ and $x_i, x_j$ values are taken into account by considering a constraint for each new variable. Therefore, the 0/1 linear model has $n + m$ variables and $m$ constraints, as shown in (11).

$$\begin{aligned} \min_{x_i, x_{ij}} Z_2 = & \sum_{i \in V} d_i x_i - \sum_{(i,j) \in E} 2a_{ij} x_{ij} + m^-_{(G)} \\ \text{s.t.} \quad & x_{ij} \le (x_i + x_j)/2 \quad \forall (i, j) \in E^+ \\ & x_{ij} \ge x_i + x_j - 1 \quad \forall (i, j) \in E^- \\ & x_i \in \{0, 1\} \; i \in V \\ & x_{ij} \in \{0, 1\} \; (i, j) \in E \end{aligned} \tag{11}$$

## 4.4 *Additional Constraints for the 0/1 Linear Model*

The structural properties of the model allow us to restrict the model by adding additional valid inequalities. Valid inequalities are utilised by our solver Gurobi as additional non-core constraints that are kept aside from the core constraints of the model. Upon violation by a solution, valid inequalities are efficiently pulled in to the model. Pulled-in valid inequalities cut away a part of the feasible space and restrict the model. Additional restrictions imposed on the model can often speed up the solver algorithm if they are valid and useful [41]. Properties of the optimal solution can be used to determine these additional constraints. Two properties we use are the non-negativity of net degrees in negative minimal graphs and the fact that in every cycle there is always an even number of edges that change sign when applying the switching function $g$.

An obvious structural property of the nodes in a negative minimal graph, $G^{g^*}$, is that their net degrees are always non-negative, i.e. $d^+(i) - d^-(i) \geq 0 \quad \forall i \in V$. This can be proved by contradiction using the definition of the switching function (4). To see this, assume a node in a negative minimal graph has a negative net degree. It follows that the negative edges incident on the node outnumber the positive edges. Therefore, switching the node decreases the total number of negative edges in a negative minimal graph which is a contradiction.

This structural property can be formulated as constraints in the problem. A *net-degree constraint* can be added to the model for each node restricting all variables associated with the connected edges. These constraints are formulated using quadratic terms of $x_i$ variables. As $x_i$ represents the colour of a node, $(1 - 2x_i)(1 - 2x_j)$ takes value $-1$ if and only if the two endpoints of edge $(i, j) \in E$ have different colours. Interpreting this based on the concept of switching sets, different values of the $x_i$ variables associated with the two endpoints of edge $(i, j) \in E$ mean that the edge should be negated (change sign) in the process of transforming to a negative minimal graph. The linearised formulation of the net-degree constraints using $x_i$ and $x_{ij}$ variables is provided in (12).

$$\sum_{j:(i,j)\in E \text{ or } (j,i)\in E} a_{ij}(1 - 2x_i - 2x_j + 4x_{ij}) \geq 0 \quad \forall i \in V \tag{12}$$

Another structural property we observe is related to the edges making a cycle. According to the definition of the switching function (4), switching one node negates all edges (changes the signs on all edges) incident on that node. Because there are two edges incident on each node in a cycle, in every cycle there is always an even number of edges that change sign when switching function $g$ is applied to signed graph $G$.

As listing all cycles of a graph is computationally intensive, this structural property can be applied to cycles of a limited length. For instance, we may apply this structural property to the edge variables making triangles in the graph. This structural property can be formulated as valid inequalities in Equation (13) in which $T = \{(i, j, k) \in V^3 \mid (i, j), (i, k), (j, k) \in E\}$ contains ordered 3-tuples of nodes whose edges form a triangle. Note that $(x_i + x_j - 2x_{ij})$ equals 1 if edge $(i, j) \in E$ is negated and equals 0 otherwise. The expression in Equation (13) denotes the total number of negated edges in the triangle formed by three edges $(i, j), (i, k), (j, k)$.

$$\begin{aligned} & x_i + x_j - 2x_{ij} + x_i + x_k - 2x_{ik} + x_j + x_k - 2x_{jk} \\ & = 0 \text{ or } 2 \quad \forall (i, j, k) \in T \end{aligned} \tag{13}$$

**Table 1** Comparison of the three optimisation models

| | QCQP (5) | UBQP (8) | 0/1 linear model (11) |
|---|---|---|---|
| Variables | $n$ | $n$ | $n+m$ |
| Constraints | $n$ | 0 | $m$ |
| Variable type | Continuous | Binary | Binary |
| Constraint type | Quadratic | – | Linear |
| Objective | Quadratic | Quadratic | Linear |

Equation (13) can be linearised to Equation (14) as follows. *Triangle constraints* can be applied to the model as four constraints per triangle, restricting three edge variables and three node variables per triangle.

$$\begin{aligned} x_i + x_{jk} &\geq x_{ij} + x_{ik} \quad \forall(i,j,k) \in T \\ x_j + x_{ik} &\geq x_{ij} + x_{jk} \quad \forall(i,j,k) \in T \\ x_k + x_{ij} &\geq x_{ik} + x_{jk} \quad \forall(i,j,k) \in T \\ 1 + x_{ij} + x_{ik} + x_{jk} &\geq x_i + x_j + x_k \quad \forall(i,j,k) \in T \end{aligned} \tag{14}$$

In order to speed up the model in (11), we consider fixing a node colour to increase the root node objective function in the solver's branch and bound process. We conjecture the best node variable to fix is the one associated with the highest unsigned node degree. This constraint is formulated in (15) which our experiments show speeds up the branch and bound algorithm by increasing the lower bound.

$$x_k = 1 \quad k = \arg\max_{i \in V} d(i) \tag{15}$$

The complete formulation of the 0/1 linear model with further restrictions on the feasible space includes the objective function and core constraints in Equation (11) and valid inequalities in Equations (12), (14), and (15). The model has $n+m$ binary variables, $m$ core constraints, and $n + 4|T| + 1$ additional constraints.

Table 1 provides a comparison of the three optimisation models based on their variables, constraints, and objective functions. In the next sections, we mainly focus on the 0/1 linear model solved in conjunction with the valid inequalities (additional constraints).

## 5 Numerical Results in Random Graphs

In this section, the frustration index of various random networks is computed by solving the 0/1 linear model (11) coupled with the additional constraints. We use Gurobi version 7 on a desktop computer with an Intel Core i5 4670 @ 3.40 GHz and

8.00 GB of RAM running 64-bit Microsoft Windows 7. The models were created using Gurobi's Python interface. All four processor cores available were used by Gurobi.

To verify our software implementation, we manually counted the number of frustrated edges given by our software's proposed node colouring for a number of test problems, and confirmed that this matched the frustration count reported by our software. These tests showed that our models and implementations were performing as expected.

## 5.1 Performance of the 0/1 Linear Model on Random Graphs

In this subsection we discuss the time performance of the branch and bound algorithms for solving the 0/1 linear model. In order to evaluate the performance of the 0/1 linear model (11) coupled with the additional constraints, we generate 10 decent-sized Erdős-Rényi random graphs [9] as test cases with various densities and percentages of negative edges. Results are provided in Table 2 in which B&B nodes stands for the number of branch and bound nodes (in the search tree of the branch and bound algorithm) explored by the solver.

The results in Table 2 show that random test cases based on Erdős-Rényi graphs with 500–600 edges can be solved to optimality in a reasonable time. The branching process for these test cases explores various numbers of nodes ranging between 0 and 12,384. These numbers also depend on the number of threads and the heuristics that the solver uses automatically.

**Table 2** Performance measures of Gurobi solving the 0/1 linear model in (11) for the random networks

| TestCase | $n$ | $m$ | $m^-$ | $\rho$ | $\frac{m^-}{m}$ | $L(G)$ | B&B nodes | Time(s) |
|---|---|---|---|---|---|---|---|---|
| 1 | 65 | 570 | 395 | 0.27 | 0.69 | 189 | 5133 | 65.4 |
| 2 | 68 | 500 | 410 | 0.22 | 0.82 | 162 | 4105 | 27.3 |
| 3 | 80 | 550 | 330 | 0.17 | 0.60 | 170 | 11,652 | 153.3 |
| 4 | 50 | 520 | 385 | 0.42 | 0.74 | 185 | 901 | 22.4 |
| 5 | 53 | 560 | 240 | 0.41 | 0.43 | 193 | 292 | 13.5 |
| 6 | 50 | 510 | 335 | 0.42 | 0.66 | 178 | 573 | 13.8 |
| 7 | 59 | 590 | 590 | 0.34 | 1.00 | 213 | 1831 | 46.0 |
| 8 | 56 | 600 | 110 | 0.39 | 0.18 | 110 | 0 | 0.4 |
| 9 | 71 | 500 | 190 | 0.20 | 0.38 | 155 | 6305 | 77.7 |
| 10 | 80 | 550 | 450 | 0.17 | 0.82 | 173 | 12,384 | 138.0 |

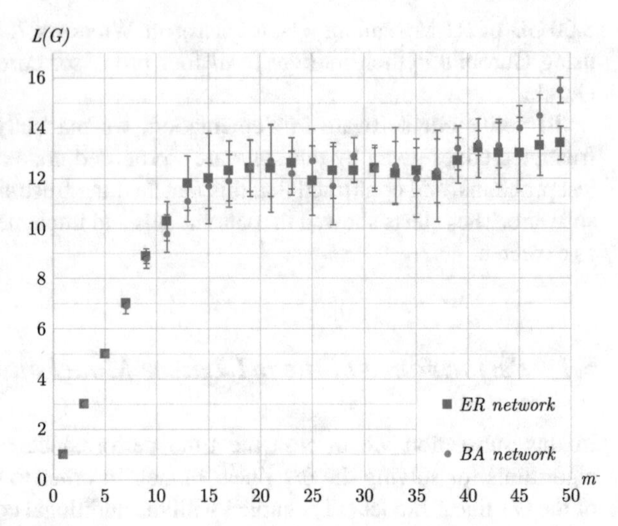


**Fig. 1** The frustration index in Erdős-Rényi (ER) networks with 15 nodes and 50 edges and Barabási-Albert (BA) networks with 15 nodes and 50 edges and various number of negative edges

## 5.2 *Impact of Negative Edges on the Frustration Index*

In this subsection we use both Erdős-Rényi and Barabási-Albert random networks [9] as synthetic data for calculation of the line index of balance. In this analysis, we use the same randomly generated graphs with different numbers of negative edges assigned by a uniform random distribution as test cases over 50 runs per experiment setting. Figure 1 shows the average and standard deviation of the line index of balance in these random signed networks with $n = 15$, $m = 50$. It is worth mentioning that we have observed similar results in other types of random graphs including small world, scale-free, and random regular graphs [9].

Figure 1 shows similar increases in the line index of balance in the two graph classes as $m^-$ increases. It can be observed that the maximum frustration index is still smaller than $m/3$ for all graphs. This shows a gap between the values of the line index of balance in random graphs and the theoretical upper bound of $m/2$. It is important to know whether this gap is proportional to graph size and density.

## 5.3 *Impact of Graph Size and Density on the Frustration Index*

In order to investigate the impact of graph size and density, 4-regular random graphs with a constant fraction of randomly assigned negative edges are analysed averaging over 50 runs per experiment setting. The frustration index is computed for 4-regular random graphs with 25%, 50%, and 100% negative edges and compared with the upper bound $m/2$. Figure 2 demonstrates the average and standard deviation of the frustration index where the degree of all nodes remains constant, but the density of the 4-regular graphs, $\rho = 4/n - 1$, decreases as $n$ and $m$ increase.

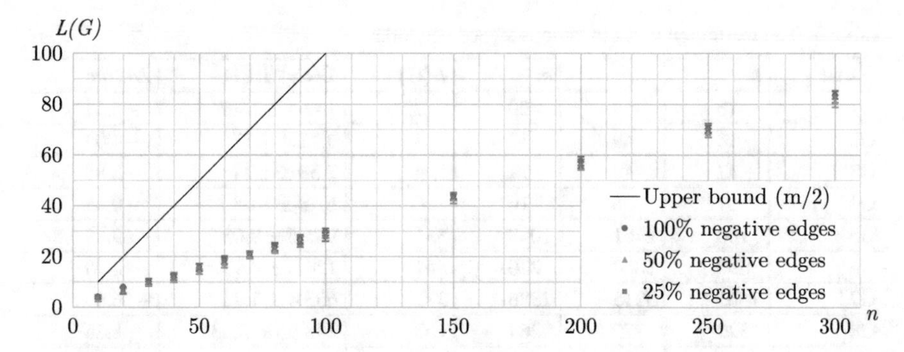


**Fig. 2** The frustration index in random 4-regular networks of different orders $n$ and decreasing densities

An observation to derive from Figure 2 is the similar frustration index values obtained for networks of the same sizes, even if they have different percentages of negative edges. It can be concluded that starting with an all-positive graph (which has a frustration index of 0), making the first quarter of graph edges negative increases the frustration index much more than making further edges negative. Future research is required to get a better understanding of how the frustration index and minimum deletion sets change when the number of negative edges is increased (on a fixed underlying structure). Another observation is that the gap between the frustration index values and the theoretical upper bound increases with increasing $n$.

# 6 Numerical Results in Real Signed Networks

In this section, the frustration index is computed in nine real networks by solving the 0/1 linear model (11) using Gurobi version 7 on a desktop computer with an Intel Core i5 4670 @ 3.40 GHz and 8.00 GB of RAM running 64-bit Microsoft Windows 7.

We use well-studied signed social network datasets representing communities with positive and negative interactions and preferences including Read's dataset for New Guinean highland tribes [57] and Sampson's dataset for monastery interactions [59] which we denote respectively by G1 and G2. We also use graphs inferred from datasets of students' choice and rejection, denoted by G3 and G4 [44, 52]. A further explanation on the details of inferring signed graphs from choice and rejection data can be found in [7]. Moreover, a larger signed network, denoted by G5, is inferred by Neal [51] through implementing a stochastic degree sequence model on Fowler's data on Senate bill co-sponsorship [23].

As well as the signed social network datasets, large-scale biological networks can be analysed as signed graphs. There are four signed biological networks analysed by [13] and [38]. Graph G6 represents the gene regulatory network of

**Table 3** The frustration index in various signed networks

| Graph | $n$ | $m$ | $m^-$ | $L(G)$ | $L(G_r) \pm$ SD | Z score |
|---|---|---|---|---|---|---|
| G1 | 16 | 58 | 29 | 7 | $14.65 \pm 1.38$ | −5.54 |
| G2 | 18 | 49 | 12 | 5 | $9.71 \pm 1.17$ | −4.03 |
| G3 | 17 | 40 | 17 | 4 | $7.53 \pm 1.24$ | −2.85 |
| G4 | 17 | 36 | 16 | 6 | $6.48 \pm 1.08$ | −0.45 |
| G5 | 100 | 2461 | 1047 | 331 | $965.6 \pm 9.08$ | −69.89 |
| G6 | 690 | 1080 | 220 | 41 | $124.3 \pm 4.97$ | −16.75 |
| G7 | 1461 | 3215 | 1336 | 371 | $653.4 \pm 7.71$ | −36.64 |
| G8 | 329 | 779 | 264 | 193 | $148.96 \pm 5.33$ | 8.26 |
| G9 | 678 | 1425 | 478 | 332 | $255.65 \pm 8.51$ | 8.98 |

*Saccharomyces cerevisiae* [12] and graph G7 is related to the gene regulatory network of *Escherichia coli* [58]. The Epidermal growth factor receptor pathway [55] is represented as graph G8. Graph G9 represents the molecular interaction map of a macrophage [54]. For more details on the four biological datasets, one may refer to [38]. The data for real networks used in this study is publicly available on the Figshare research data sharing website [5].

We use $G_r = (V, E, \sigma_r)$ to denote a reshuffled graph in which the sign function $\sigma_r$ is a random mapping of $E$ to $\{-1, +1\}$ that preserves the number of negative edges. The reshuffling process preserves the underlying graph structure. The numerical results on the frustration index of our nine signed graphs and reshuffled versions of these graphs are shown in Table 3 where, for each graph $G$, the average and standard deviation of the line index of balance in 500 reshuffled graphs, denoted by $L(G_r)$ and SD, are also provided for comparison.

Although the signed networks are not balanced, the relatively small values of $L(G)$ suggest a low level of frustration in some of the networks. Figure 3 shows how the small signed networks G1–G4 can be made balanced by negating (or removing) the edges on a minimum deletion set. Dotted lines represent negative edges, solid lines represent positive edges, and frustrated edges are indicated by dotdash lines regardless of their original signs. The node colourings leading to the minimum frustration counts are also shown in Figure 3. Note that it is pure coincidence that there are an equal number of nodes coloured black for each graph G1–G4 in Figure 3. Visualisations of graphs G1–G4 without node colours and minimum deletion sets can be found in [7].

In order to be more precise in evaluating the relative levels of frustration in G1–G9, we have implemented a very basic statistical analysis using Z scores, where $Z = (L(G) - L(G_r))/\text{SD}$. The Z scores, provided in the right column of Table 3, show how far the frustration index is from the values obtained through random allocation of signs to the fixed underlying structure (unsigned graph). Negative values of the Z score can be interpreted as a lower level of frustration than the value resulting from a random allocation of signs. G1, G2, G5, G6, and G7 exhibit a level of frustration lower than what is expected by chance, while the opposite is observed for G8 and G9. The numerical results for G3 and G4 do not allow a conclusive interpretation.

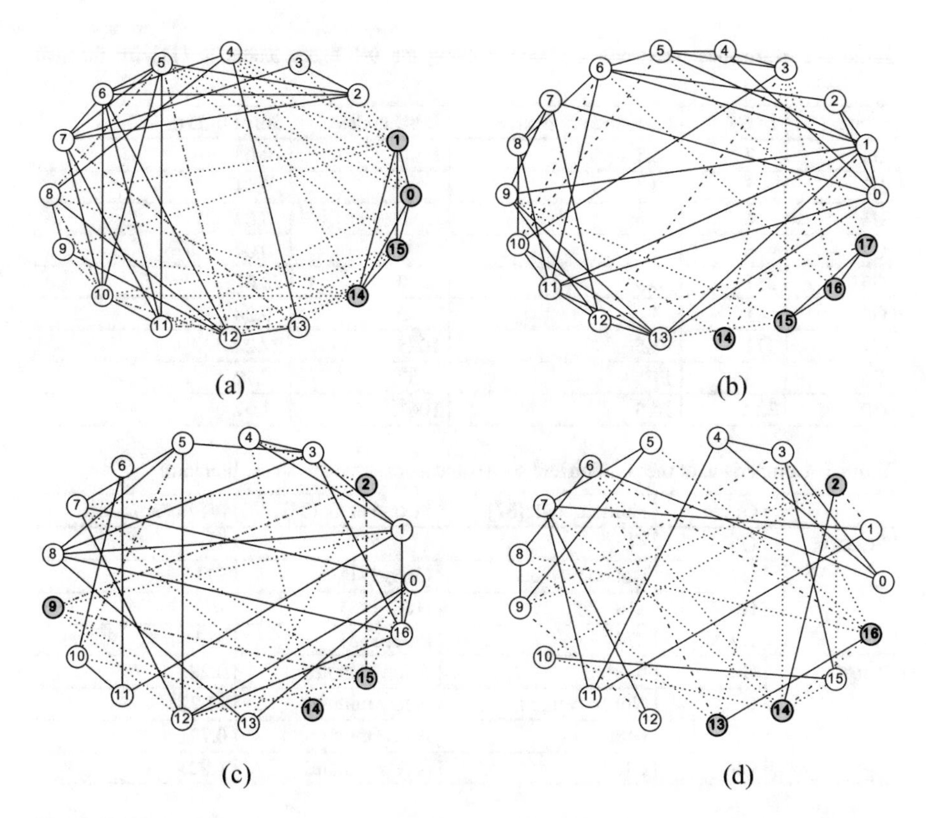


**Fig. 3** The frustrated edges represented by dotdash lines for four small signed networks inferred from the sociology datasets. (**a**) Highland tribes network (G1), a signed network of 16 tribes of the Eastern Central Highlands of New Guinea [57]. Minimum deletion set comprises 7 negative edges. (**b**) Monastery interactions network (G2) of 18 New England novitiates inferred from the integration of all positive and negative relationships [59]. Minimum deletion set comprises 2 positive and 3 negative edges. (**c**) Fraternity preferences network (G3) of 17 boys living in a pseudo-dormitory inferred from ranking data of the last week in [52]. Minimum deletion set comprises 4 negative edges. (**d**) College preferences network (G4) of 17 girls at an Eastern college inferred from ranking data of house B in [44]. Minimum deletion set comprises 3 positive and 3 negative edges

Various performance measures for the 0/1 linear model (11) coupled with the additional constraints for solving G1–G9 are provided in Table 4.

We compare the quality and solve time of our exact algorithm with that of recent heuristics and approximations implemented on the datasets. Table 5 provides a comparison of the 0/1 linear model (11) with other methods in the literature.

Hüffner et al. have previously investigated frustration in G6–G9 suggesting a data reduction scheme and (an attempt at) an exact algorithm [37]. Their suggested data reduction algorithm can take more than 5 h for G6, more than 15 h for G8, and more than 1 day for G9 if the parameters are not perfectly tuned [37]. Their algorithm coupled with their data reduction scheme and heuristic speed-ups does not converge

**Table 4** Performance measures of Gurobi solving the 0/1 linear model in (11) for the real networks

| Graph | $L(G)$ | Root node objective | B&B nodes | Solve time (s) |
|---|---|---|---|---|
| G1 | 7 | 4.5 | 0 | 0.03 |
| G2 | 5 | 0 | 0 | 0.04 |
| G3 | 4 | 2.5 | 0 | 0.02 |
| G4 | 6 | 2 | 0 | 0.04 |
| G5 | 331 | 36.5 | 0 | 78.67 |
| G6 | 41 | 3 | 0 | 0.28 |
| G7 | 371 | 21.5 | 1085 | 27.22 |
| G8 | 193 | 17 | 457 | 0.72 |
| G9 | 332 | 14.5 | 1061 | 1.92 |

**Table 5** Comparison of the solution and solve time against models in the literature

| | Graph | Hüffner et al. [37] | Iacono et al. [38] | 0/1 linear model |
|---|---|---|---|---|
| Solution | G6 | 41 | 41 | 41 |
| | G7 | Not converged | [365, 371] | 371 |
| | G8 | 210 | [186, 193] | 193 |
| | G9 | 374 | [302, 332] | 332 |
| Time | G6 | 60 s | A few minutes | 0.28 s |
| | G7 | Not converged | A few minutes | 27.22 s |
| | G8 | 6480 s | A few minutes | 0.72 s |
| | G9 | 60 s | A few minutes | 1.92 s |

for G7 [37]. In addition to these solve time and convergence issues, their algorithm provides $L(G8) = 210$, $L(G9) = 374$, both of which are incorrect based on our results.

Iacono et al. have investigated frustration in G6–G9 [38]. Their heuristic algorithm provides upper and lower bounds for G6–G9 with a 100%, 98.38%, 96.37% and 90.96% ratio of lower to upper bound, respectively. Regarding solve time, they have only mentioned that their heuristic requires a fairly limited amount of time (a few minutes on an ordinary PC).

While data reduction schemes [37] take up to 1 day for these datasets and heuristic algorithms [38] only provide bounds with up to 9% gap from optimality, our 0/1 linear model solves each of the 9 datasets to optimality in less than a minute.

## 7 Conclusion and Future Research

This study focuses on frustration index as a measure of balance in signed networks and the findings may well have a bearing on the applications of the line index of balance in the other disciplines [4]. The present study has suggested a novel method for computing a measure of structural balance that can be used for analysing dynamics of signed networks. It contributes additional evidence that suggests signed social networks and biological gene regulatory networks exhibit a relatively low level of frustration (compared to the expectation when allocating signs at random). On similar lines of research, we have undertaken a follow-up study with more focus on operations research aspects of this topic [6].

This study has a number of important implications for future investigation. *Point index of balance* is a similarly defined measure of balance based on removing minimum nodes to achieve balance [32]. The computations of this measure may also be considered as a niche point to be explored using exact and heuristic computational methods [26]. The optimisation model introduced can make network dynamics models more consistent with the theory of structural balance [3]. Many sign change simulation models that allow one change at a time use the number of balanced triads in the network as a criterion for transitioning towards balance. These models may result in stable states that are not balanced, like jammed states and glassy states [48]. This contradicts not only the instability of unbalanced states, but also the fundamental assumption that networks gradually move towards balance. Deploying decrease in the frustration index as the criterion, the above-mentioned states might be avoided resulting in a more realistic simulation of signed network dynamics that is consistent with structural balance theory and its assumptions.

**Acknowledgements** The authors are grateful for the extremely valuable comments of the anonymous reviewers that have prevented incorrect attributions in the literature review section and helped improve the discussions in this chapter.

## References

1. R.P. Abelson, M.J. Rosenberg, Symbolic psycho-logic: a model of attitudinal cognition. Behav. Sci. **3**(1), 1–13 (1958). https://doi.org/10.1002/bs.3830030102
2. J. Akiyama, D. Avis, V. Chvàtal, H. Era, Balancing signed graphs. Discret. Appl. Math. **3**(4), 227–233 (1981). https://doi.org/10.1016/0166-218X(81)90001-9
3. T. Antal, P.L. Krapivsky, S. Redner, Dynamics of social balance on networks. Phys. Rev. E **72**(3), 036121 (2005)
4. S. Aref, M.C. Wilson, Balance and frustration in signed networks. J. Complex Networks (2018, in press)
5. S. Aref, Signed networks from sociology and political science, systems biology, international relations, finance, and computational chemistry (2017). https://doi.org/10.6084/m9.figshare.5700832.v2
6. S. Aref, A.J. Mason, M.C. Wilson, An exact method for computing the frustration index in signed networks using binary programming. arXiv:1611.09030 (2017)
7. S. Aref, M.C. Wilson, Measuring partial balance in signed networks. J. Complex Networks (2018, in press). https://doi.org/10.1093/comnet/cnx044
8. F. Barahona, On the computational complexity of Ising spin glass models. J. Phys. A Math. Gen. **15**(10), 3241 (1982)
9. B. Bollobás, *Random Graphs*, 2nd edn. (Cambridge University Press, Cambridge, 2001)

10. J. Bramsen, Further algebraic results in the theory of balance. J. Math. Sociol. **26**(4), 309–319 (2002)
11. D. Cartwright, F. Harary, Structural balance: a generalization of Heider's theory. Psychol. Rev. **63**(5), 277–293 (1956)
12. M.C. Costanzo, M.E. Crawford, J.E. Hirschman, J.E. Kranz, P. Olsen, L.S. Robertson, M.S. Skrzypek, B.R. Braun, K.L. Hopkins, P. Kondu, C. Lengieza, J.E. Lew-Smith, M. Tillberg, J.I. Garrels: YPD™, PombePD™ and WormPD™: model organism volumes of the BioKnowledge™ Library, an integrated resource for protein information. Nucleic Acids Res. **29**(1), 75–79 (2001). https://doi.org/10.1093/nar/29.1.75
13. B. DasGupta, G.A. Enciso, E. Sontag, Y. Zhang, Algorithmic and complexity results for decompositions of biological networks into monotone subsystems. Biosystems **90**(1), 161–178 (2007)
14. T. Dewenter, A.K. Hartmann, Exact ground states of one-dimensional long-range random-field Ising magnets. Phys. Rev. B **90**(1), 014207 (2014)
15. P. Doreian, A. Mrvar, Structural balance and signed international relations. J. Soc. Struct. **16**, 1–49 (2015)
16. T. Došlić, D. Vukičević, Computing the bipartite edge frustration of fullerene graphs. Discret. Appl. Math. **155**(10), 1294–1301 (2007). https://doi.org/10.1016/j.dam.2006.12.003
17. P. Esmailian, S.E. Abtahi, M. Jalili, Mesoscopic analysis of online social networks: the role of negative ties. Phys. Rev. E **90**(4), 042817 (2014)
18. E. Estrada, M. Benzi, Walk-based measure of balance in signed networks: detecting lack of balance in social networks. Phys. Rev. E **90**(4), 1–10 (2014)
19. G. Facchetti, G. Iacono, C. Altafini, Computing global structural balance in large-scale signed social networks. Proc. Natl. Acad. Sci. **108**(52), 20953–20958 (2011). https://doi.org/10.1073/pnas.1109521108
20. G. Facchetti, G. Iacono, C. Altafini, Exploring the low-energy landscape of large-scale signed social networks. Phys. Rev. E **86**(3), 036116 (2012)
21. C. Flament, *Applications of Graph Theory to Group Structure* (Prentice-Hall, Upper Saddle River, 1963)
22. C. Flament, Équilibre d'un graphe: quelques résultats algébriques. Math. Sci. Hum. **8**, 5–10 (1970)
23. J.H. Fowler, Legislative cosponsorship networks in the US House and Senate. Soc. Networks **28**(4), 454–465 (2006)
24. N.G. Fytas, P.E. Theodorakis, A.K. Hartmann, Revisiting the scaling of the specific heat of the three-dimensional random-field Ising model. Eur. Phys. J. B **89**(9), 200 (2016)
25. Gurobi Optimization Inc.: Gurobi optimizer reference manual, Houston, TX (2018). www.gurobi.com/documentation/8.0/refman/index.html. Accessed 1 May 2015
26. G. Gutin, D. Karapetyan, I. Razgon, Fixed-parameter algorithms in analysis of heuristics for extracting networks in linear programs, in *International Workshop on Parameterized and Exact Computation* (Springer, Berlin, 2009), pp. 222–233
27. F. Hadlock, Finding a maximum cut of a planar graph in polynomial time. SIAM J. Comput. **4**(3), 221–225 (1975). https://dx.doi.org/10.1137/0204019
28. P.L. Hammer, Pseudo-boolean remarks on balanced graphs, in *Numerische Methoden bei Optimierungsaufgaben Band 3* (Springer, Basel, 1977), pp. 69–78
29. P. Hansen, Labelling algorithms for balance in signed graphs, in *Problèmes Combinatoires et Théorie des Graphes* (Éditions du Centre national de la recherche scientifique, Paris, 1978), pp. 215–217
30. F. Harary, On the notion of balance of a signed graph. Mich. Math. J. **2**(2), 143–146 (1953)
31. F. Harary: Structural duality. Behav. Sci. **2**(4), 255–265 (1957)
32. F. Harary, On the measurement of structural balance. Behav. Sci. **4**(4), 316–323 (1959). https://doi.org/10.1002/bs.3830040405
33. F. Harary, J.A. Kabell, A simple algorithm to detect balance in signed graphs. Math. Soc. Sci. **1**(1), 131–136 (1980)

34. F. Harary, M.H. Lim, D.C. Wunsch, Signed graphs for portfolio analysis in risk management. IMA J. Manag. Math. **13**(3), 201–210 (2002). https://doi.org/10.1093/imaman/13.3.201
35. A.K. Hartmann, Ground states of two-dimensional Ising spin glasses: fast algorithms, recent developments and a ferromagnet-spin glass mixture. J. Stat. Phys. **144**(3), 519–540 (2011)
36. F. Heider, Social perception and phenomenal causality. Psychol. Rev. **51**(6), 358–378 (1944)
37. F. Hüffner, N. Betzler, R. Niedermeier, Separator-based data reduction for signed graph balancing. J. Comb. Optim. **20**(4), 335–360 (2010)
38. G. Iacono, F. Ramezani, N. Soranzo, C. Altafini, Determining the distance to monotonicity of a biological network: a graph-theoretical approach. IET Syst. Biol. **4**(3), 223–235 (2010)
39. P.W. Kasteleyn, Dimer statistics and phase transitions. J. Math. Phys. **4**(2), 287–293 (1963). https://doi.org/10.1063/1.1703953
40. O. Katai, S. Iwai, Studies on the balancing, the minimal balancing, and the minimum balancing processes for social groups with planar and nonplanar graph structures. J. Math. Psychol. **18**(2), 140–176 (1978)
41. E. Klotz, A.M. Newman, Practical guidelines for solving difficult mixed integer linear programs. Surv. Oper. Res. Manag. Sci. **18**(1–2), 18–32 (2013). http://dx.doi.org/10.1016/j.sorms.2012.12.001
42. J. Kunegis, Applications of structural balance in signed social networks. arXiv:1402.6865 [physics] (2014)
43. J. Kunegis, S. Schmidt, A. Lommatzsch, J. Lerner, E.W. De Luca, S. Albayrak, Spectral analysis of signed graphs for clustering, prediction and visualization, in *Proceedings of the 2010 SIAM International Conference on Data Mining*, ed. by S. Parthasarathy, B. Liu, B. Goethals, J. Pei, C. Kamath, vol. 10 (Society for Industrial and Applied Mathematics, Philadelphia, 2010), pp. 559–570. https://doi.org/10.1137/1.9781611972801.49
44. T.B. Lemann, R.L. Solomon, Group characteristics as revealed in sociometric patterns and personality ratings. Sociometry **15**, 7–90 (1952)
45. J. Leskovec, D. Huttenlocher, J. Kleinberg, Signed networks in social media, in *Proceedings of the SIGCHI Conference on Human Factors in Computing Systems* (ACM, New York, 2010), pp. 1361–1370
46. L. Ma, M. Gong, H. Du, B. Shen, L. Jiao, A memetic algorithm for computing and transforming structural balance in signed networks. Knowl.-Based Syst. **85**, 196–209 (2015). http://dx.doi.org/10.1016/j.knosys.2015.05.006
47. M. Manssen, A.K. Hartmann, Matrix-power energy-landscape transformation for finding NP-hard spin-glass ground states. J. Glob. Optim. **61**(1), 183–192 (2015)
48. S.A. Marvel, S.H. Strogatz, J.M. Kleinberg, Energy landscape of social balance. Phys. Rev. Lett. **103**(19), 198701 (2009)
49. O. Melchert, A. Hartmann, Information-theoretic approach to ground-state phase transitions for two-and three-dimensional frustrated spin systems. Phys. Rev. E **87**(2), 022107 (2013)
50. M. Mézard, G. Parisi, The Bethe lattice spin glass revisited. Eur. Phys. J. B Condens. Matter Complex Syst. **20**(2), 217–233 (2001)
51. Z. Neal, The backbone of bipartite projections: inferring relationships from co-authorship, co-sponsorship, co-attendance and other co-behaviors. Soc. Networks **39**, 84–97 (2014)
52. T. Newcomb, *The Acquaintance Process* (Holt, Rinehart and Winston, New York, 1966)."The General Nature of Peer Group Influence", pp. 2–16 in College Peer Groups, ed. by T.M. Newcomb, E.K. Wilson. (Aldine Publishing Co, Chicago, 1961)
53. R.Z. Norman, F.S. Roberts, A derivation of a measure of relative balance for social structures and a characterization of extensive ratio systems. J. Math. Psychol. **9**(1), 66–91 (1972)
54. K. Oda, T. Kimura, Y. Matsuoka, A. Funahashi, M. Muramatsu, H. Kitano, Molecular interaction map of a macrophage. AfCS Res. Rep. **2**(14), 1–12 (2004)
55. K. Oda, Y. Matsuoka, A. Funahashi, H. Kitano, A comprehensive pathway map of epidermal growth factor receptor signaling. Mol. Syst. Biol. **1**(1) (2005)
56. M. Petersdorf, Einige Bemerkungen über vollständige Bigraphen. Wiss. Z. Techn. Hochsch. Ilmenau **12**, 257–260 (1966)

57. K.E. Read, Cultures of the central highlands, New Guinea. Southwest. J. Anthropol. **10**(1), 1–43 (1954)
58. H. Salgado, S. Gama-Castro, M. Peralta-Gil, E. Diaz-Peredo, F. Sánchez-Solano, A. Santos-Zavaleta, I. Martinez-Flores, V. Jiménez-Jacinto, C. Bonavides-Martinez, J.Segura-Salazar, et al., Regulondb (version 5.0): Escherichia coli k-12 transcriptional regulatory network, operon organization, and growth conditions. Nucleic Acids Res. **34**(suppl 1), D394–D397 (2006)
59. S.F. Sampson, A novitiate in a period of change. An experimental and case study of social relationships (PhD thesis), Cornell University, Ithaca, 1968
60. D. Sherrington, S. Kirkpatrick, Solvable model of a spin-glass. Phys. Rev. Lett. **35**, 1792–1796 (1975). https://doi.org/10.1103/PhysRevLett.35.1792
61. E. Terzi, M. Winkler, A spectral algorithm for computing social balance, in *Algorithms and Models for the Web Graph* (Springer, Berlin, 2011), pp. 1–13
62. I. Tomescu, Note sur une caractérisation des graphes dont le degré de déséquilibre est maximal. Math. Sci. Hum. **42**, 37–40 (1973)
63. G. Toulouse, Theory of the frustration effect in spin glasses: I. *Spin Glass Theory and Beyond: An Introduction to the Replica Method and Its Applications*, vol. 9 (World Scientific, Singapore, 1987), p. 99
64. T. Zaslavsky, Signed graphs: to: T. Zaslavsky, Discrete Applied Mathematics 4 (1982) 47–74 Erratum. Discret. Appl. Math. **5**(2), 248 (1983)
65. T. Zaslavsky, Balanced decompositions of a signed graph. J. Comb. Theory Ser. B **43**(1), 1–13 (1987)
66. T. Zaslavsky, Matrices in the theory of signed simple graphs, advances in discrete mathematics and applications, in *Proceedings of the International Conference on Discrete Mathematics, ICDM-2008*, vol. 13 (Ramanujan Mathematical Society, Mysore, 2010), pp. 207–229
67. T. Zaslavsky, A mathematical bibliography of signed and gain graphs and allied areas. Electron. J. Comb. Dynamic Surveys in Combinatorics DS8 (2012)

# Optimal Factorization of Operators by Operators That Are Consistent with the Graph's Structure

**Victoria Goncharenko, Yuri Goncharenko, Sergey Lyashko, and Vladimir Semenov**

## 1 General Factorization Problem Statement

Consider a MIMD-structure multiprocessor computer system (CS) with local interactions. We interpret it as an oriented graph $G$. The vertices of the graph are universal machines, and the arcs are one-way communication channels. If the $i$-th machine can transmit the message to the machine with the number $j$, then the graph has an arc that starts at the vertex with the number $i$ and ends at the vertex with the number $j$. If the machine has local memory, then the corresponding vertex of the graph has a loop. We call the graph $G$ the graph of inter-machine communication CS.

We denote by the symbol $\Theta_i$ the set of numbers of those vertices of the graph from which the arc leads to the $i$-th vertex. In other words, $\Theta_i$ is the set of numbers of vertices of the stellar approach graph defined by the vertex with the number $i$.

Let $X_i$ be the set of states of the $i$-th machine, then vector $\bar{x} = (x, x, \ldots, x)$, where $x \in X_i$ ( $X_i$ describes the state of the system). We denote by the symbol $D$ the set of possible states of CS ($D \subseteq \oplus_{i=1}^{n} X_i$). We distinguish the moments of state changes in the functioning of the system. It is natural to assume that the subsequent state of the machine with the number $i$ is a function of its own previous state and the state of the machines directly connected to it, i.e., machines with numbers from $\Theta_i$. Then the subsequent state of the system will be a vector of the form

$$\bar{y} = F(\bar{x}) = (f_1(\bar{x}), \ldots, f_i(\bar{x}), \ldots, f_n(\bar{x})), \tag{1}$$

V. Goncharenko · Y. Goncharenko · S. Lyashko · V. Semenov (✉)
Department of Computational Mathematics, Taras Shevchenko Kiev National University, Kiev, Ukraine
e-mail: yuragoko@mail.ru



B. Goldengorin (ed.), *Optimization Problems in Graph Theory*, Springer Optimization and Its Applications 139,
https://doi.org/10.1007/978-3-319-94830-0_4

where operator $F$ takes the set $D$ into itself, and it's coordinate functions $f_i$ depend only on the set of those coordinates of the state vector whose numbers belong to $\Theta_i$.

**Definition 1 ([3])** The operator of the form (1) called *consistent with the graph's structure* $G$ (CGS operator).

The functioning of the system can be represented as a sequence of states

$$\bar{x}(j) = \left(F_j \circ F_{j-1} \circ F_{j-2} \circ \ldots F_0\right)(\bar{x}(0)), \quad j \in \mathbb{N}, \tag{2}$$

where $F_i$ is the CGS operator; $x(0)$, $x(j)$ are vectors of the initial and subsequent states of the system. Operator $F = F_j \circ F_{j-1} \circ F_{j-2} \circ \ldots \circ F_0$, that takes $\bar{x}(0)$ into $\bar{x}(j)$ is not necessarily the CGS operator. The representation of operator $F$ in the form of composition of operators is called the *factorization*. The number of mappings in a factorization chain is called its *length*.

*Remark 1* This model of distributed computing was considered in the early work [4].

**Definition 2 ([2])** We denote as $L(D)$ some set of operators that take the set of states $D$ into itself. We call the computational system *universal in class* $L(D)$, if any operator from $L(D)$ can be represented as the composition of finite number of operators that are consistent with the structure of graph of inter-machine communication $G$.

Suppose that we want to realize the action of operator $F \in L(D)$ ($\bar{y} = F(\bar{x})$, where $\bar{x}, \bar{y} \in D$) on the CS that is universal in class $L(D)$. That means that we need $F$ to be factorized into the chain (2) of successive CGS operators. Their sequential realization on the CS transforms the system from the state $\bar{x}$ to state $\bar{y}$. The time spent by the system for the transition from the state $\bar{x}$ to state $y$ is determined by the length of the corresponding chain. Thus, the productivity of the system in the class $L(D)$ is determined by the length of the factorization chains. There are many ways to factorize the operator. The most effective is the one that leads to a chain of minimum length.

The proof of the universality of CS in the corresponding class of mappings is the content of the basic parallelization theorem for a particular CS. The construction of the factorization method leading to a factorization chain of minimal length will be called the main problem of mapping the algorithm to the structure of a given computing system.

**Definition 3 ([2])** The *factorization length* of operator $F \in L(D)$ of a CGS operator is the smallest possible length of a factorization chain

$$\min_{\sigma(F)}\{\, j \;:\; F = F_j \circ F_{j-1} \circ F_{j-1} \circ \ldots \circ F_0\},$$

where $F$ is CGS operator, $\sigma(F)$ is the set of all possible approaches of operator factorization. The *depth of factorization* for the set $L(D)$ is

$$\max_{F \in L(D)} \left\{ \min_{\sigma(F)} \{ j \ : \ F = F_j \circ F_{j-1} \circ F_{j-1} \circ \ldots \circ F_0 \} \right\}.$$

The factorization length of a particular operator $F$ characterizes the highest productivity that CS can achieve by executing the action $\bar{y} = F(\bar{x})$. The depth of the factorization serves for upper bounds and is characterized by the "worst" implementation of the operator from $L(D)$.

At a fixed speed of the physical components of the system, the increase in the productivity of CS is achieved by changing the graph of inter-machine communications and the class of operators $L(D)$. In other words, by configuring the CS architecture for a class of tasks, or by choosing the class of tasks that are optimally implemented on the CS of this architecture, or both.

We can improve the calculus of CGS operators by endowing $D$ and $L(D)$ with the properties of specific mathematical structures. Below we consider an important case for numerical analysis when the set of states of each machine is a set of real numbers, and the set of states CS is the arithmetic vector space of column vectors $\mathbb{R}^n$ ($\mathbb{C}^n$).

## 2 Factorization of Linear Operators

Let $G$ be a finite oriented graph containing $n$ vertices with a loop at each vertex (arcs and loops are single), $A$ is its adjacency matrix. Graph $G$ is *strongly connected*, if any ordered pair of its vertices is connected by an oriented path.

Suppose also that $\mathbb{R}^n$ is an $n$-dimensional arithmetic vector space over the field of real numbers. Consider the space of linear operators acting in $\mathbb{R}^n$. Further, a linear operator in $\mathbb{R}^n$ and its matrix will be denoted by one letter.

We provide an order relation on the set of square matrices with $n \times n$ dimension. We say that the matrix $B = \{b_{i,j}\}$ *precedes* the matrix $C = \{c_{i,j}\}$, if for all $i$, $j = \overline{1,n}$ from the fact that $c_{i,j} = 0$ it follows that $b_{i,j} = 0$ (denoted by $B \prec C$).

A linear operator with matrix $B$ is consistent with the structure of graph $G$ if and only if $B \prec A^T$. We note that adding a diagonal matrix does not change the consistency.

**Theorem 1 (The Factorization Criterion)** *Every linear invertible operator in* $\mathbb{R}^n$ *can be represented as a composition of operators compatible with the structure of the graph G with a loop at each vertex, if and only if G is strongly connected.*

We preface the proof by two lemmas.

**Lemma 1** *Every non-singular square matrix can be represented as a product of matrices of elementary transformations by a diagonal matrix.*

**Lemma 2** *Suppose that in a finite oriented graph G the vertices $v_i$ and $v_j$ are not connected, then there exists a numbering of the vertices of the graph G, that its adjacency matrix has a block triangular form*

$$\begin{pmatrix} A_{1,1} & A_{1,2} \\ 0_{2,1} & A_{2,2} \end{pmatrix} \tag{3}$$

*where $A_{1,1}$, $A_{2,2}$ are square matrices of dimension $p \times p$ and $q \times q$, respectively, $p+q=n$, $A_{1,2}$ is matrix of dimension $p \times q$, $0_{2,1}$ is zero matrix of dimension $q \times p$.*

*Proof* Consider two cases. Let $v_i$ be an isolated vertex. The set of vertices of the graph $V \setminus v_i$ is numbered from 1 to $n-1$, and we assign the number $n$ to the vertex $v_i$. The matrix of a graph $G$ with such a numbering has a block-diagonal form, which satisfies the hypothesis of the theorem.

Let $v_i$ be a non-isolated vertex. We denote by $V(v_i)$ the set of vertices such that the vertex $v_i$ is connected with them by outgoing ways. The set $V \setminus V(v_i)$ is nonempty since by the condition $v_i \in V \setminus V(v_i)$. Renumber the vertices from the set $V \setminus V(v_i)$ by numbers from 1 to $p$ and vertices from set $V(v_i)$ by numbers from $p+1$ to $n$. With such a numbering of vertices, the adjacency matrix of the oriented graph $G$ has a form (3).

Since the lower (upper) block-triangular matrices form a subring in the ring of square matrices, it follows that filled non-singular matrix cannot be represented in the product of the product of the upper (lower) block-triangular matrices. We return to the proof of the Theorem 1.

*Proof Sufficiency* By the hypothesis of the theorem, the adjacency matrix $A$ of the graph $G$ has a nonzero diagonal. Hence the diagonal matrix is always consistent with the structure of the graph $G$. By Lemma 1, it suffices to show that any elementary transformation can be represented as a product of elementary transformations consistent with the structure of the graph $G$. Elementary transformation when the $i$-th row of the matrix multiplied by the number $\lambda$ is added to the $j$-th row of the matrix and denoted by

$$T_{ij}(\lambda) = E + \Lambda_{ij}(\lambda),$$

where $E$ is the identity matrix, $\Lambda_{ij}(\lambda)$ is the matrix where the element with the index $ij$ is equal to $\lambda$, and all other elements are equal to zero. We note that the product $\Lambda_{ik}(a) \cdot \Lambda_{lj}(b)$ is equal to the zero matrix for $k \neq l$ and is equal to $\Lambda_{ij}(ab)$ for $l = k$. This implies that the matrices $T_{ij}(\lambda)$ and $T_{ij}(-\lambda)$ are mutually inverse.

We carry out the proof by mathematical induction on the length of the path. Suppose that the vertex $v_i$ is connected with the vertex $v_j$, then the elementary transformation $T_{ij}(\lambda)$ is consistent with the structure of the graph $G$. By the induction hypothesis, if the vertices $v_i$ and $v_k$ are connected by a path of length $m$, then the elementary transformations $T_{ik}(\lambda)$ can be decomposed into a product of matrices consistent with the structure of the graph $G$.

Let vertices $v_i$ and $v_j$ be connected by a path of length $m+1$ and the last arc of the path connects a pair of vertices $(v_k, v_j)$. A direct calculation shows that

$$T_{ij}(\lambda) = T_{ik}(-1)T_{ik}(\lambda)T_{kj}(1)T_{kj}(-\lambda),$$

which was to be shown.

*Necessity* From the opposite. Suppose that every invertible matrix can be represented as a product of matrices compatible with the structure graph $G$, but in the graph there are vertices $v_i$ and $v_j$ not connected by any path. By Lemma 2, there exists a numbering of the vertices of the graph such that its adjacency matrix has a block-triangular structure. Matrices consistent with the structure of such a graph form a subring of block-triangular matrices. This subring is closed with respect to products, and so it is impossible to be a non-singular filled matrix. The theorem is proved.

## 3 The Upper Bound of the Factorization Depth

The proof of the Theorem 1 is constructive and allows to obtain an upper bound for the depth of factorization.

**Theorem 2** *Let $d$ be the diameter of a strongly connected graph $G$ with a loop at each vertex, and $\alpha$ the factorization depth of the set of linear invertible operators in $\mathbb{R}^n$ by operators consistent with the structure of $G$. Then we have the estimation*

$$\alpha < 4dn(n-1) + 1. \tag{4}$$

*Proof* Since all elementary transformations are invertible, to restore the matrix from its diagonal matrix, it suffices to have $n(n-1)$ elementary transformations, each of which can be factored by at most $4d$ elementary transformations. We obtain the estimate (4).

The estimate (4) is rough.

**Hypothesis.** The estimate $\alpha \leq d$ holds.

Consider the problem of solving a system of equations of the form $B\bar{x} = \bar{c}$, where $B$ is a matrix consistent with the structure of the graph $G$. Its solution $\bar{x} = B^{-1}\bar{c}$ requires the inversion of the matrix $B^{-1}$, but the inverse matrix is not consistent with the structure of the graph. Then arises the question about the algorithm and the length of the factorization of non-singular matrices $B^{-1}$.

**Theorem 3 ([1])** *For any non-singular matrix $B$ consistent with the structure of the graph, the factorization length of the matrix $B^{-1}$ does not exceed $n-1$.*

*Proof* Let $B$ be a square matrix of size $n \times n$. Consider its characteristic polynomial

$$P_n(\lambda) = \det(\lambda E - B) = \lambda^n + a_1\lambda^{n-1} + a_2\lambda^{n-2} + \ldots + a_{n-1}\lambda + a_n.$$

If $B$ is a non-singular matrix, then the free term of the characteristic polynomial

$$a_n = P_n(0) = \det(-B)$$

is nonzero. By the Hamilton-Kelly theorem, the characteristic polynomial $P_\lambda$ is annihilating for the matrix $B$, so

$$B^n + a_1 B^{n-1} + a_2 B^{n-2} + \ldots + a_{n-1} B + a_n E = 0.$$

We represent the last term of this matrix equality in the form $a_n E = a_n B^{-1} B$, then

$$B^n + a_1 B^{n-1} + a_2 B^{n-2} + \ldots + a_{n-1} B = -a_n B^{-1} B.$$

Multiplying on the right by $B^{-1}$ and divide on $-a_n$ we get

$$B^{-1} = -a_n^{-1} \left( B^{n-1} + a_1 B^{n-2} + \ldots + a_{n-1} E \right).$$

Thus, the substitution of the matrix $B$ in the polynomial $(P_n(\lambda) - a_n)\,\lambda^{-1}$ gives the adjoint matrix. Denote it by

$$\widetilde{P_{n-1}}(\lambda) = \lambda^{n-1} + a_1 \lambda^{n-2} + a_2 \lambda^{n-3} + \ldots + a_{n-1} \lambda^n + a_{n-1}.$$

Let its roots (possibly complex) $\mu_1, \ldots, \mu_{n-1}$, then

$$\widetilde{P_{n-1}}(\lambda) = (\lambda - \mu_{n-1})(\lambda - \mu_{n-2})(\lambda - \mu_{n-3}) \ldots (\lambda - \mu_1).$$

So

$$B^{-1} = -a_n^{-1} (B - \mu_{n-1} E)(B - \mu_{n-2} E) \ldots (B - \mu_1 E).$$

Since the matrix $(B - \mu_{n-1} E) \prec B$, it follows that the matrix $B^{-1}$ is represented as a product of $n - 1$ matrices consistent with the structure of the graph $G$. The theorem is proved.

Thus, for the set of matrices of the inverse of the structure of the graph, the depth of factorization is $n - 1$. Estimation is not optimal. For a complete graph, every matrix is consistent with it. The proposed factorization method yields $n - 1$ factors. This estimate is exact because you can specify the graph and the matrix on which it is achieved. Indeed, let the graph $G$ contains $n$ vertices $V = \{v_1, v_2, v_3, \ldots, v_n\}$ and $2(n - 1)$ arcs $\{(v_1, v_2), (v_2, v_1), (v_2, v_3), (v_3, v_2), \ldots, (v_{n-1}, v_n), (v_n, v_{n-1})\}$. Its adjacency matrix is 3-diagonal. The elementary transformation matrix $T_{nn}(1)$ is factorized by a chain containing $n - 1$ factors, but not less than $n - 1$.

## 4 Conclusion

The notion of an operator that is consistent with the structure of a graph and the computational system that is universal in the class of operators are proposed. The model fully corresponds to the processes occurring in modern distributed computing systems. The preliminary versions of this model of distributed computations were considered in the early paper [4]. The problem of the factorization of operators by operators consistent with the structure of a graph is formulated. The criterion for the factorization of a linear invertible operator acting in a finite-dimensional linear space is proved. Upper estimations of the factorization depth of the class of linear invertible operators by linear operators compatible with the structure of the graph are obtained. The formulation of the problem of factorization and the solution of the problem of its existence are new. In our opinion, they have important practical applications for the construction of algorithms for parallel and distributed computing systems. We note that the problem of constructing an algorithm for optimal factorization for a given graph of any operator from a given class remains unresolved, even in the linear case.

An interesting problem is the construction of optimal factorizations of linear operators for concrete graphs, for example, rectangular or hexagonal lattices. This issue we will consider in one of the future works.

**Acknowledgements** This research is supported by the Ministry of Education and Science of Ukraine (project 0116U004777). Vladimir Semenov thanks the State Fund for Fundamental Researches of Ukraine for support.

## References

1. V. Yu. Goncharenko, Factorization of inverse matrices by sparse matrices. J. Comput. Appl. Math. **1**(111), 94–100 (2013)
2. V. Yu. Goncharenko, The factorization of mapping and parallelization of algorithms. Rep. Natl. Acad. Sci. Ukr., no. 2, 38–40 (1995)
3. V. Yu. Goncharenko, Some properties of operators that are consistent with the graph structure, in *Operations Research and ACS*, vol. 23, (KSU, Kiev, 1984), pp. 73–81
4. V. Yu. Goncharenko, B.B. Nesterenko, Asynchronous principles in parallel computing, part II. WP of Institute of Mathematics of the NASU, no. 82–38. Kiev, 1982

# Branching in Digraphs with Many and Few Leaves: Structural and Algorithmic Results

**Jørgen Bang-Jensen and Gregory Gutin**

## 1 Introduction

This is a survey on out-branchings with minimum and maximum number of leaves, which updates the previous one [5].[1] The reader will see, in what follows, that out-branchings with minimum and maximum number of leaves are of great relevance to several areas of graph theory and algorithms.

A subgraph $T$ of a digraph $D$ is called an **out-tree** if $T$ is an oriented tree with just one vertex $s$ of in-degree zero. The vertex $s$ is the **root** of $T$. If an out-tree $B$ is a spanning subgraph of $D$, $B$ is called an **out-branching**. A vertex $x$ of an out-branching $B$ is called a **leaf** if $d_B^+(x) = 0$. All other vertices of $B$ are called **internal vertices**. Figure 1 shows a digraph $D$ and, respectively, out-branchings with minimum and maximum number of leaves of $D$.

The problems of finding an out-branching with extremal number of leaves are of interest in practical applications; e.g., the problem of finding an out-branching with minimum number of leaves was considered in the US patent [20] by Demers and Downing, where its application to the area of database systems was described.

[1] We removed results on lower bounds on the maximum number of leaves in an out-branching as it does not seem to be of interest any longer, and added new structural and algorithmic results, especially on fixed-parameter tractable algorithms.

J. Bang-Jensen
IMADA, University of Southern Denmark, Odense M, Denmark
e-mail: jbj@imada.sdu.dk

G. Gutin (✉)
Department of Computer Science, Royal Holloway, University of London, Egham, Surrey, UK
e-mail: gutin@cs.rhul.ac.uk



B. Goldengorin (ed.), *Optimization Problems in Graph Theory*,
Springer Optimization and Its Applications 139,
https://doi.org/10.1007/978-3-319-94830-0_5

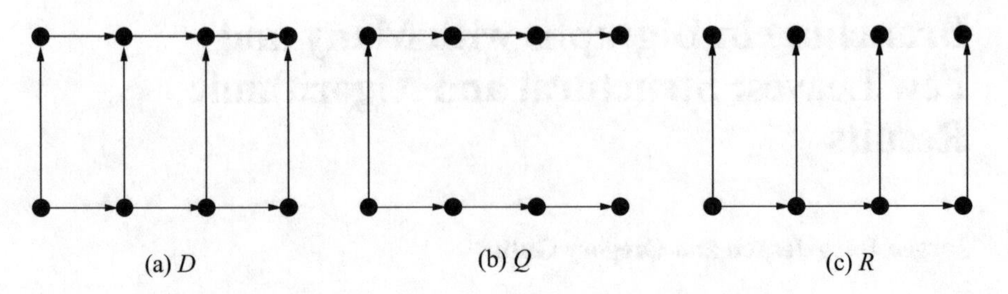


**Fig. 1** A digraph $D$ and its out-branchings with minimum and maximum number of leaves ($Q$ and $R$, respectively). (**a**) $D$, (**b**) $Q$, (**c**) $R$

Also, structural and algorithmic results on out-branchings with extremal number of leaves can be used to solve problems not directly related to the topic. For example, see the proof of Theorem 9.

For general digraphs, the problems of finding an out-branching with minimum/maximum number of leaves are $\mathcal{NP}$-hard: the problem of verifying the existence of an out-branching with just one leaf is the same as the hamiltonian path problem and the problem of finding a spanning tree with maximum number of leaves in an undirected graph is $\mathcal{NP}$-hard [27] (we may transform an undirected graph to the corresponding directed graph by replacing each edge $xy$ by two arcs $xy$ and $yx$). Thus, it is natural to consider parameterized complexities of the two problems. Let $k$ be a parameter. The problem of checking whether a digraph $D$ has an out-branching with at least $k$ leaves ($k$ internal vertices, respectively) is **fixed-parameter tractable (FPT)** which was proved by Bonsma and Dorn [14] (by Gutin et al. [28], respectively).[2] This means that each of the two parameterized problems can be solved by an algorithm of running time $O(f(k) \cdot n^{O(1)})$, where $f(k)$ is a computable function dependent on $k$ but not on $n$ and $n$ is the order of $D$. Such an algorithm is called an **FPT algorithm**; clearly, FPT algorithms generalize polynomial time algorithms, for the polynomial-time algorithms $f(k)$ can be set to a constant. Often in the parameterized algorithms and complexity literature, the running time $O(f(k) \cdot n^{O(1)})$ is written as $O^*(f(k))$, where $O^*$ hides polynomial factors.[3] We would like to note, in passing, that the problem of checking whether a digraph $D$ has an out-tree with at least $k$ leaves is also FPT [1].

Note that restricted to acyclic digraphs the problems of finding an out-branching with minimum and maximum number of leaves are of different complexities (provided $\mathcal{P} \neq \mathcal{NP}$): while the former is polynomial time solvable (see Section 3), the latter is $\mathcal{NP}$-hard (see Section 4).

[2]The algorithms of [14] and [28] have been significantly improved and we discuss the improvements in the survey.

[3]For an excellent introduction to the area of parameterized algorithms and complexity, see the monograph [16] by Cygan et al.

Our survey is organized as follows. In the next section, we will provide additional terminology and notation. Sections 3 and 4 are devoted to out-branchings with minimum and maximum number of leaves, respectively.

In this survey, we do not consider approximation or exact algorithms (with one exception in Section 4). Also, we do not consider parameterized kernelization, where some important notions and results on out-branchings with maximum number of leaves were obtained in [10]. Binkele-Raible et al. [10] proved that while the problem of deciding whether a given digraph has an out-branching rooted at a given vertex admits a polynomial-size kernel, the same problem without fixed root does not admit such a kernel subject to a well-known complexity theory hypothesis. The negative result was the first of its kind for a natural problem and led to the introduction of so-called **Turing kernels.**[4]

## 2 Terminology, Notation, and Preliminaries

For an out-branching $B$, let $\boldsymbol{L(B)}$ denote the set of leaves of $B$. For a digraph $D$ containing an out-branching, let $\boldsymbol{\ell_{\min}(D)}$ and $\boldsymbol{\ell_{\max}(D)}$ denote the minimum and maximum number of leaves in an out-branching of $D$. If $D$ has no out-branching, we can write $\ell_{\min}(D) = 0$ and $\ell_{\max}(D) = 0$.

For a digraph $D$, $\boldsymbol{\alpha(D)}$ denotes the **independence number** of $D$, i.e., the maximum size of a vertex set $X$ of $D$ such that there is no arc between any pair of vertices of $X$. A vertex $x$ of a digraph $D$ is called a **source**, if the in-degree of $x$ equals zero. The **path covering number** $\mathrm{pc}(\boldsymbol{D})$ of a digraph $D$ is the minimum number of disjoint directed paths needed to cover $V(D)$. A digraph $D$ is called **transitive** if the existence of arcs $xy$, $yz$ implies the existence of the arc $xz$, where $x$, $y$, and $z$ are distinct vertices of $D$. The underlying (undirected) graph of a digraph $D$ will be denoted by $\boldsymbol{UG(D)}$.

For more terminology and notation on digraphs, see Chapter 1 of [4].

In the study of out-branchings, the first question is when a digraph has an out-branching. This question is answered in the following well-known proposition, see, e.g., [4].

**Proposition 1** *A digraph $D$ has an out-branching if and only if $D$ has only one strongly connected component without incoming arcs.*

Since the strongly connected components of $D$ can be computed in time $O(n + m)$, where $n$ and $m$ are the number of vertices and arcs in $D$, respectively, one can decide whether a digraph has an out-branching in linear time. The same time is sufficient to decide whether $D$ has an out-branching rooted at a particular vertex $r$: just apply a breath-first search from $r$ and check whether all vertices could be reached from $r$.

[4]For more information of the area of parameterized kernels, see [16].

Later on, we will consider the question of whether a digraph contains a $k$-**leaf out-branching**, i.e. an out-branching with at least $k$-leaves. The following simple proposition helps us to answer this question. Here, a $k$-**leaf out-tree** is an out-tree with at least $k$ leaves.

**Proposition 2 ([33])** *If a digraph $D$ contains an out-branching rooted at a vertex $r$, then any $k$-leaf out-tree rooted at $r$ can be extended to a $k$-leaf out-branching rooted at $r$.*

# 3 Minimum Leaf Out-Branchings

In this subsection, we give upper bounds on $\ell_{\min}(D)$ for general and strongly connected digraphs $D$ (Section 3.1), a polynomial algorithm for computing $\ell_{\min}(D)$ for acyclic digraphs $D$ (Section 3.2), and discuss FPT algorithms for the problem to decide whether a digraph has an out-branching with at least $k$ internal vertices (Section 3.3).

## 3.1 Upper Bounds on $\ell_{\min}(D)$

Las Vergnas [34] proved the following upper bound on $\ell_{\min}(D)$ for general digraphs.

**Theorem 1 (Las Vergnas' Theorem)** *For a digraph $D$, we have $\ell_{\min}(D) \leq \alpha(D)$.*

We will prove the following proposition which immediately implies the theorem.

**Proposition 3 ([34])** *Let $B$ be an out-branching of $D$ with more than $\alpha(D)$ leaves. Then $D$ contains an out-branching $B'$ such that $L(B')$ is a proper subset of $L(B)$.*

*Proof* We will prove this claim by induction on the number $n$ of vertices in $D$. For $n \leq 2$ the result holds; thus, we may assume that $n \geq 3$ and consider an out-branching $B$ of $D$ with $|L(B)| > \alpha(D)$. Clearly, $D$ has an arc $xy$ such that $x, y$ are leaves of $B$. If the in-neighbor $p$ of $y$ in $B$ is of out-degree at least 2, then $L(B') \subset L(B)$, where $B' = B + xy - py$. So, we may assume that $d_B^+(p) = 1$. Observe that $\alpha(D - y) \leq \alpha(D) < |L(B)| = |L(B - y)|$. Hence by the induction hypothesis, $D - y$ has an out-branching $B''$ such that $L(B'') \subset L(B - y)$. Notice that $L(B - y) = L(B) \cup \{p\} \setminus \{y\}$. If $p \in L(B'')$, then observe that $L(B'' + py) \subset L(B)$. Otherwise, $L(B'' + xy) \subseteq L(B) \setminus \{x\} \subset L(B)$. □

The bound in Las Vergnas' theorem is tight as there are many digraphs $D$ for which $\ell_{\min}(D) = \alpha(D)$, see, e.g., Theorem 3. It would be interesting to find other tight upper bounds on $\ell_{\min}(D)$.

A digraph has a hamiltonian path if and only if it has an out-branching with only one leaf, so the problem of deciding whether a digraph has an out-branching with at most $k$ leaves is $\mathcal{NP}$-complete for each fixed natural number $k$. Gutin et al. [28, 29] proved that the problem of checking whether a digraph $D$ of order $n$ has an out-branching with at most $n-k$ leaves (or, equivalently, at least $k$ non-leaves) is fixed-parameter tractable.

Below we show that the well-known Gallai-Milgram theorem (Theorem 2) is a simple consequence of Las Vergnas' theorem. To do this we need the following simple result.

**Lemma 1 ([29])** *Let $D=(V,A)$ be a digraph and let $\hat{D}$ be the digraph obtained from $D$ by adding a new vertex $s$ and all possible arcs from $s$ to $V$. Then* $\mathrm{pc}(D)=\ell_{\min}(\hat{D})$.

*Proof* Since a collection of $p$ disjoint directed paths in $D$ covering $V(D)$ corresponds to an out-branching of $\hat{D}$ with $p$ leaves, we have $\mathrm{pc}(D)\ge \ell_{\min}(\hat{D})$. Let $B$ be an out-branching of $\hat{D}$ with $p$ leaves. We say that a vertex $x$ of $B$ is **branching** if $d_B^+(x)>1$. Consider a maximal directed path $Q$ of $B$ not containing branching vertices. Observe that $B-V(Q)$ has $p-1$ leaves. Thus, we can decompose the vertices of $B$ into $p$ disjoint directed paths. Deleting the vertex $s$ from this collection of paths, we see that $\mathrm{pc}(D)\le \ell_{\min}(\hat{D})$. Thus, $\mathrm{pc}(D)=\ell_{\min}(\hat{D})$. □

**Theorem 2 (Gallai-Milgram Theorem)** *[25] For every digraph $D$,* $\mathrm{pc}(D)\le \alpha(D)$.

*Proof* Consider the digraph $\hat{D}$ defined in Lemma 1. By Lemma 1 and Las Vergnas' theorem, $\mathrm{pc}(D)=\ell_{\min}(\hat{D})\le\alpha(\hat{D})=\alpha(D)$. □

The following result shows that the bound of Las Vergnas' theorem is sharp.

**Theorem 3 ([5])** *If $D$ is a transitive acyclic digraph with a unique source $s$, then* $\ell_{\min}(D)=\alpha(D)$.

*Proof* By Las Vergnas' theorem, $D$ contains an out-branching $B$ with $k\le\alpha(D)$ leaves. Observe that $B$ is rooted at $s$ and the vertices of every path in $B$ starting at $s$ and terminating at a leaf induce a clique in $UG(D)$. Thus, the vertices of $UG(D)$ can be covered by $k$ cliques and, hence, $\alpha(UG(D))\le k$. We conclude that $\ell_{\min}(D)=\alpha(D)$. □

The next theorem is equivalent to Theorem 3. Indeed, by Theorem 3 and Lemma 1, we have $\mathrm{pc}(D)=\ell_{\min}(\hat{D})=\alpha(\hat{D})=\alpha(D)$ for every transitive acyclic digraph $D$ which implies Dilworth's theorem. Since $\mathrm{pc}(D)\le\ell_{\min}(D)\le\alpha(D)$ for each transitive acyclic digraph with a unique source, Dilworth theorem implies Theorem 3.

**Theorem 4 (Dilworth's Theorem)** *[21] Every transitive acyclic digraph $D$ has* $\mathrm{pc}(D)=\alpha(D)$.

Theorems 3 and 4 give raise to the following natural research problem.

**Problem 1** Find other non-trivial digraph classes for which the equalities of Theorem 3 and/or Theorem 4 hold.

Las Vergnas proved another upper bound on $\ell_{\min}(D)$.

**Theorem 5 ([34])** *Let $D$ be a digraph on $n$ vertices such that any two distinct non-adjacent vertices have degree sum at least $2n - 2h - 1$, where $1 \leq h \leq n - 1$. Then $\ell_{\min}(D) \leq h$.*

Settling a conjecture of Las Vergnas [34], Thomassé [38] proved the following:

**Theorem 6** *If $D$ is a strong, then $\ell_{\min}(D) \leq \max\{\alpha(D) - 1, 1\}$.*

## 3.2 Acyclic Digraphs

Demers and Downing [20] suggested a heuristic approach for finding, in an acyclic digraph, an out-branching with minimum number of leaves. However, no argument or assertion has been made to provide the validity of their approach and to investigate its computational complexity. Using another approach, Gutin et al. [28] showed that a minimum leaf out-branching in an acyclic digraph can be found in polynomial time.

The following algorithm MINLEAF introduced by Gutin et al. [28] returns an out-branching with minimum number of leaves in an acyclic digraph. Observe that an acyclic digraph $D$ has an out-branching if and only if it has exactly one source. It is not difficult to prove that MINLEAF is correct and of running time $O(\sqrt{m}n^3)$, where $n$ ($m$) is the order (size) of the input digraph.

Figure 2 illustrates MINLEAF. There $M = \{rx', xy', zt'\}$ and $T = D - zy$.

The parameters directed tree-width, directed path-width, and DAG-width of digraphs are analogs of tree-width of undirected graphs; for definitions, see [26, 32]. Acyclic digraphs have directed tree-width, directed path-width, and DAG-width equal zero. It follows from one of the main results of the paper [32] by Johnson et al. that there are polynomial-time algorithms for verifying whether a digraph of bounded directed tree-width (directed path-width, DAG-width, respectively) has a Hamilton directed path. In contrast, Dankelmann et al. [19] proved that the problem

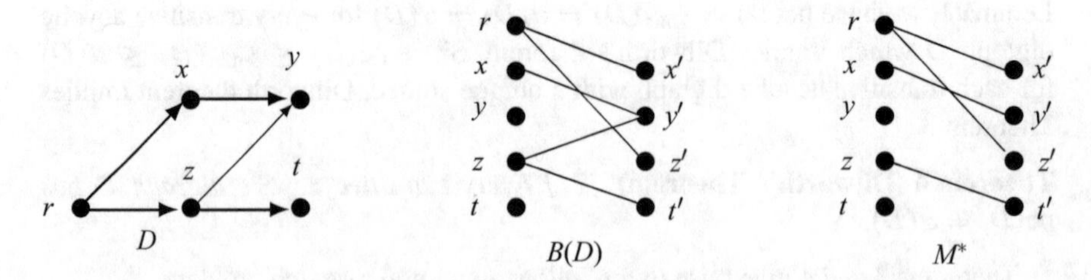


**Fig. 2** Illustration for MINLEAF

**Algorithm 1** MINLEAF

*Input:* An acyclic digraph $D$ with vertex set $V$.
*Output:* A minimum leaf out-branching $T$ of $D$ if $\ell_{\min}(D) > 0$ and "NO", otherwise.

1: Find a source $r$ in $D$. If there is another source in $D$, return "no out-branching". Let $V' = \{v' : v \in V\}$.
2: Construct a bipartite graph $B = B(D)$ of $D$ with partite sets $V$, $V' - r'$ and edge $xy'$ for each arc $xy \in A(D)$.
3: Find a maximum matching $M$ in $B$.
4: $M^* := M$. For all $y' \in V'$ not covered by $M$, set $M^* := M^* \cup \{$an arbitrary edge incident with $y'\}$.
5: $A(T) := \emptyset$. For all $xy' \in M^*$, set $A(T) := A(T) \cup \{xy\}$.
6: Return $T$.

of finding the minimum number of leaves in an out-branching of a digraph of directed tree-width (directed path-width, DAG-width, respectively) equal one is $\mathcal{NP}$-hard.

## 3.3 FPT Algorithms for General Digraphs

In this subsection we discuss FPT algorithms for the problem of deciding whether a digraph has an out-branching with at least $k$ internal vertices. This is called the $k$-Internal Out-Branching problem ($k$-IOB). The undirected version of $k$-IOB, called $k$-Internal Spanning Tree ($k$-IST), is defined similarly.

Note that $k$-IOB is a generalization of $k$-IST since the latter can easily be reduced to $k$-IOB on symmetric digraphs, i.e. digraphs in which every arc is on a directed cycle of length 2. The $k$-IOB problem a priori seems more difficult than the well-known $k$-Path problem (decide whether a given digraph has a path on $k$ vertices) since a witness of a yes-instance of $k$-path is a subgraph of size $k$, that is, a path on $k$ vertices. However, it is easy to see that a witness of a yes-instance of $k$-IOB (which has an out-branching) can be a subgraph of size $2k - 1$, that is, an out-tree with $k$ internal vertices and $k - 1$ leaves. This simple but crucial observation lies at the heart of several algorithms for $k$-IOB.

Parameterized algorithms for $k$-IST and $k$-IOB were first studied by Prieto and Sloper [36] and Gutin et al. [28], who proved that both problems are FPT. Since then several papers improved complexities of deterministic and randomized algorithms for both problems; we list these algorithms in Table 1. We also remark that approximation algorithms, exact exponential-time algorithms, and kernelization algorithms for both $k$-IOB and $k$-IST were extensively studied, but we do not overview such results in this survey.

**Table 1** FPT algorithms for $k$-IOB and $k$-IST

| Reference | Det./rand. | Space | Graph | Time$O^*(\cdot)$ |
|---|---|---|---|---|
| Prieto et al. [36] | *det* | *poly* | *Undirected* | $2^{O(k \log k)}$ |
| Gutin et al. [28] | *det* | *exp* | *Directed* | $2^{O(k \log k)}$ |
| Cohen et al. [15] | *det* | *exp* | *Directed* | $55.8^k$ |
| | *rand* | *poly* | *Directed* | $49.4^k$ |
| Fomin et al. [23] | *det* | *exp* | *Directed* | $16^{k+o(k)}$ |
| | *rand* | *poly* | *Directed* | $16^{k+o(k)}$ |
| Fomin et al. [24] | *det* | *poly* | *Undirected* | $8^k$ |
| Shachnai et al. [37] | *det* | *exp* | *Directed* | $6.855^k$ |
| Daligault [17] | *rand* | *poly* | *Directed* | $4^k$ |
| Li et al. [35] | *det* | *poly* | *Undirected* | $4^k$ |
| Zehavi [39] | *det* | *exp* | *Directed* | $5.139^k$ |
| | *rand* | *exp* | *Directed* | $3.617^k$ |
| Björklund et al. [12] | *rand* | *poly* | *Undirected* | $3.455^k$ |
| Björklund et al. [13] | *rand* | *poly* | *Directed* | $2^k$ |
| Gutin et al. [31] | *det* | *poly* | *Directed* | $3.86^k$ |
| Gutin et al. [31] | *det* | *exp* | *Directed* | $3.41^k$ |

Let us give a brief overview of the algorithms in [13] and [31] as their algorithms are currently the best among randomized and deterministic algorithms for $k$-IOB. Unlike the algorithms in the previous papers, the algorithms in [13] and [31] rely on the well-known Directed Matrix-Tree theorem.

The **(symbolic) Kirchhoff matrix** $K = K(D)$ of a directed multigraph $D$ on $n$ vertices is defined as follows, where we assume that the vertex set is $[n] := \{1, 2, \ldots, n\}$.

$$K_{ij} = \begin{cases} \sum_{\ell i \in A(D)} x_{\ell i} & \text{if } i = j, \\ -x_{ij} & \text{if } ij \in A(D), \\ 0 & \text{otherwise,} \end{cases} \quad (1)$$

where $A(D)$ is the arc set of $D$ and $x_{ij}$ are variables.

For $r \in [n]$ we denote by $K_{\bar{r}}(D)$ the matrix obtained from $K(D)$ by deleting the $r$th row and the $r$th column. Moreover, let $\mathscr{B}_r$ denote the set of out-branchings rooted at $r$. The following version of the Directed Matrix-Tree Theorem implies a natural one-to-one correspondence between the monomials of $\det(K_{\bar{r}}(D))$ and the out-branchings in $\mathscr{B}_r$.

**Theorem 7** *For every directed multigraph $D$ with symbolic Kirchhoff matrix $K(D)$ and $i \in V(D)$,* $\det(K_{\bar{r}}(D)) = \sum_{B \in \mathscr{B}_r} \prod_{ij \in A(B)} x_{ij}$.

For a proof of this theorem, see, e.g., [13].

For the algorithm in [13], Björklund et al. set $x_{ij}$ to $(1 + ty_i)z_{ij}$ and observed that $D$ has an out-branching rooted at $r$ with at least $k$ internal vertices if and only if the coefficient of $t^k$ in $\det(K_{\bar{r}}(D))$ has a monomial that is multilinear of degree $k$ in the variables $y_i$. Indeed, observe that the substitution $x_{ij} = (1 + ty_i)z_{ij}$ adds to the degree of the variable $y_i$ whether $i$ occurs as an internal vertex; the variables $z_{ij}$ make sure that distinct spanning out-branchings will not cancel each other. To compute $\det(K_{\bar{r}}(D))$ in polynomial time, we can build an arithmetic circuit $\mathscr{C}$ of polynomial size using, for example, Berkowitz's determinant circuit design [9]. To detect a multilinear monomial in $\mathscr{C}$ restricted to the coefficient of $t^k$ we can invoke Lemma 1 in [11]. This results in a randomized algorithm of running time $O^*(2^k)$.

To improve deterministic FPT algorithms for $k$-IOB, Gutin et al. [31] developed the following framework that can be used to design FPT algorithms for some other problems as well.

1. Identify a polynomial such that it has a monomial with at least $k$ distinct variables (called a *witnessing monomial*) if and only if the input instance of the problem at hand is a yes-instance. It should be possible to efficiently evaluate the polynomial (black box-access is sufficient here).
2. Color the variables of the polynomial with $k$ colors using a polynomial-delay perfect hash-family. To improve the running time of this step, we apply a problem-specific *coloring guide* to reduce the number of "random" colors. Given a $k$-coloring, we obtain a smaller polynomial by identifying all variables of the same color.
3. Use inclusion–exclusion to extract the coefficient of a colorful monomial from the reduced polynomial. By the usual color-coding arguments, if such a monomial exists, then the original polynomial contained a witnessing monomial.

While it is not possible to explain, in a short survey, all details of the implementation of the framework for $k$-IOB, let us comment on Steps 1 and 2 with respect to $k$-IOB. The polynomial used for $k$-IOB in [31] is the determinant in the Directed Matrix-Tree Theorem. We replace every variable $x_{ij}$ in (1) by $x_i$ and observe that now by Theorem 7 we have that if the polynomial $\det(K_{\bar{r}}(D))$ over variables $x_1, \ldots, x_n$ contains a monomial with at least $k$ different variables, then there exists an out-branching rooted at $r$ with at least $k$ internal vertices.

The following lemma establishes useful connections between matchings and out-trees/out-branchings; it is the key to the computation of a coloring guide in [31]. We believe it is of independent interest, too. A **matching** in a digraph is a set of arcs without common vertices.

**Lemma 2**

(1) *Let $T$ be an out-tree with $k \geq 0$ internal vertices. Then $T$ has a matching of size at least $k/2$.*
(2) *Let $D = (V, A)$ be a digraph containing an out-branching, and $M$ a matching in $D$. Then, in polynomial time, we can find an out-branching of $D$ for which no arc of $M$ has both end-vertices as leaves.*

*Proof*

(1) We prove it by induction on $k \geq 0$. The claim obviously holds for $k = 0$, so assume that $k \geq 1$. Let $k_1$ be the number of *pre-leaves*, i.e. vertices of $T$ whose only out-neighbors are leaves and $k_2$ the number of *prepre-leaves*, i.e. vertices of $T$ whose only out-neighbors are pre-leaves. Observe that $k_1 \geq k_2$ and that $T$ has a matching $M_1$ with $k_1$ edges whose vertices are some leaves and all pre-leaves. Let $T'$ be an out-tree obtained from $T$ by deleting all leaves and pre-leaves. Observe that $T'$ has $k - k_1 - k_2$ internal vertices and thus by induction hypothesis $T'$ has a matching $M_2$ of size at least $(k - k_1 - k_2)/2 \geq k/2 - k_1$. Thus, the matching $M_1 \cup M_2$ of $T$ is of size at least $k/2$.
(2) Let $B$ be an out-branching of $D$ and suppose that both end-vertices of some arc $xy$ of $M$ are leaves in $B$. Then add $xy$ to $B$ and delete $zy$ from $B$, where $z$ is the in-neighbor of $y$ in $B$. In the resulting out-branching $B'$, $x$ is an internal vertex. Notice that $zy$ does not belong to $M$. Hence, $B'$ contains one more arc of $M$ than $B$. Starting with an arbitrary out-branching and repeating the above exchange operation at most $|M|$ times, we will get an out-branching in which no arc of $M$ has both end-vertices as leaves. This process can be completed in polynomial time.

## 4 Maximum Leaf Out-Branchings

Alon et al. [2] proved the following complexity result.

**Proposition 4** *The problem of finding an out-branching of maximum number of leaves in an acyclic digraph is $\mathcal{NP}$-hard.*

*Proof* Consider a bipartite graph $G$ with bipartition $X, Y$ and a vertex $s \notin V(G)$. To obtain an acyclic digraph $D$ from $G$ and $s$, orient the edges of $G$ from $X$ to $Y$ and add all arcs $sx$, $x \in X$. Let $B$ be an out-branching in $D$. Then the set of leaves of $B$ is $Y \cup X'$, where $X' \subset X$, and for each $y \in Y$ there is a vertex $z \in Z = X \setminus X'$ such that $zy \in A(D)$. Observe that $B$ has maximum number of leaves if and only if $Z \subseteq X$ is of minimum size among all sets $Z' \subseteq X$ such that $N_G(Z') = X$. However, the problem of finding $Z'$ of minimum size such that $N_G(Z') = X$ is equivalent to the Set Cover problem ($\{N_G(y) : y \in Y\}$ is the family of sets to cover), which is $\mathcal{NP}$-hard. □

Alon et al. [2] also proved that the problem of deciding whether a digraph has an out-branching with at least $k$ leaves is FPT for acyclic and strongly connected digraphs. Bonsma and Dorn [14] extended this result to all digraphs. In [14], they presented an algorithm for the problem of running time $O^*(2^{O(k \log k)})$. Kneis et al. [33] designed an algorithm of running time $O^*(4^k)$; an algorithm of running time $O^*(3.72^k)$ was consequently obtained by Daligault et al. [18].

Since the algorithm of Kneis et al. [33] is quite simple, let us consider its short description. We will first check whether the input digraph $D$ has an out-branching rooted at vertex $r$ for each vertex $r$ of $D$. Let $R$ be the set of vertices, which are roots of out-branchings of $D$. Using Proposition 2, it suffices to check whether $D$ has an out-branching with at least $k$ leaves rooted at a vertex $r \in R$. So for each $r \in R$, we will run the following procedure. Let an out-tree $T$ initially contain only $r$ and let $r$ be active. In every iteration, choose an active vertex $v$ of $T$, declare it passive and either move to the next iteration or add to $T$ all arcs going from $v$ to vertices outside of $T$ and then move to the next iteration. Kneis et al. [33] proved that $D$ has an out-tree rooted at $r \in R$ with at least $k$ leaves if and only if the procedure obtains such a tree.

The paper [33] of Kneis et al. was a breakthrough as (a) it gave an algorithm, which was simpler and easier to analyze than that in [14], (b) the algorithm's complexity was lower than even that of algorithms to find a spanning tree with maximum number of leaves in an undirected graph. Apart from improving the running time of the algorithm in [33], Daligault et al. [18] improved the running time $O^*(2^n)$ of the following simple exact algorithm for finding an out-branching with maximum number of leaves: in a digraph $D = (V, A)$, for every $S \subset V$ delete all arcs leaving vertices of $S$ and check whether the resulting digraph has an out-branching (in which the vertices of $S$ will be a subset of leaves). By Proposition 1, the existence of an out-branching can be decided in polynomial time. The exact algorithm in [18] runs in time $O^*(1.9973^n)$; it would be interesting to design an exact algorithm with significantly smaller running time.

While the algorithmic results of Alon et al. [2] were significantly improved, their structural results are still of interest including the following theorem that was recently applied to solve an open problem, which will be discussed below.

**Theorem 8 ([2])** *Let $D$ be a strongly connected digraph. If $D$ has no out-branching with at least $k$ leaves, then the (undirected) pathwidth of $D$ is bounded by $O(k \log k)$.*

A well-known result in digraph algorithms, due to Edmonds, states that given a digraph $D$ and a positive integer $\ell$, we can decide whether $D$ has $\ell$ arc-disjoint out-branchings in polynomial time [22]. The same result holds for $\ell$ arc-disjoint in-branchings. Inspired by this fact, it is natural to ask for a "mixture" of out- and in-branchings: given a digraph $D$ and a pair $u, v$ of (not necessarily distinct) vertices, decide whether $D$ has an arc-disjoint out- branching $T_u^+$ rooted at $u$ and an in-branching $T_v^-$ rooted at $v$.

Thomassen proved (see [3]) that the problem is $\mathcal{NP}$-complete and remains $\mathcal{NP}$-complete if we add the condition that $u = v$. The same result still holds for digraphs in which the out-degree and in-degree of every vertex equals two [6].

An out-branching $T^+$ and an in-branching $T^-$ are called $k$-**distinct** if $|A(T^+) \setminus A(T^-)| \geq k$. Bang-Jensen and Yeo [7] asked whether the following problem called SINGLE-ROOT $k$-DISTINCT BRANCHINGS is FPT with respect to the parameter $k$: Given a digraph $D$; does $D$ contain a vertex $r$ and an out-branching $T^+$ and an in-branching $T^-$, both rooted at $r$ and which are $k$-distinct? Note that if $D$ has an out-branching and in-branching with the same root, then $D$ is strongly connected. Gutin et al. [30] proved that the problem is FPT and we will give their short proof below. Let us start from the following simple yet important lemma.

**Lemma 3 ([30])** *Let $D$ be a digraph containing an out-branching and an in-branching. If $D$ contains an out-branching (in-branching) $T$ with at least $k+1$ leaves, then every in-branching (out-branching) $T'$ of $D$ is $k$-distinct from $T$.*

*Proof* We will consider only the case when $T$ is an out-branching since the other case can be treated similarly. Let $T'$ be an in-branching of $D$ and let $L$ be the set of all leaves of $T$ apart from the one which is the root of $T'$. Observe that all vertices of $L$ have outgoing arcs in $T'$ and since in $T$ the incoming arcs of $L$ are the only arcs incident to $L$ in $T$, the sets of the outgoing arcs in $T'$ and incoming arcs in $T$ do not intersect.

We will use the following dynamic programming result of Bang-Jensen et al. [8].

**Lemma 4** *Let $H$ be a digraph of (undirected) treewidth $\tau$. Then* SINGLE-ROOT $k$-DISTINCT BRANCHINGS *on $H$ can be solved in time $O^*(2^{O(\tau \log \tau)})$.*

**Theorem 9 ([30])** SINGLE-ROOT $k$-DISTINCT BRANCHINGS *can be solved in time $O^*(2^{O(k \log^2 k)})$.*

*Proof* Let $D$ be an input digraph. As we noted above, we may assume that $D$ is strongly connected. Using an $O^*(3.72^k)$-time algorithm of [18] we can find an out-branching $T^+$ with at least $k+1$ leaves, or decide that $D$ has no such out-branching. If $T^+$ is found, the instance of SINGLE-ROOT $k$-DISTINCT BRANCHINGS is positive by Lemma 3 as any in-branching $T^-$ of $D$ is $k$-distinct from $T^+$. In particular, we may assume that $T^-$ has the same root as $T^+$ (a strongly connected digraph has an in-branching rooted at any vertex). Now suppose that $T^+$ does not exist. Then, by Theorem 8 the (undirected) pathwidth of $D$ is bounded by $O(k \log k)$. Thus, by Lemma 4 the instance can be solved in time $O^*(2^{O(k \log^2 k)})$.

SINGLE-ROOT $k$-DISTINCT BRANCHINGS requires that $k$-distinct in- and out-branchings must have the same root. Bang-Jensen et al. [8] considered the $k$-DISTINCT BRANCHINGS problem, where the same root requirement is removed. Bang-Jensen et al. [8] proved that the problem is FPT for strongly connected digraphs and conjectured that the result can be extended to all digraphs. This was confirmed by Gutin et al. [30], who used the approach of Theorem 9 coupled with a new digraph decomposition.

**Acknowledgements** The research of Jørgen Bang-Jensen was supported by the Danish research council, grant number 1323-00178B. The research of Gregory Gutin was supported in part by Royal Society Wolfson Research Merit Award.

## References

1. N. Alon, F. Fomin, G. Gutin, M. Krivelevich, S. Saurabh, Parameterized algorithms for directed maximum leaf problems, in *Proceedings of ICALP 2007*. Lecture Notes in Computer Science, vol. 4596 (2007), pp. 352–362
2. N. Alon, F. Fomin, G. Gutin, M. Krivelevich, S. Saurabh, Spanning directed trees with many leaves. SIAM J. Discret. Math. **23**, 466–476 (2009)
3. J. Bang-Jensen, Edge-disjoint in- and out-branchings in tournaments and related path problems. J. Combin. Theory Ser. B **51**(1), 1–23 (1991)
4. J. Bang-Jensen, G. Gutin, *Digraphs: Theory, Algorithms and Applications*, 2nd edn. (Springer, Berlin, 2009)
5. J. Bang-Jensen, G. Gutin, Out-branchings with extremal number of leaves. Ramanujan Math. Soc. Lect. Notes **13**, 91–99 (2010)
6. J. Bang-Jensen, S. Simonsen, Arc-disjoint paths and trees in 2-regular digraphs. Discret. Appl. Math. **161**(16–17), 2724–2730 (2013)
7. J. Bang-Jensen, A. Yeo, The minimum spanning strong subdigraph problem is fixed parameter tractable. Discret. Appl. Math. **156**, 2924–2929 (2008)
8. J. Bang-Jensen, S. Saurabh, S. Simonsen, Parameterized algorithms for non-separating trees and branchings in digraphs. Algorithmica **76**(1), 279–296 (2016)
9. S.J. Berkowitz, On computing the determinant in small parallel time using a small number of processors. Inf. Process. Lett. **18** 147–150 (1984)
10. D. Binkele-Raible, H. Fernau, F.V. Fomin, D. Lokshtanov, S. Saurabh, Y. Villanger, Kernel(s) for problems with n kernel: on out-trees with many leaves. ACM Trans. Algorithms **9**(4) (2011), article 39
11. A. Björklund, P. Kaski, L. Kowalik, Constrained multilinear detection and generalized graph motifs. Algorithmica **74**(2), 947–967 (2016)
12. A. Björklund, V. Kamat, L. Kowalik, M. Zehavi, Spotting trees with few leaves. SIAM J. Discret. Math. **31**(2), 687–713 (2017)
13. A. Björklund, P. Kaski, I. Koutis, Directed hamiltonicity and out-branchings via generalized Laplacians, in *Automata, Languages and Programming, 44th International Colloquium, ICALP 2017*. Leibniz International Proceedings in Informatics (LIPIcs), vol. 80 (2017), pp. 91:1–91:14
14. P. Bonsma, F. Dorn, Tight bounds and a fast FPT algorithm for directed Max-Leaf Spanning Tree. J. ACM Trans. Algorithms **7**(4), 1–19 (2011)
15. N. Cohen, F.V. Fomin, G. Gutin, E.J. Kim, S. Saurabh, A. Yeo, Algorithm for finding $k$-vertex out-trees and its application to $k$-internal out-branching problem. J. Comput. Syst. Sci. **76**, 650–662 (2010)
16. M. Cygan, F.V. Fomin, L. Kowalik, D. Lokshtanov, D. Marx, M. Pilipczuk, M. Pilipczuk, S. Saurabh, *Parameterized Algorithms* (Springer, Berlin, 2015)
17. J. Daligault, Combinatorial techniques for parameterized algorithms and kernels, with applications to multicut, PhD thesis, Universite Montpellier II, Montpellier, Herault, 2011
18. J. Daligault, G. Gutin, E.J. Kim, A. Yeo, FPT algorithms and kernels for the directed $k$-leaf problem. J. Comput. Syst. Sci. **76**, 144–152 (2010)
19. P. Dankelmann, G. Gutin, E.J. Kim, On complexity of minimum leaf out-branching problem. Discret. Appl. Math. **157**, 3000–3004 (2009)
20. A. Demers, A. Downing, minimum leaf spanning tree. US Patent no. 6,105,018, August 2000

21. R.P. Dilworth, A decomposition theorem for partially ordered sets. Ann. Math. **51**, 161–166 (1950)
22. J. Edmonds, Edge-disjoint branchings, in *Combinatorial Algorithms*, ed. by B. Rustin (Academic Press, Cambridge, 1973), pp. 91–96
23. F.V. Fomin, F. Grandoni, D. Lokshtanov, S. Saurabh, Sharp separation and applications to exact and parameterized algorithms. Algorithmica **63**, 692–706 (2012)
24. F.V. Fomin, S. Gaspers, S. Saurabh, S. Thomassé, A linear vertex kernel for maximum internal spanning tree. J. Comput. Syst. Sci. **79**, 1–6 (2013)
25. T. Gallai, A.N. Milgram, Verallgemeinerung eines graphentheoretischen Satzes von Rédei. Acta Sci. Math. Szeged **21**, 181–186 (1960)
26. R. Ganian, P. Hlineny, A. Langer, J. Obdrzalek, P. Rossmanith, Digraph width measures in parameterized algorithmics. Discret. Appl. Math. **168**, 88–107 (2014)
27. M.R. Garey, D.S. Johnson, *Computers and Intractability* (W.H. Freeman and Co., San Francisco, 1979)
28. G. Gutin, I. Razgon, E.J. Kim, Minimum leaf out-branching problems, in *AAIM'08*. Lecture Notes in Computer Science, vol. 5034 (2008), pp. 235–246
29. G. Gutin, I. Razgon, E.J. Kim, Minimum leaf out-branching and other problems. Theor. Comput. Sci. **410**, 4571–4579 (2009)
30. G. Gutin, F. Reidl, M. Wahlström, $k$-distinct in- and out-branchings in digraphs. J. Comput. Syst. Sci. **95**, 86–97 (2018)
31. G. Gutin, F. Reidl, M. Wahlström, M. Zehavi, Designing deterministic polynomial-space algorithms by color-coding multivariate polynomials. J. Comput. Syst. Sci. **95**, 69–85 (2018)
32. T. Johnson, N. Robertson, P.D. Seymour, R. Thomas, Directed tree-width. J. Combin. Theory Ser. B **82**, 138–154 (2001)
33. J. Kneis, A. Langer, P. Rossmanith, A new algorithm for finding trees with many leaves, in *ISAAC 2008*. Lecture Notes in Computer Science, vol. 5369 (2008), pp. 270–281
34. M. Las Vergnas, Sur les arborescences dans un graphe orienté. Discret. Math. **15**, 27–39 (1976)
35. W. Li, Y. Cao, J. Chen, J. Wang, Deeper local search for parameterized and approximation algorithms for maximum internal spanning tree. Inf. Comput. **252**, 187–200 (2017)
36. E. Prieto, C. Sloper, Reducing to independent set structure – the case of $k$-internal spanning tree. Nord. J. Comput. **12**, 308–318 (2005)
37. H. Shachnai, M. Zehavi, Representative families: a unified tradeoff-based approach. J. Comput. Syst. Sci. **82**, 488–502 (2016)
38. S. Thomassé, Covering a strong digraph by $\alpha - 1$ disjoint paths: a proof of Las Vergnas' conjecture. J. Combin. Theory Ser. B **83**, 331–333 (2001)
39. M. Zehavi, Mixing color coding-related techniques, in *Algorithms - ESA 2015 - 23rd Annual European Symposium*. Lecture Notes in Computer Science, vol. 9294 (2015), pp. 1037–1049

# Dominance Certificates for Combinatorial Optimization Problems

**Daniel Berend, Steven S. Skiena, and Yochai Twitto**

## 1 Introduction

One of the most active research areas in the theory of combinatorial algorithms is the design of approximation algorithms for NP-hard problems. However, while approximation ratio analysis does give some information on heuristics, it does not provide the whole picture regarding their performance in practice.

Algorithmic solutions used in practice are often some form of local improvement heuristic, based on techniques such as simulated annealing [18], HC [31], GRASP [24], tabu search [8], or genetic algorithms [17, 22]. Properly implemented, these techniques may lead to short, efficient programs which yield reasonable solutions. However, these heuristics often come with no theoretical guarantee as to the quality of the provided solution.

An $f(I)$ *combinatorial dominance guarantee* is a certificate that a solution is not worse than at least $f(I)$ solutions for a particular problem instance $I$. The intuition behind this performance measure rests on the letter of recommendation one could write on behalf of a given person, or heuristic solution. A recommendation like *"She is the best of the 75 students in my class this year"* is analogous to a combinatorial

D. Berend (✉)
Department of Mathematics and of Computer Science, Ben-Gurion University, Beer Sheva, Israel

Department of Mathematics, Rice University, Houston, TX, USA
e-mail: berend@cs.bgu.ac.il

S. S. Skiena
Department of Computer Science, Stony Brook University, Stony Brook, NY, USA
e-mail: skiena@cs.stonybrook.edu

Y. Twitto
Department of Computer Science, Ben-Gurion University, Beer Sheva, Israel
e-mail: twittoy@cs.bgu.ac.il



B. Goldengorin (ed.), *Optimization Problems in Graph Theory*,
Springer Optimization and Its Applications 139,
https://doi.org/10.1007/978-3-319-94830-0_6

dominance guarantee. It certifies the candidate as superior to a certain number of members of a given pool, with the implied assumption that this says something meaningful about the candidate's global ranking as well. The larger the number of competitors dominated by the candidate, the stronger the recommendation.

The previous body of work concerns proving existential bounds for particular problems over the space of all problem instances. In this paper, we demonstrate a general technique for awarding combinatorial dominance "certificates" to arbitrary solutions of various optimization problems. We demonstrate this technique on the TRAVELING SALESMAN and MAXIMUM SATISFIABILITY problems, and briefly experiment its usability. Observe that similar approximation ratio certificates are not forthcoming for ad-hoc solutions. Namely, given a particular solution of a problem, it is not at all clear how we can compare its quality with that of the (unknown) optimal solution.

Additionally, we describe how to simulatively estimate the number of solutions better than a given solution up to a given error with high probability. We experiment the usability of the simulative estimation for differentiating between heuristics for MAXIMUM SATISFIABILITY in terms of dominance, and compare the estimate derived from Chebyshev's inequality to simulation results.

In Section 1.1 we briefly survey previous work. The notions of combinatorial dominance guarantees are formalized in Section 1.2. In Section 2 we show how an arbitrary (and, in particular, a randomly selected) solution may be proved to have some combinatorial dominance guarantee. Brief experimental examinations are summarized in Section 3. Finally, we discuss directions for future research in Section 4.

## 1.1 Previous Work

The issue of measuring the quality of approximate solutions has been addressed by Zemel [35]. A formulation of the very basic properties expected from a function measuring the quality of approximate solutions was given, and the notion of a *proper* quality measure stated accordingly. Zemel suggested considering some measures, such as *z-approximation* [16] and *location ratio*, which is more familiar recently as *dominance ratio* [1, 14]. Both of these measures are proper.

The latter measure has been studied primarily within the operations research community. The basic notion appears to have been independently discovered several times. The primary focus has been on algorithms for TSP, specifically designing polynomial-time algorithms which dominate exponentially large neighborhoods. The first TSP heuristics with an exponential dominance number are due to Rublineckii [30] (see also Sarvanov and Doroshko [32, 33]).

The question whether there exists a polynomial-time algorithm which yields a solution dominating $(n-1)!/p(n)$ tours, where $p(n)$ is polynomial, appears to have first been raised by Glover and Punnen [9]. Dominance bounds for TSP have been most aggressively pursued by Gutin, Yeo, and Zverovich in a series of papers (cf. [10, 11]), culminating in a polynomial-time algorithm which finds a solution dominating $\Theta((n-1)!)$ tours. These bounds follow by applying certain Hamiltonian cycle decomposition theorems to the complete graph. We refer to [12] for more information.

Deineko and Woeginger [7] survey the complexity of optimizing TSP over several well-defined but exponentially large neighborhoods. Such optima by definition have large dominance numbers. Balas and Simonetti [4] perform an experimental study of certain linear-time dynamic programming algorithms for TSP, which dominate exponentially many solutions.

Gutin, Vainshtein, and Yeo [14] appear to have been the first to consider the complexity of achieving a given dominance bound. In particular, they define complexity classes of DOM-easy and DOM-hard problems. They prove that weighted MAX $k$-SAT and MAX CUT are DOM-easy while (unless P = NP) VERTEX COVER and CLIQUE are DOM-hard.

Alon, Gutin, and Krivelevich [1] provide several algorithms which achieve large dominance *ratios* for versions of INTEGER PARTITION, MAX CUT, and MAX $r$-SAT. These algorithms share a common property—they provide solutions of quality guaranteed to be not worse than the average solution value. This property has been used also in other dominance proofs [11, 14, 19, 25, 26]. Twitto [34] showed that this property by itself does not necessarily ensure good dominance.

Other works on dominance analysis include [13, 26], where it is proved that the nearest neighbor, minimum spanning tree, and greedy heuristics perform extremely poorly for symmetric and asymmetric TSP. Various combinatorial optimization problems and classical heuristics for them have been analyzed in [5, 6, 15]. In [23], a model for analyzing heuristic search algorithms (such as simulated annealing and backtracking), based on the ideas of combinatorial dominance, has been developed.

Recently, Kühn and Osthus [20] studied a polynomial-time algorithm for ATSP, and showed that it provides a dominance ratio of at least $1/2-o(1)$. In [21], together with Patel, they gave a polynomial-time algorithm with dominance ratio of $1 - n^{-1/29}$ for a special case of TSP in which the edges may take only two possible weights.

In another quite recent work, Punnen, Sripratak, and Karapetyan [27] analyzed the BBQP problem with $m + n$ variables. They proved that any solution for this problem, with quality no worse than the average, dominates at least $2^{m+n-2}$ solutions, and that this bound is the best possible. They provided an $O(mn)$ algorithm to identify such a solution.

## *1.2 Definitions*

Consider a given instance $I$ of some combinatorial optimization problem $P$. The instance is represented by a solution space $S_P(I)$ and objective function $C_P(I, x)$. The *solution space* $S_P(I)$ is the set of all combinatorial objects representing possible solutions $x$ to $I$. The objective function $C_P(I, x)$ is defined for all solutions $x \in$

$S_P(I)$. If $P$ is a maximization (minimization, resp.) problem, we seek an $x_0 \in S_P(I)$ such that $C_P(I, x_0) \geq C_P(I, x)$ ($C_P(I, x_0) \leq C_P(I, x)$, resp.) for all $x \in S_P(I)$.

A *heuristic* $H_P$ for $P$ is a procedure which, for any instance $I$, selects a solution $x \in S_P(I)$. For a given instance $I$ of $P$, denote by $F(I)$ the number of solutions that are not better than the heuristic solution $H_P(I)$. The number of all other solutions in $S_P(I)$ (which are better than $H_P(I)$) is denoted by $B(I)$.

**Definition 1** A heuristic $H_P$ offers an $F(n)$ *combinatorial dominance guarantee (dominance bound/number)* for problem $P$ if for each $n$:

1. For all instances $I$ of size $n$ of $P$, the solution $H_P(I)$ dominates at least $F(n)$ elements of $S_P(I)$.
2. There exists an instance $I'$ of size $n$ for which $H_P(I')$ dominates exactly $F(n)$ elements of $S_P(I')$.

The heuristic *blackball bound/number* of $H_P$ is $B(n) = |S_P(n)| - F(n)$.

The heuristic *dominance* (*blackball*, resp.) *ratio* is defined to be its dominance (blackball, resp.) number divided by the size of the solution space.

## 2 Certified Dominance Bounds for Arbitrary Solutions

In this section we demonstrate a general technique for awarding combinatorial dominance certificates to arbitrary solutions of various optimization problems. Such a certificate is a proof that a given solution of some instance (of some optimization problem) is not worse than at least some prescribed number of solutions of that instance. Additionally, we describe how to simulatively estimate the number of solutions better than a given solution up to a given error with high probability.

Assume we have any instance of some optimization problem, and let $X$ be its objective function. Suppose we can calculate the expected value $\mu = E(X)$ and the variance $\sigma^2 = V(X)$ of the value of the objective function at a random solution. (Note that these quantities can be calculated for many problems; see Sections 2.1 and 2.2 below for two important examples.) Now, suppose we have any solution with an objective value $x_0$ which happens to be better than $\mu$, i.e., $x_0 > \mu$ ($x_0 < \mu$, resp.) for maximization (minimization, resp.) problems. We may then assert that there is some percentage of the solutions which are not better than this solution. Indeed, denote by $x$ the objective value of a random solution, and assume, say, that we deal with a maximization problem. By the one-sided Chebyshev's inequality [29], we have

$$P(X > x_0) \leq \frac{\sigma^2}{\sigma^2 + (x_0 - \mu)^2} < 1. \tag{1}$$

One way to obtain (with probability arbitrarily close to 1) a solution significantly above the mean is to take the best of a large number of randomly selected solutions.

For example, suppose the values of the objective function are approximately normally distributed, which is true in many situations. The best solution out of approximately 40 sampled solutions is expected to be about two standard deviations above the mean, as approximately 2.5% of the solutions have this property. By Chebyshev's inequality, such a solution is in any case guaranteed to be not worse than at least four fifths of the solutions. Note that this bound holds whether the values are normally distributed or not, given the distance from the mean is verified somehow. It is the value from Chebyshev's inequality that provides the dominance certificate.

## *2.1* TSP *Certification*

In the SYMMETRIC TRAVELING SALESMAN problem (STSP), we are given an edge-weighted complete undirected graph $K_V$. We seek an ordering $p$ of the $n = |V|$ vertices, minimizing the sum of weights of the edges along the tour induced by $p$ on $K_V$. The size of the solution space is $(n-1)!/2$, and the size of the problem is taken as $n = |V|$. We apply the above technique to STSP.

Denote by $w_{ij}$ the weight of edge $(i, j)$, and let $X$ be the weight of a random tour. We have to find $E(X)$ and $V(X)$. For $1 \le i < j \le n$, put:

$$X_{ij} = \begin{cases} 1, & \text{the tour contains the edge } (i, j), \\ 0, & \text{otherwise.} \end{cases}$$

Then:

$$X = \sum_{1 \le i < j \le n} w_{ij} X_{ij}.$$

For $X_{ij}$, we have

$$E\left(X_{ij}\right) = P\left(X_{ij} = 1\right) = \frac{n}{\binom{n}{2}} = \frac{2}{n-1},$$

$$V\left(X_{ij}\right) = E\left(X_{ij}^2\right) - E^2\left(X_{ij}\right) = \frac{2(n-3)}{(n-1)^2}.$$

The covariance $\mathrm{Cov}\left(X_{ij}, X_{kl}\right)$ is given by

$$\begin{aligned}\mathrm{Cov}\left(X_{ij}, X_{kl}\right) &= E\left(X_{ij} X_{kl}\right) - E\left(X_{ij}\right) E\left(X_{kl}\right) \\ &= P\left(X_{ij} = X_{kl} = 1\right) - \left(\frac{2}{n-1}\right)^2,\end{aligned}$$

where

$$P\left(X_{ij}=X_{kl}=1\right)=\begin{cases}\frac{4}{(n-1)(n-2)}, & \{i,j\}\cap\{k,l\}=\emptyset,\\ \frac{2}{(n-1)(n-2)}, & |\{i,j\}\cap\{k,l\}|=1,\\ \frac{2}{n-1}, & \{i,j\}=\{k,l\}.\end{cases}$$

Therefore

$$E(X)=\sum_{1\le i<j\le n} w_{ij}E\left(X_{ij}\right)=\frac{2}{n-1}\sum_{1\le i<j\le n} w_{ij},$$

and

$$\begin{aligned}V(X)&=V\left(\sum_{1\le i<j\le n} w_{ij}X_{ij}\right)\\&=\sum_{1\le i<j\le n} w_{ij}^2 V\left(X_{ij}\right)+\sum_{(i,j)\neq(k,l)} w_{ij}w_{kl}\mathrm{Cov}\left(X_{ij},X_{kl}\right)\\&=\frac{2(n-3)}{(n-1)^2}\sum_{1\le i<j\le n} w_{ij}^2+\frac{4}{(n-1)^2(n-2)}\sum_{\{i,j\}\cap\{k,l\}=\emptyset} w_{ij}w_{kl}\\&\quad-\frac{2(n-3)}{(n-1)^2(n-2)}\sum_{|\{i,j\}\cap\{k,l\}|=1} w_{ij}w_{kl}.\end{aligned}$$

Having these explicit formulas for $E(X)$ and $V(X)$, we may easily, and automatically, bound the quality of any given solution for any given instance of STSP. The automation may be obtained by a program that first computes the expectation and variance by the above formulas for the given instance. Then, for any solution thereof, it computes and returns the probability given in (1) as the certified dominance bound for the solution.

## 2.2 MAXSAT *Certification*

In the MAXIMUM SATISFIABILITY problem (MAXSAT), we are given a multiset of clauses over some Boolean variables. Each clause is a disjunction of literals (a variable $x_i$ or its negation $\overline{x}_i$). We seek a true-false assignment for the variables, maximizing the number of satisfied clauses.

For disjoint sets $A, B \subseteq \{1, 2, \ldots, n\}$, denote:

$$T_{AB} = \bigvee_{i \in A} x_i \vee \bigvee_{j \in B} \overline{x}_j.$$

For example, $T_{\{1,4\}\{2\}} = x_1 \vee x_4 \vee \overline{x}_2$. Suppose the multiset, which we denote by $T$, consists of $c_{AB}$ occurrences of each $T_{AB}$. For a random assignment of values, consider the random variable $Y$—the number of satisfied clauses in $T$. We have

$$E(Y) = \sum_{A,B} c_{AB} \left(1 - 2^{-|A|-|B|}\right),$$

and

$$V(Y) = \sum_{A,B} c_{AB}^2 2^{-|A|-|B|} \left(1 - 2^{-|A|-|B|}\right)$$

$$+ \sum_{(A,B) \neq (A',B')} c_{AB} c_{A'B'} \left(P(T_{AB} = T_{A'B'} = \text{false}) - 2^{-|A|-|B|-|A'|-|B'|}\right),$$

where

$$P(T_{AB} = T_{A'B'} = \text{false}) = \begin{cases} 2^{-|A \cup A'|-|B \cup B'|}, & A \cap B' = A' \cap B = \emptyset, \\ 0, & \text{otherwise.} \end{cases}$$

Again, automating the computation of the quantities above to bound the quality of any given solution for any given instance of MAXSAT is immediate.

## 2.3 A Confidence Interval for the Blackball Ratio

Given an instance $I$ of an optimization problem $P$, and a solution $s_0$ of $I$, let:

$$g(s) = \begin{cases} 1, & s \text{ is better than } s_0, \\ 0, & \text{otherwise.} \end{cases}$$

Denote by $P_{s_0}(s)$ the probability that a random solution $s$ of $I$ is better than $s_0$.

**Lemma 1** *Let $\varepsilon, \delta \in (0, 1)$. Suppose $s_1, s_2, \ldots, s_t$ are $t \geq \left\lceil \frac{(2+\varepsilon) \ln \frac{2}{\delta}}{\varepsilon^2} \right\rceil$ uniformly and independently sampled solutions of I, and let*

$$\tilde{\mu} = \frac{1}{t} \left|\{1 \leq i \leq t : g(s_i) = 1\}\right|.$$

*Then, for a uniformly sampled solution s of I, we have*

$$\tilde{\mu} \in \left[P_{s0}(s) - \varepsilon, P_{s0}(s) + \varepsilon\right]$$

*with probability at least* $1 - \delta$.

*Proof* Consider the random variables $Y_i = g(s_i), 1 \leq i \leq t$. Note that $E(Y_i) = P(Y_i = 1) = P_{s0}(s)$. Let $Y = \sum_{i=1}^{t} Y_i$. By Chernoff's inequality [2] we have

$$\begin{aligned} P(|\tilde{\mu} - P_{s0}(s)| > \varepsilon) &= P(|Y/t - P_{s0}(s)| > \varepsilon) \\ &= P(|Y - tP_{s0}(s)| > \varepsilon t P_{s0}(s)/P_{s0}(s)) \\ &\leq 2e^{-\varepsilon^2 t/(2P_{s0}(s)+\varepsilon)} \\ &\leq 2e^{-\varepsilon^2 t/(2+\varepsilon)} \\ &\leq 2e^{-\frac{\varepsilon^2(2+\varepsilon)\ln\frac{2}{\delta}}{(2+\varepsilon)\varepsilon^2}} = \delta. \end{aligned}$$

The selection between several heuristics for a problem is often made by comparing their performance experimentally. The experiments may include applying the heuristics to random instances of the problem, or a predefined set of representative instances. Note that the above simple method might be used to obtain a single number representing the performance (in the experiments) of each of the heuristics. Namely, the method may yield an estimate of the blackball ratio of each of the heuristics in the experiments. Automation of this method is immediate. Observe that similar estimations of the approximation ratio of heuristics are not forthcoming, as the optimal objective value is usually unknown.

# 3 Experimental Results

In this section we present the results of experiments we performed in order to examine the techniques presented in Section 2.

## 3.1 *Results on the Chebyshev's Bound-Based Technique*

In these experiments, our aim is to check to what extent Chebyshev's bound gives meaningful results. To this end, we took 1000 random instances of STSP, on 20 vertices each, with edge weights selected uniformly and independently in [0, 1]. For each of the instances, we randomly selected a solution that is better than the average solution of that instance (by randomly generating solutions until obtaining a solution with this property). We compared Chebyshev's bound on the dominance

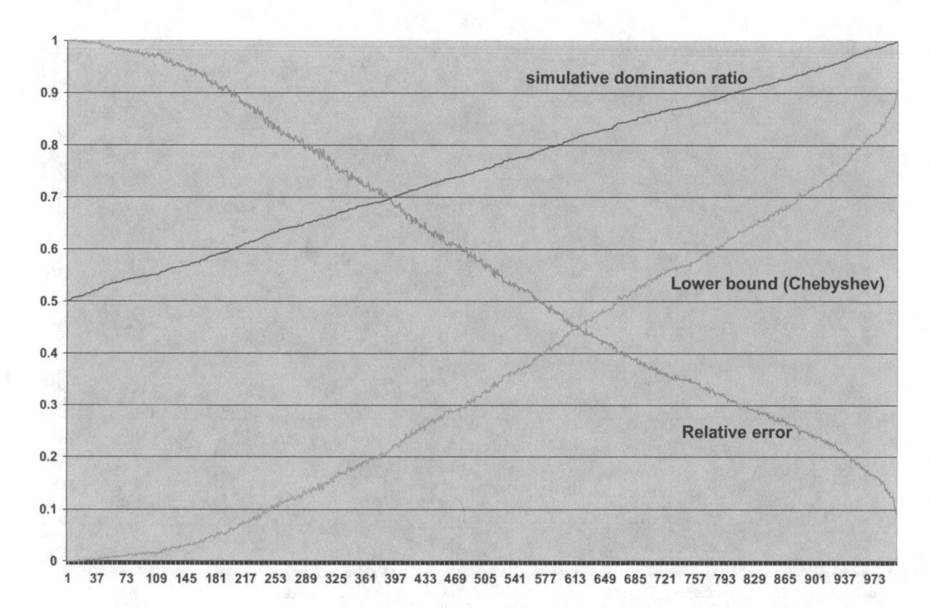


**Fig. 1** Comparing Chebyshev's lower bound to simulation results

ratio of the selected solution with an estimate of this ratio, given by simulation. The latter estimate was calculated as the percentage of solutions which outperformed our initial randomly selected solution, taken over a large number of random solutions.

Figure 1 shows the results. The instances (horizontal axis) are sorted according to their simulative estimation. The decreasing graph provides the relative error of Chebyshev's bound with respect to the (probably very accurate) estimate given by the simulation. It shows that Chebyshev's bound gets better and more meaningful as the solutions get better. For solutions close to the average solution value, Chebyshev's bound yields meaningless estimates, whereas for very good solutions it yields good estimates. A scatterplot of Chebyshev's bound (vertical axis) and the estimation provided by simulation (horizontal axis) is provided in Figure 2.

Likewise, we considered all the Euclidean instances of size of up to 1000 vertices from TSPLIB [28]. We used the following six heuristics available in the Concorde TSP Solver [3]: Greedy (GR), Boruvka (BV), Quick Boruvka (QBV), Nearest Neighbor (NN), LinKernighan (LK), and Optimal (OPT). (For more information on these heuristics consult the solver's [3] documentation.) We applied each of the heuristics on each of the instances, obtained solutions, and calculated Chebyshev's lower bounds on the dominance ratio of each of the solutions.

The graphs of the dominance of the heuristics are given in Figure 3, in which a representative part has been zoomed in. In this figure (as well as in Figure 4), the instances are arranged by their size on the horizontal axis. The vertical axis provides the dominance ratio. The location of the two coinciding graphs corresponding to the OPT and LK heuristics, above all the other graphs, shows that the dominance ratio is able to point to the better methods, and that the LK heuristic usually provides very

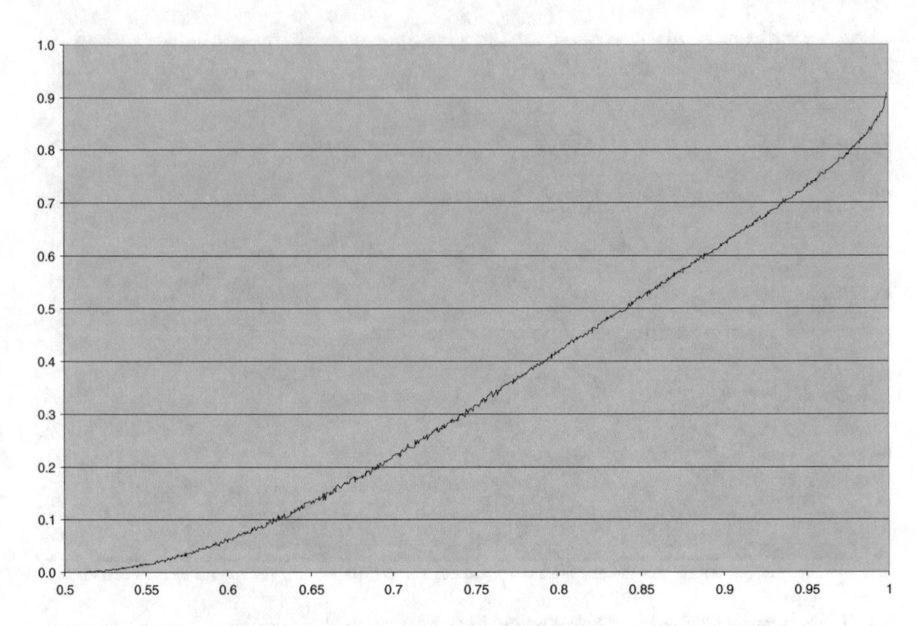


**Fig. 2** Scatterplot of Chebyshev's lower bound and simulation results

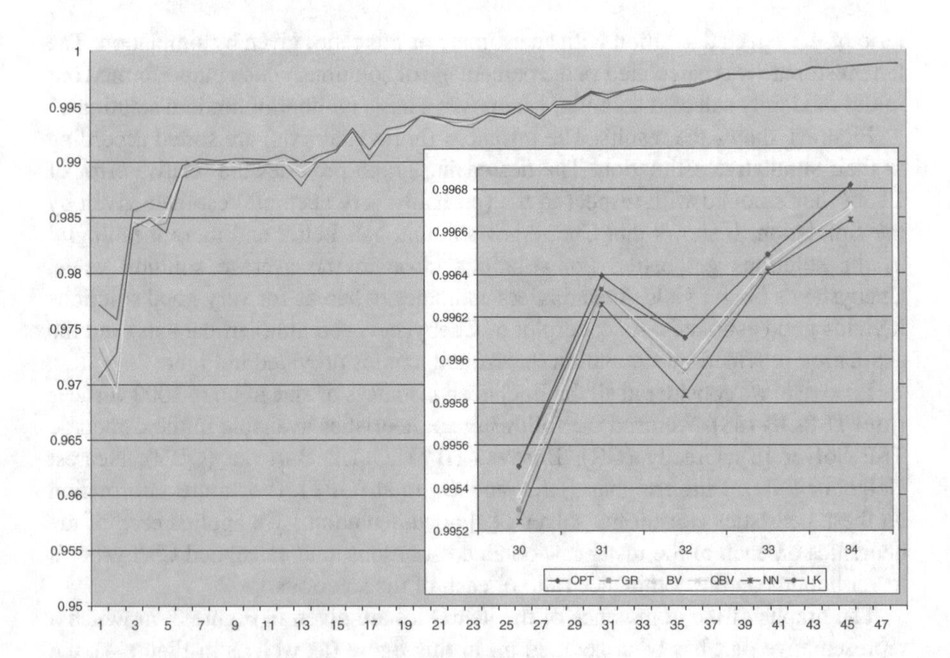


**Fig. 3** Comparison by dominance. Instances from TSPLIB

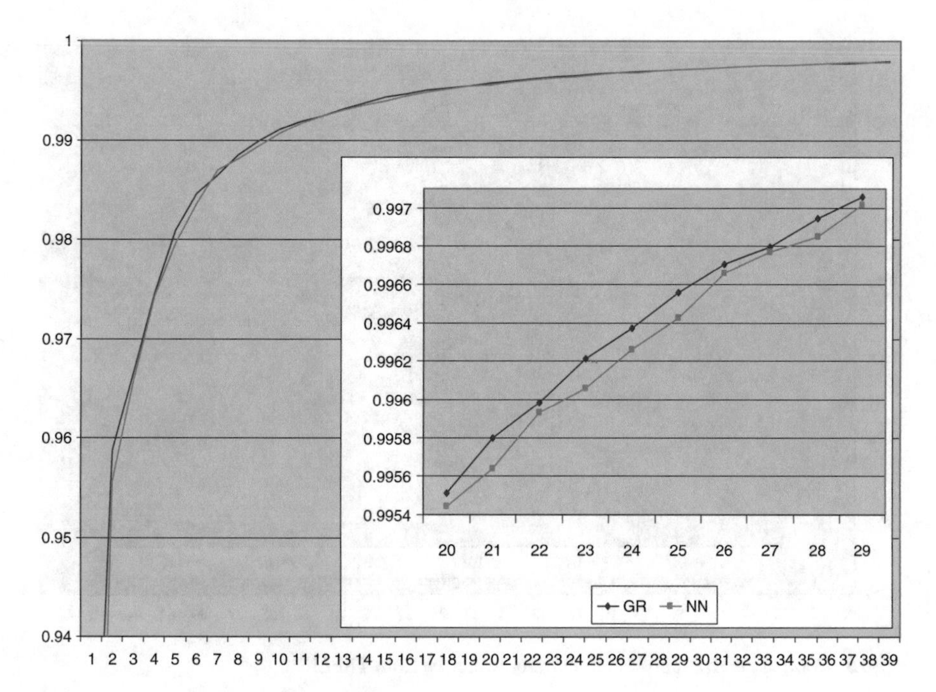


**Fig. 4** GR vs. NN. Applied on randomly generated instances

good solutions. The NN heuristic seems to be the worst all the way long. The other three heuristics (GR, BV, and QBV) are in the middle.

Similar phenomena were observed when we used Chebyshev's lower bound to compare GR and NN on randomly generated instances (Figure 4). A comparison using approximation ratio yielded similar results, as can be seen in Figure 5. To make it clearer, the instances in this figure are sorted according to the approximation ratio of the NN heuristic on them.

## 3.2 *Results on the Confidence Interval-Based Technique*

In the following, our main aim is to examine and demonstrate the usability of the technique presented in Section 2.3 as a way to compare and differentiate heuristics according to their estimated domination ratio. To this end, we compared the following heuristics for MAXSAT:

Majority Vote (MV). This heuristic assigns the value true to a variable if its number of positive occurrences is at least as large as its number of negative occurrences. Otherwise, it assigns the value false.

Step-by-step Majority Vote (SMV). This heuristic goes over the variables one-by-one according to some random order. It assigns a truth value to the current

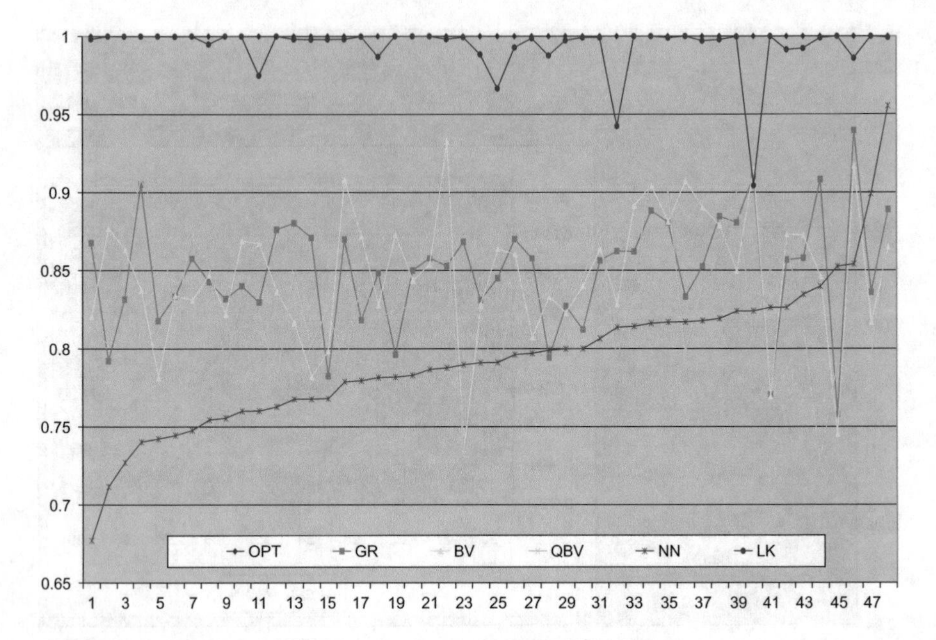


**Fig. 5** Comparison by approximation ratio. Instances from TSPLIB

variable according to the majority vote as before. However, after each assignment it discards all the clauses satisfied so far. The resulting instance is passed for the next step.

Greedy Occurrence SMV (GOSMV). Same as SMV, but at each step assigns a truth value to the currently most frequent variable. Ties are broken arbitrarily.

Greedy Unbalanced SMV (GUBSMV). Same as SMV, but at each step assigns a truth value to the variable for which the absolute value of the difference between its number of positive occurrences and number of negative occurrences is maximal at this point. Ties are broken arbitrarily.

The comparisons were done on 1000 randomly generated MAXSAT instances, on $n = 50$ variables $x_1, x_2, \ldots, x_n$ and $m = 300$ clauses. We select the number of variables to appear in each clause uniformly from the interval $[1, n]$. Then, for each variable $x_i$ we draw a random number from the interval $[0, 1/i]$, and select the ones with the largest random numbers to appear in the clause. Each of these variables appears in the clause positively or negatively with probability $1/2$.

For each such randomly generated instance we applied each of the heuristics, and obtained their solutions. The quality of a solution was assessed by Lemma 1. We have chosen $\varepsilon = 0.03$ and $\delta = 0.001$. Each variable of the random solutions generated for the assessment was set to true or false with probability $1/2$.

The estimated domination ratio of each of the heuristics on the randomly generated instances is depicted in Figure 6. On the horizontal axis, the instances are sorted in ascending order of performance of the heuristics on them. The sorting

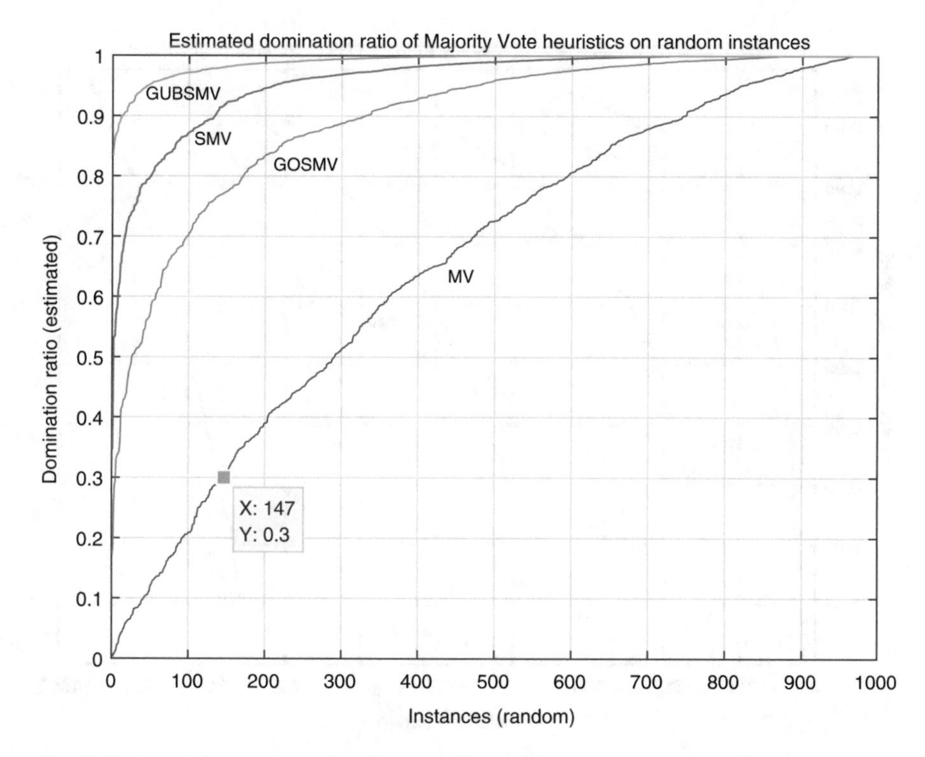


**Fig. 6** Estimated domination ratio of Majority Vote heuristics on random MAXSAT instances

is done for each of the heuristics independently, so that a specific point on the horizontal axis is likely to correspond to distinct instances for the various heuristics. The vertical axis is the estimated domination ratio at this point. For example, the point $(147, 0.3)$ marked on the graph of the MV heuristic indicates that the domination ratio of the instance ranked as the 147'th out of 1000 (from the bottom) for this heuristic is 0.3. A vertical zoom on the top echelon of the domination ratio (above 0.9) is provided in Figure 7.

A performance statistics is provided in Table 1. For each of the heuristics we give the minimum, maximum, mean, and median, estimated domination ratio measured over the 1000 instances. The standard deviation around the mean estimated domination ratio is also provided. For example, one may learn from the performance statistics that the mean estimated domination ratio of the SMV heuristic was 0.9603, whereas its worst case domination ratio was 0.3463.

Inspecting the results, one can clearly see that the MV heuristic is the worst, as expected. Better performance was shown by the GOSMV heuristics which performs relatively well on average but failed to provide good performance in the worst case. Both were inferior to the SMV heuristic. The best performance was shown by the GUBSMV heuristic, which performed well not only on average but also in the worst case; see the minimum performance in the table. The median of 1 for this heuristic

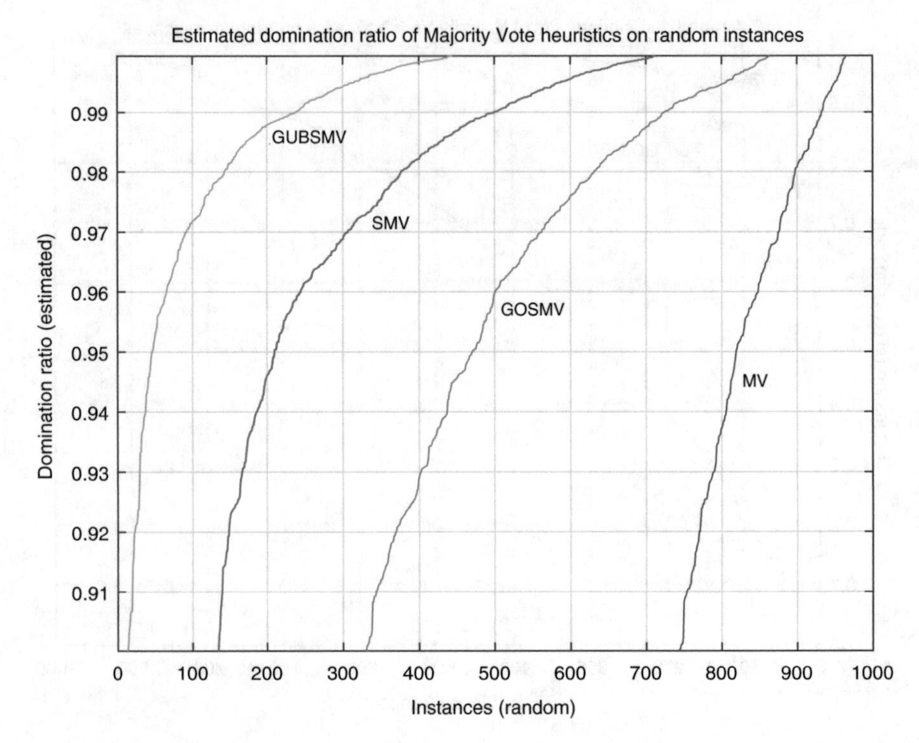


**Fig. 7** Estimated domination ratio of Majority Vote heuristics on random MAXSAT instances. A vertical zoom on the top echelon

**Table 1** Performance statistics of Majority Vote heuristics on random MAXSAT instances

| | MV | SMV | GOSMV | GUBSMV |
|---|---|---|---|---|
| min | 0.0037 | 0.3463 | 0.1716 | 0.8341 |
| max | 1 | 1 | 1 | 1 |
| mean | 0.6611 | 0.9603 | 0.8991 | 0.9908 |
| median | 0.7263 | 0.9901 | 0.9599 | 1 |
| std | 0.2811 | 0.0735 | 0.1394 | 0.0206 |

indicates that it provided a solution better than all randomly selected solutions for at least half of the instances.

It is worthwhile mentioning that, by applying the technique demonstrated in this section, we not only obtained a clear differentiation between the explored heuristics, but also gained quantitative insights regarding the performance gap between them in terms of domination ratio.

# 4 Discussion

We have demonstrated analytic and probabilistic methods to obtain a non-trivial combinatorial dominance certificate on the quality of any ad-hoc solution to a given combinatorial optimization problem on any particular instance. We have shown that these methods are easily applied to TSP and MAXSAT. We note that similar approximation ratio certificates are not forthcoming for ad-hoc solutions.

This opens up two interesting lines for investigation. The first is to apply these methods to experimentally compare heuristics for other optimization problems. These methods provide ways of identifying relatively hard instances of particular problems and certifying the quality of heuristics even in the absence of known optimal solutions. The second direction concerns theoretical investigations of the power of the Chebyshev-based method. Does the method provably yield more meaningful bounds on some problems than others? To what extent does this method apply to problems with infeasible solutions?

**Acknowledgements** The authors would like to thank Gregory Gutin and the referees for their helpful comments on this paper.

# References

1. N. Alon, G. Gutin, M. Krivelevich, Algorithms with large domination ratio. J. Algorithms **50**(1), 118–131 (2004)
2. D. Angluin, L.G. Valiant, Fast probabilistic algorithms for Hamiltonian circuits and matchings, in *Proceedings of the Ninth annual ACM Symposium on Theory of Computing (STOC)*, pp. 30–41, New York, NY (1977)
3. D. Applegate, R. Bixby, V. Chvatal, W. Cook. Concorde TSP solver (2006). See http://www.tsp.gatech.edu/concorde/
4. E. Balas, N. Simonetti, Linear time dynamic programming algorithms for some new classes of restricted TSPs: a computational study. INFORMS J. Comput. **13**(1), 56–75 (2001)
5. D. Berend, S. Skiena, Y. Twitto, Combinatorial dominance guarantees for heuristic algorithms, in *Proceedings of the International Conference on Analysis of Algorithms (AofA)*, Juan-les-Pins, France, June (2007)
6. D. Berend, S. Skiena, Y. Twitto, Combinatorial dominance guarantees for problems with infeasible solutions. ACM Trans. Algorithms **5**(1), 1–29 (2008)
7. V. Deineko, G. Woeginger, A study of exponential neighborhoods for the traveling salesman problem and the quadratic assignment problem. Math. Programm. **87**(3), 519–542 (2000)
8. F. Glover, Tabu search — Part I. ORSA J. Comput. **1**(3), 190–206 (1989)
9. F. Glover, A. Punnen, The travelling salesman problem: new solvable cases and linkages with the development of new approximation algorithms. J. Oper. Res. Soc. **48**(5), 502–510 (1997)
10. G. Gutin, A. Yeo, TSP tour domination and Hamilton cycle decompositions of regular digraphs. Oper. Res. Lett. **28**(3), 107–111 (2001)
11. G. Gutin, A. Yeo, Polynomial approximation algorithms for the TSP and the QAP with a factorial domination number. Discret. Appl. Math. **119**(1), 107–116 (2002)
12. G. Gutin, A. Yeo, A. Zverovich, Exponential neighborhoods and domination analysis for the TSP, in *The Traveling Salesman Problem and its Variations*, ed. by G. Gutin, A. Punnen (Kluwer Academic Publishers, Boston, 2002), pp. 223–256

13. G. Gutin, A. Yeo, A. Zverovich, Traveling salesman should not be greedy: domination analysis of greedy-type heuristics for the TSP. Discret. Appl. Math. **117**(1), 81–86 (2002)
14. G. Gutin, A. Vainshtein, A. Yeo, Domination analysis of combinatorial optimization problems. Discret. Appl. Math. **129**(2), 513–520 (2003)
15. G. Gutin, B. Goldengorin, J. Huang, Worst case analysis of max-regret, greedy and other heuristics for multidimensional assignment and traveling salesman problems. J. Heuristics **14**(2), 169–181 (2008)
16. R. Hassin, S. Khuller, $z$-approximations. J. Algorithms **41**(2), 429–442 (2001)
17. J.H. Holland, *Adaptation in Natural and Artificial Systems: An Introductory Analysis with Applications to Biology, Control, and Artificial Intelligence* (MIT Press, Cambridge, 1992)
18. S. Kirkpatrick, C.D. Gelatt Jr., M.P. Vecchi, Optimization by simulated annealing. Science **220**(4598), 671–680 (1983)
19. A.E. Koller, S.D. Noble, Domination analysis of greedy heuristics for the frequency assignment problem. Discret. Math. **275**(1), 331–338 (2004)
20. D. Kühn, D. Osthus, Hamilton decompositions of regular expanders: a proof of Kelly's conjecture for large tournaments. Adv. Math. **237**, 62–146 (2013)
21. D. Kühn, D. Osthus, V. Patel, A domination algorithm for $\{0, 1\}$-instances of the traveling salesman problem. Random Struct. Algorithms **48**(3), 427–453 (2016)
22. H. Mühlenbein, Genetic algorithms, in *Local Search in Combinatorial Optimization*, ed. by E. Aarts, J.-K. Lenstra (Wiley, New York, 1997), pp. 137–171
23. V. Phan, S. Skiena, P. Sumazin, A model for analyzing black-box optimization, in *Lecture Notes in Computer Science*, vol. 2748 (Springer, Berlin 2003), pp. 424–438
24. L.S. Pitsoulis, M.G.C. Resende, Greedy randomized adaptive search procedures, in *Handbook of Applied Optimization*, ed. by P.M. Pardalos, M.G.C. Resende (Oxford University Press, Oxford, 2002), pp. 178–183
25. A. Punnen, S. Kabadi, Domination analysis of some heuristics for the asymmetric traveling salesman problem. Discret. Appl. Math. **119**(1), 117–128 (2002)
26. A. Punnen, F. Margot, S. Kabadi, TSP heuristics: domination analysis and complexity. Algorithmica **35**(2), 111–127 (2003)
27. A. Punnen, P. Sripratak, D. Karapetyan, Domination analysis of algorithms for bipartite boolean quadratic programs, in *Proceedings of the International Symposium on Fundamentals of Computation Theory (FCT)*, pp. 271–282, Liverpool, August (2013)
28. G. Reinelt, TSPLIB — a traveling salesman problem library. ORSA J. Comput. **3**(4), 376–384 (1991). See also http://www.iwr.uni-heidelberg.de/groups/comopt/software/TSPLIB95/
29. S. Ross, *A First Course in Probability*, 5th edn. (Prentice Hall, Upper Saddle River, 1998)
30. V.I. Rublineckii, Estimates of the accuracy of procedures in the traveling salesman problem. Numer. Math. Comput. Technol. (in Russian) **4**, 18–23 (1973)
31. S. Russell, P. Norvig, *Artificial Intelligence: A Modern Approach*, 2nd edn. (Prentice Hall, Upper Saddle River, 2003)
32. V. Sarvanov, N. Doroshko, The approximate solution of the traveling salesman problem by a local algorithm that searches neighborhoods of exponential cardinality in quadratic time. Softw. Algorithm. Program. (in Russian) **31**, 8–11 (1981)
33. V. Sarvanov, N. Doroshko, The approximate solution of the traveling salesman problem by a local algorithm that searches neighborhoods of factorial cardinality in cubic time. Softw. Algorithm. Program. (in Russian) **31**, 11–13 (1981)
34. Y. Twitto, Dominance guarantees for above-average solutions. Discret. Optim. **5**(3), 563–568 (2008)
35. E. Zemel, Measuring the quality of approximate solutions to zero-one programming problems. Math. Oper. Res. **6**(3), 319–332 (1981)

# Conditional Markov Chain Search for the Simple Plant Location Problem Improves Upper Bounds on Twelve Körkel–Ghosh Instances

**Daniel Karapetyan and Boris Goldengorin**

## 1 Introduction

The Simple Plant Location Problem (SPLP), also known as Uncapacitated Facility Location Problem, is a classical combinatorial optimisation problem [11] with many applications in quantitative logistics [13] and flexible manufacturing systems [20]. The SPLP takes a set $I = \{1, 2, \dots, m\}$ of sites in which plants can be located, a set $J = \{1, 2, \dots, n\}$ of clients, each having a unit demand, a vector $F = (f_i)$ of fixed costs for setting up plants at sites $i \in I$, and a matrix $C = [c_{ij}]$ of transportation costs from $i \in I$ to $j \in J$ as input. It computes a set $P^\star$, $\emptyset \subset P^\star \subseteq I$, at which plants can be located so that the total cost of satisfying all client demands is minimal. The costs involved in meeting the client demands include the fixed costs of setting up plants, and the transportation cost of supplying clients from the plants that are set up.

Historical roots of the SPLP can be found in pioneering Weber's publication [39], and a modern formulation of the SPLP as a Mixed Integer Linear Programming (MILP) problem can be read in [3].

A detailed introduction to this problem has appeared in [11]. The SPLP forms the underlying model in several combinatorial problems, such as set covering, set partitioning, information retrieval, simplification of logical Boolean expressions, airline crew scheduling, vehicle despatching, and is a subproblem for various location analysis problems (see [21] and the references within).

D. Karapetyan (✉)
Institute for Analytics and Data Science, University of Essex, Colchester, UK

B. Goldengorin
Department of Information Systems and Decision Science, Merrick School of Business, University of Baltimore, Baltimore, MD, USA



B. Goldengorin (ed.), *Optimization Problems in Graph Theory*, Springer Optimization and Its Applications 139,
https://doi.org/10.1007/978-3-319-94830-0_7

The SPLP is $\mathcal{NP}$-hard [11], and several exact and heuristic algorithms for solving it have been discussed in the literature. Most of the exact algorithms are based on a mathematical programming formulation of the SPLP (see, for example, Cornuejols and Thizy [12], Morris [33], and Schrage [36]). Polyhedral results for the SPLP polytope have been reported in Trubin [38], Balas and Padberg [2], Cho et al. [9], Cho et al. [10], Farias [14], Cánovas et al. [8], and Galli et al. [17]. In theory, these results allow us to solve the SPLP by applying the simplex algorithm to the strong linear programming relaxation, with the additional stipulation that a pivot to a new extreme point is allowed only when this new extreme point is integral. However, efficient implementations of this pivot rule are not available. Beasley [6] reported computational experiments with Lagrangian heuristics for SPLP instances. Körkel [28] proposed algorithms based on refinements to a dual-ascent heuristic procedure to solve the dual of a linear programming relaxation of the SPLP combined with the use of the complementary slackness conditions to construct primal solutions [15]. Barahona and Chudak [4, 5] have reported optimal solutions to some SPLP instances with $m = n = 3000$ and paid attention to computationally difficult SPLP instances with large fixed costs and several opened sites in an optimal solution and easy solvable SPLP instances with small fixed costs and almost all opened sites in an optimal solution.

Since the SPLP is NP-hard, an essential number of publications are devoted to approximation and heuristic algorithms (see, e.g., [35]). For example, Guha and Khuller [22] have established a lower bound of 1.463 for the approximation factor, under some widely believed assumptions. Another heuristic by Jain et al. [26] has a performance guarantee of only 1.61, but in computational experiments returns good quality SPLP solutions within 2% of their optimality. In practice, these heuristics tend to be much closer to optimality for non-pathological instances. There is a long list of heuristics without any theoretically proven approximation ratio for the found feasible solutions which return high quality SPLP solutions. Among them constructive and local search heuristics rooted from the pioneering work of Kuehn and Hamburger [30], and successfully continued by simulated annealing [1] and [40], genetic algorithms [29], complete local search with memory [18], and tabu search [32] as well as Sun [37]. Dual-based methods such as Erlenkotters [15] dual ascent, Guignard's [23] Lagrangian dual ascent, and the volume algorithm by Barahona and Chudak [4] have also shown promising results. An experimental comparison of some state-of-the-art heuristics is presented by Hoefer [25] with a recommendation that tabu search finds the highest quality heuristic solutions within reasonable CPU time.

Researchers found that many SPLP instance families are relatively easy to solve. For example, one can solve all Beaslye SPLP benchmark instances just by two Khumawala preprocessing rules combined with a few branchings on variables with the largest violation within a fraction of a second [19]. Letchford and Miller [31] designed preprocessing rules which are effective for the SPLP instances with facilities and clients located at points on the Euclidean plane. In 2003, Ghosh [18] proposed a class of computationally hard instances which are now known as Körkel–Ghosh (KG) instances since these instances are modified Körkel instances [28].

Fischetti et al. [16] explain the computational intractability of the KG instances because they have a large number of near-optimal solutions, which makes it hard to identify variables that could not be in an optimal KG SPLP instance solution. Since on average at least 80% of all sites should be closed in an optimal solution to the KG instances most of the preprocessing approaches are not successful in their efforts to find high quality solutions to the KG instances. The KG instance library includes three classes of instances, namely A, B, and C. In class A, the fixed costs $f_i$ are drawn uniformly from $[100, 200]$, in class B—from $[1000, 2000]$, and in class C—from $[10000, 20000]$. The transportation costs $c_{ij}$ are always drawn uniformly from $[1000, 2000]$. Symmetric and asymmetric instances are included, where symmetric instances satisfy $c_{ij} = c_{ji}$. The KG library includes instances of size $m \times n = 250 \times 250$, $m \times n = 500 \times 500$, and $m \times n = 750 \times 750$.

In the recent decade, many different heuristics (see, e.g., [37]) as well as exact approaches by Beltran-Royo et al. [7], Posta et al. [34], and Fischetti et al. [16] were applied to improve the best known upper bounds for the KG instances. A recent attack on the KG benchmark showed that the upper bounds for many of the instances can still be improved [16] but this takes a significant computational effort. Fischetti et al. [16] conclude that 50 KG instances still remain out of reach for existing exact methods. Nevertheless, they have been able to improve the best known upper bounds for 22 KG instances solutions and matched the other 21 within 3600 s. After increasing the CPU time budget to 7200 s they slightly improved their results by keeping 22 strictly improved and one more matched (now 22 matched) solutions. For the remaining six instances (out of 50) their upper bounds are worse than the best known.

The purpose of our paper is to present the next step in finding better solutions to the KG SPLP benchmark instances. While all the previous attempts to tackle SPLP were based on human-designed algorithms, we applied automated heuristic generation to produce an effective method for SPLP.

The main idea behind automated heuristic generation is that (meta)heuristic design is a labour-intensive process in which an expert is required to use their skills and intuition about the domain to combine available components into an algorithm with complex behaviour. Automated generation of (meta)heuristics, also known as generating hyper-heuristics, is meant to make the design process cheaper and quicker, and avoid the subjective judgement of the expert that usually affects the algorithm architecture. The completely automated algorithm design is not yet available; however, a recent approach called Conditional Markov Chain Search (CMCS) enables one to automatically compose a metaheuristic from a set of given domain-specific routines in [27].

The CMCS gives a flexible framework capable of describing a wide range of metaheuristics using a set of parameters. Each specific combination of parameter values is called a *configuration*. In other words, a configuration is a specific composition of a metaheuristic from the available domain-specific routines. By selecting one of the top performing configurations, we generate an effective metaheuristic.

We use this approach to automatically design a simple yet effective metaheuristic for SPLP. An important contribution of the paper is a refined CMCS generation method. Observe that the problem of selecting the best performing configuration out of several candidates is not well-defined, mainly because there is unlikely to be a single configuration performing better than every other configuration in every test. We propose an approach to selecting the best CMCS configuration from the space of all feasible configurations. We further apply several rules to significantly reduce the space of CMCS configurations and use a brute-force-like algorithm to choose the best of them.

While searching for the best performing CMCS configuration, we use a training dataset consisting of small instances, and use short running times. Nevertheless, the performance of our selected configuration scales well to the size of large KG instances. In particular, we show that our automatically generated metaheuristic clearly outperforms previous state-of-the-art heuristics including the most recent computational records in [16]. Moreover, in our experiments it improved 12 (and matched 38 remaining) best known values among the 50 yet unsolved KG SPLP instances, and have not returned any worse solution for all previously solved 90 KG benchmark instances keeping the total CPU time budget not more than 1 s!

The paper is structured as follows. The SPLP-specific parts of the algorithm, i.e. the data structures and algorithmic components, are described in Section 2. The CMCS framework, the generation procedure, and the best performing CMCS configurations are discussed in Section 3. The computational results of applying the best performing CMCS configurations to the benchmark instances are reported in Section 4. The concluding remarks and future work are discussed in Section 5.

# 2 SPLP Components

In this section we describe the domain-specific components that will later be used within our CMCS configurations. All of these components are well-known from the SPLP literature, or are variations of standard algorithms.

## 2.1 Data Structures

Our data structure is based on the ideas previously proposed in the SPLP literature, see, e.g., [24]. We store the list of opened sites in two forms: a vector $y \in \{0, 1\}^m$, where $y_i$ indicates whether site $i$ is opened, and a set of indices $P \subseteq I$ of opened sites. In addition, for each client $j \in J$, we store the closest opened site $p(j) \in P$ and the second closest opened site $q(j) \in P$. Thus, the objective value of a solution can be efficiently computed as

$$\sum_{i \in P} f_i + \sum_{j \in J} c_{p(j),j} \, .$$

In practice, we never need to compute the objective value as we store it in a variable $v$ and maintain its value while manipulating the solution.

Our data structure requires that at least two sites are opened, which is a reasonable assumption for our test problems, and so we enforce this constraint in every component. If for some other problem instances such an assumption would be too strong, one can start the search with evaluating all the $m$ solutions containing exactly one opened site, which would take only $O(mn)$ time.

We also precompute a matrix $\pi = [\pi(i, j)]$ of size $m \times n$, where $\pi(i, j)$ is the index of the $j$th closest site for the client $i$. In other words, $c_{\pi(i,1)} \leq c_{\pi(i,2)} \leq \cdots \leq c_{\pi(i,m)}$. We will need the matrix $\pi$ for efficient exploration of a neighbourhood, see Section 2.4.

This data structure allows efficient procedures for opening or closing a site. For details see Algorithms 1 and 2. The worst case time complexity of opening a site is $O(n)$ and of closing—$O(n|P|)$.

## 2.2 *Open Random (k)*

The first two components we discuss are mutation operators, i.e. components that make random changes to the solution, usually applied to escape a local minimum by worsening its quality. The 'Open Random ($k$)' component opens $k$ randomly selected sites. More specifically, the component selects $k$ distinct sites, opening those of them that are not currently opened (we assume that the number of opened sites is relatively small and thus the probability of hitting a site that is already opened is relatively small).

The time complexity of the 'Open Random ($k$)' component is $O(kn)$.

**Algorithm 1** Opening a site

**input** : Site $i^* \in I \setminus P$ to be opened
1 **forall the** $j \in J$ **do**
2   **if** $c_{i^*,j} < c_{p(j),j}$ **then**
3     $v \leftarrow v - c_{p(j),j} + c_{i^*,j}$ $q(j) \leftarrow p(j)$ $p(j) \leftarrow i^*$
4 $v \leftarrow v + f_{i^*}$ $P \leftarrow P \cup \{i^*\}$ $y_{i^*} \leftarrow 1$

**Algorithm 2** Closing a site

**input** : Site $i^* \in P$ to be closed
5 $P \leftarrow P \setminus \{i^*\}$ $y_{i^*} \leftarrow 0$ **forall the** $j \in J$ **do**
6   **if** $p(j) = i^*$ **then**
7     $v \leftarrow v - c_{p(j),j} + c_{q(j),j}$ $p(j) \leftarrow q(j)$ $q(j) \leftarrow \arg\min_{i \in P \setminus \{p(j)\}} c_{i,j}$
8   **else if** $q(j) = i^*$ **then**
9     $q(j) \leftarrow \arg\min_{i \in P \setminus \{p(j)\}} c_{i,j}$
10 $v \leftarrow v - f_{i^*}$

## 2.3 Close Random (k)

The 'Close Random $(k)$' component is another mutation operator; it closes $k$ randomly selected sites. More specifically, the component selects $\min\{k, |P| - 2\}$ currently opened sites and closes them. The worst case time complexity of the 'Close Random $(k)$' component is $O(kn|P|)$.

## 2.4 Open Best

The 'Open Best' component is a local search procedure that opens a single site if that improves the solution. The cardinality of the corresponding neighbourhood is $m - |P|$, and the procedure chooses the best candidate. A naive implementation of the 'Open Best' local search would take $O(mn)$ time. We reduce this to $O(m + \sum_{j \in J} p(j))$. (Observe that $p(j) \leq m$ and hence $\sum_{j \in J} p(j) \leq mn$, while in practice the sum is considerably smaller.) This is achieved by gradually building a vector $\delta_i$, $i \in I$, where $\delta_i$ is the change in the objective value if the site $i$ is to be opened. We initialise $\delta_i \leftarrow f_i$. Then, for each client $j \in J$, we scan through the sites $i$ that are closer than $p(j)$, i.e. through the sites that, if opened, will improve the transportation cost for that client, and update $\delta_i \leftarrow \delta_i + c_{i,j} - c_{p(j),j}$. This operation is implemented efficiently by utilising the precomputed $\pi$ matrix, see Section 2.1. At the end, we choose $i^* = \arg\min_{i \in I} \delta_i$. If $\delta_{i^*} < 0$, then we open site $i^*$. Otherwise we leave the solution unchanged.

## 2.5 Close Best

The 'Close Best' component is a local search procedure that closes a single site if that improves the solution. The corresponding neighbourhood consists of $|P|$ solutions, and the procedure chooses the best out of them.

A naive implementation of the 'Close Best' local search would take $O(|P|n)$ time, whereas we reduce the exploration time to $O(m + n)$ time. We gradually build a vector $\delta_i$, $i \in I$, where $\delta_i$ is the change in the objective value caused by closing site $i$. Initially we set $\delta_i = -f_i$ for every $i \in P$. Then, for each client $j \in J$, we increase $\delta_{p(j)}$ by $c_{q(j),j} - c_{p(j),j}$. At the end, we choose $i^* = \arg\min_{i \in P} \delta_i$, and if $\delta_{i^*} < 0$, then we close site $i^*$. Otherwise we leave the solution unchanged.

## 2.6 Exchange Best

The 'Exchange Best' component is a local search procedure that closes one site and opens another one if that improves the solution. It chooses the best candidate out of the $(m - |P|)|P|$ solutions in the neighbourhood.

A native implementation would take $O(mn|P|)$ time to explore the 'Exchange Best' neighbourhood. By following the logic of the 'Open Best' local search implementation and building a matrix $\delta$ of size $|P| \times n$, we reduce this to $O(mn)$.

## 2.7 *Exchange Half Fixed*

Observe that the 'Exchange Best' local search is relatively slow comparing to the other two local search procedures. We propose a local search that explores only a fraction of the 'Exchange Best' neighbourhood but runs much quicker. Our 'Exchange Half Fixed' local search randomly selects the site $i^* \in P$ to be closed, and then searches for the best site to be opened. This exploration takes $O(\sum_{j \in J} p(j) + m\gamma)$ time, where $\gamma$ is the number of clients $j$ for which $p(j) = i^*$. For a random instance, the expected value of $\gamma$ is $n/|P|$.

# 3 Conditional Markov Chain Search

Conditional Markov Chain Search (CMCS) was first introduced by Karapetyan et al. [27] as a framework to enable automated generation of metaheuristics. It gives a flexible way of composing a metaheuristic from a set of domain-specific components, with the behaviour of the control mechanism defined by numerical parameters.

Let $\mathcal{H}$ be an ordered set of available domain-specific components, which we call a *solution pool*. By component we mean a black box algorithm that takes the problem instance and a solution as an input and outputs a new (modified) solution. A component could implement, for example, a local search procedure or a random move (mutation).

CMCS is a single-point metaheuristic. It applies the components to the current solution, one at a time, in a certain sequence. The output of the previous component is an input of the next component in the sequence. For example, if the sequence is

$$\mathcal{H}_1, \mathcal{H}_1, \mathcal{H}_3, \mathcal{H}_3, \mathcal{H}_2, \mathcal{H}_1,$$

and the initial solution is $S_0$, the CMCS will proceed as follows:

$$\begin{aligned}
S_1 &\leftarrow \mathcal{H}_1(S_0),\\
S_2 &\leftarrow \mathcal{H}_1(S_1),\\
S_3 &\leftarrow \mathcal{H}_3(S_2),\\
S_4 &\leftarrow \mathcal{H}_3(S_3),\\
S_5 &\leftarrow \mathcal{H}_2(S_4),\\
S_6 &\leftarrow \mathcal{H}_1(S_5).
\end{aligned}$$

CMCS saves the best solution found so far. Hence, at the end of iteration $i$, it stores two solutions: current solution $S_i$ and the best of $S_0, S_1, \ldots, S_i$.

Each component modifies the solution according to its internal logic. The change may improve or worsen the solution quality; a component may also leave the solution intact.

The components are stateless, and are independent (do not communicate with each other). A component may be randomised or be deterministic.

Given a fixed set of components, the behaviour of CMCS is defined by the sequence in which the components are executed. This sequence is decided online. The decision of which component to execute in iteration $i$ is made at the end of iteration $i-1$. In particular, this decision depends on two factors: (i) which component was executed at iteration $i-1$, and (ii) whether $S_{i-1}$ is better than $S_{i-2}$. Hence, the sequence of components is a Markov chain, with the state consisting of the last executed component and a Boolean variable indicating whether the last executed component has improved the solution.

The specific logic of the next component selection is called *configuration*. While CMCS is a framework, a CMCS configuration is a fixed metaheuristic algorithm. Observe that a CMCS configuration can be completely defined by two transition matrices, $M^{\text{succ}}$ and $M^{\text{fail}}$, each of size $|\mathscr{H}| \times |\mathscr{H}|$. The matrix $M^{\text{succ}}$ is used when the solution is improved by the last executed component, and $M^{\text{fail}}$ is used otherwise (when either the objective value has not changed, or has worsened). To keep the paper self-contained, we include a pseudo-code of CMCS in Algorithm 3, a close copy from [27].

CMCS does not have any acceptance criteria, i.e. it never backtracks any changes made by the components. (Backtracking can be implemented within a domain-specific component, e.g. inside a local search procedure, however, once the component execution is finished, the change, in general, cannot be undone.) There-

**Algorithm 3** Conditional Monte-Carlo search

**input** : Components pool $\mathscr{H}$;
**input** : Matrices $M^{\text{succ}}$ and $M^{\text{fail}}$ of size $|\mathscr{H}| \times |\mathscr{H}|$;
**input** : Objective function $f(S)$ to be minimised;
**input** : Instance data $\mathscr{I}$;
**input** : Initial solution $S_0$;
**input** : Termination time *terminate-at*;

11 $S^* \leftarrow S_0$ $f^* \leftarrow f(S_0)$ $f_{\text{prev}} \leftarrow f^*$ $h \leftarrow 1$ $i \leftarrow 1$ **while** *now* < *terminate-at* **do**
12 $S_i \leftarrow \mathscr{H}_h(\mathscr{I}, S_{i-1})$ $f_{\text{cur}} \leftarrow f(S_i)$ **if** $f_{cur} < f_{prev}$ **then**
13 $h \leftarrow RouletteWheel(M^{\text{succ}}_{h,1}, M^{\text{succ}}_{h,2}, \ldots, M^{\text{succ}}_{h,|\mathscr{H}|})$
14 **if** $f_{cur} < f^*$ **then**
15 $S^* \leftarrow S_i$ $f^* \leftarrow f_{\text{cur}}$
16 **else**
17 $h \leftarrow RouletteWheel(M^{\text{fail}}_{h,1}, M^{\text{fail}}_{h,2}, \ldots, M^{\text{fail}}_{h,|\mathscr{H}|})$
18 $f_{\text{prev}} \leftarrow f_{\text{cur}}$ $i \leftarrow i+1$
19 **return** $S^*$

fore, the only source of the improvement pressure in CMCS is the improvement pressure generated by some of the components. As a result, it is necessary to include in the component pool at least one component that would be biased towards good solutions, such as a local search procedure. It is equally important to include at least one component capable of worsening the solution, such as a mutation operator, to escape local minima.

Observe that CMCS is completely domain-independent as all the knowledge of the domain is incorporated in the components treated as black boxes. Hence, both the control mechanism and the configuration generation routines are domain-independent and reusable. Such a reusability is a long-standing goal in the area of optimisation algorithm design.

We proceed by discussing in Section 3.1 how to restrict CMCS to leave only a finite manageable number of configurations, and also how to enumerate them, and then, in Section 3.2, we discuss how to choose the best of the available configurations.

## *3.1 Deterministic CMCS*

Recall that a CMCS configuration is specified by two matrices, $M^{\text{succ}}$ and $M^{\text{fail}}$. Each value in a transition matrix defines the probability of the corresponding transition, hence the space of configurations is continuous. Searching in this space is particularly hard due to the roughness of the landscape, typical in parameter tuning. However, as shown in [27], discretisation of the search space allows one to use brute force to optimise some special cases of CMCS.

In this project, we restrict CMCS to the deterministic case, i.e. to the case where each row of $M^{\text{succ}}$ and $M^{\text{fail}}$ contains exactly one non-zero element. (Note that the resultant configuration is not necessarily a deterministic SPLP algorithm; it is only the transition mechanism that is deterministic.) This leaves us with $k^{2k}$ feasible configurations.

Out of these configurations, some are equivalent. Consider the example in Figure 1. As $\mathscr{H}_3$ is unreachable in either of the two configurations, the last row of the matrices can be ignored (it does not affect the behaviour of the configurations). Then the two configurations, formally different, are equivalent.

To exclude such 'duplicates', we follow a two-step procedure:

1. At first, we generate all non-empty subsets $H \subseteq \mathscr{H}$.
2. For each subset $H$, we generate all the configurations that use every component $h \in H$. By 'use' we mean that there exists a non-zero probability of transition from any $h' \in H$ to $h$, perhaps within several iterations. This can be formalised using a directed graph $G = (H, E)$ with a node set $H$ and arc set $E$ which includes an arc $(h, h') \in E$ if and only if $M^{\text{succ}}_{h,h'} + M^{\text{fail}}_{h,h'} > 0$. We say that the configuration uses all the components $H$ if and only if graph $G$ is strongly connected. One may note that some component $h$ may not be reachable from

$$M^{\text{succ}} = \begin{bmatrix} & \mathcal{H}_1 & \mathcal{H}_2 & \mathcal{H}_3 \\ \mathcal{H}_1 & 1 & 0 & 0 \\ \mathcal{H}_2 & 1 & 0 & 0 \\ \mathcal{H}_3 & 0 & 1 & 0 \end{bmatrix} \quad M^{\text{fail}} = \begin{bmatrix} & \mathcal{H}_1 & \mathcal{H}_2 & \mathcal{H}_3 \\ \mathcal{H}_1 & 0 & 1 & 0 \\ \mathcal{H}_2 & 0 & 1 & 0 \\ \mathcal{H}_3 & 1 & 0 & 0 \end{bmatrix}$$

(a)

$$M^{\text{succ}} = \begin{bmatrix} & \mathcal{H}_1 & \mathcal{H}_2 & \mathcal{H}_3 \\ \mathcal{H}_1 & 1 & 0 & 0 \\ \mathcal{H}_2 & 1 & 0 & 0 \\ \mathcal{H}_3 & 0 & 0 & 1 \end{bmatrix} \quad M^{\text{fail}} = \begin{bmatrix} & \mathcal{H}_1 & \mathcal{H}_2 & \mathcal{H}_3 \\ \mathcal{H}_1 & 0 & 1 & 0 \\ \mathcal{H}_2 & 0 & 1 & 0 \\ \mathcal{H}_3 & 1 & 0 & 0 \end{bmatrix}$$

(b)

**Fig. 1** Configurations A (**a**) and B (**b**) are equivalent. Indeed, they are only different in the last row of $M^{\text{succ}}$, i.e. in transition from $\mathcal{H}_3$. However, $\mathcal{H}_3$ is unreachable. Hence, the last row of either of the transition matrices does not affect the behaviour of the configurations

some other components, however be executed during the first iterations of the algorithm, for example, if it is the entry point. We assume here that the effect of $h$ in such a case is likely to be negligible after a large number of iterations.

We can further eliminate some configurations by imposing several constraints:

- At least one of the components in $H$ needs to generate improvement pressure. In practice, this usually means that at least one of the components is a local search.
- At least one of the components in $H$ needs to be able to worsen the solution, as otherwise the search will quickly converge to a local minimum and stop there. In practice, this usually means that as least one of the components is a mutation.
- If component $H_h$ is a classic local search, i.e. it explores some deterministic neighbourhood and makes the move if and only if it improves the solution, then we can fix $M^{\text{fail}}_{h,h} = 0$.

We say that a configuration that satisfies all the above conditions is *meaningful*. This still leaves us with a considerable number of meaningful configurations. For example, for a set of six components, where three of them are deterministic local search procedures and three are mutations, the number of meaningful configurations is approximately $3.4 \cdot 10^8$. This is a significant improvement over the number of feasible configurations $k^{2k} \approx 2 \cdot 10^9$, but still impractical even for such a small number of components. Thus, we introduce an additional constraint; we only consider configurations for $|H| = \lambda$, where $\lambda$ is a parameter. Then we can ask a question of the form 'what is the best configuration composed of exactly $\lambda$ components' or 'what is the best configuration composed of at most $\Lambda$ components'.

The parameter $\lambda$ greatly reduces the number of configurations. The number of feasible $\lambda$-component configurations is

$$\binom{|\mathcal{H}|}{\lambda} \cdot \lambda^{2\lambda} \text{ configurations.} \tag{1}$$

Given three local searches and ten mutation, there are only $2.1 \cdot 10^5$ three-component and $1.2 \cdot 10^3$ two-component feasible configurations. By using the conditions discussed above, we end up with $3.7 \cdot 10^4$ three-component and $1.8 \cdot 10^2$ two-component meaningful configurations (there are no meaningful configurations with one component, as we require that both local searches and mutations are included into $H$, see the above constraints). Compare this to the overall $9.2 \cdot 10^{28}$ feasible configurations.

We recognise that the parameter $\lambda$ greatly restricts the complexity of the CMCS configurations, however, this restriction allows us to include many components and let the CMCS generator, rather than a human expert, choose which component combinations are most efficient.

## 3.2 CMCS Generator

CMCS generator is a procedure that finds the best (or some very good) CMCS configuration. In this project, as we restrict the set of configurations to deterministic configurations, our CMCS generator aims at selecting the best of all the meaningful configurations.

To evaluate a configuration, we use a training dataset $\mathcal{T}$. Each element of $\mathcal{T}$ is a triple $(\mathit{Inst}, S_0, t)$, where *Inst* is the SPLP instance, $S_0$ is the initial solution, and $t$ is the time budget. Let $f(C, \mathit{Inst}, S_0, t)$ be the objective value of a solution obtained by solving instance *Inst* with the initial solution $S_0$ and the time budget $t$ by the CMCS configuration $C$. Then we can interpret the problem of selecting the best configurations as a multi-dimensional optimisation problem, with the objective functions $f(C, \mathit{Inst}, S_0, t)$, $(\mathit{Inst}, S_0, t) \in \mathcal{T}$. With a large number of dimensions, one may assume that the majority of the configurations would be Pareto optimal. However, in practice many configurations demonstrate very poor solution quality and as a result are dominated by top ranked configurations. Hence, the Pareto domination approach is sufficient to filter out the majority of poorly performing configurations.

The approach taken in [27] was to run all the tests for each configuration. Here we improve this by filtering out the least promising configurations after the first few tests. More specifically, if a configuration $C$ performs strictly worse than some other configuration $C^*$ in the first seven tests, we can use the sign test condition to conclude that $C^*$ is superior to $C$ with significance level 99%. This heuristic approach will not allow us to select the best performing configuration but it will let us quickly focus on the most promising configurations.

Selection of the best performing configuration out of the most promising candidates requires multiple-criteria decision-making. The standard methods, such as the Analytic hierarchy process or ELECTRE, are designed to tackle problems with hard to quantify and compare attributes. In our problem, all the attributes (the $f(C, \mathit{Inst}, S_0, t)$ values) have equal weights, and are inherently easy to quantify. Thus, we use the simple weighted sum model, with equal weights.

To summarise, our CMCS generator performs in two stages:

1. Form a set $\mathscr{C}$ of all meaningful configurations and run the first seven tests $\mathscr{T}_1, \mathscr{T}_2, \ldots, \mathscr{T}_7$ for each configuration $C \in \mathscr{C}$. Select non-dominated configurations and save them to $\mathscr{C}'$, i.e.

$$\mathscr{C}' = \{C \in \mathscr{C} : \nexists C^* \in \mathscr{C} \text{ such that } C^* \text{ dominates } C\}. \tag{2}$$

   $\mathscr{C}'$ usually includes only a small fraction of all the meaningful configurations.
2. Run the remaining tests for each configuration in $\mathscr{C}'$. As the scale of objective values may vary between tests, then, for each $(Inst, S^0, t) \in \mathscr{T}$, normalise $f(C, Inst, S^0, t)$ by scaling it to the [0, 1] interval:

$$f'(C, Inst, S^0, t) = \frac{f(C, Inst, S^0, t) - \min_{C' \in \mathscr{C}'} f(C', Inst, S^0, t)}{\max_{C' \in \mathscr{C}'} f(C', Inst, S^0, t) - \min_{C' \in \mathscr{C}'} f(C', Inst, S^0, t)}.$$

   Finally, select a configuration $C \in \mathscr{C}'$ that minimises

$$\sum_{(Inst, S^0, t) \in \mathscr{T}} f(C, Inst, S^0, t). \tag{3}$$

# 4 Computational Results

We first describe in Section 4.1 our set up for the CMCS configuration generation, as well as the produced configurations. We then apply in Section 4.2 these configurations to the KG instances to obtain upper bounds for the yet unsolved instances and compare our results to the state-of-the-art heuristics from the literature. We also check the performance of our CMCS configurations on already solved instances.

All our algorithms were implemented in C#, and the experiments were conducted on a Windows machine with two Intel Xeon E5-2690 v4 (2.6 GHz) CPUs and HyperThreading enabled. Our implementation of CMCS does not use concurrency. We used the multiple cores to run the experiments in parallel, but not more than one experiment per physical CPU core.

## 4.1 CMCS Configuration Generation for SPLP

In our experiments, we restricted the set of configurations to two- and three-component configurations, i.e. set $\Lambda = 3$. Our pool $\mathscr{H}$ of components consisted of:

- 'Open Best', 'Close Best', 'Exchange Best', and 'Exchange Half Fixed' local searches;
- 'Open Random $(k)$' and 'Close Random $(k)$' mutations for $k = 1, 2, 3, 4$.

The training dataset included 200 tests $(Inst, S_0, t)$. The instances *Inst* were generated using the KG instance generator, with the instance size $n = m$ selected uniformly at random between 300 and 400. The size and the type of the instance ('a', 'b', or 'c') was also drawn uniformly at random. The initial solution $S_0$ was generated by opening $r$ random sites, where $r \in [2, \lfloor 0.1n \rfloor]$ was chosen uniformly at random. Note that an optimal solution to a KG instance is likely to have more than two but less than $\lfloor 0.1n \rfloor$ opened sites; hence, we exercise both situations when the number of opened sites needs to be increased and decreased. The time budget $t$ for each test was set to 0.5 s. This time budget was selected to allow a CMCS configuration to run a sufficiently large number of iterations (about 50) to reveal its long-term behaviour.

Some of the data on CMCS generation is summarised in Table 1. One can see that the value of $\lambda$ significantly affects the number of meaningful configurations, as well as the number of configurations selected at the first stage of the generation. As a result, the wall time taken by the generator for $\lambda = 2$ was around 4 min whereas for $\lambda = 3$ it was more than 12 h. This shows that the current generator is not well suited for more complex configurations or significantly larger component pools, however, even this limited set of configurations yields good results, as we will show in our computational study.

The configurations generated for $\lambda = 2$ and $\lambda = 3$ are shown in Figure 2. Each node in these diagrams corresponds to a component, and each arc to a transition. Blue arcs show transitions when the last component execution was successful

**Table 1** CMCS generation data

| | $\lambda = 2$ | $\lambda = 3$ |
|---|---|---|
| Feasible deterministic configurations (see (1)) | 1056 | 160,380 |
| Meaningful deterministic configurations | 216 | 43,326 |
| Second stage configurations | 28 | 4901 |
| Overall generation time, sec (wall time) | 293 | 43,326 |

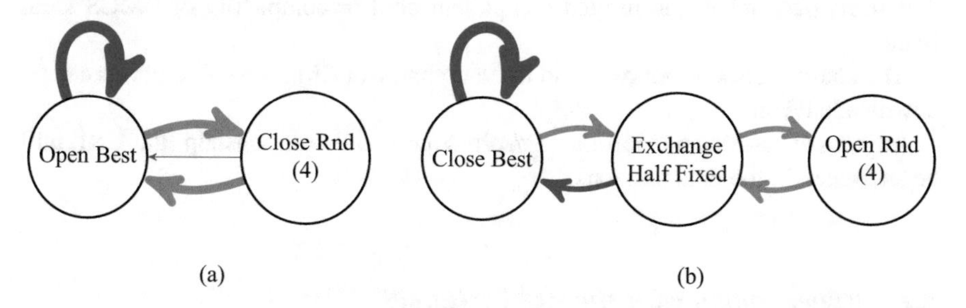


**Fig. 2** Best performing CMCS configurations. The blue arcs correspond to successful transitions (after the solution was improved), and the red arcs correspond to the unsuccessful transitions (after the solution was not improved). The thickness of an arc indicates the frequency of the corresponding transition. (**a**) Best performing two-component CMCS configuration. (**b**) Best performing three-component CMCS configuration

(improved the solution) and red—when unsuccessful (the solution quality was worsened or has not changed). The thickness of the arc indicates the frequency of that transition; it is proportional to square root of that frequency.

The strategy of the two-component configuration is easy to explain. The configuration opens sites, in a greedy manner, as long as this improves the solution. Then it closes four sites randomly and gets back to adding new sites. This strategy, in fact, exactly replicates the behaviour of iterated local search.

The three-component configuration is more complex, but its logic can still be interpreted. The algorithm closes sites, in a greedy manner, until it reaches a local minimum. Then it attempts to replace some site with another one ('Exchange Half Fixed'). If successful, there is a chance some other site got redundant which is why it returns to the 'Close Best' component. If, however, 'Exchange Half Fixed' fails, a mutation is applied. In particular, four random sites are opened and the control is passed back to 'Exchange Half Fixed'. There is a high chance that 'Exchange Half Fixed' will be able to improve the solution, and then the control will be returned to 'Close Best'. Note that this strategy closely resembles variable neighbourhood search, with three neighbourhoods. As soon as the search in one neighbourhood fails, the next neighbourhood is used. It is non-typical though that one of the local searches ('Exchange Half Fixed') explores only a randomly chosen area of the neighbourhood.

While we can explain the behaviour of the three-component CMCS configuration and even show its similarity to a well-known metaheuristic, the point of the automated generation is that this strategy was produced without prior knowledge of existing metaheuristics. Moreover, many decisions were taken automatically, such as which components to include in the metaheuristic, and in which order to use them. The whole design process is completely unbiased, hence the generated configuration is objectively one of the most effective ones possible within the framework. (The exact definition of effectiveness may vary, but then the CMCS generation procedure can also be adjusted accordingly.) Thus, we can expect that the generated configuration performs at least as well as any human-designed metaheuristics, unless the limited component pool or complexity of CMCS is an issue.

The source codes of our two- and three-component CMCS configurations can be downloaded from

http://csee.essex.ac.uk/staff/dkarap/splp-source-and-solutions.zip (the URL will be shortened in the final version).

## *4.2 Experiments with the KG Instances*

In this section we solve the KG instances with the two- and three-component CMCS configurations discussed above. We first solve the 50 instances to which optimal solutions are not yet known. We use time budget 7200 s, same as in [16]. (The test machine used in [16] is based on Intel Xeon E3-1220V2 CPU (3.10 GHz), which is comparable to our CPUs. However, Fischetti et al. [16] utilised four CPU cores, effectively increasing computational power fourfold. By assuming that one time unit in the experiments of [16] is equal to one time unit in our experiments, we give [16] advantage.

Our results are reported in Table 2. These new upper bounds are the best solutions we found in our experiments (as the reader will see later, all these solutions were obtained by the three-component CMCS configuration with time budget 1000 s). We improved 12 best known upper bounds and matched all others. We further tested our solvers on the instances for which optimal solutions are known. Our three-component CMCS configuration could solve any of those instances to optimality within 1 s. While not being a formal proof, this suggests that we, perhaps, have also reached optimal solutions for most of the yet unsolved instances.

**Table 2** Previous and new upper bounds for the yet unsolved KG instances

| Instance | Previously best known | Our best | Difference |
|---|---|---|---|
| ga500a-1 | 511,383 | 511,383 | 0 |
| ga500a-2 | 511,255 | 511,255 | 0 |
| ga500a-3 | 510,810 | 510,810 | 0 |
| ga500a-4 | 511,008 | 511,008 | 0 |
| ga500a-5 | 511,239 | 511,226 | −13 |
| ga500b-1 | 538,060 | 538,060 | 0 |
| ga500b-2 | 537,850 | 537,850 | 0 |
| ga500b-3 | 537,924 | 537,921 | −3 |
| ga500b-4 | 537,925 | 537,925 | 0 |
| ga500b-5 | 537,482 | 537,482 | 0 |
| ga750a-1 | 763,528 | 763,520 | −8 |
| ga750a-2 | 763,653 | 763,623 | −30 |
| ga750a-3 | 763,697 | 763,684 | −13 |
| ga750a-4 | 763,945 | 763,941 | −4 |
| ga750a-5 | 763,786 | 763,786 | 0 |
| ga750b-1 | 796,454 | 796,454 | 0 |
| ga750b-2 | 795,963 | 795,963 | 0 |
| ga750b-3 | 796,130 | 796,130 | 0 |
| ga750b-4 | 797,013 | 797,013 | 0 |
| ga750b-5 | 796,387 | 796,312 | −75 |
| ga750c-1 | 902,026 | 902,026 | 0 |
| ga750c-2 | 899,651 | 899,651 | 0 |
| ga750c-3 | 900,010 | 900,010 | 0 |
| ga750c-4 | 900,044 | 900,044 | 0 |
| ga750c-5 | 899,235 | 899,235 | 0 |
| gs500a-1 | 511,188 | 511,187 | −1 |
| gs500a-2 | 511,179 | 511,179 | 0 |

(continued)

**Table 2** (continued)

| Instance | Previously best known | Our best | Difference |
|---|---|---|---|
| gs500a-3 | 511,112 | 511,106 | −6 |
| gs500a-4 | 511,137 | 511,137 | 0 |
| gs500a-5 | 511,293 | 511,293 | 0 |
| gs500b-1 | 537,931 | 537,931 | 0 |
| gs500b-2 | 537,763 | 537,763 | 0 |
| gs500b-3 | 537,854 | 537,854 | 0 |
| gs500b-4 | 537,742 | 537,742 | 0 |
| gs500b-5 | 538,270 | 538,270 | 0 |
| gs750a-1 | 763,671 | 763,671 | 0 |
| gs750a-2 | 763,548 | 763,548 | 0 |
| gs750a-3 | 763,727 | 763,702 | −25 |
| gs750a-4 | 763,887 | 763,887 | 0 |
| gs750a-5 | 763,614 | 763,614 | 0 |
| gs750b-1 | 797,026 | 797,026 | 0 |
| gs750b-2 | 796,170 | 796,170 | 0 |
| gs750b-3 | 796,589 | 796,589 | 0 |
| gs750b-4 | 796,734 | 796,709 | −25 |
| gs750b-5 | 796,365 | 796,365 | 0 |
| gs750c-1 | 900,363 | 900,363 | 0 |
| gs750c-2 | 897,886 | 897,886 | 0 |
| gs750c-3 | 901,656 | 901,089 | −567 |
| gs750c-4 | 901,239 | 901,239 | 0 |
| gs750c-5 | 900,216 | 900,216 | 0 |

Our best solutions can be downloaded from http://csee.essex.ac.uk/staff/dkarap/splp-source-and-solutions.zip (the URL will be shortened in the final version), and are also reported in Appendices 1 and 2.

In Table 3, we compare our two- and three-component CMCS configurations to the results of the previous attack on the KG instances [16]. The time budget of each method is given in the second row of the table. Observe that either of the two CMCS configurations clearly outperforms [16] being given only 100 s, whereas the time budget in [16] is 7200 s. Moreover, being given 1000 s, the three-component CMCS configuration matches or outperforms [16] on every instance, finding all the new best solutions. Hence, the three-component CMCS is faster than Fischetti et al. [16] by two orders of magnitude, and it is capable of achieving higher solution quality.

We also note here that the three-component CMCS configuration performs better, on average, than the two-component one. For example, the three-component CMCS configuration given 1000 s achieves the same solution quality as the two-component CMCS configuration given 7200 s. The two-component configuration was also less successful on the solved KG instances; even given 7200 s per instances, it could not reach the optimal solution for one of them. This demonstrates the importance of configuration complexity and diversity of components; setting $\Lambda = 2$ could be

**Table 3** Comparison of the CMCS configurations to [16]

| Solver: Budget, sec: | Fischetti et al. [16] | Two-component configuration | | | | | Three-component configuration | | | | |
|---|---|---|---|---|---|---|---|---|---|---|---|
| | 7200 | 1 | 10 | 100 | 1000 | 7200 | 1 | 10 | 100 | 1000 | 7200 |
| ga500a-1 | 511,383 | 53 | 7 | 0 | 0 | 0 | 0 | 0 | 0 | 0 | 0 |
| ga500a-2 | 511,255 | 7 | 0 | 0 | 0 | 0 | 21 | 0 | 0 | 0 | 0 |
| ga500a-3 | 510,810 | 3 | 0 | 0 | 0 | 0 | 7 | 0 | 0 | 0 | 0 |
| ga500a-4 | 511,008 | 37 | 30 | 0 | 0 | 0 | 48 | 30 | 30 | 0 | 0 |
| ga500a-5 | 511,239 | 95 | 0 | −13 | −13 | −13 | 50 | 1 | −13 | −13 | −13 |
| ga500b-1 | 538,060 | 0 | 0 | 0 | 0 | 0 | 0 | 0 | 0 | 0 | 0 |
| ga500b-2 | 537,850 | 7 | 0 | 0 | 0 | 0 | 0 | 0 | 0 | 0 | 0 |
| ga500b-3 | 537,924 | −3 | −3 | −3 | −3 | −3 | 166 | −3 | −3 | −3 | −3 |
| ga500b-4 | 537,925 | 43 | 0 | 0 | 0 | 0 | 69 | 0 | 0 | 0 | 0 |
| ga500b-5 | 537,482 | 0 | 0 | 0 | 0 | 0 | 0 | 0 | 0 | 0 | 0 |
| ga750a-1 | 763,528 | 141 | 0 | 0 | −8 | −8 | 150 | −8 | 18 | −8 | −8 |
| ga750a-2 | 763,653 | 63 | 8 | −30 | −30 | −30 | 59 | −19 | −17 | −30 | −30 |
| ga750a-3 | 763,697 | 196 | 46 | −13 | −13 | −13 | 170 | 59 | −7 | −13 | −13 |
| ga750a-4 | 763,945 | 200 | 34 | −4 | 19 | −4 | 180 | 103 | −4 | −4 | −4 |
| ga750a-5 | 763,786 | 260 | 12 | 4 | 8 | 0 | 212 | 63 | 0 | 0 | 0 |
| ga750b-1 | 796,454 | 337 | 0 | 0 | 0 | 0 | 26 | 0 | 0 | 0 | 0 |
| ga750b-2 | 795,963 | 190 | 0 | 0 | 0 | 0 | 248 | 0 | 0 | 0 | 0 |
| ga750b-3 | 796,359 | 132 | −216 | −229 | −229 | −229 | 90 | −229 | −229 | −229 | −229 |
| ga750b-4 | 797,013 | 128 | 0 | 0 | 0 | 0 | 0 | 0 | 0 | 0 | 0 |
| ga750b-5 | 796,549 | −104 | −237 | −237 | −237 | −237 | −65 | −214 | −237 | −237 | −237 |
| ga750c-1 | 902,026 | 0 | 0 | 0 | 0 | 0 | 0 | 0 | 0 | 0 | 0 |
| ga750c-2 | 899,651 | 81 | 0 | 0 | 0 | 0 | 81 | 0 | 0 | 0 | 0 |
| ga750c-3 | 900,019 | −9 | −9 | −9 | −9 | −9 | −9 | −9 | −9 | −9 | −9 |
| ga750c-4 | 900,044 | 0 | 0 | 0 | 0 | 0 | 0 | 0 | 0 | 0 | 0 |
| ga750c-5 | 899,235 | 0 | 0 | 0 | 0 | 0 | 0 | 0 | 0 | 0 | 0 |
| gs500a-1 | 511,188 | 12 | −1 | −1 | −1 | −1 | 114 | 41 | −1 | −1 | −1 |
| gs500a-2 | 511,179 | 0 | 0 | 0 | 0 | 0 | 0 | 0 | 0 | 0 | 0 |
| gs500a-3 | 511,112 | 0 | 25 | −6 | −6 | −6 | 31 | 0 | −6 | −6 | −6 |
| gs500a-4 | 511,137 | 117 | 0 | 0 | 0 | 0 | 139 | 0 | 0 | 0 | 0 |
| gs500a-5 | 511,293 | 81 | 27 | 27 | 0 | 0 | 88 | 0 | 0 | 0 | 0 |
| gs500b-1 | 537,931 | 64 | 0 | 0 | 0 | 0 | 0 | 0 | 0 | 0 | 0 |
| gs500b-2 | 537,763 | 16 | 0 | 0 | 0 | 0 | 48 | 0 | 0 | 0 | 0 |
| gs500b-3 | 537,854 | 72 | 0 | 0 | 0 | 0 | 0 | 0 | 0 | 0 | 0 |
| gs500b-4 | 537,742 | 0 | 0 | 0 | 0 | 0 | 0 | 0 | 0 | 0 | 0 |
| gs500b-5 | 538,270 | 82 | 0 | 0 | 0 | 0 | 82 | 0 | 0 | 0 | 0 |
| gs750a-1 | 763,671 | 63 | 5 | 0 | 0 | 0 | 159 | 17 | 0 | 0 | 0 |
| gs750a-2 | 763,548 | 199 | 15 | 15 | 0 | 0 | 157 | 15 | 14 | 0 | 0 |
| gs750a-3 | 763,727 | 155 | 46 | 21 | −25 | −25 | −12 | 11 | 21 | −25 | −25 |
| gs750a-4 | 763,922 | 58 | 6 | −35 | −35 | −35 | 53 | 22 | −3 | −8 | −35 |
| gs750a-5 | 763,614 | 102 | 27 | 2 | 0 | 0 | 87 | 18 | 2 | 0 | 0 |

(continued)

**Table 3** (continued)

| Solver: Budget, sec: | Fischetti et al. [16] 7200 | Two-component configuration 1 | 10 | 100 | 1000 | 7200 | Three-component configuration 1 | 10 | 100 | 1000 | 7200 |
|---|---|---|---|---|---|---|---|---|---|---|---|
| gs750b-1 | 797,329 | −138 | −303 | −303 | −303 | −303 | −303 | −303 | −303 | −303 | −303 |
| gs750b-2 | 796,170 | 31 | 25 | 0 | 0 | 0 | 31 | 31 | 0 | 0 | 0 |
| gs750b-3 | 796,589 | 0 | 0 | 0 | 0 | 0 | 534 | 0 | 0 | 0 | 0 |
| gs750b-4 | 797,020 | −311 | −286 | −311 | −311 | −311 | −178 | −286 | −311 | −311 | −311 |
| gs750b-5 | 796,365 | 0 | 0 | 0 | 0 | 0 | 0 | 0 | 0 | 0 | 0 |
| gs750c-1 | 900,363 | 0 | 0 | 0 | 0 | 0 | 0 | 0 | 0 | 0 | 0 |
| gs750c-2 | 897,886 | 0 | 0 | 0 | 0 | 0 | 187 | 0 | 0 | 0 | 0 |
| gs750c-3 | 901,656 | −567 | −567 | −567 | −567 | −567 | −567 | −567 | −567 | −567 | −567 |
| gs750c-4 | 901,239 | 0 | 0 | 0 | 0 | 0 | 0 | 0 | 0 | 0 | 0 |
| gs750c-5 | 900,216 | 0 | 0 | 0 | 0 | 0 | 0 | 0 | 0 | 0 | 0 |
| Improved | | 6 | 8 | 14 | 15 | 16 | 6 | 9 | 14 | 16 | 16 |
| Same | | 14 | 28 | 31 | 33 | 34 | 16 | 29 | 31 | 34 | 34 |
| Worse | | 30 | 14 | 5 | 2 | 0 | 28 | 12 | 5 | 0 | 0 |

considered as a minimal option, which restricts the performance that can be achieved by corresponding configurations. On the other hand, we have evidence that $\Lambda = 3$ is sufficient to achieve outstanding performance when compared to human-designed algorithms. A more efficient CMCS generation routine will let us verify if a four-component configuration can achieve an even better performance.

## 5 Conclusions

In this paper, we discussed automated generation of CMCS configurations for the SPLP, and have shown the success of our approach. In particular, we clearly outperformed the previous state-of-the-art solver and improved the best known upper bounds for 12 out of 50 yet unsolved KG instances. The outstanding performance of our SPLP heuristic is attributed to that it was generated automatically.

The automated generation of the algorithm has several obvious advantages. One is that it saves labour and human expertise required for (meta)heuristic design. Also, automated generation is significantly quicker than a manual design process, hence the entire algorithm design can be completed within a few days. Finally, the computer is capable of testing more combinations than a human and objectively selecting the best of them. This lack of bias means, among other things, that the computer does test strategies that a human would usually rule out, and in our experience such unusual strategies often demonstrate unexpectedly good performance.

In the spirit of the no free lunch theorem, we note here that the selected configuration performs best only under certain circumstances such as a specific instance family or certain time budget. By correctly selecting the training dataset, we can obtain an algorithm that is best suited for our particular case. Moreover, it is easy to obtain several algorithms for various circumstances and requirements and then use the most appropriate one for each job.

In this project, we limited CMCS configurations to deterministic strategies, and also restricted the number of components to be included in a configuration. These simple measures greatly reduced the number of candidate configurations allowing us to enumerate all of them and choose the best performing one. We use a combination of Pareto dominance and sign test to quickly rule out less promising configurations, and then apply a multi-criteria optimisation method to choose a single best candidate.

We leave for future work investigation of more efficient CMCS generation procedures, which will allow one to include more components into the component pool and consider more sophisticated configurations. Also, it would be interesting to select not a single best configuration but several well-performing configurations with complementary properties, and select the most appropriate one at run time. Finally, we are interested to apply the CMCS approach to other classes of allocation and clustering instances as well as to routing and scheduling optimisation problems.

## *Appendix 1: Optimal Solutions for Instances Solved to Optimality*

| Instance | Obj. v. | Opened sites |
|---|---|---|
| ga250a-3 | 257953 | 22 35 39 46 57 66 76 86 97 100 105 112 114 116 121 124 126 127 144 154 155 176 192 196 200 207 211 219 223 227 229 237 246 249 |
| ga250a-5 | 258190 | 13 17 29 35 40 43 49 55 60 63 79 82 110 126 135 139 150 157 161 174 178 179 198 201 204 208 211 230 232 241 248 |
| ga500c-5 | 621313 | 4 75 183 259 360 491 |
| gs500c-3 | 621204 | 98 195 216 245 333 429 |
| gs500c-5 | 623180 | 22 51 276 355 439 444 |
| ga250a-1 | 257957 | 21 32 38 47 53 56 58 84 94 100 101 103 111 129 136 139 144 146 149 150 168 170 175 178 203 204 219 224 234 238 239 250 |
| ga250a-2 | 257502 | 11 37 55 62 64 84 88 99 100 103 107 114 115 116 118 132 146 157 158 160 171 191 200 211 213 217 218 221 237 238 240 250 |
| ga250a-4 | 257987 | 4 5 7 30 31 37 53 55 69 73 74 75 84 92 93 103 108 115 119 127 129 153 163 168 173 174 188 199 200 203 208 213 219 236 250 |
| ga250b-1 | 276296 | 10 56 60 94 106 129 149 150 170 203 219 250 |
| ga250b-2 | 275141 | 37 55 88 103 135 141 158 191 211 213 231 |
| ga250b-3 | 276093 | 1 18 22 35 39 50 97 192 200 229 246 |
| ga250b-4 | 276332 | 5 7 36 56 77 92 124 160 228 236 250 |

(continued)

| Instance | Obj. v. | Opened sites |
|---|---|---|
| ga250b-5 | 276404 | 40 57 79 110 157 161 183 184 208 211 241 246 |
| ga250c-1 | 334135 | 100 154 175 231 |
| ga250c-2 | 330728 | 45 55 88 99 |
| ga250c-3 | 333662 | 22 97 127 138 |
| ga250c-4 | 332423 | 5 124 143 188 |
| ga250c-5 | 333538 | 74 110 157 247 |
| ga500c-1 | 621360 | 127 269 378 403 430 |
| ga500c-2 | 621464 | 28 107 212 315 344 456 |
| ga500c-3 | 621428 | 68 187 314 326 370 474 |
| ga500c-4 | 621754 | 56 97 307 350 436 |
| gs250a-1 | 257964 | 4 10 12 25 27 30 47 51 63 71 119 123 126 132 137 143 145 155 161 163 169 176 177 178 203 214 232 234 236 238 245 246 249 |
| gs250a-2 | 257573 | 9 24 25 40 43 46 52 74 77 86 87 88 95 96 98 100 101 113 114 120 130 139 154 160 161 165 166 184 191 241 245 250 |
| gs250a-3 | 257626 | 9 20 33 34 37 38 55 60 67 69 71 72 91 110 120 121 132 139 144 148 166 172 174 177 187 189 190 199 204 209 223 229 234 |
| gs250a-4 | 257961 | 3 20 31 36 46 54 101 102 104 115 118 128 139 143 144 159 160 163 168 179 188 193 195 207 208 217 221 226 233 237 |
| gs250a-5 | 257896 | 18 33 36 47 49 60 76 77 89 98 104 114 118 122 124 133 137 156 161 168 172 189 204 207 209 212 213 217 223 227 228 230 235 250 |
| gs250b-1 | 276761 | 8 27 47 63 71 113 137 145 170 178 229 232 |
| gs250b-2 | 275675 | 25 43 52 69 77 87 120 139 149 160 221 245 |
| gs250b-3 | 275710 | 32 55 57 60 67 69 82 166 172 174 210 |
| gs250b-4 | 276114 | 13 19 31 35 97 106 139 144 157 177 191 247 |
| gs250b-5 | 275916 | 18 36 118 122 124 137 166 172 177 204 209 230 |
| gs250c-1 | 332935 | 63 170 176 232 |
| gs250c-2 | 334630 | 25 52 83 144 |
| gs250c-3 | 333000 | 57 60 69 166 |
| gs250c-4 | 333158 | 52 144 157 191 |
| gs250c-5 | 334635 | 18 84 114 186 |
| gs500c-1 | 620041 | 29 102 112 242 440 |
| gs500c-2 | 620434 | 70 286 424 439 495 |
| gs500c-4 | 620437 | 96 247 283 316 390 |

## *Appendix 2: Best Known Solutions for Instances Not Yet Solved to Optimality*

The objective values of these solutions can be found in Table 2, column 'Our best'.

| Instance | Opened sites |
|---|---|
| ga500a-1 | 22 28 52 59 65 70 73 79 86 90 100 103 111 126 142 152 156 173 177 189 199 205 208 219 221 234 245 246 260 265 269 275 280 299 301 303 313 375 378 397 410 419 430 463 475 486 490 494 |
| ga500a-2 | 34 51 54 70 84 116 120 122 126 144 155 158 169 177 188 193 195 204 212 213 218 222 223 238 255 289 313 315 321 333 338 345 360 372 391 397 399 401 413 415 437 450 466 470 476 478 485 487 500 |
| ga500a-3 | 12 33 36 37 44 49 55 65 75 85 88 92 94 96 114 132 145 148 150 166 178 184 185 187 192 196 201 220 241 252 257 278 285 288 290 303 340 364 367 370 375 378 386 393 431 451 474 |
| ga500a-4 | 18 19 29 35 39 42 43 49 56 65 74 78 102 119 138 140 144 155 188 197 204 214 267 273 280 281 282 293 317 329 340 346 350 360 364 371 377 388 404 411 417 419 430 436 448 456 466 484 496 |
| ga500a-5 | 4 11 14 22 34 36 38 40 47 51 55 95 100 120 123 125 127 133 155 174 181 183 199 216 229 283 284 301 316 321 326 328 332 336 348 369 371 380 382 387 390 397 399 424 429 487 488 491 497 |
| ga500b-1 | 34 100 127 153 156 176 184 189 199 236 277 375 378 379 410 430 470 |
| ga500b-2 | 28 34 51 137 212 213 225 238 241 245 249 268 315 336 338 344 459 478 |
| ga500b-3 | 36 66 92 94 166 185 187 189 241 290 300 303 326 340 370 393 431 474 |
| ga500b-4 | 24 138 204 282 293 329 330 343 360 388 396 436 448 451 456 466 484 496 |
| ga500b-5 | 2 14 55 123 124 135 142 147 181 183 231 258 259 349 360 382 399 414 |
| gs500a-1 | 6 22 42 53 55 58 94 95 115 116 120 121 126 127 129 144 149 154 164 171 173 212 239 252 270 285 294 300 320 321 327 335 336 343 352 377 379 384 385 389 399 420 426 429 434 442 464 490 |
| gs500a-2 | 9 50 53 62 70 99 109 110 115 124 168 169 175 185 198 202 204 218 229 233 241 247 269 276 289 290 294 295 301 316 333 335 336 356 358 376 383 394 400 422 426 437 439 453 457 459 460 463 464 470 |
| gs500a-3 | 7 17 28 38 41 55 65 67 74 84 86 110 117 147 152 153 162 173 212 219 223 244 256 259 269 271 273 287 300 301 308 310 365 369 371 377 381 385 394 401 413 417 421 437 453 456 493 494 |
| gs500a-4 | 9 10 14 18 56 67 68 84 87 93 95 123 124 136 137 161 165 173 180 189 194 196 202 217 229 231 258 273 277 281 290 350 356 359 363 371 378 380 390 391 435 438 453 458 464 484 490 491 495 |
| gs500a-5 | 3 4 7 13 15 30 40 47 60 69 78 86 122 123 136 153 156 159 160 171 174 185 192 231 233 234 235 250 251 257 268 281 304 312 316 322 331 338 384 391 411 424 431 444 459 460 481 498 |
| gs500b-1 | 22 29 45 82 102 112 116 193 215 257 258 313 385 410 440 468 470 471 |
| gs500b-2 | 24 70 85 95 115 168 233 247 329 356 358 382 383 408 437 439 457 488 |
| gs500b-3 | 7 41 116 216 235 245 255 269 273 279 287 299 308 333 347 371 392 429 |
| gs500b-4 | 76 91 103 120 132 142 165 171 173 212 230 351 380 406 437 438 447 491 |

(continued)

| Instance | Opened sites |
|---|---|
| gs500b-5 | 3 7 13 40 65 105 153 185 226 235 319 322 372 384 431 444 460 486 |
| ga750a-1 | 2 18 35 50 52 67 69 71 74 105 110 111 117 118 127 152 155 209 219 233 234 242 263 280 296 305 309 316 330 335 346 381 430 431 435 446 449 457 478 484 494 512 538 540 548 559 564 587 616 640 644 647 650 667 680 689 711 713 716 729 738 745 |
| ga750a-2 | 1 18 40 45 71 102 104 109 110 114 120 126 131 144 150 154 168 170 180 183 211 214 227 234 235 237 239 259 283 288 289 296 301 316 351 352 357 359 362 369 375 382 387 436 447 455 491 511 528 530 539 550 581 582 615 642 644 645 673 684 710 722 |
| ga750a-3 | 49 68 71 75 88 95 101 109 113 127 183 202 206 212 254 266 295 298 334 339 349 356 361 379 389 400 415 419 428 433 436 439 446 447 450 464 465 483 506 560 561 565 570 575 580 581 585 590 606 626 627 645 657 666 669 679 682 694 698 731 |
| ga750a-4 | 25 54 79 87 99 100 104 108 112 149 154 160 163 169 176 195 224 226 258 266 271 279 291 292 303 305 319 324 326 363 386 400 413 416 418 420 437 454 468 487 496 534 536 551 561 568 628 635 636 652 656 669 672 684 693 695 703 718 733 737 748 |
| ga750a-5 | 18 34 35 67 75 77 80 87 93 117 119 146 161 167 168 187 189 195 224 228 235 246 247 271 315 320 325 329 359 365 367 373 389 411 421 424 426 429 452 456 475 476 507 522 523 524 532 553 562 565 578 588 655 678 702 706 709 713 734 736 747 749 |
| ga750b-1 | 58 67 71 100 117 184 214 335 346 386 478 484 512 559 589 593 616 647 662 711 720 745 746 |
| ga750b-2 | 1 45 109 110 144 168 182 214 235 237 239 283 288 296 308 329 375 505 637 644 645 712 |
| ga750b-3 | 53 68 101 202 206 215 254 298 334 356 404 408 464 560 570 575 596 604 669 673 726 728 |
| ga750b-4 | 47 52 55 104 115 128 149 154 202 232 266 303 358 434 468 551 635 639 672 704 733 |
| ga750b-5 | 29 42 49 87 99 182 193 194 218 224 351 363 373 376 380 456 473 523 667 685 700 746 |
| ga750c-1 | 214 418 476 587 593 644 711 |
| ga750c-2 | 1 170 182 235 237 564 590 |
| ga750c-3 | 68 101 173 215 439 548 616 |
| ga750c-4 | 128 144 154 279 413 456 704 |
| ga750c-5 | 14 344 376 456 577 659 685 |
| gs750a-1 | 2 20 52 64 75 89 93 124 128 130 143 147 177 199 205 218 232 236 237 240 264 279 281 289 305 317 320 335 336 374 393 403 404 425 453 458 466 468 482 487 496 518 538 541 545 554 555 564 602 609 637 651 658 659 663 669 672 681 682 706 713 729 743 |
| gs750a-2 | 1 19 24 62 64 70 92 93 94 100 108 113 134 137 146 157 178 180 186 213 225 265 268 278 281 325 334 336 341 348 362 393 397 398 411 444 451 460 462 493 494 498 504 554 574 575 594 595 620 625 628 636 639 646 650 661 721 724 735 |

(continued)

| Instance | Opened sites |
|---|---|
| gs750a-3 | 6 22 35 38 46 66 96 110 117 119 133 135 171 182 192 230 250 287 288 291 317 354 357 360 367 374 385 392 396 409 412 425 437 450 458 463 465 487 499 510 528 554 561 562 583 596 609 612 624 640 677 688 702 710 715 721 728 732 746 |
| gs750a-4 | 2 4 24 26 34 56 78 80 81 95 107 115 117 123 136 139 145 146 151 172 174 190 241 243 258 266 269 294 305 323 332 346 377 388 399 412 434 443 459 473 477 498 500 522 530 536 537 551 558 573 603 617 640 641 656 666 669 673 721 722 740 749 |
| gs750a-5 | 3 12 31 46 54 65 74 88 91 96 102 104 112 114 119 135 150 182 198 202 207 231 234 288 302 317 356 359 386 394 397 410 411 417 421 425 483 579 582 587 607 612 638 644 654 662 663 669 672 680 693 707 716 720 721 725 731 744 |
| gs750b-1 | 41 45 67 75 128 130 213 232 237 242 279 281 313 428 487 538 573 658 698 699 725 743 |
| gs750b-2 | 1 10 108 126 191 213 214 257 265 314 336 341 348 367 444 450 494 508 617 636 639 650 661 |
| gs750b-3 | 22 26 64 135 171 192 281 291 304 317 446 462 484 553 561 583 702 706 715 727 728 |
| gs750b-4 | 4 59 81 95 107 113 165 174 179 190 238 248 261 294 305 355 431 459 616 641 654 721 |
| gs750b-5 | 3 31 74 90 104 200 281 302 359 392 394 427 487 540 549 568 593 627 635 693 707 709 |
| gs750c-1 | 67 112 128 428 479 603 639 |
| gs750c-2 | 29 92 108 265 336 494 639 |
| gs750c-3 | 6 44 304 583 624 680 698 |
| gs750c-4 | 26 139 151 287 522 628 721 |
| gs750c-5 | 104 198 302 557 607 635 707 |

# References


1. M.L. Alves, M.T. Almeida, Simulated annealing algorithm for the simple plant location problem: a computational study. Revista Invest. Oper. **12**, (1992)
2. E. Balas, M.W. Padberg, On the set covering problem. Oper. Res. **20**, 1152–1161 (1972)
3. M. Balinski, Integer programming: methods, uses, computations. Manag. Sci. **12**, 253–313 (1965)
4. F.F. Barahona, F.N.A. Chudak, Solving large scale uncapacitated facility-location problems, in *Approximation and Complexity in Numerical Optimization*, ed. by P.M. Pardalos (Kluwer Academic Publishers, Norwell, MA, 1990), pp. 48–62
5. F.F. Barahona, F.N.A. Chudak, Near-optimal solutions to large-scale facility location problems. Discret. Optim. **2**, 35–50 (2005)
6. J.E. Beasley, Lagrangian heuristics for location problems. Eur. J. Oper. Res. **65**, 383–399 (1993)
7. C. Beltran-Royo, J.-P. Vial, A. Alonso-Ayuso, Semi-Lagrangian relaxation applied to the uncapacitated facility location problem. Comput. Optim. Appl. **51**, 387–409 (2012)
8. L. Cánovas, M. Landete, A. Marin, On the facets of the simple plant location packing polytope. Discret. Appl. Math. **23**, 27–53 (2002)

9. D.C. Cho, E.J. Johnson, M.W. Padberg, M.R. Rao, On the uncapacitated location problem I: valid inequalities and facets. Math. Oper. Res. **8**, 579–589 (1983)
10. D.C. Cho, E.J. Johnson, M.W. Padberg, M.R. Rao, On the uncapacitated location problem II: facets and lifting theorems. Math. Oper. Res. **8**, 590–612 (1983)
11. G. Cornuejols, G. Nemhauser, L.A. Wolsey, The uncapacitated facility location problem, in *Discrete Location Theory*, ed. by P.B. Mirchandani, R.L. Francis (Wiley-Interscience, New York, 1990)
12. G. Cornuejols, J.-M. Thizy, A primal approach to the simple plant location problem. SIAM J. Algebraic Discret. Methods **3**(4), 504–510 (1982)
13. M.S. Daskin, *Network and Discrete Location: Models, Algorithms, and Applications*, 2nd edn. (Wiley, New York, 2013)
14. I.R. de Farias, A family of facets for the uncapacitated p-median polytope. Oper. Res. Lett. **28**, 161–167 (2001)
15. D. Erlenkotter, A dual-based procedure for uncapacitated facility location. Oper. Res. **26**, 992–1009 (1978)
16. M. Fischetti, I. Ljubić, M. Sinnl, Redesigning benders decomposition for large-scale facility location. Manag. Sci. **63**(7), 2146–2162 (2017)
17. L. Galli, A.N. Letchford, S.J. Miller, New valid inequalities and facets for the simple plant location problem. Eur. J. Oper. Res. **269**(3), 824–833 (2018)
18. D. Ghosh, Neighborhood search heuristics for the uncapacitated facility location problem. Eur. J. Oper. Res. **150**(4), 150–162 (2003)
19. B. Goldengorin, Data correcting approach for routing and location in networks, in *Handbook of Combinatorial Optimization*, ed. by P.M. Pardalos, D.-Z. Du, R.L. Graham (Springer, New York, 2013), pp. 929–993
20. B. Goldengorin, D. Krushinsky, P.M. Pardalos, *Cell Formation in Industrial Engineering: Theory, Algorithms and Experiments* (Springer, New York, 2013)
21. B. Goldengorin, G.A. Tijssen, D. Ghosh, G. Sierksma, Solving the simple plant location problems using a data correcting approach. J. Glob. Optim. **25**, 377–406 (2003)
22. S. Guha, S. Khuller, Greedy strikes back: improved facility location algorithms. J. Algorithms **31**, 228–248 (1999)
23. M. Guignard, A Lagrangean dual ascent algorithm for simple plant location problems. Eur. J. Oper. Res. **33**, 193–200 (1988)
24. P. Hansen, N. Mladenović, Variable neighborhood search for the p-median. Locat. Sci. **5**(4), 207–226 (1997)
25. M. Hoefer, Experimental comparison of heuristic and approximation algorithms for uncapacitated facility location, in *Experimental and Efficient Algorithms* (Springer, Berlin, 2003), pp. 165–178
26. K. Jain, M. Mahdian, E. Markakis, A. Saberi, V.V. Vazirani. Greedy facility location algorithms analyzed using dual fitting with factor-revealing LP. J. ACM **60**, 795–824 (2003)
27. D. Karapetyan, A.P. Punnen, A.J. Parkes, Markov chain methods for the bipartite boolean quadratic programming problem. Eur. J. Oper. Res. **260**(2), 494–506 (2017)
28. M. Körkel, On the exact solution of large-scale simple plant location problems. Eur. J. Oper. Res. **39**, 157–173 (1989)
29. J. Kratica, D. Tosic, V. Filipović, I. Ljubić, Solving the simple plant location problem by genetic algorithm. RAIRO Oper. Res. **35**, 127–142 (2001)
30. A.A. Kuehn, M.J. Hamburger, A heuristic program for locating warehouses. Manag. Sci. **9**(4), 643–646 (1963)
31. A. Letchford, S. Miller, An aggressive reduction scheme for the simple plant location problem. Eur. J. Oper. Res. **234**, 674–682 (2014)
32. L. Michel, P. Van Hentenryck, Solving the simple plant location problem by genetic algorithm. RAIRO Oper. Res. **35**, 127–142 (2001)
33. J.G. Morris, On the extent to which certain fixed charge depot location problems can be solved by LP. J. Oper. Res. Soc. **29**, 71–76 (1978)
34. M. Posta, J.A. Ferland, P. Michelon, An exact cooperative method for the uncapacitated facility location problem. Math. Programm. Comput. **6**, 199–231 (2014)

35. M.G.C. Resende, R. Werneck, A hybrid multistart heuristic for the uncapacitated facility location problem. Eur. J. Oper. Res. **174**, 54–68 (1975)
36. L. Schrage, Implicit representation of variable upper bounds in linear programming. Math. Programm. Study **4**, 118–132 (1975)
37. M. Sun, Solving uncapacitated facility location problems using tabu search. Comput. Oper. Res. **33**, 2563–2589 (2006)
38. V.A. Trubin, On a method of solution of integer programming problems of a special kind. Soviet Math. Doklady **10**, 1544–1546 (1969)
39. A. Weber, Theory of the Location of Industries. The University of Chicago Press Chicago, Illinois (1929). English Edition, with Introduction and Notes by Carl J. Friedrich.
40. V. Yigit, M.E. Aydin, O. Turkbey, Evolutionary simulated annealing algorithms for uncapacitated facility location problems, in *Adaptive Computing in Design and Manufacture VI*, ed. by I.C. Parmee (Springer, New York, 2004), pp. 185–194

# An Algorithmic Answer to the Ore-Type Version of Dirac's Question on Disjoint Cycles

**H. A. Kierstead, A. V. Kostochka, T. Molla, and D. Yager**

*Dedicated to Gregory Gutin on the occasion of his 60th birthday*

## 1 Introduction

For a multigraph $G = (V, E)$, let $|G| = |V|$, $\|G\| = |E|$, and $\alpha(G)$ be the independence number of $G$. Also, for $S, T \subseteq V$, let $||S, T|| = |\{uv \in E : u \in S, v \in T\}|$. The *minimum degree of* $G$, denoted $\delta(G)$, is the minimum number of edges incident to a vertex where edges are counted according to their multiplicities. For a simple graph $G$, let $\overline{G}$ denote the complement of $G$. For multigraphs $G$ and $H$, let $G \cup H$ denote the multigraph with $V(G \cup H) = V(G) \cup V(H)$ and $E(G \cup H) = E(G) \cup E(H)$. For disjoint graphs $G$ and $H$, let $G \vee H$ denote $G \cup H$ together with all edges from $V(G)$ to $V(H)$.

Let $K(X)$ be the complete graph with vertex set $X$, and $K_t(X) = K(X)$ indicate that $|X| = t$.

If we only want to specify one vertex $v$ of $K_t$ we write $K_t(v)$.

H. A. Kierstead
School of Mathematical and Statistical Sciences, Arizona State University, Tempe, AZ, USA
e-mail: kierstead@asu.edu

A. V. Kostochka (✉)
Department of Mathematics, University of Illinois, Urbana, IL, USA

Sobolev Institute of Mathematics, Novosibirsk, Russia
e-mail: kostochk@math.uiuc.edu

T. Molla
Department of Mathematics and Statistics, University of South Florida, Tampa, FL, USA
e-mail: molla@usf.edu

D. Yager
Department of Mathematics, University of Illinois, Urbana, IL, USA
e-mail: yager2@illinois.edu



B. Goldengorin (ed.), *Optimization Problems in Graph Theory*,
Springer Optimization and Its Applications 139,
https://doi.org/10.1007/978-3-319-94830-0_8

The problem of finding the maximum number of disjoint cycles in a graph is $NP$-hard, since even some partial cases of it are:

**Theorem 1 ([6], p. 68)** *Determining whether a* $3n$*-vertex graph has* $n$ *disjoint triangles is an* $NP$*-complete problem.*

On the other hand, Bodlaender [1] and independently Downey and Fellows [4] showed that this problem is *fixed parameter tractable*:

**Theorem 2 ([1, 4])** *For every fixed* $k$*, the question whether an* $n$*-vertex graph has* $k$ *disjoint cycles can be resolved in linear (in* $n$*) time.*

Since the general problem is hard, it is natural to look for sufficient conditions that ensure the existence of "many" disjoint cycles in a graph. One well-known result of this type is the following theorem of Corrádi and Hajnal [2] from 1963:

**Theorem 3 ([2])** *Let* $k \in \mathbb{Z}^+$*. Every graph* $G$ *with* $|G| \geq 3k$ *and* $\delta(G) \geq 2k$ *contains* $k$ *disjoint cycles.*

The hypothesis $\delta(G) \geq 2k$ is best possible, as shown by the $3k$-vertex graph $H = \overline{K}_{k+1} \vee K_{2k-1}$, which has $\delta(H) = 2k - 1$ but does not contain $k$ disjoint cycles. The proof yields a polynomial algorithm for finding $k$ disjoint cycles in the graphs satisfying the conditions of the theorem.

Theorem 3 was refined and generalized in several directions. Enomoto [5] and Wang [15] generalized the Corrádi–Hajnal Theorem in terms of the *minimum Ore-degree* $\sigma_2(G) := \min\{d(x) + d(y) : xy \notin E(G)\}$:

**Theorem 4 ([5],[15])** *Let* $k \in \mathbb{Z}^+$*. Every graph* $G$ *with* $|G| \geq 3k$ *and*

$$\sigma_2(G) \geq 4k - 1$$

*contains* $k$ *disjoint cycles.*

Kierstead, Kostochka, and Yeager [9] refined Theorem 3 by characterizing all simple graphs that fulfill the weaker hypothesis $\delta(G) \geq 2k-1$ and contain $k$ disjoint cycles. This refinement depends on an extremal graph $\mathbf{Y}_{k,k,k}$ where $\mathbf{Y}_{h,s,t} = \overline{K_h} \vee (K_s \cup K_t)$ and $\mathbf{Y}_{h,s,t}(X_0, X_1, X_2) = \overline{K_h}(X_0) \vee (K_s(X_1) \cup K_t(X_2))$ (Figure 1).

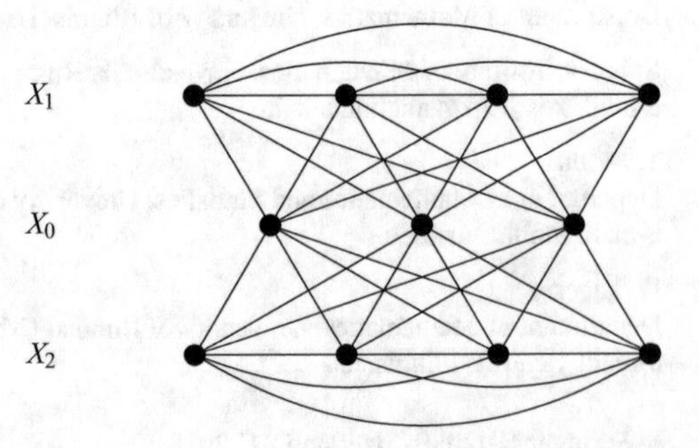


**Fig. 1** $\mathbf{Y}_{h,t,s}$, shown with $h = 3$ and $t = s = 4$

**Theorem 5 ([9])** *Let $k \geq 2$. Every simple graph $G$ with $|G| \geq 3k$ and $\delta(G) \geq 2k-1$ contains $k$ disjoint cycles if and only if:*

*(i)* $\alpha(G) \leq |G| - 2k$;
*(ii) if $k$ is odd and $|G| = 3k$, then $G \neq \mathbf{Y}_{k,k,k}$; and*
*(iii) if $k = 2$ then $G$ is not a wheel.*

Theorem 4 was refined in a similar way in [9] and [8] (see Theorem 13 in the next section).

Dirac [3] described all 3-connected multigraphs that do not have two disjoint cycles and posed the following question:

*Question 1 ([3])* Which $(2k-1)$-connected multigraphs[1] do not have $k$ disjoint cycles?

Kierstead, Kostochka, and Yeager [10] used Theorem 5 to answer Question 1 (see Theorem 11 in Section 2). The goal of this paper is to resolve the Ore-type version of Question 1 for multigraphs in an algorithmic way. In Theorem 14, we consider the class $\mathscr{D}\mathscr{O}_k$ of multigraphs $G$ whose underlying simple graph $\underline{G}$ satisfies $d_{\underline{G}}(x)+d_{\underline{G}}(y) \geq 4k-3$ for all nonadjacent vertices $x$ and $y$, and describe all graphs in $\mathscr{D}\mathscr{O}_k$ that do not have $k$ disjoint cycles. Using this description we construct a polynomial time algorithm that for every multigraph $G$ in $\mathscr{D}\mathscr{O}_k$ decides whether $G$ has $k$ disjoint cycles or not.

In the next section, we introduce notation and discuss existing results to be used later on. In Section 3 we state our main results, Theorems 14 and 15. In the next two sections, we prove Theorem 14, and in the last section we prove Theorem 15.

## 2 Preliminaries and Known Results

### 2.1 *Notation*

A *loop* is an edge consisting of a single vertex, and a *strong edge* is an edge with multiplicity greater than one. For every multigraph $G$, let $V_1 = V_1(G)$ be the set of vertices in $G$ incident to loops, and $V_2 = V_2(G)$ be the set of vertices in $G - V_1$ incident to strong edges. Let $F = F(G)$ be the simple graph with $V(F) = V_2$ and $E(F)$ consisting of the strong edges in $G - V_1$. We define $\alpha' = \alpha'(F)$ to be the size of a maximum matching in $F$. Let $\underline{G}$ denote the *underlying simple graph of* $G$, i.e., the simple graph on $V(G)$ such that two vertices are adjacent in $\underline{G}$ if and only if they are adjacent in $G$. For $e \notin E(G)$, let $G+e$ denote the graph with $V(G+e) = V(G)$ and $E(G+e) = E(G) \cup \{e\}$. For a path $P$ with $P \cap G = \emptyset$, let sd$(G, e, P)$ be the result of subdividing $e$ with $P$.

[1] Dirac used the word *graphs*, but in [3] this appears to mean *multigraphs*.

Recall that $K_t(X) = K(X)$ denotes the complete graph with vertex set $X$ where $|X| = t$. Similarly, $K(Y, Z)$ is the complete $Y, Z$-bigraph. We also extend this notation to the case that $Y$ is a graph. Then, $K(Y, Z)$ is $K(V(Y), Z) \cup Y$.

For $S \subseteq V(G)$, let $\overline{S} = V(G) - S$ and let $N_G(S) = \bigcup_{v \in S} N_G(v)$. For a matching $M$, let $W = W(M)$ denote the set of vertices saturated by $M$, and $G' = G'(M) = G - W(M)$. If $|F| = 2\alpha'$, then $G'(M) = G'(M')$ for all perfect matchings $M$ and $M'$ in $F$.

For $v \in V(G)$, we define $s(v) = |N(v)|$ to be the *simple degree* of $v$, and we say that $\mathscr{S}(G) = \min\{s(v) : v \in V\}$ is the *minimum simple degree* of $G$. Similarly, $\mathscr{S}\mathscr{O}(G) = \min\{s(v) + s(u) : v, u \in V, v \neq u$ and $uv \notin E(\underline{G})\}$. Let $c(G)$ be the maximum number of disjoint cycles contained in $G$.

We define $\mathscr{D}_k$ to be the family of multigraphs $G$ with $\mathscr{S}(G) \geq 2k - 1$ and $\mathscr{D}\mathscr{O}_k$ to be the family of multigraphs $G$ with $\mathscr{S}\mathscr{O}(G) \geq 4k - 3$. For a graph $G \in \mathscr{D}\mathscr{O}_k$, call a vertex $v \in V(G)$ *low* if $d_G(v) \leq 2k - 2$. Let $\mathscr{B}_k = \{G \in \mathscr{D}_k : c(G) < k\}$, and let $\mathscr{B}\mathscr{O}_k = \{G \in \mathscr{D}\mathscr{O}_k : c(G) < k\}$.

If $G \in \mathscr{D}\mathscr{O}_k$ is an $n$-vertex multigraph and $\alpha(G) \geq n - 2k + 2$, then for any distinct $v_1, v_2$ in a maximum independent set $I$, $s(v_1)+s(v_2) \leq (2k-2)+(2k-2) < 4k-3$. Thus, $\alpha(G) \leq n-2k+1$ for every $n$-vertex $G \in \mathscr{D}\mathscr{O}_k$; so, we call $G \in \mathscr{D}\mathscr{O}_k$ *extremal* if $\alpha(G) = n - 2k + 1$. If $G \in \mathscr{D}\mathscr{O}_k$ is extremal, and $v_1$ and $v_2$ are distinct vertices in a maximum independent set $I$, then $s(v_1)+s(v_2) \leq (2k-1)+(2k-1) = 4k - 2$. Since $\mathscr{S}\mathscr{O}(G) \geq 4k - 3$, this means that for some $v \in \{v_1, v_2\}$ we have $s(v) = 2k - 1$ and $I$ is exactly $V(G) - N(v)$. Thus, to check whether $G$ is extremal it is enough to check for every $v \in V(G)$ with $s(v) = 2k - 1$ whether the set $V(G) - N(v)$ is independent. If $I$ is a maximum independent set in an extremal $G \in \mathscr{D}\mathscr{O}_k$, then since $\mathscr{S}\mathscr{O}(G) \geq 4k - 3$,

*at most one vertex in $I$ has nonneighbors in $V(G) - I$, and any such vertex has at most one nonneighbor in $V(G) - I$.* (1)

We call all such maximum independent sets in an extremal graph *big sets*. On the other hand, if $x$ is a common vertex of big sets $I$ and $J$, then $s(x) \leq |G| - |I \cup J| \leq 2k - 1 - |J - I|$. Hence, for every $y \in I - x$, $s(x) + s(y) \leq 4k - 2 - |J - I|$, and so $|J - I| \leq 1$. Furthermore, if $|J - I| = 1$ and there is $x' \in J \cap I - x$, then $s(x) + s(x') \leq 2(n - \alpha(G) - 1) = 4k - 4$, a contradiction. Thus, in this case $\alpha(G) = 2$. This yields the following.

Let $G$ be extremal. If $|G| > 2k + 1$, then every two distinct big sets in $G$ are disjoint. If $|G| = 2k + 1$, sets $I, J \subset V(G)$ are big and $x \in I \cap J$, then $s(x) = 2k - 2$. (2)

## 2.2 Gallai–Edmonds Theorem

We will use the classical Gallai–Edmonds Theorem on the structure of graphs without perfect matchings. Recall that a graph $H$ is *odd* if $|H|$ is odd, and that $o(H)$ denotes the number of odd components of $H$. For a matching $M$ and $uv \in M$, we say that $u$ is the *M-mate* of $v$. For a graph $H$ and $S \subseteq V(H)$, the *deficiency* $\operatorname{def}(S)$ is $o(H - S) - |S|$. Next, $\operatorname{def}(H) := \max\{\operatorname{def}(S) : S \subseteq V(H)\}$. For each graph $H$, $\operatorname{def}(F) \geq 0$, since $\operatorname{def}(\emptyset) = o(H) \geq 0$.

**Theorem 6 (Gallai–Edmonds)** *Let $H$ be a graph and $D$ be the set of $v \in V(H)$ such that there is a maximum matching in $H$ not covering $v$. Let $A$ be the set of the vertices in $V(H) - D$ that have neighbors in $D$, and let $C = V(H) - D - A$. Let $H_1, \ldots, H_k$ be the components of $H[D]$. If $M$ is a maximum matching in $H$, then all of the following hold:*

*(a) $C \cup A \subseteq W(M)$ and the $M$-mates of $A$ are in distinct components of $H[D]$.*
*(b) For each $H_i$ and every $v \in V(H_i)$, $H_i - v$ has a perfect matching.*
*(c) If $\emptyset \neq S \subseteq A$, then $N(S)$ intersects at least $|S| + 1$ components of $H[D]$.*
*(d) $\operatorname{def}(H) = \operatorname{def}(A) = k - |A|$.*

We refer to $(D, A, C)$ as the Gallai–Edmonds decomposition (GE-decomposition) of $H$.

## 2.3 Results for $\mathscr{D}_k$

Since every cycle in a simple graph has at least three vertices, the condition $|G| \geq 3k$ is necessary in Theorem 3. However, it is not necessary for multigraphs, since loops and multiple edges form cycles with fewer than three vertices. Theorem 3 can easily be extended to multigraphs, although the statement is no longer as simple:

**Theorem 7** *For $k \in \mathbb{Z}^+$, let $G$ be a multigraph with $\mathscr{S}(G) \geq 2k$, and set $F = F(G)$, $V_1 = V_1(G)$, and $\alpha' = \alpha'(F)$. Then, $G$ has no $k$ disjoint cycles if and only if*

$$|V(G)| - |V_1| - 2\alpha' < 3(k - |V_1| - \alpha'), \tag{3}$$

*that is, $|V(G)| + 2|V_1| + \alpha' < 3k$.*

*Proof* If (3) holds, then $G$ does not have enough vertices to contain $k$ disjoint cycles. If (3) fails, then we choose $|V_1|$ cycles of length one and $\alpha'$ cycles of length two from $V_1 \cup V(F)$. By Theorem 3, the remaining (simple) graph contains $k - |V_1| - \alpha'$ disjoint cycles. □

Theorem 7 yields the following.

**Corollary 1** *Let $G$ be a multigraph with $\mathscr{S}(G) \geq 2k - 1$ for some integer $k \geq 2$, and set $F = F(G)$, $V_1 = V_1(G)$, and $\alpha' = \alpha'(F)$. Suppose $G$ contains at least one loop. Then, $G$ has no $k$ disjoint cycles if and only if $|V(G)| + 2|V_1| + \alpha' < 3k$.*

Since acyclic graphs are exactly forests, Theorem 5 can be restated as follows:

**Theorem 8** *For $k \in \mathbb{Z}^+$, let $G$ be a simple graph in $\mathscr{D}_k$. Then, $G$ has no $k$ disjoint cycles if and only if one of the following holds:*

($\alpha$) $|G| \leq 3k - 1$;
($\beta$) $k = 1$ *and* $G$ *is a forest with no isolated vertices;*
($\gamma$) $k = 2$ *and* $G$ *is a wheel;*
($\delta$) $\alpha(G) = n - 2k + 1$; *or*
($\epsilon$) $k > 1$ *is odd and* $G = \mathbf{Y}_{k,k,k}$.

Dirac [3] described all 3-connected multigraphs that do not have two disjoint cycles:

**Theorem 9 ([3])** *Let* $G$ *be a 3-connected multigraph. Then,* $G$ *has no two disjoint cycles if and only if one of the following holds:*

*(A)* $\underline{G} = K_4$ *and the strong edges in* $G$ *form either a star (possibly empty) or a 3-cycle;*
*(B)* $G = K_5$;
*(C)* $\underline{G} = K_5 - e$ *and the strong edges in* $G$ *are not incident to the ends of* $e$;
*(D)* $\underline{G}$ *is a wheel, where some spokes could be strong edges; or*
*(E)* $G$ *is obtained from* $K_{3,|G|-3}$ *by adding non-loop edges between the vertices of the (first) 3-class.*

Going further, Lovász [12] described *all* multigraphs with no two disjoint cycles. To state his result, let a *bud* be a vertex incident to at most one edge. Also, let $W_n = K_1 \vee C_n$ be the wheel with $n + 1$ vertices and $\mathbf{W}^+_\mathbf{n}$ be obtained from $W_n$ by replacing each spoke with a strong edge. Similarly, let $\mathbf{K}^+_{\mathbf{3,n-3}}$ be the $n$-vertex multigraph obtained from $K_{3,n-3}$ by adding strong edges connecting all pairs of the vertices of the (first) 3-class. Then, each multigraph described by Theorem 9(A) above is contained either in $\mathbf{W}^+_\mathbf{3}$ or in $\mathbf{K}^+_{\mathbf{3,1}}$.

Lovász [12] observed that any connected multigraph can be transformed into a multigraph with minimum degree at least 3 or a multigraph with exactly one vertex without affecting the maximum number of disjoint cycles in it by using a sequence of operations of the following two types: (1) deleting a bud; (2) replacing a vertex $v$ of degree 2 that has neighbors $x$ and $y$ (where $v \notin \{x, y\}$ but possibly $x = y$) by a new (possibly parallel) edge connecting $x$ and $y$.

He also proved the following:

**Theorem 10 ([12])** *Let* $H$ *be a multigraph with* $\delta(H) \geq 3$. *Then,* $H$ *has no two disjoint cycles if and only if :*

*(L1)* $H = K_5$;
*(L2)* $H \subseteq \mathbf{W}^+_{\mathbf{|H|-1}}$;
*(L3)* $H \subseteq \mathbf{K}^+_{\mathbf{3,|H|-3}}$; *or*
*(L4)* $H$ *is obtained from a forest* $T$ *and vertex* $x$ *with possibly some loops at* $x$ *by adding edges linking* $x$ *to* $T$.

Say that a multigraph $G$ *has the 2-property* if the vertices of degree at most 2 form a clique $Q(G)$ (possibly with some multiple edges). Let $G \in \mathscr{D}\mathscr{O}_2$ with no two disjoint cycles. Then, $G$ has the 2-property. By Lovász's observation above, $G$ can be transformed into a multigraph $H$ that has exactly one vertex or is of type (L1)–(L4) by a sequence of deleting buds and/or contracting edges. Note that if a multigraph $G'$ has the 2-property, then the multigraph obtained from $G'$ by deleting a bud or contracting an edge also has the 2-property. Thus, $H$ and all the intermediate multigraphs have the 2-property. Reversing this transformation, $G$ can be obtained from $H$ by adding buds and subdividing edges. If $H$ has exactly one vertex and at most one edge, then any multigraph with the 2-property that can be obtained from $H$ this way has maximum degree at most 2 or is a path with a single loop at one end. Hence, $G$ is a $K_i$ for $i \leq 3$, is a path on at most 3 vertices with a loop at an endpoint, or forms a strong edge. If $\delta(H) \geq 3$, then the clique $Q := Q(G)$ cannot have more than 2 vertices: by the definition of $Q(G)$, $|Q| \leq 3$, and if $|Q| = 3$, then $Q$ induces a $K_3$-component of $G$ and $\delta(G-Q) \geq 3$; thus, $G-Q$ has another cycle. Let $Q' := V(G) - V(H)$. By above, $Q \subseteq Q'$. If $Q' \neq Q$, then $Q$ consists of a single leaf in $G$ with a neighbor of degree 3, so $G$ is obtained from $H$ by subdividing an edge and adding a leaf to the vertex of degree 2. If $Q' = Q$, then $Q$ is a component of $G$, or $G = H + Q + e$ for some edge $e \in E(H, Q)$, or at least one vertex of $Q$ subdivides an edge $e \in E(H)$. In the last case, when $|Q| = 2$, $e$ is subdivided twice by $Q$.

In case (L4), because $\delta(H) \geq 3$, either $T$ has at least two buds, each linked to $x$ by multiple edges, or $T$ has one bud linked to $x$ by an edge of multiplicity at least 3. So, this case cannot arise from $G$. Also, $\delta(H) = 3$, unless $H = K_5$, in which case $\delta(H) = 4$. So, $Q$ is not an isolated vertex, lest deleting $Q$ leave $H$ with $\delta(H) \geq 5 > 4$; and if $Q$ has a vertex of degree 1, then $H = K_5$. Else all vertices of $Q$ have degree 2, and $Q$ consists of the subdivision vertices of one edge of $H$. This yields the following characterization of multigraphs in $G \in \mathscr{D}\mathscr{O}_2$ with no two disjoint cycles.

Set $Z_t = \{z_1, \ldots, z_t\}$, and define $\mathbf{S_3} = K(Z_5) \cup z_1xy$, $\mathbf{S_4} = \mathrm{sd}(K(Z_5), z_1z_2, x) \cup xy$, and $\mathbf{S_5} = \mathrm{sd}(K(Z_5), z_1z_2, xy)$ (see Figure 2).

**Corollary 2** *All $G \in \mathscr{D}\mathscr{O}_2$ with $|G| \geq 4$ and no 2 disjoint cycles satisfy one of:*

*(Y1)* $G \subseteq \mathbf{S_3}$*;*
*(Y2)* $G \subseteq \mathbf{S_4}$*;*
*(Y3)* $G = \mathbf{S_5}$*;*
*(Y4)* $G \in \{H, \mathrm{sd}(H, e, x), \mathrm{sd}(H, e, xy)\}$*, where* $W_{|H|-1} \subseteq H \subseteq \mathbf{W}^+_{|\mathbf{H}|-\mathbf{1}}$*;*

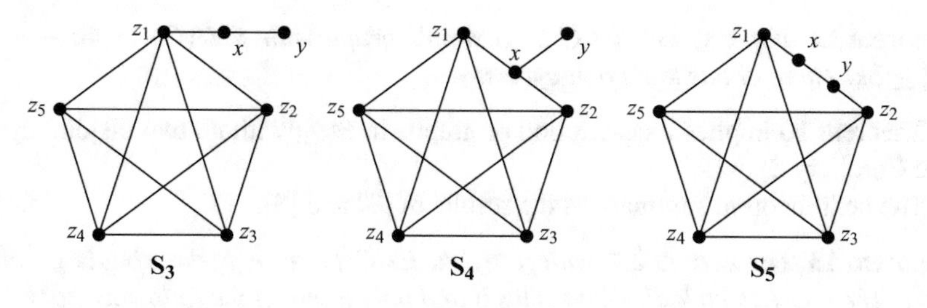


**Fig. 2** Graphs $\mathbf{S_3}$, $\mathbf{S_4}$, and $\mathbf{S_5}$

*(Y5)* $G \in \{H, \mathrm{sd}(H, e, x), \mathrm{sd}(H, e, xy)\}$, *where* $H \subseteq \mathbf{K}^+_{3,|H|-3}$ *and* $H$ *contains* $K_{3,|H|-3}$ *minus an edge.*

By Corollary 1, in order to describe the multigraphs in $\mathscr{D}_k$ not containing $k$ disjoint cycles, it is enough to describe such multigraphs with no loops. Recently, Kierstead, Kostochka, and Yeager [10] proved the following:

**Theorem 11 ([10])** *Let* $k \geq 2$ *and* $n \geq k$ *be integers. Let* $G$ *be an* $n$*-vertex graph in* $\mathscr{D}_k$ *with no loops. Set* $F = F(G)$, $\alpha' = \alpha'(F)$, *and* $k' = k - \alpha'$. *Then,* $G$ *does not contain* $k$ *disjoint cycles if and only if one of the following holds:*

*(a)* $n + \alpha' < 3k$;
*(b)* $|F| = 2\alpha'$ *(i.e.,* $F$ *has a perfect matching) and either*

  *(i)* $k'$ *is odd and* $G - F = \mathbf{Y}_{k',k',k'}$, *or*
  *(ii)* $k' = 2 < k$ *and* $G - F = W_5$;

*(c)* $G$ *is extremal and either*

  *(i) some big set is not incident to any strong edge, or*
  *(ii) for some two distinct big sets* $I_j$ *and* $I_{j'}$, *all strong edges intersecting* $I_j \cup I_{j'}$ *have a common vertex outside of* $I_j \cup I_{j'}$ *and if* $v \in I_j \cap I_{j'}$ *(this may happen only if* $k' = 2$*), then* $v$ *is not incident with a strong edge;*

*(d)* $n = 2\alpha' + 3k'$, $k'$ *is odd, and there is* $S = \{v_0, \ldots, v_s\} \subseteq V(F)$ *such that* $F[S]$ *is a star with center* $v_0$, $F - S$ *has a perfect matching and either*

  *(i)* $G - (F - S + v_0) = \mathbf{Y}_{k'+1,k',k'}$, *or*
  *(ii)* $s = 2$, $v_1v_2 \in E(G)$, $G - F = \mathbf{Y}_{k'-1,k',k'}$ *and* $G$ *has no edges between* $\{v_1, v_2\}$ *and the set* $X_0$ *in* $G - F$;

*(e)* $k = 2$ *and* $W_{n-1} \subseteq G \subseteq \mathbf{W}^+_{\mathbf{n-1}}$;
*(f)* $k' = 2$, $|F| = 2\alpha' + 1 = n - 5$, *and* $G - F = C_5$.

## 2.4 *Results for* $\mathscr{DO}_k$

Theorem 4 can be restated as follows.

**Theorem 12** *For* $k \in \mathbb{Z}^+$, *let* $G$ *be a simple graph with* $\mathscr{SO}(G) \geq 4k - 1$ *and* $|G| \geq 3k$. *Then,* $G$ *has* $k$ *disjoint cycles.*

Theorem 10 implies a description of graphs in $\mathscr{DO}_2$ with no two disjoint cycles (see Corollary 2).

The next theorem summarizes the results of [9] and [8].

**Theorem 13** *For* $k, n \in \mathbb{Z}^+$ *with* $n \geq 3k$, *let* $G$ *be an* $n$*-vertex simple graph in* $\mathscr{DO}_k$. *Then,* $G$ *has no* $k$ *disjoint cycles if and only if one of the following holds:*

*(S1)* $k = 1$ *and* $G$ *is a forest with at most one isolated vertex;*

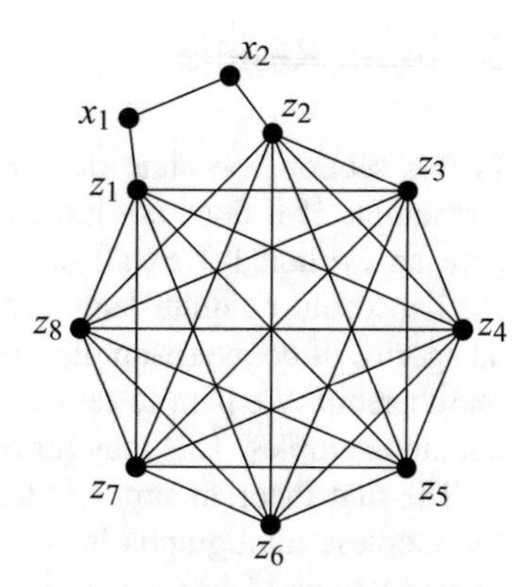

**Fig. 3** Graph $\mathbf{F_1}$

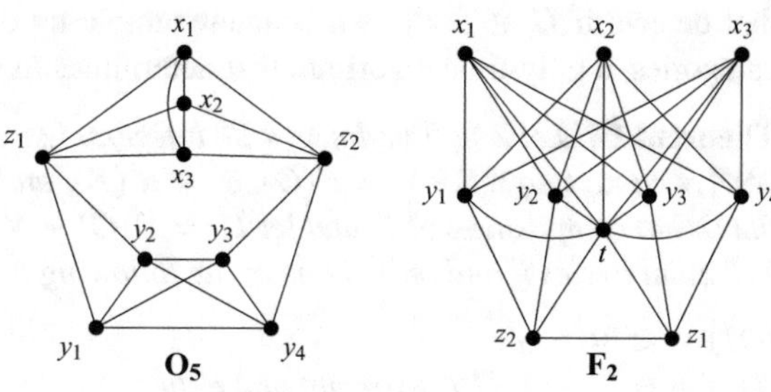

**Fig. 4** Graphs $\mathbf{O_5}$ and $\mathbf{F_2}$

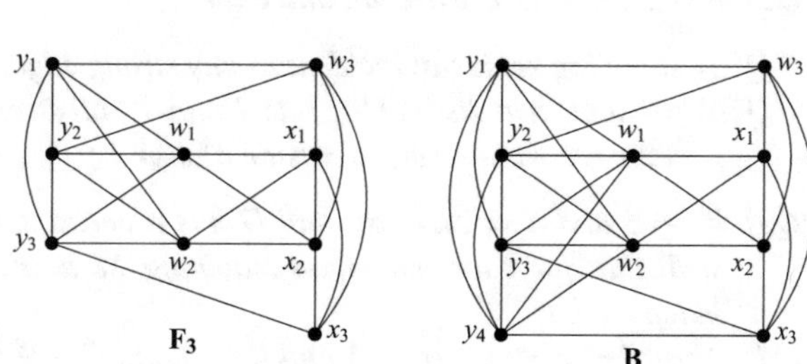

**Fig. 5** Graphs $\mathbf{F_3}$ and $\mathbf{B}$

*(S2)* $k = 2$ *and* $G$ *satisfies the conditions of Corollary 2;*
*(S3)* $\alpha(G) = n - 2k + 1$*;*
*(S4)* $k = 3$ *and* $G = \mathbf{F}_1$ *(see Figure 3);*
*(S5)* $k = 3$ *and* $G = \mathbf{F}_2 = \{t\} \vee \overline{O}_5$ *where* $\mathbf{O}_5$ *is the 5-chromatic graph in Figure 4;*
*(S6)* $k = 3$ *and* $G$ *is the graph* $\mathbf{F}_3$ *in Figure 5;*
*(S7)* $k \geq 3, n = 3k, \alpha(G) \leq k$*, and* $\chi(\overline{G}) > k$*;*
*(S8)* $k \geq 3, n = 3k$*, and* $G \subseteq \mathbf{Y}_{k,c,2k-c}$ *for some odd* $1 \leq c \leq 2k - 1$*;*
*(S9)* $k \geq 3, n = 3k$*, and* $G = \mathbf{Y}_{k-1,1,2k}$*.*

*Remark 1 The result of Rabern [14] (see also [7, 11]) implies that if (S7) holds, then* $k \leq 4$*.*

## 3 Main Results

In this section, we state our main results. We call a graph $G \in \mathcal{DO}_k$ a counterexample if it does not have $k$ disjoint cycles. We believe that it is possible to give an explicit list of all counterexamples in the style of previous results, but the list would be quite long and complicated. Here, we are content to give broad categories of counterexamples together with a poly-time algorithm that determines membership. We plan to revisit this problem again and give a more explicit list of counterexamples. This may (or may not) lead to a faster algorithm.

The first theorem supports the algorithmic problem by proving that for $k \geq 5$ the loopless multigraphs in $\mathcal{DO}_k$ are counterexamples if and only if they belong to at least one of five categories. The second theorem gives a poly-time algorithm that detects if $G \in \mathcal{DO}_k$ is a counterexample by describing, for each of the five categories, a poly-time algorithm that determines membership.

**Theorem 14** *Let $k \geq 5$ and $n \geq k$ be integers. Let $G$ be an $n$-vertex multigraph in $\mathcal{DO}_k$ with no loops. Set $F = F(G)$, $\alpha' = \alpha'(F)$, and $k' = k - \alpha'$. Let $(D, A, C)$ be the GE-decomposition of $F$ and let $D' = V(G) - V(F)$. Then, $G$ does not contain $k$ disjoint cycles if and only if one of the following holds:*

*(Q1)* $n < 3k - \alpha'$;
*(Q2)* $n > 2k + 1$, *$G$ is extremal and either*

- *(Q2a) some big set is not incident to any strong edge, or*
- *(Q2b) for some two distinct big sets $J$ and $J'$, all strong edges intersecting $J \cup J'$ have a common vertex outside of $J \cup J'$;*

*(Q3)* $k' \geq 5$ *and* $n = 3k - \alpha'$, *and $G$ has a vertex $x \in D'$ of degree $k + \alpha' - 1$ such that for each maximum matching $M$ in $F$, the set $N(x) - W(M)$ is independent;*
*(Q4)* $3k - \alpha' \leq n \leq 3k - \alpha' + 1$ *and* $k' \leq 4$ *and* $\underline{G} - W(M)$ *has no* $k - |M|$ *disjoint cycles for all (possibly nonmaximum) matchings $M$ in $F$; or*
*(Q5)* $k' \geq 5$ *and* $n = 3k - \alpha'$, $|F| - 2\alpha' \in \{0, |D| - 2, |D| - 1\}$ *and for all maximum matchings $M$ in $F$ either* $\alpha(G'(M)) = k' + 1$ *or* $G'(M) \subseteq \mathbf{Y}_{k',c,2k'-c}$ *for some odd* $c \leq k'$.

**Theorem 15** *There is a polynomial time algorithm that for every multigraph $G \in \mathcal{DO}_k$ decides whether $G$ has $k$ disjoint cycles or not.*

## 4 Proof of Theorem 14: Sufficiency

We will prove that if $G$ contains a set $\mathcal{C} = \{C_1, \ldots, C_k\}$ of $k$ disjoint cycles, then all of the conditions (Q1)–(Q5) fail. Given such $\mathcal{C}$, let $M \subseteq \mathcal{C}$ be the set of cycles in $\mathcal{C}$ that are strong edges, $m = |M|$ and $\mathcal{C}' = \mathcal{C} - M$. Since $m \leq \alpha'$ and each cycle that is not a strong edge has at least 3 vertices, $n \geq 2m + 3(k - m) = 3k - m \geq 3k - \alpha'$; so (Q1) does not hold.

If (Q2) holds, then $G$ is extremal. Every big set $J$ satisfies $|V(G) - J| < 2k$. So some cycle $C_J \in \mathscr{C}$ has at most one vertex in $V(G) - J$. Since $J$ is independent, $C_J$ has at most one vertex in $J$. Thus $C_J$ is a strong edge and (Q2a) fails. Suppose there are big sets $J$ and $J'$ satisfying (Q2b). Then, $|W(M) \cap (J \cup J')| \leq 1$, and so for some $I \in \{J, J'\}$, $I \subseteq V - W(M)$. By this fact and independence, each cycle in $\mathscr{C}$ has at least two vertices outside of $I$, and so $|I| \leq n - 2k$ contradicting the definition of a big set. So (Q2b) also fails.

Note that the $k - |M|$ cycles in $\mathscr{C}'$ correspond to disjoint cycles in $\underline{G} - W(M)$, so (Q4) does not hold.

If $n = 3k - \alpha'$, then $m = \alpha'$, every cycle in $\mathscr{C}'$ is a triangle and every vertex in $G - W(M)$ belongs to exactly one triangle in $\mathscr{C}'$. Therefore, (Q3) and (Q5) do not hold.

## 5 Proof of Theorem 14: Necessity

Suppose $G$ does not have $k$ disjoint cycles and that none of the conditions (Q1)–(Q5) hold. Because (Q4) does not hold, either $n \geq 3k - \alpha' + 2 = 2k + k' + 2$, or $k' \geq 5$; the latter implies that $n \geq 3k - \alpha' = 2k + k' \geq 2k + 5$. Therefore,

$$n \geq 2k + 3. \tag{4}$$

Among the maximum matchings in $F$, choose a matching $M$ such that

$$\alpha(G - W) \text{ is minimum, where } W = W(M). \tag{5}$$

Then, $|M| = \alpha'$, $G' = G - W$ is simple, and $\mathscr{SO}(G') \geq 4k - 3 - 4\alpha' = 4k' - 3$. So $G' \in \mathscr{DO}_{k'}$. Let $n' := |V(G')| = n - 2\alpha'$. Since (Q1) does not hold,

$$n' \geq 3k'. \tag{6}$$

If $n' = 3k'$, then $G'$ is quite dense, so sometimes it will be convenient to consider the complement of $\underline{G}$. For $v \in V(G)$, let $\overline{N}[v] = V(G) - N[v]$ and $\overline{s}(v) = |\overline{N}[v]| = n - 1 - s(v)$. When $n' = 3k'$, we have $n = 2k + k'$ and thus the inequality $s(v) + s(u) \geq 4k - 3$ can be written as

$$\overline{s}(v) + \overline{s}(u) \leq 2k' + 1 \quad \text{for all } vu \notin E(G). \tag{7}$$

Since $G'$ has no $k'$ disjoint cycles and (6) holds, one of (S1)–(S9) in Theorem 13 holds for $G'$ with $k'$ in place of $k$. We will now show that each of (S1)–(S9) will lead to one of (Q1)–(Q5) holding or some other contradiction.

*Case 1* (S4), (S5), or (S6) **hold for** $G'$.
Then, $k' \le 3$ and $3k' \le |G'| \le 3k' + 1$, so (Q4) holds, because $G$ has no $k$ disjoint cycles.

*Case 2* (S3) **holds for** $G'$.
Then, $n > 2k + 1$ and $G'$ is extremal. Let $J$ be a big set in $G'$. Then, $|J| = n' - 2k' + 1 = n - 2k + 1$. So, $G$ is extremal and $J$ is a big set in $G$. Since (Q2a) fails, some $w \in J$ has a strong neighbor $v$. Let $vu$ be the edge in $M$ containing $v$. In $F$, consider the maximum matching $M' = M - vu + wv$, and set $G'' = G - W(M')$. By (5), $G''$ contains a big set $J'$, and $J'$ is big in $G$. Since $w \notin J'$, $J' \neq J$. So by (2), $J' \cap J = \emptyset$ (possibly, $u \in J'$). Since (Q2a) fails, some $w' \in J'$ has a strong neighbor $v'$. Possibly, $v' = v$, but then since (Q2b) fails, some $w'' \in J \cup J'$ has a strong neighbor $v'' \neq v$. Thus, we can choose notation so that $v' \neq v$. As $M'$ is maximum, there is an edge $v'u' \in M'$. Set $M'' = M' + w'v' - v'u'$ and $G^* := G - W(M'')$. Again by (5), $G^*$ contains a big set $J''$. Since $w, w' \notin J''$, we have $J'' \notin \{J, J'\}$. So by (2), $J'' \cap (J \cup J') = \emptyset$. Thus, since $V(G^*) \supseteq (J - w) \cup (J' - w') \cup J''$,

$$n' \ge 3|J| - 2 = 3(n' - 2k' + 1) - 2 = 3n' - 6k' + 1,$$

which yields $2n' \le 6k' - 1$, a contradiction to (6). Hence, (Q2) holds.

*Case 3* (S7) **holds for** $G'$.
So $k' \ge 3$, $|G'| = 3k'$, $\alpha(G') \le k'$, and $\chi(\overline{G'}) > k'$. Since $|G'| = 3k'$, (7) must hold. Since $\chi(\overline{G'}) > k'$, $G'$ contains an induced subgraph $G_0$ such that $\overline{G_0}$ is a vertex-$(k' + 1)$-critical graph. By (7),

$$\text{for every } xy \in E(\overline{G_0}), \text{ the sum of the degrees of } x \text{ and } y \text{ in } \overline{G_0} \text{ is at most } 2k' + 1. \tag{8}$$

The $(k' + 1)$-critical graphs satisfying (8) were studied recently. If $k' \ge 5$, then by results in [7] and [14], $\overline{G_0} = K_{k'+1}$, which means $\alpha(G') \ge k' + 1$, a contradiction to the case. If $k' \le 4$, then (Q4) holds.

*Case 4* (S1) **holds for** $G'$.
So $k' = 1$ and $G'$ is a forest with at most one isolated vertex. Since $k \ge 5$, $|M| \ge 4$. Let $xz, x'z', x''z''$ be three strong edges in $M$.

*Case 4.1:* $G'$ has at least two non-singleton components, say $H_1$ and $H_2$. Then, $n' \ge 4$. For $i = 1, 2$, let $P_i$ be a longest path in $H_i$, and let $u_i$ and $w_i$ be the ends of $P_i$. As $\mathscr{SO}(G) \ge 4k - 3$, at most two edges between $W$ and $\{u_1, u_2, w_1, w_2\}$ are missing in $G$. So, we may assume that at most one edge between $\{x, z\}$ and $\{u_1, u_2, w_1, w_2\}$ is missing in $G$. By symmetry, we assume that among these edges only $xu_1$ could be missing in $G$. Then, the $\alpha' - 1$ strong edges of $M - xz$ and the cycles $xu_2w_2x$ and $zu_1w_1z$ form $k$ disjoint cycles in $G$, a contradiction.

*Case 4.2:* $G'$ has a unique non-singleton component $H$, and this $H$ is not a star. Let $P = y_1 \dots y_t$ be a longest path in $H$. Since $H$ is not a star, $t \ge 4$. Then, $y_1$ is a leaf in $G'$, and either $d_{G'}(y_2) = 2$ or $y_2$ is adjacent to a leaf $l \neq y_1$. Let $y_1' = y_2$

if $d_H(y_2) = 2$ and $y_1' = l$ otherwise. Similarly, either $d_{G'}(y_{t-1}) = 2$ or $y_{t-1}$ is adjacent to a leaf $l' \neq y_t$. Let $y_t' = y_{t-1}$ if $d_H(y_{t-1}) = 2$ and $y_t' = l'$ otherwise. Since $y_1 y_t', y_1' y_t \notin E(G)$ and $G \in \mathcal{DO}_k$,

*the number of missing edges between* $\{y_1, y_1', y_t, y_t'\}$ *and* $W$ *in* $G$ *is at most* $q + r$, *where* $q = |\{y_1', y_t'\} \cap \{y_2, y_{t-1}\}|$ *and* $r$ *is the number of low vertices in* $\{y_1, y_1', y_t, y_t'\}$. (9)

Since $q \leq 2$, $r \leq 2$, and $|M| \geq 3$, we can assume that at most one edge between $\{x, z\}$ and $\{y_1, y_1', y_t, y_t'\}$ is missing in $G$. So, we get a contradiction as at the end of Case 4.1.

*Case 4.3:* The unique non-singleton component $H$ of $G'$ is a star. The leaves of the star form an independent set of size $n' - 1 = n' - 2k' + 1$. By (4), we are in Case 2.

*Remark* The proof of the next case works even if (5) does not hold, and we will use this in Case 6.

*Case 5* (S9) **holds for** $G'$.
So $n' = 3k'$ and $G' \subseteq \mathbf{Y}_{k'-1,1,2k'}(Y, \{x\}, Z)$. If $k' \leq 4$, then (Q4) holds. So, below we assume

$$k' \geq 5. \tag{10}$$

Since $n' = 3k'$, we will often use (7). Since each $y \in Y$ has $k' - 2$ nonneighbors in $Y$, (7) yields

$$|\overline{N}[y] - Y| + |\overline{N}[y'] - Y| \leq 5 \quad \text{for all distinct } y, y' \in Y. \tag{11}$$

Since $x$ is not adjacent to any of the $2k'$ vertices in $Z$, by (7)

$$N(x) = V(G) - Z - x \text{ and } N(z) = V(G) - x - z \text{ for each } z \in Z. \tag{12}$$

If $x$ has a strong neighbor $v_0$ with the $M$-mate $u_0$, then we construct $k$ disjoint cycles in $G$ as follows. First, take the $\alpha'$ strong edges in $M - v_0u_0 + v_0x$. By (12), $G[Z] = K_{2k'}$ and each $y \in Y + u_0$ is adjacent to all of $Z$. So, we take $k'$ 3-cycles each of which contains one vertex in $Y+u_0$ and two vertices in $Z$. This contradiction shows that $x \in D'$.

Since $x \in D'$ and $d(x) = k+\alpha'-1$, if (Q3) does not hold, then $F$ has a maximum matching $M'$ such that

$$\text{there are } u_1, u_2 \in V(G) - W(M') - Z \text{ with } u_1u_2 \in E(G). \tag{13}$$

For $i = 1, 2$, the symmetric difference $M \triangle M'$ contains a path $P_i$ of an even length an end of which is $u_i$. Since the other end $w_i$ of $P_i$ is not covered by $M$, $w_i \in V(G') \cap D$. Also by definition, none of the vertices in $G'$ is an internal vertex in

$P_i$. In particular, $x \notin V(P_i)$. Let $M''$ be the maximum matching in $F$ such that $M \triangle M'' = P_1 \cup P_2$. Then, $V(G) - W(M'') = V(G') - \{w_1, w_2\} \cup \{u_1, u_2\}$. If $|\{w_1, w_2\} \cap Z| = \ell_Z$ and $|\{w_1, w_2\} \cap Y| = \ell_Y$, then we can renumber the vertices in $Z - \{w_1, w_2\}$ and $Y - \{w_1, w_2\}$ as $z_1, \ldots, z_{2k'-\ell_Z}, y_1, \ldots, y_{k'-1-\ell_Y}$ and construct $k$ disjoint cycles in $G$ as follows. Take the $k - k'$ strong edges in $M''$, then take the cycle $xu_1u_2x$ and for $j = 1, \ldots, k' - 1 - \ell_Y$ take the cycle $(y_j, z_{2j-1}, z_{2j})$. Finally, if $\ell_Y \geq 1$, then $|Z - \{z_1, \ldots, z_{2(k'-1-\ell_Y)}, w_1, w_2\}| = 3\ell_Y$, then we simply take $\ell_Y$ triangles in the remaining complete graph $G[Z - \{z_1, \ldots, z_{2(k'-1-\ell_Y)}, w_1, w_2\}]$. Hence, (Q3) holds.

*Case 6* (S8) **holds for** $G'$.
So $n' = 3k'$ and $G' \subseteq \mathbf{Y}_{k',c,2k'-c}(Y, X, Z)$ for $k' \geq 3$ and some odd $1 \leq c \leq k'$. If $k' \leq 4$, then (Q4) holds. So, as in the previous case, we assume $k' \geq 5$.

Let $M'$ be an arbitrary maximum matching in $F$. Since $G$ has no $k$ disjoint cycles, $G'(M')$ does not have $k'$ disjoint triangles. Therefore, by (6), (10), and Theorem 13 (with Remark 1), we have that one of (S3), (S8), or (S9) holds in $G'(M')$. By the remark before Case 5,

$$\textit{if (S9) holds in } G'(M'), \textit{ then } (Q3) \textit{ holds.} \tag{14}$$

If (S3) holds in $G'(M')$, then $\alpha(G'(M')) = n' - 2k' + 1 = k' + 1$. Therefore, if we assume (Q3) does not hold, to show that (Q5) holds, we only need to show that $|F| - 2\alpha' \in \{0, |D| - 2, |D| - 1\}$ which is true when $|W| \leq 2 + |A| + |C|$.

Since $n' = 3k'$, we will often use (7). Since each $y \in Y$ has $k' - 1$ nonneighbors in $Y$, (7) yields

$$|\overline{N}[y] - Y| + |\overline{N}[y'] - Y| \leq 3 \quad \text{for all } y, y' \in Y. \tag{15}$$

By (15),

$$\text{there is } y_0 \in Y \text{ such that } |\overline{N}[y] - Y| \leq 1 \text{ for every } y \in Y - y_0. \tag{16}$$

Since each $x \in X$ has $2k' - c$ nonneighbors in $Z$, if $x$ has a nonneighbor $y \in Y$, then by (7),

$$2k' + 1 \geq \overline{s}(x) + \overline{s}(y) \geq (2k' - c + 1) + (k' - 1 + 1) = 3k' - c + 1,$$

which yields $c = k'$. Moreover, if in this case some $z \in Z$ also has a nonneighbor $y' \in Y$, then again by (7), $2k' + 1 \geq \overline{s}(x) + \overline{s}(z) \geq (k' + 1) + (k' + 1) = 2k' + 2$, a contradiction. Thus, we may assume (by possibly switching the roles of $X$ and $Z$ when $c = k'$) that

$$|\overline{N}[x] \cap W| \leq 1 \text{ and } \overline{N}[x] \cap \overline{W} = Z \text{for each } x \in X, \tag{17}$$

and

$$|\overline{N}[z] - X| \leq 1 \text{ for each } z \in Z, \text{ and if } c = k', \text{ then } G[Z] = K_c. \tag{18}$$

We will need the following observation.

**Lemma 1** *Let $t \geq 2$ and $\epsilon \in \{0, 1\}$. Let $H$ be a graph with a partition $V(H) = R \cup Q$ such that $|R| = 2t + \epsilon$, $|Q| = 3t - |R| = t - \epsilon$, and let $y_0 \in Q$. If*

1. *each $u \in R$ has at most one nonneighbor in $H$ and*
2. *each $y \in Q - y_0$ has at most $1 + \epsilon$ nonneighbors in $R$ and*
3. *$y_0$ has at most two nonneighbors in $R$ and has only $1 + \epsilon$ nonneighbors if $t = 2$,*

*then $H$ contains $t$ vertex-disjoint triangles.*

*Proof* Using induction, note that the lemma holds for $t = 2$. If $t \geq 3$, then $H$ has a triangle $T = y_0 z_1 z_2 y_0$ with $z_1, z_2 \in R$. By induction, $H' := H - T$ has $t - 1$ disjoint triangles. □

*Claim 1* Let $G' \subseteq \mathbf{Y}_{k',c,2k'-c}(Y, X, Z)$ for $k' \geq 4$ and an odd $c \leq k'$. Suppose there are $w \in V(G')$ and $u \in W$ such that $F$ has an $M$-alternating $u, w$-path $P$.

(A) If $w \in Y \cup Z$, then $u$ has no neighbor in $Y - w$ or no neighbor in $X$.
(B) If $w \in X$, then $u$ has no neighbor in $Y$ or no neighbor in $Z$.

*Proof* Let $M'$ be the matching obtained from $M$ by switching edges on $P$. Then, $W(M') = W(M) - w + u$. Set $t = (2k' - c - 1)/2$. Since $1 \leq c \leq k'$ and is odd, by (10),

$$|Z| = 2k' - c \geq 5 \text{ and } k' - 1 \geq t \geq 2. \tag{19}$$

Arguing by contradiction, we assume that the lemma fails and construct $k$ disjoint cycles.

*Case 1* $w \in Y \cup Z$. Since (A) does not hold, $u$ has neighbors $x \in X$ and $y \in Y - w$.

Pick $y \in N(u) \cap Y - w$ with $s(y)$ minimum. Then, for $y_0$ defined in (16), we have

$$\text{if } y_0 \in Y - w - y, \text{ then } y_0 u \notin E(G), \text{ and so by (15), } |\overline{N}[y_0] \cap Z| \leq 2. \tag{20}$$

By (17), $T := uxyu \subseteq G$. Set $\epsilon := 0$ if $w \in Z$; else $\epsilon := 1$. Partition $Y - y - w$ as $\{Q, \overline{Q}\}$ so that $|Q| = t - \epsilon$, $|\overline{Q}| = \frac{c-1}{2}$, and $y_0 \in \overline{Q} \cup \{w, y\}$ if $c > 1$. So $t \geq 3$, if $y_0 \in Q$. Regardless, by (16), (18), and (20), $Q$ and $R := Z - w$ satisfy the conditions of Lemma 1. Thus, $Q \cup R$ contains $t$ disjoint triangles. By (17), $(X - x) \cup \overline{Q}$ contains $\frac{c-1}{2}$ disjoint triangles. Counting these $k' - 1$ triangles, $T$, and $k - k'$ strong edges of $M'$ gives $k$ disjoint cycles.

*Case 2* $w \in X$. Since (B) fails, there are $z \in N(u) \cap Z$ and $y \in N(u) \cap Y$. Our first goal is to show that there is an edge with ends in $N(u) \cap Y$ and $N(u) \cap Z$. If

$N(u) \cap N(z) \cap Y \neq \emptyset$, then we are done. Else, by (18), $N(z) \cap Y = Y - y = \overline{N}[u] \cap Y$. Let $y' \in Y - y$. By (15) applied to $y$ and $y'$, $|\overline{N}[y] \cap Z| \leq 2$. By (7) applied to $u$ and $y'$, $|\overline{N}[u] \cap Z| \leq 2$. By (19), $|Z| \geq 5$, so there is $z' \in Z \cap N(u) \cap N(y)$, and we are done.

Pick $yz \in E$ with $y \in N(u) \cap Y$ and $z \in N(u) \cap Z$ so that $s(y)$ is minimum and let $T := uzyu$. Then, for $y_0$ defined in (16), using (15),

$$\text{if } y_0 \in Y - y \text{ then } |\overline{N}[y_0] \cap (Z - z)| \leq 2, \tag{21}$$

since $y_0u \notin E(G)$ or $y_0z \notin E(G)$.

Partition $Y - y$ as $\{Q, \overline{Q}\}$ so that $|Q| = t$, $|\overline{Q}| = \frac{c-1}{2}$, and $y_0 \in \overline{Q} + y$ if $c > 1$. So $t \geq 3$, if $y_0 \in Q$. Regardless, by (16), (18), and (21), $Q$ and $R := Z - z$ satisfy the conditions of Lemma 1. Thus, $Q \cup R$ contains $t$ disjoint triangles. By (17), $(X - w) \cup \overline{Q}$ contains $\frac{c-1}{2}$ disjoint triangles. Counting these $k' - 1$ triangles, $T$, and $k - k'$ strong edges of $M'$ gives $k$ disjoint cycles. □

*Claim 2* Let $G' \subseteq \mathbf{Y}_{k',c,2k'-c}(Y, X, Z)$ for $k' \geq 4$ and an odd $c \leq k'$. Then, $|D \cap W| \leq 2$.

*Proof* Suppose $u \in D \cap W$. Then, there is a matching $M'$ and vertex $w_u \in V(G')$ such that $W(M') = W(M) + w_u - u$ and there is an $M, M'$-alternating path from $u$ to $w_u$. By Claim 1, $u$ has no neighbors in $Y - w_u$ or in $X$ or in $Z$.

By degree condition (7), there is at most one $u \in D \cap W$ with no neighbor in $X$ or no neighbor in $Z$: otherwise, for any $x \in X$ and $z \in Z$, we have the contradiction

$$\|\{x, z\}, W\| \leq 4\alpha' - 2 \text{ and so } s(x) + s(z) \leq 4k' - 2 + 4\alpha' - 2 \leq 4k - 4.$$

Similarly, there is at most one $u \in D \cap W$ with at most one neighbor in $Y$: otherwise, as $k' \geq 4$, there are two $y, y' \in Y$ with

$$\left\|\{y, y'\}, W\right\| \leq 4\alpha' - 4 \text{ and so } s(y) + s(y') \leq 4k' + 4\alpha' - 4 \leq 4k - 4.$$

Thus, $|D \cap W| \leq 2$. □

Lemma 2 yields that $|W| \leq 2 + |A| + |C|$. Thus, (Q5) holds.

*Case 7* (S2) **holds for** $G'$.
So, $n' \geq 3k'$ and $k' = 2$ and $G'$ satisfies one of (Y1)–(Y5) from Corollary 2. If $n' \leq 7$, then (Q4) holds, so assume $n' \geq 8$. This implies that $G'$ satisfies either (Y4) or (Y5). As $k \geq 5$, $|M| = \alpha' = k - k' \geq 3$.

Define a vertex $v \in \overline{W}$ to be *$i$-acceptable* if $|N(v) \cap W| \geq 2\alpha' - i$, *acceptable* if it is 1-acceptable, and *good* if it is 0-acceptable. Let $u, v \in \overline{W}$ with $uv \notin E$. If $i$ and $j$ are minimum natural numbers such that $u$ is $i$-acceptable and $v$ is $j$-acceptable, then

$$i + j \leq d_{G'}(u) + d_{G'}(v) - 5. \tag{22}$$

*Case 7.1:* $G'$ satisfies (Y4), i.e., $G' \in \{H, \text{sd}(H, e, x), \text{sd}(H, e, xy)\}$, where $W_{|H|-1} \subseteq H \subseteq \mathbf{W}^+_{|\mathbf{H}|-\mathbf{1}}$. Set $t = |H| - 1$. Let $H$ have center $v_0$ and rim $v_1 \dots v_t v_1$, and let $\mathbf{W}'_\mathbf{t}$ be the result of adding a parallel edge between $v_0$ and $v_1$ in $W_t$. Since $G'$ is simple, we may assume $H \in \{W_t, \mathbf{W}'_\mathbf{t}\}$. If $G' \neq H$ then we may assume that the subdivided edge $e$ is incident to $v_1$. As $n' \geq 8, t \geq 5$.

*Case 7.1.1:* $t = 5$. The subdividing vertex $x$ exists. By (22), the subdividing vertices and $v_3, v_4, v_5$ are all good, $v_2$ is acceptable, and $v_1$ is 2-acceptable. As $|M| \geq 3$, there is an edge $ab \in M$ with $av_1, bv_2 \in E$. Then, there are $k$ disjoint cycles $v_0v_4v_5v_0$, $av_1xa$, $bv_2v_3b$, and $|M - ab|$ strong edges, contradicting $G \in \mathscr{BO}_k$.

*Case 7.1.2:* $t \geq 6$. By (22), the rim vertices $v_3, v_4, v_5, v_6$ are all acceptable. As $|M| \geq 3$, there is an edge $ab \in M$ such that $av_3v_4a$ and $bv_5v_6b$ are cycles. Let $C$ be the smallest cycle containing $v_0, v_1, v_2$ (and any subdividing vertices). Then, there are $k$ disjoint cycles $C, av_3v_4a, bv_5v_6b$ and $\alpha' - 1$ strong edges, contradicting $G \in \mathscr{BO}_k$.

*Case 7.2:* $G'$ satisfies (Y5), i.e., $G' \in \{H, \text{sd}(H, e, x), \text{sd}(H, e, xy)\}$, where

$$K_{3,|H|-3}(Y, Z_t) - e' \subseteq H \subseteq \mathbf{K}^+_{\mathbf{3},|\mathbf{H}|-\mathbf{3}}(Y, Z_t) \text{ where } Y = \{y_1, y_2, y_3\},$$
$$Z_t = \{z_1, \cdots, z_t\}.$$

As $n' \geq 8, t \geq 3$. If $\alpha(G') \geq n' - 2k' + 1$, then (Q2) holds and so (S3) holds which falls under Case 2. So, assume that the subdividing vertex $x$ exists in $G'$.

*Case 7.2.1:* $e = y_hy_i$, where $\{h, i, j\} = [3]$. Since $\alpha(G') \leq n' - 2k'$ and $Z + x$ is independent, $e$ is subdivided twice. As $d_{G'}(x) = 2$, every vertex of $Z$ is adjacent to every vertex of $Y$ (and no other vertex of $G'$). Thus, $G' = \text{sd}(H, e, xy)$ and the vertices of $Z + x + y$ are all good.

Suppose $t = 3$. Then, $d_{G'}(y_j) \leq 5$. By (22), $y_j$ is 2-acceptable. As $|M| \geq 3$, there is an edge $ab \in M$ with $ay_j \in E$. Thus, there are $k$ disjoint cycles $ay_jz_1a$, $bxyb$, $z_2y_hz_3y_iz_2$, and $\alpha' - 1$ strong edges, contradicting $G \in \mathscr{BO}_k$.

Otherwise, $t \geq 4$. Then, for every $ab \in M$, there are $k$ disjoint cycles $axya$, $bz_1y_1z_2b$, $z_3y_2z_4y_3z_3$, and $\alpha' - 1$ other strong edges, contradicting $G \in \mathscr{BO}_k$.

*Case 7.2.2:* $e \in E(Y, Z_t)$. Now, $H$ is simple. Say $e = y_1z_1$ and $e' = y'z'$. If $e' \notin E(H)$, then $y' \neq y_1$. By degree conditions $xz' \in E$, so $z' = z_1$. As $xz_i, z_1z_i \notin E$ for $i \geq 2$, (22) implies that all vertices of $Z - z_1$ and all subdividing vertices are good, $z_1$ is acceptable, and $z_1$ is good if $e' \notin H$.

*Case 7.2.2.1* $t \geq 4$. Let $ab \in M$ with $a \in N(z_1)$. If $t \geq 5$, then there are $k$ disjoint cycles, $az_1xa$, $bz_2y_1z_3b$, $z_4y_2z_5y_3z_4$, and $\alpha' - 1$ strong edges, contradicting $G \in \mathscr{BO}_k$. Else, $t = 4$. Since $d_{G'}(y_2) \leq 6$ and $xy_2 \notin E$, (22) implies $y_2$ is 3-acceptable. As $z_1$ is acceptable and $|M| \geq 3$, there is an edge $ab \in M$ with $az_1, by_2 \in E$. As $x$ and $z_2$ are good, this yields $k$ disjoint cycles $az_1xa$, $by_2z_2b$, $z_3y_1z_4y_3z_3$, and $\alpha' - 1$ strong edges, contradicting $G \in \mathscr{BO}_k$.

*Case 7.2.2.2* $t = 3$ and $z_1y_1$ is subdivided twice with $z_1x, yy_1 \in E$. Then, $x$ and $y$ are both good. Since $d_{G'}(y_1) \le 5$ and $xy_1 \notin E$, $y_1$ is 2-acceptable. As $z_1$ is acceptable, there is an edge $ab \in M$ with $az_1, by_1 \in E$. Thus, there are $k$ disjoint cycles $az_1xa$, $by_1yb$, $y_2z_2y_3z_3y_2$, and $\alpha' - 1$ strong edges, contradicting $G \in \mathscr{BO}_k$.

*Case 7.2.2.3* $t = 3$ and $z_1y_1$ is subdivided once. Suppose there is an edge $y_iy_j \in E$, where $[3] = \{i, j, h\}$. Then, $d_{G'}(y_h) \le 5$ and either $y_hx \notin E$ or $y_hz_1 \notin E$. By (22), $y_h$ is 3-acceptable. As $|M| \ge 3$, there is an edge $ab \in M$ with $az_1, by_h \in M$. Thus, there are $k$ disjoint cycles $az_1xa$, $by_hz_2b$, $y_iz_3y_jz_3$, and $\alpha' - 1$ strong edges, contradicting $G \in \mathscr{BO}_k$. So, assume $\|G[Y]\| = 0$.

If $|F| = 2\alpha'$, then (Q4) holds. Else, there are edges $ab, a'b' \in M$ and a vertex $u \in \overline{W}$ with $au \in E(F)$. All vertices of $G'$ are good except one of $y_1, z_1$ might only be acceptable. Choose notation so that $\{b, a', b'\} = \{c_1, c_2, c_3\}$ and $|N(c_1) \cap \overline{W}| \ge 6$ and $|N(c_2) \cap \overline{W}|, |N(c_3) \cap \overline{W}| \ge 7$. By inspection, $G' - u$ contains a perfect matching $\{e_1, e_2, e_3\}$ with $e_1 \subseteq N(c_1)$. Thus, $G$ contains $k$ disjoint cycles, $c_1e_1c_1$, $c_2e_2c_2$, $c_3e_3c_3$, $aua$ and $\alpha' - 2$ other strong edges, contradicting $G \in \mathscr{BO}_k$.

□

# 6 Proof of Theorem 15

To construct the algorithm, we first describe several subroutines.

**Lemma 2** *Let $k \ge 4$ and $n = 3k$. There is a subroutine that for any simple graph $G = (V, E) \in \mathscr{DO}_k$ with $|G| = n$ checks whether $G \subseteq \mathbf{Y}_{k,s,2k-s}$ for some $s \le k$, and in this case constructs the representation $G \subseteq \mathbf{Y}_{k,s,2k-s}(Y, X, Z)$, all in $O(n^3)$ time.*

*Proof* Note that if $G \subseteq \mathbf{Y}_{k,s,2k-s}(Y, X, Z)$ for some $s \le k$, then we can choose notation so that every $x \in X$ is adjacent to every other vertex in $X \cup Y$ by (17). Search for a vertex $x$ such that $d(x) \le 2k - 1$ and $N[x]$ can be partitioned as $\{Q, R\}$ so that $Q = \{v \in N[x] : N[v] = N[x]\}$ and $R$ is independent with $|R| = k$. This takes $O(n^3)$ time. If we find such an $x$, then $G \subseteq \mathbf{Y}_{\mathbf{k,s,2k-s}}(R, Q, Z)$, $Q = X$, and $R = Y$. Otherwise, $G \nsubseteq \mathbf{Y}_{k,s,2k-s}$ for any $s \le k$. □

**Lemma 3** *There are subroutines that for any simple graph $F$ with $|F| = n$ construct:*

1. *a maximum matching of $F$ in $O(n^{2.5})$ time;*
2. *a GE-decomposition $(D, A, C)$ of $F$ in $O(n^{3.5})$ time.*

*Proof* For (1), see [13]. For (2), using (1), find the sizes of a maximum matching in $F$ and in all the graphs $(F - v)$ with $v \in V(F)$. This can be done in $O(n^{3.5})$ time. Set $D = \{v \in V(F) : \alpha'(F - v) = \alpha'(F)\}$, $A = N_F(D) - D$, and $C = V(F) - D - A$.

□

Now, we are ready to define our algorithm.

**Setup** We are given a positive integer $k$ and a multigraph $G \in \mathcal{DO}_k$. By Corollary 1, we may assume that $G$ is loopless. Construct the simple graph $F$ induced by the strong edges of $G$ and the GE-decomposition $(D, A, C)$ of $F$ in $O(n^{3.5})$ time. Set $|G| = n$, $D' = V(G) - V(F)$, $\alpha' = \alpha'(F)$, and $k' = k - \alpha'$.

If $k \leq 4$, then construct $G_1$ from $G$ by subdividing each edge. Then, $G_1$ is a simple graph with $n + \|G\|$ vertices and $2\|G\|$ edges, and the number of disjoint cycles in $G_1$ equals that in $G$. By Theorem 2, we can determine whether $G_1$ has $k$ disjoint cycles in linear time in $n + \|G\|$. So, in total this step takes $O(n^2)$ time. Thus, below we assume $k \geq 5$ and apply Theorem 14 to $G$. Checking (Q1) is trivial, so it remains to show how to check (Q2)–(Q5).

**Check (Q2)** First check whether $n > 2k+1$ and $G$ is extremal: $\mathcal{DO}_k$. As observed in Section 2.1, every big set $J \subseteq G$ has the form $J = V(G) - N(v)$ for some vertex $v$ with $s(v) = 2k - 1$. We can find all such sets in $O(n^3)$ time by checking whether $V(G) - N(v)$ is independent for each $v \in V(G)$ with $s(v) = 2k - 1$. If $n \leq 2k+1$ or there are no such sets, then (Q2) fails. Otherwise, let $I_1, \ldots, I_q$ be the big sets in $G$. As $n > 2k + 1$, (2) implies they are disjoint, so $q < n$. For each $j \in [q]$, check whether $I_j$ has no strong neighbors or has a unique strong neighbor $w(j)$. This takes $O(n^2)$ time. If at least one $I_j$ has no strong neighbors or $w(j) = w(j')$ for some distinct $j, j' \in [q]$, then (Q2) holds; otherwise, (Q2) does not hold.

**Check (Q3)** First confirm that $k' \geq 5$ and $n = 3k - \alpha'$. Next, construct the set $U$ of vertices $v \in D'$ with $s(v) = k + \alpha' - 1$. Then, test each $v \in U$ to see if (*) for some adjacent pair $\{x, y\} \in N(v)$ there is an $\alpha'$-matching contained in $F - x - y$. This uses $O(n^{5.5})$ steps. Now, (Q3) holds if and only if (*) fails.

**Check (Q4)** First confirm that $3k - \alpha' \leq n \leq 3k - \alpha' + 1$ and $k' \leq 4$. If so, then we still need to check whether $\underline{G} - W(M)$ has no $k - |M|$ disjoint cycles for all matchings $M$ in $F$. If $|\underline{G} - W(M)| \leq 3(k - |M|) - 1$, then $\underline{G} - W(M)$ does not have enough vertices to have $k - |M|$ disjoint cycles. So, it suffices to check for every $W \subseteq V(G)$ with $2(\alpha' - 1) \leq |W| \leq 2\alpha'$ whether: (i) $F[W]$ has a perfect matching and (ii) $\underline{G} - W$ has no $k - |W|/2$ disjoint cycles. Then, (Q4) holds if and only if (i) implies (ii) for all such $W$. As $n - |W| \leq 3(k - \alpha' + 1) \leq 15$, there are $O(n^{15})$ sets to test. Testing (i) takes $O(n^{2.5})$ time and testing (ii) takes $O(1)$ time. So altogether, we use $O(n^{17.5})$ time.

**Check (Q5)** First confirm that $k' \geq 5$, $n = 3k - \alpha'$ and $|F| - 2\alpha' \in \{0, |D| - 2, |D| - 1\}$. If so, then we still need to check that for all maximum matchings $M$ either: (i) $\alpha(G - W(M)) = k' + 1$ or (ii) $G - W(M) \subseteq \mathbf{Y}_{k',c,2k'-c}$ for some odd $c \leq k'$. We do this by checking certain subsets $W \subseteq V(G)$ to see if $F[W]$ has a perfect matching $M$ satisfying (i) and (ii). If $|F| - 2\alpha' = 0$, then $W := W(M) = V(F)$; else, using $|F| - 2\alpha' \in \{|D| - 2, |D| - 1\}$, $V(F) = A \cup C \cup D$ and $W = A \cup C \cup (W \cap D)$, we have

$$|D - W| = |F| - 2\alpha' \geq |D| - 2 = |D - W| + |W \cap D| - 2,$$

so $|W \cap D| \leq 2$. Thus, we only need to check $O(n^2)$ sets $W$. By Lemma 2 and the argument in Check (Q2), each check takes $O(n^3)$ time, so all together we use $(n^5)$ time.

This completes our description of the algorithm. □

**Acknowledgements** We thank a referee for a number of helpful comments. Research of A. Kostochka is supported in part by NSF grant DMS-1600592 and by grants 18-01-00353A and 16-01-00499 of the Russian Foundation for Basic Research. Research of T. Molla is supported in part by NSF grant DMS-1500121. Research of D. Yager is supported by the Campus Research Board of the University of Illinois.

## References

1. H.L. Bodlaender, On disjoint cycles. Int. J. Found. Comput. Sci. **5**, 59–68 (1994)
2. K. Corrádi, A. Hajnal, On the maximal number of independent circuits in a graph. Acta Math. Acad. Sci. Hungar. **14**, 423–439 (1963)
3. G. Dirac, Some results concerning the structure of graphs. Can. Math. Bull. **6**, 183–210 (1963)
4. R.G. Downey, M.R. Fellows Fixed-parameter tractability and completeness. Congr. Numer. **87**, 161–178 (1992)
5. H. Enomoto, On the existence of disjoint cycles in a graph. Combinatorica **18**(4), 487–492 (1998)
6. M.R. Garey, D.S. Johnson, Computers and intractability, in *A Guide to the Theory of NP-Completeness. A Series of Books in the Mathematical Sciences* (W. H. Freeman and Co., San Francisco, 1979), x+338 pp. (p. 68)
7. H.A. Kierstead, A.V. Kostochka, Ore-type versions of Brooks' theorem. J. Comb. Theory Ser. B **99**, 298–305 (2009)
8. H. Kierstead, A. Kostochka, T. Molla, E.C. Yeager, Sharpening an Ore-type version of the Corrádi-Hajnal theorem. Abhandlungen aus dem Mathematischen Seminar der Universität Hamburg, published online (2016)
9. H.A. Kierstead, A.V. Kostochka, E.C. Yeager, On the Corrádi-Hajnal theorem and a question of dirac. J. Comb. Theory Ser. B **122**, 121–148 (2017)
10. H.A. Kierstead, A.V. Kostochka, E.C. Yeager, The $(2k-1)$-connected multigraphs with at most $k-1$ disjoint cycles. Combinatorica **37**(1), 77–86 (2017)
11. A.V. Kostochka, L. Rabern, M. Stiebitz, Graphs with chromatic number close to maximum degree. Discret. Math. **312**, 1273–1281 (2012)
12. L. Lovász, On graphs not containing independent circuits, (Hungarian. English summary) Mat. Lapok **16**, 289–299 (1965)
13. S. Micali, V. Vazirani, An $O(\sqrt{|V|} \cdot |E|)$ algorithm for finding a maximum matching in general graphs, in *Proceedings of Twenty-first Annual Symposium on Foundations of Computer Science* (IEEE Berkeley, California, 1980), pp. 17–27
14. L. Rabern, A-critical graphs with small high vertex cliques. J. Comb. Theory Ser. B **102**, 126–130 (2012)
15. H. Wang, On the maximum number of disjoint cycles in a graph. Discret. Math. **205**, 183–190 (1999)

# Combinatorial and Graph-Theoretical Problems and Augmenting Technique

Ngoc C. Lê

## 1 Introduction

Berge's lemma [3] states that a matching $M$ (a set of edges without common vertices) of a graph $G$ is maximum (contains the largest number of edges) if and only if there is no augmenting path (a path that starts and ends on free (unmatched) vertices, and alternates between edges in and not in the matching) with $M$. Edmonds [11] used this idea to develop *Blossom Algorithm* for this problem. This idea was used first for the Maximum Independent Set problem, i.e. the problem asks for a largest number of vertices set without edges among them, by Sbihi [33] and Minty [28]. Clearly, a matching in a graph $G$ corresponds to an independent set in the line graph of $G$. Hence, we can use Edmonds' algorithm to find Maximum Independent Set for line graphs. Sbihi and Minty extended this idea for a more general graph class, say claw-free graph, by showing that an independent set $S$ of a graph $G$ is maximum if and only if there is no augmenting path (a path that starts and ends on vertices not lies in the independent set and alternates between vertices in and not in the independent set) with $S$. This technique was extended for more general graph classes by using the augmenting graph concept as described as follows.

**Definition 1 ([17])** Given a graph $G$ and an independent set $S$, an induced bipartite subgraph $H = (W, B, E)$ of $G$ is called an augmenting graph for $S$ if (**i**) $W \subseteq S$, $B \subseteq V(G)\backslash S$, (**ii**) $N(B) \cap (S\backslash W) = \emptyset$, and (**iii**) $|B| > |W|$.

An augmenting graph $H$ is called minimal if it does not contain any augmenting graph as a proper induced subgraph.

---

N. C. Lê (✉)
School of Applied Mathematics and Informatics, Hanoi University of Science and Technology, Hanoi, Vietnam



B. Goldengorin (ed.), *Optimization Problems in Graph Theory*, Springer Optimization and Its Applications 139,
https://doi.org/10.1007/978-3-319-94830-0_9

**Theorem 1 ([17])** *An independent set $S$ in a graph $G$ is maximum if and only if there is no augmenting graph for $S$.*

This theorem suggests the following general approach to find a maximum independent set in a graph $G$. Begin with any independent set $S$ (may be empty) in $G$ and as long as $S$ admits an augmenting graph $H$, exchange white and black vertices of $H$. Clearly, the problem of consecutively finding augmenting graphs and of applying these augmentations is generally NP-hard, as the MIS problem is NP-hard. Moreover, we can restrict ourselves in minimal augmenting graph only. Hence, for a polynomial time solution to some graph class, one has to solve the two following problems:

**(P1)** Find a complete list of (minimal) augmenting graphs.
**(P2)** Develop polynomial time algorithms for detecting (minimal) augmenting graphs.

So far, characterizations of (minimal) augmenting graphs mainly followed the two following directions. In the first approach, augmenting graphs in ($S_{1,2,k}$,banner)-free graphs are characterized based on the observation that a banner-free bipartite graph is either $C_4$-free or complete. The most general result follows this direction described by Lozin and Milanič [23] for ($S_{1,2,5}$,banner)-free graphs.

In the second approach, augmenting graphs of subclasses of $P_5$-free graphs are characterized based on the observation showed indepedently by many researchers (e.g., [29]) that every connected $P_5$-free bipartite graph is a *bipartite-chain* graph, i.e. the vertices of each part can be ordered under inclusion of their neighborhood. Based on this property, polynomial solutions were obtained for some subclasses of $P_5$-free graphs [4, 16, 25, 29, 30]. It is also worth to notice that the MIS problem is shown polynomially solvable in $P_5$-free graphs [22].

In this paper, we try to combine the two above approaches to a subclass of (banner$_2$,domino)-free graphs (see Figure 1). In particular, we obtain the following theorem.

**Theorem 2** *Given integers $m, l$, the MIS problem is polynomially solvable in ($S_{2,2,5}$,banner$_2$,domino,$M_m$, $K_{m,m} - e$,$R_l^1$, $R_l^2$, $R_l^3$)-free graphs. (See Figure 4.)*

Obviously, banner and domino are two natural generalizations of $P_5$ and banner (banner$_1$), $R_l^1$, $R_l^2$, $R_l^3$ are generalizations of $S_{1,2,5}$. Hence, our result is a generalization of some previous known results for ($S_{1,2,5}$, banner)-free graphs [23], ($P_5$, $K_{3,3} - e$)-free graphs [16, 25] for ($P_5$, $K_{2,m} - e$)-free graphs, and some subclasses of $S_{1,2,2}$-free graphs [20].

The organization of the paper is as follows. Augmenting graphs for some subclasses of $S_{2,2,l}$-free graphs are characterized in Section 2, i.e. to solve Problem P1. Methods for finding such augmenting graphs are described in Section 3, i.e. to solve Problem V2. In Section 4, we summarize some results in using technique for other combinatorial and graph-theoretical problem. Section 5 is a discussion about the issue. Many of long proofs are put in the Appendix part.

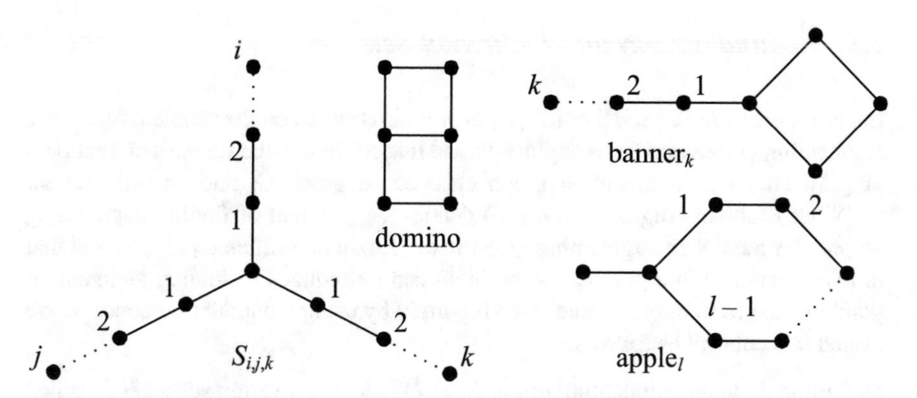


**Fig. 1** $S_{i,j,k}$, domino, banner$_k$, and apple$_l$

Here, we want to collect most of the terminology and notations used in the paper. For those not given here, they will be defined when needed. For those not given, we refer the readers to [5]. Given a graph $G = (V, E)$, for a vertex $u$, we denote by $N(u) := \{v \in V : uv \in E\}$ the *neighborhood* of $u$ in $G$. For a subset $U \subset V(G)$, we denote by $N(U) := (\bigcup_{u \in U} N(u)) \setminus U$ the neighborhood of $U$. If $W, U$ are two vertex subsets of $G$, then $N_U(W) := N(W) \cap U$. Also, $N_U(v) := N(v) \cap U$ for a vertex $v$. Given a graph $G = (V, E)$ and a vertex subset $U$, we denote by $G - U$ the graph obtained from $G$ by deleting all vertices (together with adjacent edges) in $U$. For two vertices $u, v \in V$, we write $u \sim v$ if $uv \in E$. For a vertex $u$, we denote by $d(u) := |N(u)|$, the degree of $u$ in $G$. We also denote by $G[U] := G - (V(G) \setminus U)$, the subgraph of $G$ induced by $U$.

## 2 Augmenting Graphs in Subclasses of ($S_{2,2,l}$,banner$_l$)-Free Graphs

Hertz and Lozin [17] obtained the following observation about minimal augmenting graphs.

**Lemma 1 ([17])** *If $H = (B, W, E)$ is a minimal augmenting graph for an independent set $S$ of a graph $G$, then*

1. *$H$ is connected;*
2. $|W| = |B| - 1$;
3. *for every subset $U \subseteq W$, $|U| < |N_B(U)|$.*

## 2.1 Redundant Sets and Reduction Sets

Let us report from Section 3 of [23] a general observation on the problem of finding augmenting graphs and let us slightly extend it according to the Remark of Section 3 of [23]. Given an augmenting graph class $\mathscr{A}$, a graph $G$, and an independent set $S$, let Problem Augmentation ($\mathscr{A}$) denote the problem of finding augmenting graphs if $S$ admits an augmenting graph in $\mathscr{A}$. Lozin and Milanič [23] showed that in ($S_{1,2,5}$,banner)-free graphs, the problem can be reduced to finding augmenting graphs of the form tree$^1$,...,tree$^6$ (see Figure 2) by using redundant set concept. We extend this concept as follows.

**Definition 2** In an augmenting graph $H = (W, B, E)$, a vertex subset $U$ is called redundant if

1. $|U \cap W| = |U \cap B|$ and
2. for every vertex $b \in B \setminus U$, $N_{W\setminus U}(U \cap B) \subseteq N_{W\setminus U}(b)$.

**Theorem 3** *Let $\mathscr{A}_1$ and $\mathscr{A}_2$ be two classes of augmenting graphs. If there exists a constant k such that, for every augmenting graph $H = (W, B, E) \in \mathscr{A}_2$, there exists a redundant subset U of size at most k such that $H - U \in \mathscr{A}_1$, then Problem Augmentation($\mathscr{A}_2$) is polynomially reducible to the problem Augmentation($\mathscr{A}_1$).*

*Proof* Assume that Algorithm $Augment_1(G, S)$ outputs a subset $V' \subseteq V(G)$ such that $G[V']$ is augmenting for $S$ whenever $S$ admits an augmenting graph from $\mathscr{A}_1$ (and perhaps even if this is not the case). The procedure also returns $\emptyset$ if no augmenting graph is found.

Assume that $S$ admits an augmenting graph $H = (B, W, E) \in \mathscr{A}_2$. Then by the theorem's assumption, $H$ contains a redundant set $U$ of size at most $k$ such that $H - U \in \mathscr{A}_1$. It is obvious that the graph $H - U$ is augmenting for $S\setminus U$. Moreover, since $U$ is redundant, $G''$ contains every vertex of $H - U$, i.e. Steps 1 and 2 have not removed any vertex of $H - U$. Therefore, Algorithm $Augment_1$ must output a non-empty set $T$. Consequently, Algorithm $Augment_2$ also outputs a non-empty set $U \cup T$.

We show that $G[U \cup T]$ is augmenting for $S$. Indeed, by Step 1, $G[U \cup T]$ is a bipartite graph. Since $T$ is augmenting for $S\setminus U$ in $G''$, $|T \cap S\setminus U| < |T \cap V(G'')|$. Moreover, since $|U \cap S| = |U \cap V(G)\setminus S|$, $|(T \cup U) \cap S| < |(T \cup U) \cap V(G)\setminus S|$. By Step 2, $N_S(U\setminus S) \subseteq T \cap S$, i.e. $N_S((T \cup U)\setminus S) \subseteq (T \cup U) \cap S$. Hence, the graph $G[U \cup T]$ is augmenting for $S$, even if $G[T]$ does not coincide with $H - U$. Therefore, whenever $S$ admits an augmenting graph in $\mathscr{A}_2$, Algorithm $Augment_2$ finds an augmenting graph.

To this end, the procedure inspects polynomially many subsets of vertices of the input graph, which results in polynomially many calls of Algorithm $Augment_1$. The construction of the graph $G''$ also is performed in polynomial time. Hence, Problem Augmentation($\mathscr{A}_2$) is polynomially reducible to Problem Augmentation ($\mathscr{A}_1$).

Note that Problem Augmentation($\mathscr{A}_1$) becomes Problem (P2) when $\mathscr{A}_1$ is the class of all (possible) augmenting graphs.

**Algorithm 1** $Augment_2(G, S)$ (Version 1)

**Input:** A graph $G$ and an independent set $S$ of $G$

**Output:** A subset $V' \subseteq V(G)$ such that $G[V']$ is augmenting for $S$ whenever $S$ admits an augmenting graph from $\mathscr{A}_2$. Return $\emptyset$ if no augmenting graph is found.

1: **for all** $U \subseteq V(G)$ of size at most $k$ such that

1. $B_0 := U \cap (V(G)\backslash S)$ is independent in $G$,
2. $|B_0| = |U \cap S|$

**do**

2: $G' := G - N_G(B_0) \cap (V(G)\backslash S)$ {Remove the (black) neighbors of $B_0$ in $V(G)\backslash S$};

3: $G'' := G' - \{b \in V(G')\backslash S : N_{S\backslash U}(B_0)\backslash N_{S\backslash U}(b) \neq \emptyset\}$ {Remove the (black) vertices of $V(G')\backslash S$ whose neighborhood in $S\backslash U$ does not cover the neighborhood of $B_0$ in $S\backslash U$};

4: $T := Augment_1(G'' - U, S\backslash U)$;

5: **if** $T \neq \emptyset$ **then**

6: **return** $U \cup T$ {We have an augmenting graph for $S$}

7: **end if**

8: **end for**

9: **return** $\emptyset$

Moreover, we can also extend the redundant set concept further as follows. If Algorithm $Augment_1$ starts with some initialization process (see Algorithm 2), which computes some finite vertex set $C$ such that $N_{S\backslash U}(U\backslash S) \subseteq N_S(C\backslash S)$, then we can process this initialization procedure in $Augment_2$ as in Version 2 and remove the condition that every neighbor in $S\backslash U$ of black vertices in $B\backslash U$ covers the neighbor of $U$ in $S\backslash U$ (see Algorithm 3). More precisely, we have the following definition.

**Definition 3** Let $\mathscr{A}_1$ and $\mathscr{A}_2$ be the two augmenting graph classes. Given an integer $k$. Assume that there exists a polynomial time procedure finding an augmenting graph in $\mathscr{A}_1$ (or deciding such augmenting graph does not exist) and such a procedure has a form as in Algorithm 2, i.e. starts by generating some candidates and from each candidate $C$, builds up augmenting graphs ($Generate_1(C, G, S)$). In an augmenting graph $H = (B, W, E) \in \mathscr{A}_2$, a vertex subset $U$ is called a reduction set associated with some key set $B^* \subseteq B \cap C$ if $|U \cap B| = |U \cap W|$ and $N_{W\backslash U}(U \cap B) \subseteq N_{W\backslash U}((B^*\backslash U) \cap B)$.

And by the above arguments, we have the following observation.

**Theorem 4** *Let $\mathscr{A}_1$ and $\mathscr{A}_2$ be the two augmenting graph classes. Then Problem Augmentation($\mathscr{A}_2$) is polynomially reducible to Problem Augmentation($\mathscr{A}_1$) if there are two integers $k_1, k_2$ such that for every augmenting graph $H = (B, W, E) \in \mathscr{A}_2$, there is a reduction set $U$ of size at most $k_1$ associated with a key set $B^*$ of size at most $k_2$ such that $H - U \in \mathscr{A}_1$.*

**Algorithm 2** $Augment_1(G, S)$

**Input:** A graph $G$ and an independent set $S$ of $G$

**Output:** A subset $V' \subseteq V(G)$ such that $G[V']$ is augmenting for $S$ whenever $S$ admits an augmenting graph from $\mathscr{A}_1$. Return $\emptyset$ if no augmenting graph is found.

1: Generate Candidates;
2: **for all** Candidates $C$ **do**
3: $\quad T := Generate_1(C, G, S)$;
4: $\quad$ **if** $T \neq \emptyset$ **then**
5: $\quad\quad$ **return** $T$ {We have an augmenting graph for $S$}
6: $\quad$ **end if**
7: **end for**
8: **return** $\emptyset$

**Algorithm 3** $Augment_2(G, S)$ (Version 2)

**Input:** A graph $G$ and an independent set $S$ of $G$.

**Output:** A subset $V' \subseteq V(G)$ such that $G[V']$ is augmenting for $S$ whenever $S$ admits an augmenting graph from $\mathscr{A}_2$. Return $\emptyset$ if no augmenting graph is found.

1: **for all** $U \subseteq V(G)$ of size at most $k$ such that

1. $B_0 := U \cap (V(G) \backslash S)$ is independent in $H$,
2. $|B_0| = |U \cap S|$

**do**

2: $\quad G' := G - N_G(B_0) \cap (V(G) \backslash S)$ {Remove the (black) neighbors of $B_0$ in $V(G) \backslash S$};
3: $\quad$ Generate Candidates;
4: $\quad$ **for all** Candidates $C$ of $G'$ such that $N_{S \backslash U}(B_0) \subseteq N_{S \backslash U}(C \cap (V(G' - U) \backslash S))$ **do**
5: $\quad\quad T := Generate_1(C, G' - U, S \backslash U)$;
6: $\quad\quad$ **if** $T \neq \emptyset$ **then**
7: $\quad\quad\quad$ **return** $U \cup T$ {We have an augmenting graph for $S$}
8: $\quad\quad$ **end if**
9: $\quad$ **end for**
10: **end for**
11: **return** $\emptyset$

## 2.2 Augmenting Graphs in Subclasses of $S_{2,k,l}$-Free Graphs

The following corollary is a consequence of Lemma 1 and was obtained in [23].

**Corollary 1 ([23])** *Let $H = (B, W, E)$ be a minimal augmenting graph for an independent set $S$ of a graph $G$. Then for every vertex $b \in M$, there exists a perfect matching between $B \backslash \{b\}$ and $W$ in $H$, i.e. a matching consists of every vertex of $B \backslash \{b\}$ and $W$.*

*Remark 1* By the above corollary, from now on, given a minimal augmenting graph $H = (B, W)$ and a black vertex $b \in B$, we denote by $M$ such a perfect matching and for every vertex $u$ of $H$ different from $b$ and by $\mu(u)$ the matched vertex of $u$ in $M$. For a subset $U \subseteq V(H)$, we also denote $\mu(U) := \{\mu(u) : u \in U\}$.

**Corollary 2** *Let $H = (B, W)$ be a minimal augmenting graph. Then every white vertex of $H$ is of degree at least two.*

We say that $G$ is an $(k, m)$-extended-chain if $G$ is a tree and contains two vertices $a, b$ such that there exists an induced path $P \subset G$ connecting $a, b$, every vertex of $G - P$ is of distance at most $k - 1$ from either $a$ or $b$, and every vertex of $G - P$ has no neighbor in $P$ except possibly $a$ or $b$ and every vertex of $G$ is of degree at most $m - 1$. The following observation is an extension of Theorem 8 of [17]. The result was announced in [21] without full proof.

**Lemma 2 ([21])** *For any three integers $k, l$, and $m$ such that $4 \leq 2k \leq l$ and $m \geq 3$, in ($S_{2,2k,l}$,apple$^l_4$, apple$^l_6$,. . .,apple$^l_{2k+2}$, $K_{1,m}$)-free graphs, there are only finitely many minimal augmenting graphs different from augmenting $(2k, m)$-extended-chains and not of the form apple$_{2p}$. Moreover, if $H$ is of the form augmenting $(2k, m)$-extended-chain, then every white vertex is of degree two.*

Note that in an augmenting graph of the form apple$_{2p}$ (or augmenting apple for short), the vertex of degree three is white. However, given an augmenting apple $H = (B, W, E(H))$, where $b$ is the black vertex of degree one and $w$ is the white vertex of degree three. Then $U := \{b, w\}$ is a redundant set such that $H - U$ is an augmenting chain, a special case of augmenting $(k, m)$-extended-chain.

## 2.3 Augmenting Graphs in Subclasses of $S_{2,2,5}$-Free Graphs

Now, we try to omit $K_{1,m}$ from the list of forbidden induced subgraphs by considering ($S_{2,2,5}$,banner$_2$,domino)-free augmenting graphs. We extend the consideration of Section 4 in [23]

**Lemma 3** *Given a graph $G$ and an ($S_{2,2,5}$,banner$_2$,domino)-free minimal augmenting graph $H = (B, W, E)$ for an independent set $S$, at least one of the following statements is true:*

1. *$H$ belongs to some finite set of augmenting graphs;*
2. *$H$ is an augmenting chain or an augmenting apple (see Figure 1);*
3. *$H$ is an augmenting graph of the form tree$^1$, tree$^2$, . . . , tree$^7$ (see Figure 2) or can be reduced by a redundant set containing at most 32 vertices to an augmenting graph of the form tree$^1$, tree$^2$, . . . , tree$^7$;*
4. *there is a vertex $b \in B$ such that $b$ is adjacent to all vertices of $W$.*

Such $b$ of Case 4 is called the *augmenting vertex* of $S$, as in [29, 30]. We also call augmenting graphs of the form tree$^1$, tree$^2$, . . . , tree$^7$ as *augmenting trees*. For Case 4 of Lemma 3, we show that under some restrictions, these augmenting graphs have structural properties similar to $P_5$-free augmenting graphs, i.e. being a bipartite-chain by the following observation.

**Lemma 4** *Given a (domino,banner$_2$)-free graph $G$, an integer $m \geq 3$, and an $M_m$-free (see Figure 3) minimal augmenting graph $H = (B, W, E)$ for an independent set $S$ such that there exists some black vertex $b \in B$ adjacent to every white vertex of $W$, and $|W| \geq 2m + 1$, at least one of the following statements is true.*

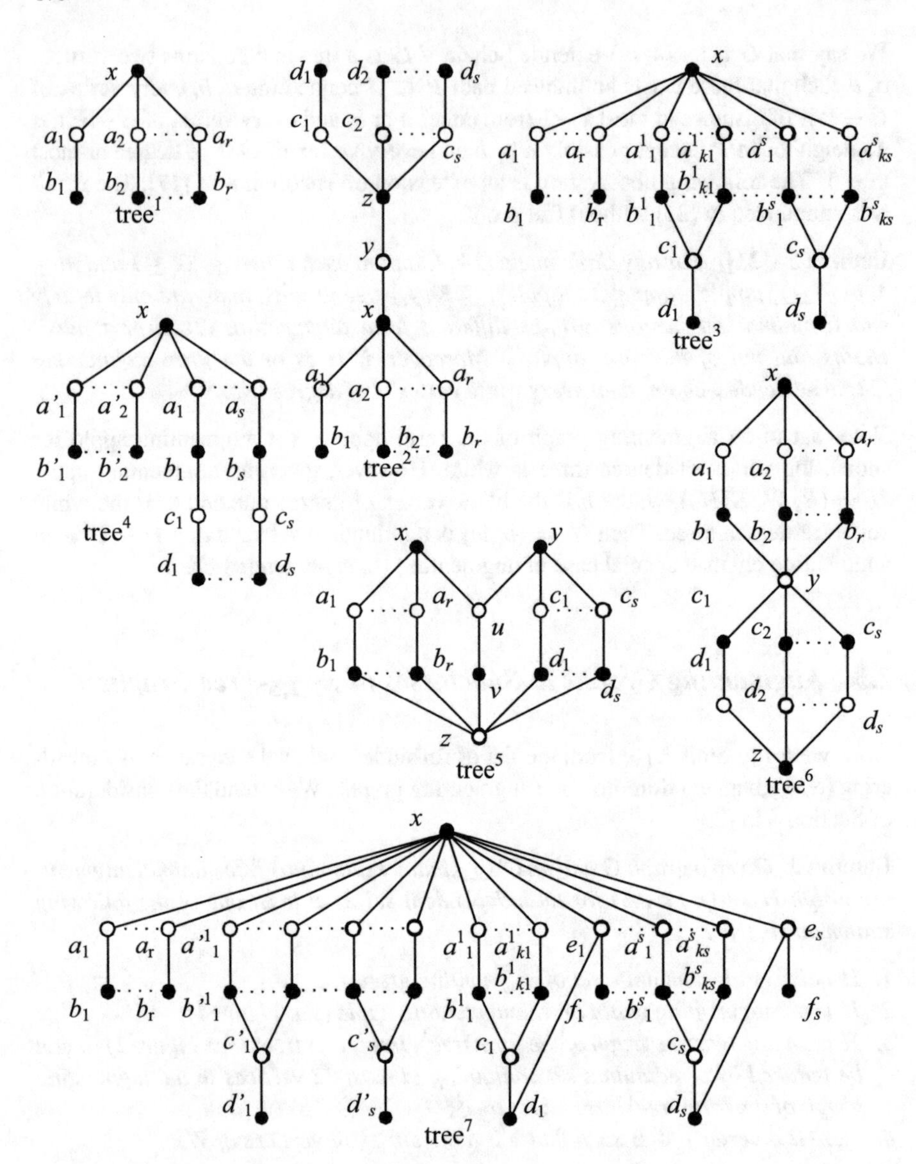


**Fig. 2** Augmenting trees

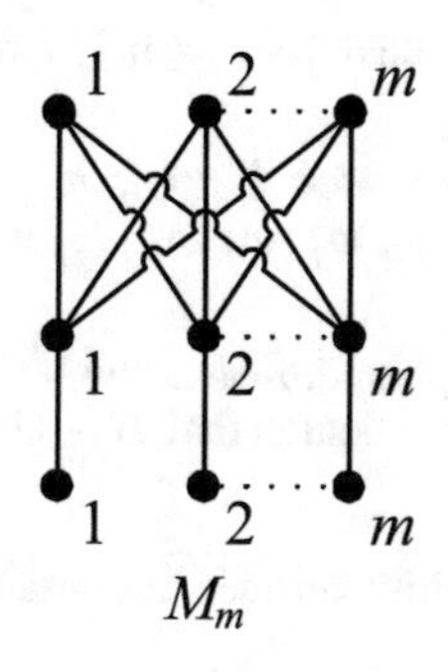


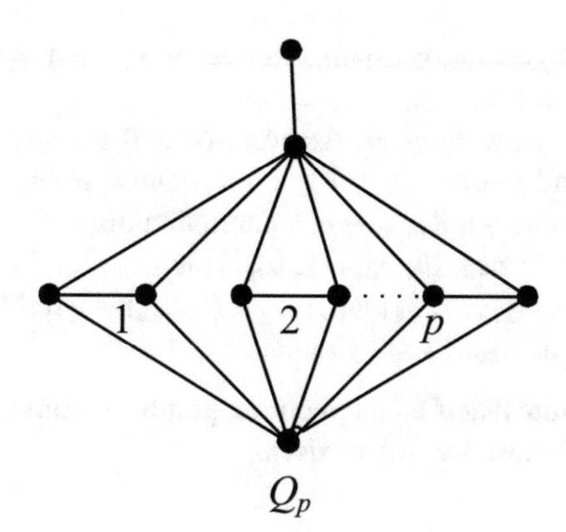


**Fig. 3** $M_m$ and $Q_p$

1. *H is of the form tree*[1] *or there exists a reduction set U of size at most* $2m-2$ *associated with a key set of size one such that* $H-U$ *is of the form tree*[1].
2. *H is a bipartite-chain or there exists a redundant set U of size at most* $2m-2$ *such that* $H-U$ *is a bipartite-chain.*

*Proof* We refer to Lemma 10 for the procedure finding tree[1] and note that such a procedure starts by finding a candidate containing $b$, i.e. $b$ is adjacent to every white vertex in the augmenting tree[1] and we have the key set $B^* := \{b\}$.

Let $B = \{b, b_1, \ldots, b_q\}$, $b$ be the vertex $b$ in Corollary 1, $p$ be the integer $p$ in Lemma 8 such that $N_W(b_i) \supseteq N_W(b_j)$ for every $1 \le i \le p, i < j \le q$ and $|N_W(b_i)| = 1$ for every $i \ge p+1$.

If $p \le m-1$, then $U = \{b_1, \ldots, b_p, \mu(b_1), \ldots, \mu(b_p)\}$ is a reduction set of size at most $2m-2$ associated with $B^*$ such that $H-U$ is of the form tree[1].

If $p \ge q-m+1$, then $U = \{b_{p+1}, \ldots, b_q, \mu(b_{p+1}), \ldots, \mu(b_q)\}$ is a redundant set of size at most $2m-2$ such that $H-U$ is a bipartite-chain.

If $m \le p \le q-m$, then $\{b, b_1, \ldots, b_{k-1}, b_{q-k+1}, \ldots, b_q, \mu(b_{q-k+1}), \ldots, \mu(b_q)\}$ induces an $M_m$, a contradiction.

The following observation is a generalization of Lemma 10 in [4] and Theorem 1 in [16] about augmenting graphs in $(P_5, K_{2,m}-e)$-free graphs and $(P_5, K_{3,3}, -e)$-free graphs, respectively.

**Lemma 5** *Given a graph G, an independent set S of G, an integer m, and a* $(K_{m,m}-e)$*-free minimal augmenting bipartite-chain* $H = (B, W, E)$*, either*

1. *H has at most* $2m-2$ *white vertices; or*
2. *H is of the form* $K_{l,l+1}$ *or there is a redundant set of size at most* $2m-4$ *such that* $H-U$ *is of the form* $K_{l,l+1}$*, for some l.*

*Proof* Assume that $|W| = p \ge 2m-1$. Let $W = \{w_1, w_2, \ldots, w_p\}$ and $B = \{b_1, b_2, \ldots, b_p, b_{p+1}\}$. Assume that $N_W(b_i) \subseteq N_W(b_j)$ for $i < j$. Moreover, by Corollary 1, there exists a perfect matching between $B \backslash \{b_{p+1}\}$ and $W$. Without loss of generality, assume that $b_i \sim w_i$ for $1 \leq i \leq p$. Then we have $|N_W(b_i)| \geq i$ for $i = 1, 2, \ldots$.

Now, $b_i \sim w_j$ for every $b_i \in B$ and $w_j \in W$ such that $p - m + 4 \geq i \geq m - 1$ and $p - m + 3 \geq j \geq i + 1$, otherwise $\{b, b_p, \ldots, b_{p-m+3}, b_i, w_j, w_{m-1}, \ldots, w_1\}$ induces a $K_{m,m} - e$, a contradiction.

Hence, $\{b, b_p, \ldots, b_{m-1}, w_{p-m+1}, \ldots, w_1\}$ induces a $K_{p-m+3,p-m+2}$ and $U := \{b_{m-2}, \ldots, b_1, w_p, \ldots, w_{p-m+2}\}$ is a redundant of size $2m - 4$ such that $H - U$ is a $K_{p-m+3,p-m+2}$.

Note that if an augmenting graph contains at most $2m - 2$ white vertices, it contains at most $4m - 3$ vertices.

## 3 Finding Augmenting Graphs

Now, we consider Problem (P2), i.e. the problem of finding augmenting graphs characterized in Section 2. Remind that we can enumerate all augmenting graphs of bounded size in polynomial time. Moreover, Hertz and Lozin [17] described a method of finding augmenting graphs of the form $K_{m,m+1}$ in banner$_2$-free graphs. Besides, it is obvious that augmenting apples can be reduced to augmenting chains by a redundant set of size two. Hence, we have to find augmenting extended-chains and augmenting trees.

### *3.1 Augmenting Extended-Chain and Augmenting Trees*

The method for finding augmenting chains in ($S_{1,2,j}$,banner)-free graphs has been described by Hertz, Lozin, and Schindl [18]. We have extended this method and obtain the following result, which was published in [21] without proof.

**Lemma 6 ([21])** *Given integers $l$ and $m$, where $l$ is even, an ($S_{2,l,l}$,banner$_l$, $R_l^1$, $R_l^2$, $R_l^3$, $R_l^4$, $R_l^5$)-free graph $G$, and an independent set $S$ in $G$, one can determine whether $S$ admits an augmenting $(l, m)$-extended-chain in polynomial time (Figure 4).*

By extending the techniques presented in [23] (finding augmenting trees of the form tree$^1, \ldots,$ tree$^6$ in ($S_{1,2,5}$, banner)-free graphs), we obtain the following result.

**Lemma 7** *An augmenting graph of the form tree$^1$, tree$^2$, ..., tree$^7$ can be found in ($S_{2,2,5}$,banner$_2$)-free graphs in polynomial time.*

Together with the method of Lozin and Hertz [17] for finding augmenting graphs of the form $K_{p,p+1}$ in banner$_2$-free graphs, it leads us to the following result.

**Corollary 3** *Given an integer $m$, the MIS problem is polynomially solvable in ($S_{1,2,5}$,banner$_2$,domino,$M_m$, $K_{m,m} - e$)-free graphs.*

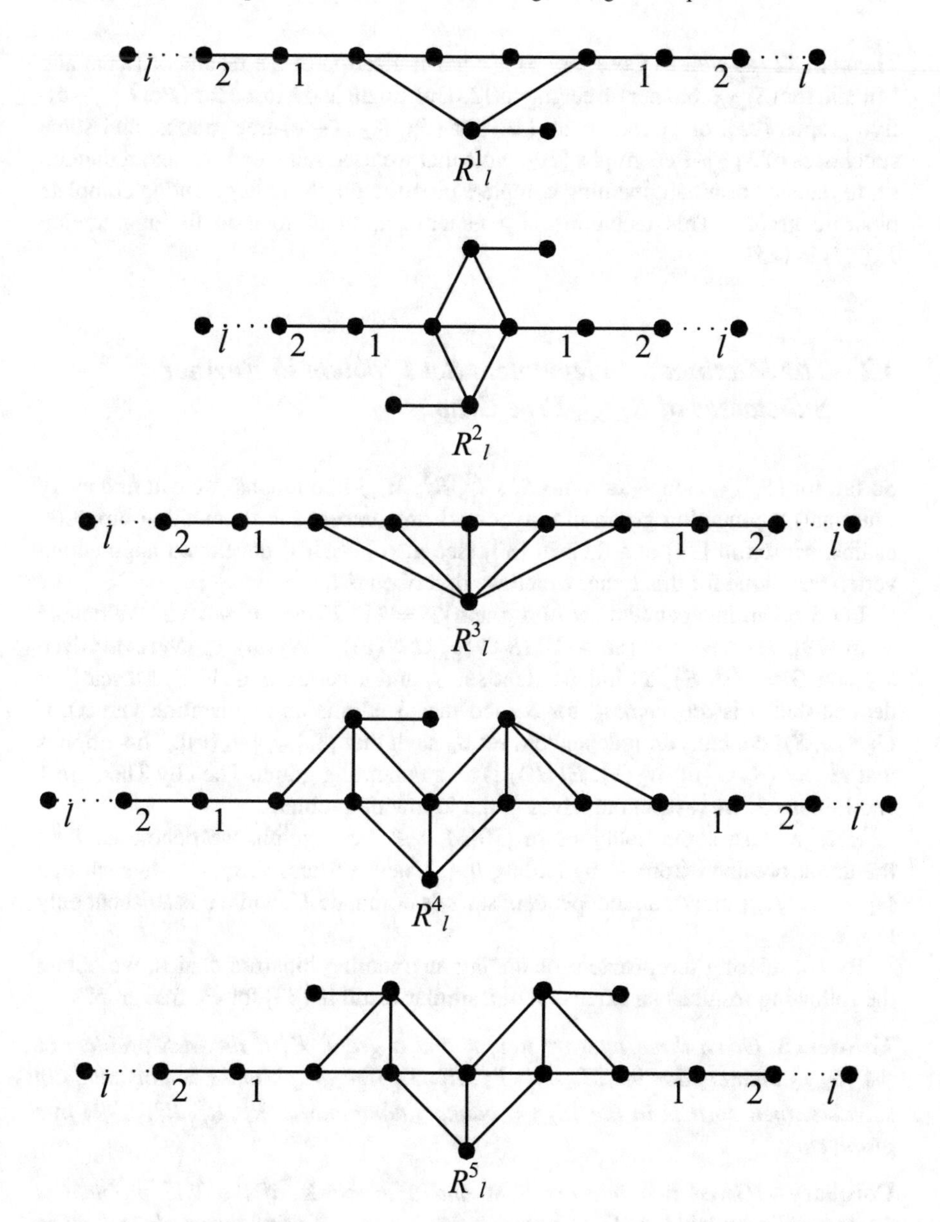


**Fig. 4** $R_l^1$, $R_l^2$, $R_l^3$, $R_l^4$, and $R_l^5$

Theorem 42 (as well as Corollary 3) is a generalization of the results of Lozin and Milanič for ($S_{1,2,5}$, banner)-free graphs [23], of Lozin and Mosca for ($P_5$, $K_{3,3}-e$)-free graphs [25], of Gerber et al. [16] for ($P_5$, $K_{2,m} - e$)-free graphs, and some subclasses of $S_{1,2,2}$-free graphs [20]. Note that we used redundant set and reduction set to reduce "near" augmenting complete bipartite graphs to augmenting complete bipartite graphs. This technique is a generalization of method for augmenting $K^+_{m,m}$'s in [25].

## 3.2 *The Maximum Independent Set Problem in Further Subclasses of $S_{2,2,5}$-Free Graphs*

So far, for ($S_{2,2,5}$,banner$_2$,domino,$R^1_l$, $R^2_l$, $R^3_l$, $M_m$)-free graphs, we can find every (minimal) augmenting graph in polynomial time except for augmenting bipartite-chains. Mosca in [29] and then in [30] (see also [14, 16]) developed augmenting vertex technique for this issue, which we describe next.

Let $S$ be an independent set of a graph $G = (V, E)$ and $v \in V\backslash S$. We denote as in [29], $H(v, S) := \{w \in V\backslash(S \cup \{v\} \cup N(v)) : N_S(w) \subset N_S(v)\}$. Given a graph $G = (V, E)$, an independent set $S$, and a vertex $v \in V\backslash S$, Mosca [29] defined that $v$ is *augmenting* for $S$ (and that $S$ admits an augmenting vertex), if $G[H(v, S)]$ contains an independent set $S_v$ such that $|S_v| \geq |N_S(v)|$. This implies that $H' := (S_v \cup \{v\}, N_S(v), E(H'))$ is an augmenting graph. Then by Theorem 1 and Lemma 3, we restrict ourselves in the following problem.

Here we use some notations in [30]. Let $K$ be a graph, we denote as $K^{(h)}$ the graph obtained from $K$ by adding $h + 1$ new vertices $v, s_1, \ldots, s_h$ such that $\{s_1, s_2, \ldots, s_h\}$ induce an independent set, $s_i$'s dominate $K$, while $v$ is adjacent only to $s_i$'s.

By considering the problem of finding augmenting bipartite chains, we obtain the following result as an extension of a similar result in [30] for $P_5$-free graphs.

**Theorem 5** *Given three integers $h, l, m$ and a graph $K$, if the MIS problem in the ($S_{2,2,5}$,banner$_2$,domino,$M_m$, $R^1_l$, $R^2_l$, $R^3_l$, $K$)-free graph class is polynomially solvable, then so it is in the ($S_{2,2,5}$, banner$_2$,domino,$M_m$, $R^1_l$, $R^2_l$, $R^3_l$, $K^{(h)}$)-free graph class.*

**Corollary 4** *Given two integers $h, m$ and a graph $K$, if the MIS problem is polynomially solvable in ($S_{1,2,5}$,banner$_2$,domino,$M_m$, $K$)-free graph class, then so it is the ($S_{1,2,5}$,banner$_2$,domino,$M_m$,$K^{(h)}$)-free graph class.*

Especially, Theorem 5 leads to some interesting polynomially solvable graph classes of the MIS problem. Remind that the MIS problem was proved to be polynomially solvable in $P_5$-free graphs [22], ($P_2$+claw)-free graphs [24], $2P_3$-free graphs [26], and $pK_2$-free graphs [1], we have the following consequence.

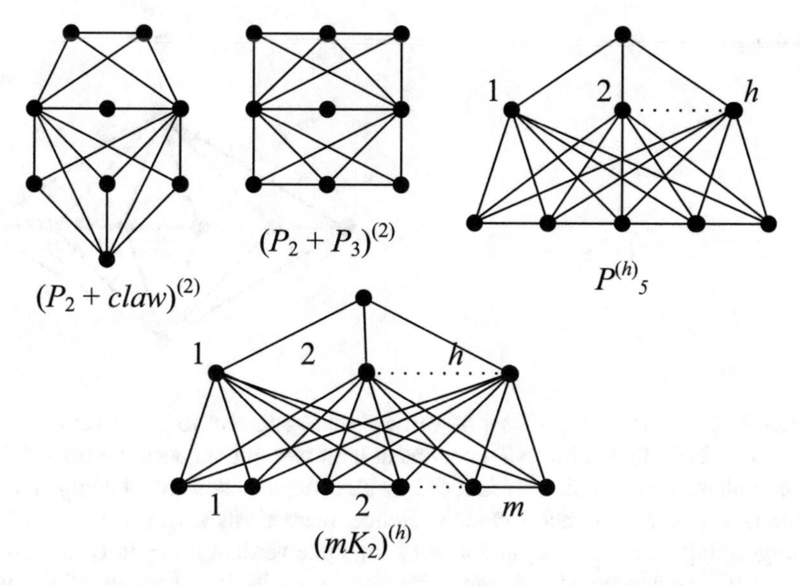


**Fig. 5** Special graphs in Corollary 5

**Corollary 5** *Given four integers $h, l, m, p$, the MIS problem is polynomially solvable in the following graph classes (see Figure 5):*

1. *$(S_{2,2,5}$,banner$_2$,domino,$M_m$, $R_l^1$, $R_l^2$, $R_l^3$, $P_5^{(h)}$)-free graphs,*
2. *$(S_{2,2,5}$,banner$_2$,domino,$M_m$, $R_l^1$, $R_l^2$, $R_l^3$, $(P_2 + claw)^{(2)}$)-free graphs,*
3. *$(S_{2,2,5}$,banner$_2$,domino,$M_m$, $R_l^1$, $R_l^2$, $R_l^3$, $(2P_3)^{(2)}$)-free graphs, and*
4. *$(S_{2,2,5}$,banner$_2$,domino,$M_m$, $R_l^1$, $R_l^2$, $R_l^3$, $(pK_2)^{(h)}$).*

Let tree$_r$ be the graph of the form tree[1] with parameter $r$ (Figure 2). Let $G = (V, E)$ be a graph, $U$ be a subset of $V$ and $u$ be a vertex of $G$ outside $U$. We say that $u$ *distinguishes* $U$ if $u$ has both a neighbor and a non-neighbor in $U$. A subset $U \subseteq V(G)$ is called a *module* in $G$ if it is indistinguishable for any vertex outside $U$. A module $U$ is *trivial* if $U$ is a single vertex or $V$ itself, otherwise it is *non-trivial*. A graph whose each module is trivial is called *prime*. It has been shown (for example in [27]) that if the problem is polynomially solvable for every prime graph of a graph class $\mathscr{X}$, then it is also polynomial solvable in $\mathscr{X}$. Using the modular decomposition technique described in [6] for $P_5$-free graphs we can extend Case 4., the case $h = 2$ of the above corollary, as follows.

**Corollary 6** *Given four integers $l$, $m$, $p$, and $r$, the MIS problem is polynomially solvable in $(S_{2,2,5}$,banner$_2$,domino,$M_m$, tree$_r$,$R_l^1$, $R_l^2$, $R_l^3$, $Q_p$)-free graph class (see Figure 3).*

*Proof* We show that a prime ($Q_p$,tree$_r$)-free graph is $((2p + r - 2)K_2)^{(2)}$-free. Indeed, let $G$ be a prime ($Q_p$,tree$_r$)-free graph, and suppose that $G$ contains an induced subgraph $Q'$ isomorphic to $((2p + r - 2)K_2)^{(2)}$.

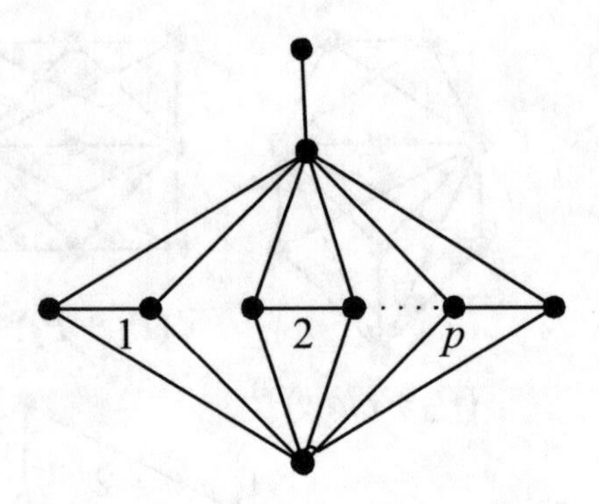


**Fig. 6** $Q_p$

Let $T \subseteq V(G)$ be the subset of vertices of $G$ adjacent to every vertex of the $(2p+r-2)K_2$ of $Q'$. Since $T$ contains at least two non-adjacent vertices, $\bar{G}[T]$, the complement subgraph of $G$ induced by $T$, contains a non-trivial component $C$. Since $G$ is prime, $C$ is not a module. Hence, there exists a vertex $v \in V(G) \backslash C$ distinguishing $C$, i.e. $v \sim c_1$ and $v \nsim c_2$ for some vertices $c_1, c_2$ in $C$. Moreover, since $\bar{G}[C]$ is connected, we can substitute $c_1, c_2$ by two vertices of the path connecting them and can assume that $c_1 \nsim c_2$ in $G$.

If $v$ is adjacent to every vertex of the $(2p+r-2)K_2$ of $Q'$, then $v \in T$ and since $v \nsim c_2$, $v \in C$, a contradiction. Hence, there exists a vertex $c'$ of the $(2p+r-2)K_2$ of $Q'$ such that $c' \nsim v$.

Since $G$ is tree$_r$-free, $v$ distinguishes at most $r-1$ edges of the $(2p+r-2)K_2$ of $Q'$. Then we have the two following cases.

*Case 1* $v$ is adjacent to both end-vertices of at least $p$ edges of the $(2p+r-2)K_2$ of $Q'$. Then $\{v, c', c_2\}$ together with these $p$ edges induce a $Q_p$, a contradiction.

*Case 2* $v$ is non-adjacent to both end-vertices of at least $p$ edges of the $(2p+r-2)K_2$ of $Q'$. Then $\{v, c_1, c_2\}$ together with these $p$ edges induce a $Q_p$, a contradiction (Figure 6).

## 4 Augmenting Graphs in Other Problems

In [19], we have extended the augmenting graph approach for a more more general combinatorial and graph-theoretical problem, say Maximum $\Pi$-set Problem. Given a graph $G$, the problem asks for a maximum vertex subset such that the induced subgraph satisfies some give properties $\Pi$. Here are some examples for the property $\Pi$ and related problems.

**Maximum $k$-Independent Set.** [13] $\Pi$: Every vertex is of degree at most $k-1$. Note that the Maximum Independent Set is the case $k = 1$.

**Maximum $k$-Path Free Set.** $\Pi$: The graph contains no path (not necessarily induced) of $k$ vertices ($k \geq 2$), also called $k$-path free. This problem is a dual version of the Minimum Vertex $k$-Path Cover problem [7].

**Maximum Forest.** $\Pi$: The graph contains no cycle. This problem is a dual version of the Minimum Feedback Vertex Cover problem [12].

**Maximum Induced Bipartite Subgraph.** $\Pi$: The graph contains no cycle of odd length.

**Maximum $k$-Acyclic Set.** $\Pi$: The graph contains no cycle of length at most $k$.

**Maximum $k$-Chordal Set.** $\Pi$: The graph contains no cycle of length larger than $k$.

**Maximum $k$-Cycle Free Set.** $\Pi$: The graph contains no cycle of length $k$ ($k \geq 3$), also called $k$-cycle free. This problem is a dual version of the Minimum Vertex $k$-Cycle Cover problem.

**Maximum Induced Matching.** [8] $\Pi$: Every vertex is of degree one.

**Maximum $k$-Regular Induced Subgraph.** [9] $\Pi$: Every vertex is of degree $k$.

**Maximum $k$-Regular Induced Bipartite Subgraph.** [9] $\Pi$: The graph is bipartite and every vertex is of degree $k$.

**Maximum Induced $k$-Cliques.** $\Pi$: Every connected component is a $k$-clique. This problem is a generalization of Maximum Induced Matching problem ($k = 2$).

We have considered two special cases of the problem. First, the property $\Pi$ is hereditary (i.e., if a graph $G$ satisfies $\Pi$, then every induced subgraph of $G$ satisfies $\Pi$) and additive (i.e., a graph $G$ satisfies $\Pi$ if and only if every connected component of $G$ satisfies $\Pi$). Second, $\Pi$ is of the form $\mathscr{F}$-induced subgraph, i.e. every connected component of $G$ belongs to some graph set $\mathscr{F}$. In both cases, we have defined the augmenting graphs and the key theorem, says the $\Pi$-set is maximum if and only if there exists no augmenting graph. We also have considered a simple case, says the ($S_{1,2,l}$,banner$_l$,$K_{1,m}$)-free minimal augmenting graph either belongs to a finite set or is augmenting extended-chain. By showing that we can find augmenting extended-chain in polynomial time for the above problem, we obtained polynomial algorithms for these non-trivial problems in ($S_{1,2,l}$,banner$_l$,$K_{1,m}$)-free graphs.

## 5 Conclusion

In this paper, we have combined the methods applied for $P_5$-free graphs and ($S_{1,2,5}$,banner)-free graphs to generalize some known results. By extending the method of Lozin and Milanič [23] for ($S_{1,2,5}$,banner)-free graphs, we show that the problem can be restricted to finding augmenting chains and augmenting bipartite-chains in ($S_{2,2,5}$,banner$_2$,domino, $M_m$)-free graphs by using concepts of redundant sets (in the extended senses). It leads us to some generalizations of results about ($P_5$, $K_{2,m} - e$)-free graphs [4], ($P_5$, $K_{3,3} - e$)-free graphs [16], and augmenting vertex in $P_5$-free graphs [14, 16, 29, 30]. It also leads to some interesting results in ($S_{2,2,5}$,banner$_2$,domino,$M_m$)-free graphs, e.g. Corollaries 5 and 6.

Note that $S_{1,1,2}$ (fork) and $S_{0,1,3}$ ($P_5$) are the largest single known forbidden subgraphs, for which the MIS problem is polynomially solvable. For $S_{i,j,k}$ such that $i + j + k \geq 5$, even for subclasses, to the best of our knowledge, there are still not many known results except in some subclasses of $P_6$-free graphs, graphs of bounded maximum degree, planar graphs, ($S_{1,2,5}$,banner)-free graphs [23], ($S_{1,1,3}$, $K_{p,p}$)-free graphs [10], and ($S_{1,2,l}$,banner$_l$,$K_{1,m}$)-free graphs [19]. Combining different techniques is a potential approach helping us extend these results to tackling the general question about complexity of the MIS problem in $S_{i,j,k}$-free graphs.

Besides, by applying a technique, which has been used for $P_5$-free graphs, for a larger graph class, e.g. $S_{2,2,5}$-free graphs, we believe that it is possible to apply other techniques, which were used in $P_5$-free graphs, in $S_{2,2,l}$-free graphs.

The augmenting graph technique is also very potential in many other non-trivial combinatorial and graph-theoretical problems.

**Acknowledgements** This research is supported by National Foundation for Science and Technology Development (NAFOSTED) of Vietnam, project code: 101.99-2016.20. We would like to express our special thanks to an anonymous reviewer for his/her very useful suggestions and comments.

## Appendix 1: Proof of Lemma 2

*Proof* (of Lemma 2) Let $H = (B, W, E)$ be a minimal augmenting graph. If $\Delta(H) = 2$, then $H$ is a cycle or a chain. Since $H$ is bipartite and $|B| = |W| + 1$, $H$ cannot be a cycle. Now, assume that $H$ is not a chain. We show that either (i) there exists some vertex $a$ such that there is no vertex of distance $2k + l + 1$ from $a$ or (ii) $H$ is an augmenting extended-chain or augmenting apple. Note that, every vertex of $H$ is of degree at most $m - 1$, otherwise an induced $K_{1,m}$ appears, a contradiction. Since $H$ is connected, if we have (i), then

$$|V(H)| \leq \sum_{i=0}^{2k+l+1} (m-1)^i = \frac{1-(m-1)^{2k+l+2}}{2-m},$$

i.e., $H$ belongs to some finite set of augmenting graphs.

If a white vertex $w \in W$ has two black neighbor $b_1, b_2$ of degree one, then $\{b_1, a, b_2\}$ is an augmenting $P_3$, a contradiction. Hence, we have the following observation.

*Claim 1* Every white vertex of $H$ has at most one black neighbor of degree one. In particular, if a white vertex $w$ is of degree at least four, then there are at least three neighbors of $w$ of degree two.

*Claim 2* Either $H$ contains a vertex, say $a$, of degree at least three and $a$ has at least three neighbors of degree at least two or $H$ is an augmenting apple.

*Proof* Since $H$ is neither a chain or a cycle, there exists at least one vertex of degree at least three.

By Corollary 2, every white vertex of $H$ is of degree at least two, i.e. every white neighbor of a black vertex has another black neighbor. Hence, if $H$ contains a black vertex of degree three, then this vertex is a desired vertex $a$.

Hence, we assume that **(1)** every black vertex of $H$ is of degree at most two. If there exist two black vertices of degree one, then by (1), the path connecting these two black vertices is an augmenting chain, a contradiction. Hence, we assume that **(2)** there exists at most one black vertex of degree one.

By Claim 1, there exists no white vertex of degree four or we have a desired vertex $a$. Moreover, if there exist two white vertices of degree three, then either one of them has three neighbors of degree two, i.e. we have a desired vertex $a$, or we have two black vertex of degree one.

Now, if every white vertex of $H$ is of degree two except one of degree three whose one black neighbor is of degree one, then $H$ is an augmenting apple.

Let $a$ be a vertex in the conclusion of the above claim. Denote by $V_i$ the subset of vertices of $H$ of distance $i$ from $a$. Let $a_p$ be the vertex of maximum distance from $a$ and assume that $p \geq 2k + l + 1$. Let $P := (a_0, a_1, \ldots, a_p)$, where $a_i \in V_i$, be a shortest path connecting $a = a_0$ and $a_p$. Let $V_1 = \{a_1, b_{1,1}, b_{1,2}, \ldots\}$, and $b_{i+1,j}$ be a vertex of $N_{V_{i+1}}(b_{i,j})$, if such one exists. By the assumption about $a$, $b_{2,1}$, and $b_{2,2}$ exist (note that they may coincide).

We show that $a_i \nsim b_{i+1,1}$ and $a_{i+1} \nsim b_{i,1}$ for $i = 1, 2, \ldots, 2k$ by induction. Note that it also implies that $b_{i,j} \neq a_i$ for every $i$, $j$.

If $a_2 \sim b_{1,1}$, then $\{b_{1,1}, a, a_1, a_2, a_3, \ldots, a_{l+2}\}$ induces a banner$_l$, a contradiction.

If $a_1 \sim b_{2,1}$, then either $\{b_{2,1}, b_{1,1}, a, a_1, a_2, \ldots, a_{l+1}\}$ or $\{b_{2,1}, a_1, a_2, a_3, a_4, \ldots, a_{l+3}\}$ induces a banner$_l$ depending on $a_3 \sim b_{2,1}$ or not, a contradiction.

Now, by induction hypothesis, consider $2 \leq i \leq k$. If $a_i \sim b_{i+1,1}$, then either $\{b_{i+1,1}, a_i, a_{i+1}, a_{i+2}, a_{i+3}, \ldots, a_{i+l+2}\}$ induces a banner$_l$ or $\{b_{i+1,1}, b_{i,1}, \ldots, b_{1,1}, a, a_1, \ldots, a_i, a_{i+1}, \ldots, a_{i+l}\}$ induces an apple$^l_{2i+2}$ depending on $a_{i+2} \sim b_{i+1}$ or not, a contradiction. If $a_{i+1} \sim b_{i,1}$ for $2 \leq i \leq k$, then $\{b_{i,1}, b_{i-1,1}, \ldots, b_{1,1}, a, a_1, a_2, \ldots, a_{i+1}, a_{i+2}, \ldots, a_{i+l+1}\}$ induces an apple$^l_{2i+2}$, a contradiction.

Again, by induction hypothesis, consider $k + 1 \leq i \leq 2k$. If $a_i \sim b_{i+1,1}$, then either $\{b_{i+1,1}, a_i, a_{i+1}, a_{i+2}, a_{i+3}, \ldots, a_{i+l+2}\}$ induces a banner$_l$ or $\{a_{i-1}, a_{i-2}, \ldots, a_1, a, b_{1,1}, b_{2,1}, \ldots, b_{i+1,1}, a_i, a_{i+1}, a_{i+2}, \ldots, a_{i+l}\}$ induces an $S_{2,2k,l}$ depending on $a_{i+1} \sim b_{i,1}$ or not, a contradiction. If $a_{i+1} \sim b_{i,1}$, then $\{a_i, a_{i-1}, a_{i-2}, \ldots, a_1, a, b_{1,1}, b_{2,1}, \ldots, b_{i,1}, a_{i+1}, a_{i+2}, \ldots, a_{i+l+1}\}$ induces an $S_{2,2k,l}$, a contradiction.

Hence, $a_i$ has only one neighbor, say $a_{i+1}$, in $V_{i+1}$ and only one neighbor, say $a_{i-1}$, for $i = 1, 2, \ldots, 2k$.

If $b_{i,1} \sim b_{i+1,2}$ for some $1 \le i \le 2k-1$ (if such two vertices exist), then $\{b_{1,1}, \ldots, b_{i,1}, b_{i+1,2}, b_{i,2}, \ldots, b_{1,2}, a, a_1, \ldots, a_l\}$ induces an $\text{apple}^l_{2i+2}$, a contradiction. Hence, $b_{i,j}$ (if such vertex exists) has at most one neighbor in $V_{i+1}$ for $1 \le i \le 2k-1$. It also implies that $b_{i,j} \ne b_{i,k}$ for every $1 \le i \le 2k$ and $j \ne k$ if such vertices exist.

If $V_{2k}$ contains at least two vertices, say $a_{2k}$ and, without loss of generality, $b_{2k,1}$, then $\{b_{2,2}, b_{1,2}, a, b_{1,1}, b_{2,1}, \ldots, b_{2k,1}, a_1, a_2, \ldots, a_l\}$ induces an $S_{2,2k,l}$, a contradiction.

To summarize, $V_{2k} = \{a_{2k}\}$, every vertex of $V_i$ has only one neighbor in $V_{i-1}$, for every $1 \le i \le p$.

Let $T$ be the connected component of $H - a_1$ containing $a$. Then $T$ is a tree by the above arguments. We show that $a$ is black. Indeed, for contradiction, suppose that $a$ is white. Let $a_1$ be the black vertex $b$ of Corollary 1. Then there is a perfect matching between $B \cap T$ and $W \cap T$. Let $b$ be a leaf of $T$. Then by Corollary 2, $b$ is black and hence $\mu(b)$ be the (only) white neighbor of $b$. It also implies that $\mu(b)$ has only one neighbor being a leaf. Indeed, if $\mu(b)$ has another black neighbor being a leaf $b'$, then there exists no $\mu(b')$, a contradiction. Then by induction on $T$, $a$ has only one black neighbor in $T$, a contradiction to $a$ is of degree at least three. Hence, we have the following claim.

*Claim 3* If $a$ is a vertex of the conclusion of Claim 2, then $a$ is black. Moreover, there exists a neighbor $w$ of $a$ such that the connected component of $H - w$ containing $a$ is a tree $T$, every vertex of $T$ is of distance at most $2k-2$ to $a$, and every white vertex of $T$ is of degree two.

Let $a$ be the black vertex $b$ of Corollary 1. Then there is a perfect matching between $B \cap T \backslash \{a\}$ and $W \cap T$, i.e. $|B \cap T| = |W \cap T| + 1$. Claims 1 and 3 lead to the following observation.

*Claim 4* Every white vertex $w$ of $H$ is either of degree two or three. Moreover, in the latter case, exactly one black neighbor of $w$ is of degree one.

Let $j$ be the largest number such that $|V_j| \ge 2$. Then $2 \le j \le 2k-2$. Moreover, $j$ is even, since every leaf of $T$ is black.

Note that every black vertex $a_q$ such that $2k - j < q < p - 2k$ is of degree two, otherwise $a_q$ becomes a vertex of the conclusion of Claim 2 and there exist at least two vertices of degree $2k$ from $a_q$, a contradiction to Claim 3.

Let $T_1$ and $T_2$ be the two connected component of $H - a_{2k-j+1} - a_{p-2k-1}$ containing $a_{2k-j}$ and $a_{p-2k}$, respectively. Then by Claim 3, $T_1$ and $T_2$ are trees such that the most distance between a vertex of $T_1$ (respectively, $T_2$) to $a_{2k-j}$ (respectively, $a_{p-2k}$) is $2k-2$. Moreover $|W \cap (T_1 + T_2)| + 2 = |B \cap (T_1 + T_2)|$.

Now, every white vertex $a_q$, where $2k - j < q < p - 2k$, is of degree two or three, and in the later case a black neighbor of $a_q$ different from $a_{q-1}$ and $a_{q+1}$ is of degree one. Hence, every such white vertex is of degree two, otherwise we have a contradiction to $|W| + 1 = |B|$.

Thus, $H$ is an augmenting $(2k-1, m)$-extended-chain.

## Appendix 2: Proof of Lemma 3

We go through the Proof by first obtaining some results related to the cases when the considered augmenting graph contains a $K_{1,m}$ as an induced subgraph.

**Lemma 8** *Let $G = (X, Y, E)$ be a bipartite graph such that there exists a vertex $x \in X$ and $N_Y(x) = Y$. Assume that $|X| = m+1$. Then at least one of the following statements is true.*

1. *$H_2$ contains a banner$_2$ or a domino.*
2. *We can linearly order $X = (x, x_1, x_2, \ldots, x_m)$ so that there exists a natural $p$, with $0 \leq p \leq m$, such that (i) $N_Y(x_1) \supseteq \ldots \supseteq N_Y(x_p)$ and $|N_Y(x_i)| = 1$ for every $i \geq p+1$. Moreover, if $p \geq m-1$, then $G$ is a bipartite-chain.*

*Proof* First, assume that Case 1 does not happen. We linearly order $X$ by the construction method.

Assume that we already have chosen $x_1, \ldots, x_p$. Let $U = X \backslash \{x, x_1, \ldots, x_p\}$. Let $x_{p+1} \in U$ be a vertex such that $|N_Y(x_{p+1})|$ is largest among vertices in $U$. Suppose that $|N_Y(x_{p+1})| \geq 2$ and there exists a vertex $x_i \in U \backslash \{x_{p+1}\}$ such that $x_i \sim y_i$ and $x_{p+1} \nsim y_i$ for some $y_i \in Y$. By the choice of $x_{p+1}$, $x_i \nsim y_j$ for some $y_j \in N_Y(x_{p+1})$. Then $\{x, y_k, y_i, y_j, x_{p+1}, x_j\}$ induces a domino or a banner$_2$ for some $y_k \in N_Y(x_{p+1}) \backslash \{y_j\}$ depending on $x_i \sim y_k$ or not and $x$ is a vertex of degree three in both cases, a contradiction.

If $p \geq m-1$, then $N_Y(x) \supseteq N_Y(x_i) \supseteq N_Y(x_j)$ for every $1 \leq i < j \leq m$. We show that for $y_i, y_j \in Y$, either $N_X(y_i) \subseteq N_X(y_j)$ or $N_X(y_j) \subseteq N_X(y_i)$. Indeed, suppose that $y_i \sim x_i$ and $y_j \sim x_j$ for some $x_i \in X \backslash N(y_j)$ and $x_j \in X \backslash N(y_i)$. Then $N_Y(x_i) \nsubseteq N_Y(x_j)$ and $N_Y(x_j) \nsubseteq N_Y(x_i)$, a contradiction.

**Lemma 9** *If an ($S_{2,2,5}$,banner$_2$,domino)-free minimal augmenting graph $H$ contains no black vertex of degree more than $k$ ($k \geq 2$), then the degree of each white vertex is at most $k^2+k+2$.*

*Proof* Suppose that $H$ contains a white vertex $w$ of degree more than $k^2+k+2$. Denote by $V_j$ the set of vertices of $H$ at distance $j$ from $w$. Hence, $|V_1| \geq k^2+k+3$.

*Claim 5* $|V_2| \geq k^2+k+1$, $V_2$ contains at least $k^2+1$ vertices having only one neighbor in $V_1$, i.e. having a neighbor in $V_3$, and $|V_3| \geq k+1$.

*Proof* Suppose that $V_3 = \emptyset$. Then by Lemma 1, $|V_2| = |V_1| - 2 \geq k^2+k+1$. Let $p$ be the $p$ in 2. of Lemma 8. Note that $p \leq k$, otherwise there exists a black vertex in $V_1$ having at least $k$ neighbors in $H$, a contradiction. Hence, by Lemma 8, there exists a white vertex in $V_2$ having only one neighbor in $V_1$, i.e. only one black neighbor. This contradiction (with Corollary 2) implies that $V_3 \neq \emptyset$.

Then $|V_2| \geq k^2+k+1$, otherwise $H[\{b\} \cup V_1 \cup V_2]$ is an augmenting graph, a contradiction. Again, by Lemma 8 and condition that there is no black vertex of degree larger than $k$, $V_2$ contains at least $k^2+1$ vertices having only one neighbor in $V_1$, i.e. having a neighbor in $V_3$ by Corollary 2. Since every black vertices of $V_3$ has at most $k$ neighbors in $V_2$, $|V_3| \geq k+1$. □

*Claim 6* $V_4 = \emptyset$, i.e. $|V_3| + |V_1| = |V_2| + 2$.

*Proof* Suppose that $V_4$ contains a (white) vertex $x$ and let $y$ be its neighbor in $V_3$. Assume that $y \sim w_1 \in V_2$ and $w_1 \sim b_1 \in V_1$.

If $w_1 \sim b_2$ for some $b_2 \in V_1 \backslash \{b_1\}$, then $\{b_1, w, b_2, w_1, y, x\}$ induces a banner$_2$, a contradiction. Hence, $N_{V_1}(w_1) = \{b_1\}$.

By Corollary 2, $x$ has at least one more black neighbor, named $z$ ($z \in V_3$ or $z \in V_5$). Now, let $b_1$ be the $b$ in Corollary 1. We have $|\mu(V_1 \backslash \{b_1, \mu(w)\})| \geq k^2 + k + 1$. Since $d(y), d(z) \leq k$, $V_1$ contains at least two vertices, named $b_2, b_3$ whose the neighbors, say $w_2 = \mu(b_2)$, $w_3 = \mu(b_3) \in V_2$, respectively, adjacent to neither $y$ nor $z$.

If $w_2 \sim b_1$, then $\{w, w_1, w_2, b_1, b_2, y\}$ induces a domino or a banner$_2$, depending on $y \sim w_2$ or not, a contradiction. If $w_2 \sim b_4$ for some $b_4 \in V_1 \backslash \{b_1, b_2\}$, then $\{b_2, w_2, b_4, w, b_1, w_1\}$ induces a banner$_2$, a contradiction. Hence, $w_1$, $w_2$, and $w_3$, each has only one neighbor in $V_1$. Moreover, $z \nsim w_1$, otherwise, $\{y, x, z, w_1, b_1, w\}$ induces a banner$_2$, a contradiction. Now, $\{w_3, b_3, w, w_2, b_2, b_1, w_1, y, z, x\}$ induces an $S_{2,2,5}$, a contradiction. Therefore, $V_4$ is empty and $|V_3| + |V_1| = |V_2| + 2$ by Lemma 1. □

Let $b \in V_3$ be the vertex $b$ in Corollary 1. Since $\mu(w)$ has at most $k - 1$ neighbors in $\mu(V_3 \backslash \{b\})$, there exists a vertex $d_1 \in V_3$ such that $\mu(d_1) \nsim \mu(w)$.

*Claim 7* $\mu(d_1)$ has a neighbor $a_1$ in $V_1$ such that $\mu(a_1)$ has no neighbor in $V_1$ other than $a_1$.

*Proof* Let $a_1$ be a neighbor of $\mu(d_1)$ in $V_1$, i.e. $\mu(a_1) \neq w$. If $\mu(a_1)$ has no neighbor in $V_1$ other than $a_1$, then we have the statement of the claim. Now, let $a_2$ be a neighbor of $\mu(a_1)$ in $V_1$. Then $\mu(d_1) \sim a_2$, otherwise $\{w, a_2, \mu(a_1), a_1, \mu(d_1), d_1\}$ induces a domino or a banner$_2$ depending on $d_1 \sim \mu(a_1)$ or not, a contradiction. It implies that $\mu(a_2) \neq w$. We continue considering $\mu(a_2)$. Since $V_1$ is finite, this process must stop, i.e. we have the claim. □

Note that $d_1 \nsim \mu(a_1)$, otherwise $\{\mu(a_1), d_1, \mu(d_1), a_1, w, \mu(w)\}$ induces a domino or a banner$_2$ depending on $\mu(w) \sim \mu(a_1)$ or not, a contradiction. Since $\mu(a_1)$ has no neighbor in $V_1$ other than $a_1$, by Corollary 2, $\mu(a_1)$ has a neighbor $d_2 \in V_3$.

Since $|N_{V_2}(\{d_1, d_2\})| \leq 2k$, there is a vertex $a \in V_1 \backslash \{\mu(w)\}$ such that $\mu(a)$ is not adjacent to $d_1, d_2$. Then $\mu(a) \nsim a_1$, otherwise $\{w, a, \mu(a), a_1, \mu(a_1), d_2\}$ induces a banner$_2$, a contradiction. If $\mu(a) \sim a_2$ for some $a_2 \in V_1$, then $\{a, \mu(a), a_2, w, a_1, \mu(a_1)\}$ induces a banner$_2$, a contradiction. Hence, $\mu(a)$ has only one neighbor in $V_1$ and has a neighbor, named $d_3 \in V_3$, by Corollary 2.

Then $\mu(d_1) \nsim a$, otherwise $\{w, a_1, a, \mu(d_1), d_3, \mu(a)\}$ induces a domino or a banner$_2$ depending on $d_3 \sim \mu(d_1)$ or not, a contradiction. Moreover $\mu(d_3) \nsim a$, otherwise $\{w, a_1, a, \mu(a), d_3, \mu(d_3)\}$ induces a domino or a banner$_2$ depending on $a_1 \sim \mu(d_3)$ or not, a contradiction.

We show that $d_1 \nsim \mu(d_3)$. Indeed, if $d_1 \sim \mu(d_3)$, then $\mu(d_1) \nsim d_3$, otherwise $\{\mu(d_3), d_1, \mu(d_1), d_3, \mu(a), a\}$ induces a banner$_2$, a contradiction. If $\mu(d_1) \sim a_2$ for some $a_2 \in V_1 \backslash \{a_1\}$, then $\{w, a_1, \mu(d_1), a_2, d_1, \mu(d_3)\}$ induces a domino or a banner$_2$ depending $\mu(d_3) \sim a_2$ or not, a contradiction. If $\mu(d_3)$ has two neighbors $a_2, a_3 \in V_1\backslash\{a_1\}$, then $\{a_2, w, a_3, \mu(d_3), d_1, \mu(d_1)\}$ induces a banner$_2$, a contradiction. Hence, $\mu(d_1)$ has only one neighbor in $V_1$ and $\mu(d_3)$ has at most one neighbor in $V_1$ different from $a_1$. Thus, because $|N_{V_2}(d_1, d_2)| \leq 2k$, there exist two vertices $b_1, b_2 \in V_1\backslash\{\mu(w)\}$ such that $\mu(b_1), \mu(b_2)$ each has only one neighbor in $V_1$ and is not adjacent to $d_1, d_3$. Now, $\{\mu(b_1), b_1, w, b_2, \mu(b_2), a_1, \mu(d_1), d_1, \mu(d_3), d_3\}$ induces an $S_{2,2,5}$, a contradiction.

Similarly, $d_3$ is not adjacent to $\mu(d_1)$, $\mu(a_1)$, and $\mu(d_3) \nsim d_2$. Moreover $\mu(d_1) \nsim d_2$, otherwise $\{w_1, d_2, \mu(d_1), a_1, w, \mu(w)\}$ induces a banner$_2$, a contradiction. Similarly, $\mu(a_1) \nsim d_1$.

Now, $\{d_2, \mu(a_1), a_1, \mu(d_1), d_1, w, a, \mu(a), d_3, \mu(d_3)\}$ induces an $S_{2,2,5}$, a contradiction. □

*Proof (of Lemma 3)* We proof by contradiction. Let $b \in B$ such that $|N_W(b)|$ is largest. If every black vertex is of degree one, then $H$ is an augmenting $P_3$. If $N_W(b) = W$, then we have 4. By Lemma 9, if every black vertex of $H$ is of degree bounded by a given number $k$, then every white vertex of $H$ is of degree bounded by $k^2 + k + 2$, i.e. $H$ is $K_{1,m}$-free for $m = k^2 + k + 3$. In this case, by Lemma 2, we have 1. or 2.

Now, we assume that $10 \leq |N_W(b)| \leq |W| - 1$. Let $b$ be the vertex $b$ of Corollary 1. Let $A = N(b) = \{w_1, w_2, \ldots, w_k\}$ $(k \geq 10)$, $C = W\backslash A$, i.e. $C \neq \emptyset$. Let $b_i = \mu(w_i)$. Let $C_1$ denote the set of vertices in $C$ having at least one neighbor in $\mu(A)$ and $C_0 = C\backslash C_1$. By the connectivity of $H$, one can choose $\mu(A)$ in order that $C_1 \neq \emptyset$. We have the following observations.

*Claim 8* $H[A \cup \mu(A)]$ is an induced sub-matching of $M$.

*Proof* We show that $b_i \nsim w_j$ for every pair $i, j$ such that $i \neq j$, $1 \leq i, j \leq k$. Let $z \in C_1$ and without loss of generality, assume that $z \sim b_1 \in \mu(A)$.

By the choice of $b$, $b_1$ is not adjacent to all $w_i$'s, without loss of generality, assume that $b_1 \nsim w_2$.

Now, $b_2 \nsim w_1$, otherwise $\{b, b_1, b_2, w_1, w_2, z\}$ induces a domino or a banner$_2$ depending on $b_2 \sim z$ or not, a contradiction.

Moreover, $b_2 \nsim w_i$ for every $i > 2$, otherwise $\{b, b_1, b_2, w_1, w_2, w_i\}$ induces a domino or a banner$_2$ depending on $b_1 \sim w_i$ or not, a contradiction.

Now, $b_1 \nsim w_i$, for every $i > 2$, otherwise $\{w_1, b_1, w_i, b, w_2, b_2\}$ induces a banner$_2$, a contradiction.

Hence, $b_i \nsim w_1$ for $i > 2$, otherwise $\{b, w_i, b_i, w_1, b_1, z\}$ induces a domino or a banner$_2$, depending on $z \sim b_i$ or not, a contradiction.

Thus, $b_i \nsim w_2$ for $i > 2$, otherwise $\{w_2, b_i, w_i, b, w_1, b_1\}$ induces a banner$_2$, a contradiction.

Moreover $b_i \nsim w_j$, for any $j \neq i$ and $i, j > 2$, otherwise $\{w_j, b_i, w_i, b, w_1, b_1\}$ induces a banner$_2$, a contradiction.

*Claim 9* There exists no vertex pair $z_1, z_2 \in C_1$ sharing two neighbors in $\mu(A)$.

*Proof* Suppose that there exists a vertex pair $z_1, z_2 \in C_1$ sharing two neighbors in $\mu(A)$, without loss of generality, assume that they are $b_1, b_2$. Then $\{z_1, b_2, z_2, b_1, w_1, b\}$ induces a banner$_2$, a contradiction. □

*Claim 10* Given $z \in C_1$, $z \sim b_j$ for some $b_j \in \mu(A)$, a black neighbor $c$ of $z$ different from $b_j$, a black neighbor $\mu(t)$ of $z$ for some $t \in C$, and another white neighbor $y \in C$ of $\mu(t)$ different from $z$, the following statements are true:

1. $c \nsim w_j$;
2. $y \nsim b_j$ and $\mu(y) \nsim z$; and
3. if $\mu(t) \sim w_i$ for some $i \neq j$, then $y, z$ are not adjacent to $b_i$ and $\mu(y) \nsim w_i$;
4. in particular, $\mu(y)$ and $\mu(t)$ cannot share a same neighbor in $A$.

*Proof* Suppose that $c \sim w_j$. Then $c \sim w_i$ for every $i \neq j$, otherwise $\{b_j, z, c, w_j, b, w_i\}$ induces a banner$_2$, a contradiction. But now, we have a contradiction to the choice of $b$.

Now, $y \nsim b_j$, otherwise $\{z, \mu(t), y, b_j, w_j, b\}$ induces a banner$_2$, a contradiction. Moreover, $\mu(y) \nsim z$, otherwise $\{w_j, b_j, z, \mu(t), y, \mu(y)\}$ induces a domino or a banner$_2$ depending on $\mu(y) \sim w_j$ or not, a contradiction.

Assume that $\mu(t) \sim w_i$ for some $i \neq j$. Then $z \nsim b_i$, otherwise $\{\mu(t), w_i, b_i, z, b_j, w_j\}$ induces a banner$_2$, a contradiction. Hence, $y \nsim b_i$, otherwise $\{b_i, y, \mu(t), w_i, b, w_j\}$ induces a banner$_2$, a contradiction. Now, $\mu(y) \nsim w_i$, otherwise $\{w_i, \mu(y), y, \mu(t), z, b_j\}$ induces a banner$_2$, a contradiction. □

*Claim 11* Every black vertex different from $b$ has at most one neighbor in $A$.

*Proof* Clearly, every black vertex of $\mu(A)$ has only one neighbor in $A$ by Claim 8. Now, suppose that there exists some black vertex $y \in B \backslash (\{b\} \cup \mu(A))$ having two neighbors, without loss of generality, assume that they are $w_1, w_2 \in A$. Then $y$ is adjacent to every vertex $w_i \in A \backslash \{w_1, w_2\}$, otherwise $\{w_1, y, w_2, b, w_i, b_i\}$ induces a banner$_2$, contradiction. Now, $y$ is adjacent to every vertex of $A$ and $\mu(y)$, a contradiction to the choice of $b$. □

*Claim 12* There exists no vertex $b_j \in \mu(A)$ having two neighbors $z_1, z_2 \in C_1$ sharing another black neighbor, named $c \neq b_j$.

*Proof* Indeed, otherwise, by Claim 10, $c \nsim w_j$, then $\{z_1, c, z_2, b_j, w_j, b\}$ induces a banner$_2$, a contradiction. □

*Claim 13* Given a vertex $b_j \in \mu(A)$, let $C(b_j)$ be the set of vertices of $C_1$ adjacent to $b_j$. Then $H[C(b_j) \cup \mu(C(b_j))]$ is an induced sub-matching of $M$.

*Proof* For contradiction, without loss of generality, suppose that $z_1, z_2 \in C$ are two neighbors of $b_j$ and $z_1 \sim \mu(z_2)$. By Claim 10, $\mu(z_2) \nsim w_j$. Hence, $\{z_1, \mu(z_2), z_2, b_j, w_j, b\}$ induces a banner$_2$, a contradiction. □

*Claim 14* If $H$ contains a vertex $y \in C_1$ adjacent to at least $k-3$ vertices of $\mu(A)$, then either $H$ is of the form tree$^5$ or tree$^6$ or $H$ contains a redundant set $U$ of size at most 32, such that $H - U$ is of the form either tree$^1$, tree$^4$, tree$^5$, or tree$^6$.

*Proof* Let $D_1$ be the subset of vertices of $C_1$ sharing some neighbor in $\mu(A)$ with $y$, $A_1$ be the vertex subset of $A$ such that $\mu(A_1) = N_{\mu(A)}(y)$, $A_2 = A\backslash A_1$, $E_1$ be the vertices subset of $C_1$ adjacent to some vertex in $\mu(A_2)$. Without loss of generality, assume that $w_1, w_2, \ldots, w_{k-3} \in A_1$. We have the following observations.

**(1)** $y$ has no neighbor in $\mu(D_1)$ and $\mu(y)$ has no neighbor in $A_1 \cup D_1$. Indeed, by Claim 10, $\mu(y)$ has no neighbor in $A_1$. If for some $z \in D_1$, without loss of generality, assume that $z \sim b_1$, $y \sim \mu(z)$, then $y \nsim b_1$, by Claim 10, a contradiction. Moreover, since $\mu(y) \nsim w_1$, $\mu(y) \nsim z$, otherwise $\{z, \mu(y), y, b_1, w_1, b\}$ induces a banner$_2$, a contradiction.

**(2)** By Claim 9, every vertex of $D_1$ has exactly one neighbor in $\mu(A_1)$. In particular, every vertex of $C_1\backslash\{y\}$ has at most four neighbors in $\mu(A)$. Moreover, there exists only one vertex $y \in C_1$ adjacent to at least $k-3$ vertices in $\mu(A)$.

**(3)** Any two vertices of $D_1$ have different neighbors in $\mu(A_1)$. Indeed, without loss of generality, suppose that $z_1, z_2 \in D_1$ both are adjacent to $b_1$. By Claim 11, and since $|A_1| = k-3 \geq 7$, there exist $w_i, w_j \in A_1$ different from $w_1$ and not adjacent to $\mu(z_1), \mu(z_2)$. By (2) and Claim 13, $\{\mu(z_1), z_1, b_1, z_2, \mu(z_2), y, b_i, w_i, b, w_j\}$ induces an $S_{2,2,5}$, a contradiction.

**(4)** Similar to Claim 13, let $C(y)$ be the subset of vertices of $C_0$ adjacent to $\mu(y)$. Then $H[C(y) \cup \mu(C(y))]$ is an induced sub-matching of $M$.

**(5)** Similarly to (3) (using (4)), there is at most one vertex of $C_0$ adjacent to $\mu(y)$.

**(6)** $H[(C_1\backslash\{y\}) \cup \mu(C_1\backslash\{y\})]$ is an induced sub-matching of $M$. Indeed, suppose that for a couple of vertices $z_1, z_2 \in C_1\backslash\{y\}$, $z_1 \sim \mu(z_2)$. Without loss of generality, assume that $z_1, z_2$ are adjacent to $b_{i_1}, b_{i_2} \in \mu(A)$, respectively. Then by Claim 10, $\mu(z_2) \nsim w_{i_2}$. Hence, $z_1 \nsim b_{i_2}$, otherwise $\{z_2, \mu(z_2), z_1, b_{i_2}, w_{i_2}, b\}$ induces a banner$_2$, a contradiction. By (**2**) and Claim 11, there exists a pair of vertices $b_i, b_j \in \mu(A)$ not adjacent to $z_1, z_2$ such that $w_i$ and $w_j$ are not adjacent to $\mu(z_1)$, $\mu(z_2)$. Now, $\{b_i, w_i, b, w_j, b_j, w_{i_2}, b_{i_2}, z_2, \mu(z_2), z_1\}$ induces an $S_{2,2,5}$, a contradiction.

**(7)** There exists no vertex $t \in C\backslash\{y\}$ having a neighbor in $\mu(C_1\backslash\{y, \mu(t)\})$. Indeed, if $t \in C$ is adjacent to $\mu(z)$ for some $z \in C_1\backslash\{y, t\}$, then for the vertex $b_j$ adjacent to $z$, $t \nsim b_j$ by Claim 10. By (2) and Claim 13, there exists a pair of vertices $w_i, w_l$ non-adjacent to $\mu(z)$ such that $b_i, b_l$ non-adjacent $z, t$. Now, $\{b_i, w_i, b, w_l, b_l, w_j, b_j, z, \mu(z), t\}$ induces an $S_{2,2,5}$, a contradiction.

**(8)** Similarly, there exists no vertex $t \in C_1\backslash\{y\}$ having a neighbor in $\mu(C\backslash\{y, \mu(t)\})$.

**(9)** If $C_0 = \{z\}$, then $z \sim \mu(y)$. If $|C_0| \geq 2$, then there exists a vertex $x \in C_0$ such that $x \sim \mu(z)$. For every such vertex $x$, the following statements are true: $y \sim \mu(x)$, $\mu(x) \nsim z$, and $\mu(x) \nsim w_i$ for $w_i \in A_1$. Moreover, if $|C_0| \geq 2$, then $A_2 = \emptyset$, i.e. $y$ is adjacent to every vertex of $\mu(A)$.

Indeed, if $C_0 \neq \emptyset$, then by (7) and the minimality of $H$, there exists a vertex $z \in C_0$ such that $z \sim \mu(y)$, otherwise $|C_0| = |N_H(C_0)|(= |\mu(C_0)|)$, a contradiction. Moreover, no other vertex of $C_0$ is adjacent to $\mu(y)$ by (5). Hence, if $|C_0| \geq 2$, then, again by (7) and the minimality of $H$, there exists a vertex $x \in C_0$ such that $x \sim \mu(z)$.

Let $x \in C_0$ such that $x \sim \mu(z)$. Since $\mu(z) \nsim y$ by Claim 10, $x \nsim \mu(y)$, otherwise $\{z, \mu(z), x, \mu(y), y, b_1\}$ induces a banner$_2$, a contradiction. Thus, $\mu(x) \nsim z$, otherwise $\{y, \mu(y), z, \mu(z), x, \mu(x)\}$ induces a domino or a banner$_2$, depending on $\mu(x) \sim y$ or not, a contradiction. Now, if $y \nsim \mu(x)$, then by Claim 11, there exists a pair of vertices $b_i, b_j \in \mu(A_1)$ such that $w_i$ and $w_j$ are not adjacent to $\mu(x), \mu(z)$ and $\{w_i, b_i, y, b_j, w_j, \mu(y), z, \mu(z), x, \mu(x)\}$ induces an $S_{2,2,5}$, a contradiction. Then $\mu(x) \nsim w_i$ for any $w_i \in A_1$, otherwise $\{y, b_i, w_i, \mu(x), x, \mu(t)\}$ induces a banner$_2$, a contradiction.

Assume that $|C_0| \geq 2$, we show that $A_2 = \emptyset$. Indeed, without loss of generality, assume that $y \nsim b_k$. Let $x \in C_0$ be a vertex such that $x \sim \mu(z)$. Then $\mu(y)$ or $\mu(z)$ is not adjacent to $w_k$, otherwise since $z \nsim w_k$ by Claim 10, $\{z, \mu(z), w_k, \mu(y), y, b_1\}$ induces a banner$_2$, a contradiction. Similarly, $\mu(x)$ or $\mu(z)$ is not adjacent to $w_k$. Now, $\mu(y) \nsim w_k$, otherwise since there exists a pair of vertices $w_i, w_j \in A_1$ not adjacent to $\mu(y), \mu(z)$ by Claim 11, $\{b_i, w_i, b, w_j, b_j, w_k, \mu(y), z, \mu(z), x\}$ induces an $S_{2,2,5}$, a contradiction. By similar reasons, $\mu(x) \nsim w_k$. Now, by Claim 11, there exists a vertex $w_i \in A_1$ not adjacent to $\mu(x)$ and $\{z, \mu(y), y, \mu(x), x, b_i, w_i, b, w_k, b_k\}$ induces an $S_{2,2,5}$, a contradiction.

**(10)** If $|D_1| \geq 2$, then no vertex of $\mu(D_1)$ has a neighbor in $A$. Indeed, by (3), without loss of generality, let $z_1, z_2 \in D_1$ be adjacent to $b_1, b_2$, respectively. To the contrary, suppose that $\mu(z_1)$ has a neighbor $w_i \in A$. By Claim 10, $w_i \neq w_1$. If $w_i = w_2$, then by (1), (6), and Claims 10, 11, $\{z_2, b_2, w_2, b, w_j, \mu(z_1), z_1, b_1, y, \mu(y)\}$ induces an $S_{2,2,5}$ for some vertex $w_j \neq w_1, w_2$ such that $w_j \nsim \mu(z_1)$, a contradiction. If $w_i \neq w_1, w_2$, then by (1) and (6), $\{w_2, b, w_i, \mu(z_2), z_2, \mu(z_1), z_1, b_1, y, \mu(y)\}$ induces an $S_{2,2,5}$ in the case that $\mu(z_2) \sim w_i$, or $\{\mu(z_2), z_2, b_2, y, \mu(y), w_2, b, w_i, \mu(z_1), z_1\}$ induces an $S_{2,2,5}$ in the case that $\mu(z_2) \nsim w_i$, a contradiction.

**(11)** If there exist two vertices $z_1, z_2 \in C_1$ sharing a neighbor in $\mu(A_2)$, then either $H$ is of the form tree$^5$ or there is a redundant set $U$ containing at most four vertices such that $H - U$ is of the form tree$^2$ or tree$^5$.

First, since $A_2 \neq \emptyset$, $|C_0| \leq 1$ by (9). Without loss of generality, assume that $z_1, z_2$ share a neighbor $b_k \in \mu(A_2)$.

If $z_2$ has another neighbor, say $b_l \in \mu(A)$, then since by (2), there exists a pair of vertices $b_i, b_j \in \mu(A_1)$ not adjacent to $z_1, z_2$, one has that $\{b_i, w_i, b, w_j, b_j, w_l, b_l, z_2, b_k, z_1\}$ induces an $S_{2,2,5}$, a contradiction. Thus, $b_k$ is the only one neighbor in $\mu(A)$ for any vertex $z \in C_1$ adjacent to $b_k$.

Note that, for any such $z$, $\mu(z) \nsim w_k$ by Claim 10. Moreover, $\mu(z) \nsim w_j \in A$ for $w_j \neq w_k$, otherwise $\{b_i, w_i, b, b_l, w_l, w_j, \mu(z), z, b_k, z'\}$ induces an $S_{2,2,5}$ for $z'$ be another neighbor of $b_k$ in $C_1$ different from $z$; by Claim 11 and (2), $b_i, b_l$ not adjacent to $z, z'$; and $w_i, w_l$ not adjacent to $\mu(z)$, a contradiction.

Now, $y$ is adjacent to at least one vertex among $\mu(z_1), \mu(z_2)$, otherwise by (6), $\{\mu(z_1), z_1, b_k, z_2, \mu(z_2), w_k, b, w_1, b_1, y\}$ induces an $S_{2,2,5}$, a contradiction. Without loss of generality, assume that $y \sim \mu(z_1)$. Then $y \sim \mu(z_2)$, otherwise by

(6), $\{w_1, b_1, y, b_2, w_2, \mu(z_1), z_1, b_k, z_2, \mu(z_2)\}$ induces an $S_{2,2,5}$, a contradiction. Hence, $y$ is adjacent to every vertex $z \in C_1$ adjacent to $b_k$.

That also implies that $y$ has no other non-neighbor than $b_k$ in $\mu(A)$. Indeed, without loss of generality, suppose that $y \nsim b_{k-1}$. Then $\{z_1, \mu(z_1), y, \mu(z_2), z_2, b_1, w_1, b, w_{k-1}, b_{k-1}\}$ induces an $S_{2,2,5}$, a contradiction.

Moreover, $\mu(y) \nsim z$ for every vertex $z \in C_1$ adjacent to $b_k$, otherwise $\{\mu(y), z, \mu(z), y, b_1, w_1\}$ induces a banner$_2$, a contradiction.

Besides, $D_1 = \emptyset$. Indeed, without loss of generality, suppose that there exists some vertex $t \in D_1$ such that $t \sim b_1$. Then $t \nsim b_k$, otherwise an $S_{2,2,5}$ arises. Moreover, $t \nsim \mu(z)$ for any $z \in C_1$ adjacent to $b_k$, otherwise $\{t, \mu(z), y, b_1, w_1, b\}$ induces a banner$_2$, a contradiction. Now, by (6), $\{\mu(z_1), z_1, b_k, z_2, \mu(z_2), w_k, b, w_1, b_1, t\}$ induces an $S_{2,2,5}$, a contradiction.

We consider the two following cases.

**Case 1.** $C_0 = \emptyset$. Then

$$U := \{y, \mu(y)\}$$

is a redundant set of size two such that $H - U$ is of the form tree$^2$ in the case that $\mu(y) \nsim w_k$, or $H$ is of the form tree$^5$ in the case that $\mu(y) \sim w_k$.

**Case 2.** $C_0 = \{x\}$ and $x \sim \mu(y)$ by (9). Then $\mu(x) \nsim w_k$, otherwise $\{x, \mu(x), w_k, \mu(y), y, b_1\}$ induces a banner$_2$ or $\{w_1, b_1, y, b_2, w_2, \mu(y), x, \mu(x), w_k, b_k\}$ induces an $S_{2,2,5}$ depending on $\mu(y) \sim w_k$ or not, a contradiction. Thus, $\mu(x) \nsim z$ for any $z \in C_1$ adjacent to $b_k$, otherwise, by Claim 11, there exists a pair of vertices $w_i, w_j \neq w_k$ not adjacent to $\mu(x)$ and hence, $\{b_i, w_i, b, w_j, b_j, w_k, b_k, z, \mu(x), x\}$ induces an $S_{2,2,5}$, a contradiction. Moreover, $\mu(x) \nsim w_i$ for any $w_i \in A_1$, otherwise $\{z_1, \mu(z_1), y, \mu(z_2), z_2, \mu(y), x, \mu(x), w_i, b\}$ induces an $S_{2,2,5}$, a contradiction. Now,

$$U := \{y, \mu(y), x, \mu(x)\}$$

is a redundant set of size at most four such that $H - U$ is of the form tree$^2$, in the case that $\mu(y) \nsim w_k$, or

$$U := \{x, \mu(x)\}$$

is a redundant set of size at most two such that $H - U$ is of the form tree$^5$, in the case that $\mu(y) \sim w_k$.

From now on, we assume the following statement.

**(11')** Two different vertices in $C_1 \backslash \{y\}$ share no common neighbor in $\mu(A)$. This also implies that $|E_1| \leq 3$.

**(12)** If $D_1 = \emptyset$, then there exists a redundant set $U$ of size at most 24 such that $H - U$ is of the form tree$^1$. Indeed, if in addition, $C_0 = \emptyset$, then by Claim 11,

$$U := \{y, \mu(y)\} \cup A_2 \cup \mu(A_2) \cup E_1 \cup \mu(E_1) \cup N_A(\mu(E_1)) \cup \mu(N_A(\mu(E_1)))$$

is a redundant set of size at most 20 such that $H - U$ is of the form tree$^1$. Now, we consider the two following cases.

**Case 1.** $C_0 = \{z\}$. Then by (9) and Claim 11,

$$U := \{y, \mu(y), z, \mu(z)\} \cup A_2 \cup \mu(A_2) \cup E_1 \cup \mu(E_1) \cup$$
$$\cup N_A(\mu(E_1) \cup \{\mu(z)\}) \cup \mu(N_A(\mu(E_1) \cup \{\mu(z)\}))$$

is a redundant set of size at most 24 such that $H - U$ is of the form tree$^1$.

**Case 2.** $|C_0| \geq 2$. Then $y$ is adjacent to every vertex of $\mu(A)$ by (2). Let $z$ be the (only) vertex of $C_0$ adjacent to $\mu(y)$. Denote by $C_0'$ the set of vertices of $C_0\backslash\{z\}$ adjacent to $\mu(z)$ and let $C_0'' := C_0\backslash(C_0' \cup \{z\})$. Then $C_0' \neq \emptyset$, otherwise $|C_0\backslash\{z\}| = |N_H(C_0\backslash\{z\})|$, a contradiction to the minimality of $H$. Moreover, for every $x \in C_0'$, $\mu(x) \sim y$, $\mu(x)$ is not adjacent to any vertex of $A_1$, and $x \nsim \mu(y)$ by (9).

**2.1.** $C_0'' = \emptyset$. Then $H$ is of the form tree$^5$ or tree$^6$ depending on $\mu(z)$ has a neighbor in $A$ or not.

**2.2.** $C_0'' \neq \emptyset$. Then it must contain a vertex $t \sim \mu(x)$ for some $x \in C_0'$, otherwise $|N(C_0'')| = |C_0''|$, a contradiction to the minimality of $H$. Now, $\mu(t) \nsim x$, otherwise $\{z, \mu(z), x, \mu(x), t, \mu(t)\}$ induces a domino or a banner$_2$ depending on $\mu(t) \sim z$ or not, a contradiction. Thus, $\mu(t) \nsim y$, otherwise $\{y, \mu(t), t, \mu(x), x, \mu(z)\}$ induces a banner$_2$, a contradiction. Now, by Claim 11, there exists a pair of vertices $w_i, w_j$ is not adjacent to $\mu(x), \mu(t), \mu(z)$ and hence, $\{\mu(t), t, \mu(x), x, \mu(z), y, b_i, w_i, b, w_j\}$ induces an $S_{2,2,5}$, a contradiction.

From now on, we assume the following statement.

**(12')** $D_1 \neq \emptyset$.

**(13)** If $|C_0| \geq 2$, then $H$ contains a redundant set $U$ of size at two such that $H - U$ is of the form tree$^5$.

By (9), $y$ is adjacent to every vertex of $\mu(A)$. Let $z$ be the (only) vertex of $C_0$ adjacent to $\mu(y)$ and $x \in C_0$ be adjacent to $\mu(z)$. Also by (9), for every such vertex $x$, $\mu(x) \sim y$, $\mu(x) \nsim z$. Moreover, by Claim 10, $z$ has no neighbor in $\mu(A)$.

Since $D_1 \neq \emptyset$, without loss of generality, assume that there exists a vertex $z_1 \in D_1$ adjacent to $b_1$. Now, $\mu(z) \sim w_1$, otherwise $\{\mu(z_1), z_1, b_1, w_1, b, y, \mu(y), z, \mu(z), x\}$ induces an $S_{2,2,5}$, a contradiction. Moreover, by (3) and Claim 11, $D_1 = \{z_1\}$. We consider the two following cases.

**Case 1.** $z$ has a neighbor $\mu(t) \in \mu(C_0)$ for some $t \in C_0$ different from $z$. Then by (7), (8), and Claim 10, $\mu(t) \sim w_1$, otherwise $\{\mu(z_1), z_1, b_1, w_1, b, y, \mu(y), z, \mu(t), t\}$ induces an $S_{2,2,5}$, a contradiction. But now, $\{\mu(z), w_1, \mu(t), z, \mu(y), y\}$ induces a banner$_2$, a contradiction.

**Case 2.** $z$ has no neighbor in $\mu(C_0)$ other than $\mu(z)$. Let $x$ be a vertex in $C_0$ adjacent to $\mu(z)$ and $C_0'$ be the set of vertices of $C_0$ different from $z$ and not adjacent to $\mu(z)$. If $C_0' \neq \emptyset$, then by (7) and (8), there exists a vertex $t \in C_0'$ adjacent to $\mu(x)$, otherwise $|C_0'| = |N_H(C_0')|$, a contradiction to the minimality of $H$. Now, $t \nsim \mu(z)$, otherwise $\{\mu(y), z, \mu(z), x, \mu(x), t\}$ induces a domino or a banner$_2$ depending on $t \sim \mu(y)$ or not, a contradiction. Now, by Claim 11, there exists a pair of vertices $w_i, w_j$ different from $w_1$ not adjacent to $\mu(x)$ and hence, $\{b_i, w_i, b, w_j, b_j, w_1, \mu(z), x, \mu(x), t\}$ induces an $S_{2,2,5}$, a contradiction.

From above considerations, every vertex $x \in C_0$ different from $z$ is adjacent to $\mu(z)$ and $\mu(x)$ is adjacent to $y$. Now,

$$U := \{z_1, \mu(z_1)\}$$

is a redundant set of size two, such that $H - U$ is of the form tree$^5$.

From now on, we assume the following statement.

**(13')** $|C_0| \leq 1$.

**(14)** If $|D_1| \geq 2$, then by (10) and (13'),

$$\begin{aligned} U := \{y, \mu(y)\} \cup C_0 \cup \mu(C_0) \cup E_1 \cup \mu(E_1) \cup \\ \cup N_A(\mu(E_1) \cup \mu(C_0)) \cup \mu(N_A(\mu(E_1) \cup \mu(C_0))) \cup \\ \cup N_{D_1}(\mu(N_A(\mu(E_1) \cup \mu(C_0)))) \cup \\ \cup \mu(N_{D_1}(\mu(N_A(\mu(E_1) \cup \mu(C_0))))) \end{aligned}$$

is a redundant set of size at most 26 such that $H - U$ is of the form tree$^4$.

**(15)** If $|D_1| = 1$, then

$$\begin{aligned} U := \{y, \mu(y)\} \cup C_0 \cup \mu(C_0) \cup D_1 \cup \mu(D_1) \cup E_1 \cup \mu(E_1) \cup \\ \cup N_A(\mu(D_1) \cup \mu(E_1) \cup \mu(C_0)) \cup \mu(N_A(\mu(D_1) \cup \mu(E_1) \cup \mu(C_0))) \cup \\ \cup N_{D_1}(\mu(N_A(\mu(D_1) \cup \mu(E_1) \cup \mu(C_0)))) \\ \cup \mu(N_{D_1}(\mu(N_A(\mu(D_1) \cup \mu(E_1) \cup \mu(C_0))))) \end{aligned}$$

is a redundant set of size at most 32 such that $H - U$ is of the form tree$^1$.

All the above observations ((1)–(15)) finish the proof of the claim. □

From now on, assume that every vertex of $C_1$ has at least four non-neighbors in $\mu(A)$.

*Claim 15* $C_0 = \emptyset$, i.e. $C = C_1$.

*Proof* Suppose that $C_0 \neq \emptyset$. Then there exists some vertex $z \in C_1$, without loss of generality, assume that $z \sim b_1$, and $y \in C_0$ such that $y \sim \mu(z)$, otherwise $|C_0| = |N_H(C_0)|$, a contradiction to the minimality of $H$. Thus, $\{b_i, w_i, b, w_j, b_j, w_1, b_1, z, \mu(z), y\}$ induces an $S_{2,2,5}$, for $b_i, b_j$ not adjacent to $z$ and $w_i, w_j$ not adjacent to $\mu(z)$, a contradiction. □

*Claim 16* If $|C| \leq 4$, then $H$ contains a redundant set $U$ of size at most 16 such that $H - U$ is of the form tree$^1$.

*Proof* Assume that $|C| \leq 4$, i.e. $|\mu(C)| \leq 4$. Note that every (black) vertex of $\mu(C)$ has at most one neighbor in $A$ by Claim 11, i.e. $|N_A(\mu(C))| \leq 4$. Then

$$U := C \cup \mu(C) \cup N_A(\mu(C)) \cup \mu(N_A(\mu(C)))$$

is a redundant set of size at most 16 such that $H - U$ is of the form tree$^1$. □

*Claim 17* Assume that $|C| \geq 5$. Then the following statements are true.

**Case 1.** If there exist vertices $z_1, z_2 \in C$ sharing some neighbor in $\mu(A)$, then $H$ is of the form tree$^2$.

**Case 2.** If for any two vertices $y, z \in C$, $y, z$ share no neighbor in $\mu(A)$, then $H$ is of the form tree$^3$ or tree$^7$ or $H$ contains a redundant set $U$ of size at most six such that $H - U$ is of the form tree$^3$.

*Proof* We consider the two above cases.

**Case 1.** Without loss of generality, assume that $z_1, z_2 \in C$ share a neighbor $b_1 \in \mu(A)$. Let us consider the following occurrences which are exhaustive by symmetry.

**1.1.** $z_2$ has another neighbor, say $b_2 \in \mu(A)$. Note that then $b_2 \nsim b_1$ since otherwise a banner$_2$ arises. Assume that there exist two vertices, without loss of generality, assume that they are $b_3, b_4$, not adjacent to $z_1, z_2$. Then $\{b_3, w_3, b, b_4, w_4, w_2, b_2, z_2, b_1, z_1\}$ induces an $S_{2,2,5}$, a contradiction. Hence, $|N_{\mu(A)}(\{z_1, z_2\})| \geq k - 1$. Since both $z_1$ and $z_2$ have at most $k - 4$ neighbors in $\mu(A)$, each of them has at least four neighbors in $\mu(A)$.

Let $z_3 \in C$ be adjacent to some vertex $b_i \in N_{\mu(A)}(\{z_1, z_2\})$. Then $z_3$ has at least four neighbors in $\mu(A)$. Hence, $z_3$ shares two neighbors in $\mu(A)$ with $z_1$ or $z_2$, a contradiction to Claim 9. So, there exists no other vertex in $C$ (than $z_1, z_2$) having a neighbor in $N_{\mu(A)}(\{z_1, z_2\})$. Together with $|C| \geq 5$, this implies that $|N_{\mu(A)}(\{z_1, z_2\})| \leq k - 1$, i.e. $|N_{\mu(A)}(\{z_1, z_2\})| = k - 1$.

Without loss of generality, assume that $z_1, z_2$ are not adjacent to $b_k$. Since $|C| \geq 5$, there exist $z_3, z_4 \in C$ such that $z_3, z_4$ are adjacent to $b_k$. Moreover, $z_3, z_4$ have no other neighbor in $\mu(A)$. By Claim 11, there exists a vertex $b_i$ such that $b_i \sim z_1$ and $w_i$ is not adjacent to $\mu(z_3), \mu(z_4)$. Hence, by Claim 13, $\{\mu(z_3), z_3, b_k, z_4, \mu(z_4), w_k, b, b_i, w_i, z_1\}$ induces an $S_{2,2,5}$, a contradiction.

**1.2.** Every vertex of $C$ adjacent to $b_1$ has only one neighbor ($b_1$) in $\mu(A)$. Note that, for every such vertex $z$, $\mu(z) \nsim w_1$ by Claim 10. Moreover, $\mu(z) \nsim w_i \in A$ for $w_i \neq w_1$, otherwise since by Claim 11, there exists a pair of vertices $w_j, w_l \neq w_1$ and non-adjacent to $\mu(z)$ and one has that $\{b_j, w_j, b, w_l, b_l, w_i, \mu(z), z, b_1, z'\}$ induces an $S_{2,2,5}$ for $z'$ be another neighbor of $b_1$ in $C$ different from $z$, a contradiction.

Now, let $C_{11}$ be the set of vertices of $C_1$ adjacent to $b_1$ and $C_{12} := C_1 \backslash C_{11}$. If $C_{12} = \emptyset$, then $H$ is of the form tree$^2$. Then assume that $C_{12} \neq \emptyset$ and let $y \in C_{12}$ and, without loss of generality, assume that $y \sim b_2 \in \mu(A)$. If $y$ is not adjacent to two vertices, say $\mu(z_1), \mu(z_2) \in \mu(C_{11})$, then $\{\mu(z_1), z_1, b_1, z_2, \mu(z_2), w_1, b, w_2, b_2, y\}$ induces an $S_{2,2,5}$, a contradiction.

If $y$ is adjacent to two vertices $\mu(z_1), \mu(z_2) \in \mu(C_{11})$, then $y$ is adjacent to every vertex $b_i \in \mu(A)$ different from $b_1$, otherwise $\{z_1, \mu(z_1), y, \mu(z_2), z_2, b_2, w_2, b, w_i, b_i\}$ induces an $S_{2,2,5}$, a contradiction.

Now, $y$ has at least $k-1$ neighbors in $\mu(A)$, a contradiction. Hence, $C_{11} = \{z_1, z_2\}$ and every vertex $y \in C_{12}$ is adjacent to exactly one vertex of $\mu(C_{11})$.

If $\mu(z_1)$ is adjacent to two vertices $y_1, y_2 \in C_{12}$, then $\{y_1, \mu(z_1), y_2, b_i, w_i, b\}$ induces a banner$_2$ in the case that $y_1, y_2$ share the same neighbor $b_i \in \mu(A)$ by Claim 10 or $\{b_{i_1}, y_1, \mu(z_1), y_2, b_{i_2}, z_1, b_1, w_1, b, w_i\}$ induces an $S_{2,2,5}$ for $b_{i_1}, b_{i_2}$ be (different) neighbors of $y_1, y_2$ in $\mu(A)$, respectively, and $w_i \in A$ different from $w_1, w_{i_1}, w_{i_2}$, a contradiction. Hence, each $\mu(z_1), \mu(z_2)$ has at most one neighbor in $C_{12}$. It implies that $|C_{12}| \leq 2$ and thus, $|C| \leq 4$, a contradiction.

**Case 2.** If for every vertex $\mu(z) \in \mu(C_1)$, $z$ is the only neighbor of $\mu(z)$, then $H$ is of the form tree$^3$.

Then assume that there is a vertex $\mu(z) \in \mu(C_1)$ such that $z$ is not the only neighbor of $\mu(z)$. First we show that for every pair $z_1, z_2 \in C$, $\mu(z_1) \nsim z_2$. Indeed, for contradiction, suppose that $\mu(z_1) \sim z_2$. Without loss of generality, assume that $z_1, z_2$ are adjacent to $b_1, b_2$, respectively. Then $\mu(z_2) \nsim z_1$, otherwise by Claim 10, $\{\mu(z_2), z_1, \mu(z_1), z_2, b_2, w_2\}$ induces a banner$_2$, a contradiction.

Moreover, $N_{\mu(A)}(\{z_1, z_2\}) \geq k-2$, otherwise by Claim 11, there exists a pair of vertices $w_i, w_j$ not adjacent to $\mu(z)$ such that $b_i, b_j$ not adjacent to $z_1, z_2$, and hence, $\{b_i, w_i, b, w_j, b_j, w_2, b_2, z_2, \mu(z_1), z_1\}$ induces an $S_{2,2,5}$, a contradiction.

Hence, the non-neighbors of $z_1, z_2$ in $\mu(A)$ have at most two neighbors in $C$, i.e. $|C| \leq 4$, a contradiction.

Then there exists some vertex $z \in C$, such that $\mu(z)$ is adjacent to some vertex of $A$. Without loss of generality, assume that $z \sim b_1$ and $\mu(z) \sim w_2$. Then $b_2 \nsim z$, by Claim 10. We consider the two following subcases.

**2.1.** $b_2 \sim y$ for some $y \in C$. Then for every $x \in C \backslash \{y, z\}$, $\mu(x) \sim w_2$, otherwise $\{z, \mu(z), w_2, b_2, y, b, w_i, b_i, x, \mu(x)\}$ induces an $S_{2,2,5}$ for $b_i \sim x$, a contradiction. By Claim 11, that also implies that $\mu(y)$ is not adjacent to any vertex $w_i \in A$ such that $b_i \sim x$ for some $x \in C_1$ different from $y$, otherwise $|C| = 2 < 5$, a contradiction. Now,

$$U := \{w_2, b_2, y, \mu(y)\} \cup N_A(\mu(y)) \cup \mu(N_A(\mu(y)))$$

is a redundant set containing at most six vertices such that $H - U$ is of the form tree$^3$.

**2.2.** $N_C(b_2) = \emptyset$. Assume that there exists some vertex $y \in C$, without loss of generality, assume that $y \sim b_3$ and $\mu(y) \sim w_2$. Then for every $x \in C$ different from $y, z$, $\mu(x) \sim w_2$, otherwise $\{z, \mu(z), w_2, \mu(y), y, b, w_i, b_i, x, \mu(x)\}$ induces an $S_{2,2,5}$ for $b_i \sim x$, a contradiction. Now,

$$U := \{w_2, b_2\}$$

is a redundant set of size two such that $H - U$ is of the form tree$^3$.

Now, if there exists no vertex pair $y, z \in C$, such that $\mu(y), \mu(z)$ share the same neighbor in $A$, then $H$ is of the form tree[7]. □

All above claims finish the proof.

## Appendix 3: Proof of Lemma 6

*Proof (of Lemma 6)* To simplify the proof, we start with a pre-processing consisting in detecting augmenting $(l, m)$-extended-chains whose path-part is of length at most $2l$ since such an augmenting $(l, m)$-extended-chain contains at most $\frac{1-(m-1)^l}{2-m}+2l+1$ vertices and can be enumerated in polynomial time.

In order to determine whether $S$ admits an augmenting $(l, m)$-extended-chain whose path-part is of length at least $2l + 2$, we first find a candidate, i.e. a pair $(L, R)$, where $L$ and $R$ are disjoint trees consisting induced paths $x_0, x_1, \ldots, x_l$ and $x_{2p-l}, x_{2p-l+1}, \ldots, x_{2p}$, respectively ($p \geq l+1$) and every vertex outside that path of $L$ ($R$, respectively) is of distance at most $l-1$ from $x_0$ ($x_{2p}$, respectively) and not adjacent to any vertices among $\{x_1, x_2, \ldots, x_l, x_{2p-l}, x_{2p-l+1}, \ldots, x_{2p}\}$. If such a candidate does not exist, then there is no augmenting $(l, m)$-extended-chain whose path-part is of length at least $2l + 2$ for $S$. Moreover, since such candidates contain only finite vertices, we can enumerate them in polynomial time.

Our purpose is to find an alternating chain connecting $x_l$ and $x_{2p-l}$. Evidently, if there are no such chains, then there is no augmenting $(l, m)$-extended-chain whose path-part is of length at least $2l + 2$ for $S$ containing $L$ and $R$.

Having found a candidate $(L, R)$, we have the following observations about vertices of $G$ in the sense that the vertices not satisfying these assumptions can be simply removed from the graph, since they cannot occur in any valid alternating chain connecting $x_l$ and $x_{2p-l}$. Let $P := (x_0, x_1, \ldots, x_{2p})$ be the path part of a desired $(l, m)$-extended-chain.

*Claim 18*

1. Each white vertex has at least two black neighbors.
2. Each black vertex lying outside $L$ and $R$ has exactly two white neighbors.
3. No black vertex outside $L$ and $R$ has a neighbor in $L$ or $R$.
4. No white vertex outside $L$ and $R$ has a neighbor in $L$ or $R$, except such a neighbor is $x_l$ or $x_{2p-l}$.
   Moreover, no white vertex outside $P$ has a neighbor in $P$.

*Proof* 1. and 2. are obvious since a vertex not satisfying these conditions cannot occur in any augmenting extended-chain containing $L$ and $R$ as sub-extended-chains.

Note that $x_l$ and $x_{2p-l}$ are black vertices. Hence, if a black vertex outside $L$ and $R$ has a neighbor in $L$ or $R$, then clearly such a vertex cannot belong to the desired augmenting chain, similar for a white vertex outside $L$ and $R$.

If a white vertex outside $P$ has a neighbor in $P$, then clearly such a neighbor is black and hence it has at least three white neighbors, a contradiction.

From the conditions of the above claim, we have the following observation.

*Claim 19* If $S$ admits an augmenting $(l, m)$-extended-chain containing $L$ and $R$, then no vertex of $P \backslash (L \cup R)$ is the center of an induced claw.

*Proof* By contradiction, suppose that $G$ contains a claw $G[C]$, where $C = \{a, b, c, d\}$, whose center $a$ (i.e., the vertex of degree three) is a vertex $x_j$ on $P$. Without loss of generality, we choose a claw such that $|\{b, c, d\} \backslash P|$ is minimal and, among such claws, choose a claw such that $j$ is minimum. Note that, since there exists at least one vertex of $\{b, c, d\}$ lying outside $P$, together with 3. of Claim 18, $l + 1 \leq j \leq 2p - l - 1$. Moreover, since every black vertex of $P$ has all its white neighbors lying in $P$, every vertex of $C \backslash P$ is black.

We shall use the following convention: for a black vertex $v$ outside $P$, if only one of the two white neighbors of $v$ is defined explicitly, then the other is denoted as $\bar{v}$. Also, for a vertex $v$ of $C$ not belonging to $P$ such that $N(v) \cap P \neq \emptyset$, we denote by $r(v)$ the largest index in $\{j, j+1, \ldots, 2p - l - 1\}$ and by $s(v)$ the smallest index in $\{l + 1, l + 2, \ldots, j\}$ such that $v$ is adjacent to $x_{r(v)}, x_{s(v)}$.

We now analyze three cases: exactly one (C1), two (C2), or three (C3) vertex/vertices of $\{b, c, d\}$ do(es)n't belong to $P$.

**Case (C1).** Without loss of generality, assume that $b = x_{j-1}$ and $c = x_{j+1}$. Then we have the following observations.

(1) $d$ is not adjacent to $x_{j-2}, x_{j+2}$. Indeed, if $d \sim x_{j-2}$ (similar for the case $d \sim x_{j+2}$), then $\{x_{j-2}, x_{j-1}, x_j, d, x_{r(d)}, x_{r(d)+1}, \ldots, x_{r(d)+l-1}\}$ induces a banner$_l$ in the case $r(d) \geq j+2$ or $\{d, x_{j-2}, x_{j-1}, x_j, x_{j+1}, \ldots, x_{j+l}\}$ induces a banner$_l$ in the case $r(d) = j$, a contradiction.

(2) $r(d) = j$ or $s(d) = j$. Indeed, by (1), suppose that $r(d) \geq j + 3$ and $s(d) \leq j - 3$. Then $\{x_{j-1}, x_j, d, x_{s(d)}, x_{s(d)-1}, \ldots, x_{s(d)-l+1}, x_{r(d)}, x_{r(d)+1}, \ldots, x_{r(d)+l-1}\}$ induces an $S_{2,l,l}$, a contradiction.

(3) $s(d) \geq j-3$ and $r(d) \leq j+3$. Indeed, suppose that $s(d) \leq j-4$ (similar for the case $r(d) \geq j+4$). Then by (2), $\{x_{j-2}, x_{j-1}, x_j, x_{s(d)}, x_{s(d)-1}, \ldots, x_{s(d)-l+1}, x_{j+1}, x_{j+2}, \ldots, x_{j+l-1}\}$ induces an $S_{2,l,l}$, a contradiction.

(4) $r(d) = s(d) = j$. Indeed, by (2) and (3), suppose that $r(d) = j + 3$ and $s(d) = j$ (similar for the case $s(d) = j - 3$ and $r(d) = j$). Among $\{x_j, x_{j+3}\}$, there exists at most one white vertex. Hence, $\{x_{j+2}, x_{j+1}, \bar{d}, d, x_{j+3}, x_{j+4}, x_{j+5}, \ldots, x_{j+l+3}, x_j, x_{j-1}, \ldots, x_{j-l}\}$ induces an $R_l^1$, a contradiction.

Now, since $r(d) = s(d) = j$, $\{\bar{d}, d, x_j, x_{j-1}, x_{j-2}, \ldots, x_{j-l}, x_{j+1}, x_{j+2}, \ldots, x_{j+l}\}$ induces an $S_{2,l,l}$, a contradiction.

**Case (C2).** Without loss of generality, assume that $b = x_{j-1}$ and $c$ and $d$ are outside $P$. Then we have the following observations.

(1) $x_{j+1}$ is adjacent both to $c$ and $d$ to avoid (C1).

(2) Also to avoid (C1), $c$ is adjacent to $x_{s(c)+1}, x_{r(c)-1}$, similarly for $d$.
(3) It cannot happen that $s(c) = s(d) \leq j-2$ or $r(c) = r(d) \geq j+2$. Indeed, say if $s(c) = s(d) \leq j-2$, then $\{c, x_{j+1}, d, x_{s(c)}, x_{s(c)-1}, \ldots, x_{s(c)-l}\}$ induces a banner$_l$, a contradiction.
(4) Similarly, if $s(c) = s(d) = j$, then there exists no common neighbor $x_i$ of $c$ and $d$ for $i \geq j+2$ and if $r(c) = r(d) = j+1$, then there exists no common neighbor $x_i$ of $c$ and $d$ for $i \leq j-2$. And in both cases, $c$ and $d$ have no common neighbor outside $P$.
(5) $c$ and $d$ are not adjacent to $x_{j-2}$. Indeed, suppose that $c \sim x_{j-2}$ (similar for the case $d \sim x_{j-2}$). Then $r(c) = j+1$ (similarly, $r(d) = j+1$), otherwise $\{x_j, x_{j-1}, x_{j-2}, c, x_{r(c)}, x_{r(c)+1}, \ldots, x_{r(c)+l-1}\}$ induces a banner$_l$, a contradiction, and $s(c) = j-3$, otherwise $\{x_j, x_{j-1}, x_{j-2}, c, x_{s(c)}, x_{s(c)-1}, \ldots, x_{s(c)-l+1}\}$ induces a banner$_l$, a contradiction. Moreover, $d$ is neither adjacent to $x_{j-2}$ nor $x_{j-3}$ also by (4). Hence, $s(d) = j$, otherwise $\{x_{j-1}, x_{j-2}, c, x_j, d, x_{s(d)}, x_{s(d)-1}, \ldots, x_{s(d)-l+1}\}$ induces a banner$_l$, a contradiction. Now, among $\{x_j, x_{j+1}\}$, there exists exactly one white vertex. Moreover, $c \nsim \bar{d}$ by (4). Now, $\{d, \bar{d}, x_{j+1}, c, x_{j-3}, x_{j-4}, \ldots, x_{j-l-2}, x_{j+2}, x_{j+3}, \ldots, x_{j+l+1}\}$, induces an $S_{2,l,l}$, a contradiction.
(6) By (2) and (5), if $s(c) \leq j-3$, then $s(c) \leq j-4$.
(7) $s(c) = j$ or $r(c) = j+1$. Similarly, $s(d) = j$ or $r(d) = j+1$. Indeed, by (5) and (6), if $s(c) \leq j-4$ and $r(c) \geq j+2$, then $\{x_{j-1}, x_j, c, x_{s(c)}, x_{s(c)-1}, \ldots, x_{s(c)-l+1}, x_{r(c)}, x_{r(c)+1}, \ldots, x_{r(c)+l-1}\}$ induces an $S_{2,l,l}$, a contradiction.
(8) $s(c) = j$ or $r(d) = j+1$ (similarly, $s(d) = j$ or $r(c) = j+1$). Indeed, by (5) and (6), without loss of generality, suppose that $s(c) \leq j-4$ and $r(d) \geq j+2$. Then by (7), $r(c) = j+1$ and $s(d) = j$. Hence, $\{x_{j-2}, x_{j-1}, x_j, c, x_{s(c)}, x_{s(c)-1}, \ldots, x_{s(c)-l+2}, d, x_{r(d)}, x_{r(d)+1}, \ldots, x_{r(d)+l-2}\}$ induces an $S_{2,l,l}$, a contradiction.
(9) $s(c) = j$ or $s(d) = j$. Indeed, by (5) and (6), without loss of generality, suppose that $s(c), s(d) \leq j-4$. Then $r(c) = r(d) = j+1$, by (7). Now, by (3), without loss of generality, assume that $s(c) < s(d)$. Then by (4), $\{x_{s(d)+1}, d, x_{j+1}, c, x_{s(c)}, x_{s(c)-1}, \ldots, x_{s(c)-l+2}, x_{j+2}, x_{j+3}, \ldots, x_{j+l+1}\}$ induces an $S_{2,l,l}$, a contradiction.
(10) $r(c) = j+1$ or $r(d) = j+1$. Indeed, if $r(c), r(d) \geq j+2$, then by (7), $s(c) = s(d) = j$. Without loss of generality, by (2) and (4), assume that $r(c) > r(d)+1$. Then $\{x_{r(d)}, d, x_j, c, x_{r(c)}, x_{r(c)+1}, \ldots, x_{r(c)+l-2}, x_{j-1}, x_{j-2}, \ldots, x_{j-l}\}$ induces an $S_{2,l,l}$, a contradiction.
(11) $s(c) = s(d) = j$. Indeed, by (5) and (6), suppose that $s(c) \leq j-4$ (similar for the case that $s(d) \leq j-4$). Then by (9), (8), and (7), $s(d) = j, r(d) = r(c) = j+1$. Note that, among $\{x_j, x_{j+1}, x_{s(c)}, x_{s(c)+1}\}$, neighbors of $c$, there exist exactly two white vertices and hence, $c \nsim \bar{d}$. Now, $\{\bar{d}, d, x_{j+1}, c, x_{s(c)}, x_{s(c)-1}, \ldots, x_{s(c)-l+2}, x_{j+2}, x_{j+3}, \ldots, x_{j+l+1}\}$ induces an $S_{2,l,l}$, a contradiction.
(12) $r(c) = r(d) = j+1$. Indeed, by (10), suppose that $r(c) = j+1$ and $r(d) \geq j+2$. Among $x_j, x_{j+1}$, there exists only one white vertex and $d \nsim \bar{c}$ by (4).

Then $\{\bar{c}, c, x_j, x_{j-1}, x_{j-2}, \ldots, x_{j-l}, d, x_{r(d)}, x_{r(d)+1}, \ldots, x_{r(d)+l-2}\}$ induces an $S_{2,l,l}$, a contradiction.

Now, $\{\bar{c}, c, x_j, d, \bar{d}, x_{j-1}, x_{j-2}, \ldots, x_{j-l}, x_{j+1}, x_{j+2}, \ldots, x_{j+l+1}\}$ induces an $R_l^2$, a contradiction.

**Case (C3).** We have the following observations.

(1) First, note that, $r(b)$, $r(c)$, and $r(d)$ (and similarly, $s(b)$, $s(c)$, and $s(c)$) are three mutually different integers. Otherwise, suppose that $r(b) = r(c)$. Then we have the claw $\{x_{r(c)}, x_{r(c)+1}, b, c\}$, i.e. (C2).

(2) To avoid (C1), if $b \sim x_i$ for some $i$, then $b$ is adjacent to at least one vertex among $x_{i-1}, x_{i+1}$. It implies $b$ is adjacent to $x_{s(b)+1}, x_{r(b)-1}$. Similarly for $c$ and $d$.

(3) Moreover, by the minimality of $j$ and to avoid (C2), we know that $x_{j-1}$ has exactly two neighbors in $\{b, c, d\}$, say $b$ and $c$. To avoid (C1) and (C2), we conclude that $x_{j+1}$ is adjacent to $d$ and has at least one neighbor in $\{b, c\}$, say $c$. Moreover, $b \nsim x_{j+1}$. Indeed, if $b \sim x_{j+1}$, then $r(b), r(c), r(d) \leq j + 2$, otherwise $\{x_{j-1}, b, x_{j+1}, c, x_{r(c)}, x_{r(c)+1}, \ldots, x_{r(c)+l-1}\}$ or $\{x_{j-1}, c, x_{j+1}, b, x_{r(b)}, x_{r(b)+1}, \ldots, x_{r(b)+l-1}\}$ or $\{b, x_{j-1}, c, x_{j+1}, d, x_{r(d)}, x_{r(d)+1}, \ldots, x_{r(d)+l-2}\}$ induces a banner$_l$ depending on which is the largest index among $r(b)$, $r(c)$, $r(d)$, a contradiction. But now, $j + 1 \leq r(c), r(b), r(d) \leq j + 2$, a contradiction with the mutual difference of $r(b)$, $r(c)$, and $r(d)$.

(4) It also implies that at least one of $s(b)$, $s(c)$ is less than $j - 1$ and at least one of $r(d)$, $r(c)$ is greater than $j + 1$.

(5) $b \nsim x_{j+1}$, together with $b \sim x_{r(b)-1}$, it implies that if $r(b) \geq j + 2$, then $r(b) \geq j + 3$. Similarly, if $s(d) \leq j - 2$, then $s(d) \leq j - 3$.

(6) In a pair of consecutive vertices of $P$, there is a black vertex and a white vertex. Hence, $b, c, d$ are not adjacent to three pairs of consecutive vertices of $P$, otherwise we have a black vertex with three white neighbors, a contradiction. Together with $c$ is adjacent to $x_{s(c)+1}$ and $x_{r(c)-1}$, it leads to either $r(c) \leq j+2$ or $s(c) \geq j - 2$. Moreover, if $c$ is adjacent to $x_{j-2}, x_{j+2}$, then $s(c) = j - 2$ and $r(c) = j + 2$. Similarly, we have the following observations: $r(b) = j$ or $s(b) \geq j - 2$, $s(d) = j$ or $r(d) \leq j + 2$.

(7) $c$ and $b$ cannot share a neighbor $x_i$ for some $i \leq j - 2$, otherwise $\{x_i, c, x_j, b, x_{r(b)}, \ldots, x_{r(b)+l-1}\}$, $\{b, x_i, c, x_j, d, x_{r(d)}, \ldots, x_{r(d)+l-2}\}$, or $\{x_i, b, x_j, c, x_{r(c)}, \ldots, x_{r(c)+l-1}\}$ induces a banner$_l$ depending on which is the largest index among $r(b)$, $r(c)$, $r(d)$ (note that at least one of these integers is bigger than $j + 1$ and they are mutually different by (1)), a contradiction. Moreover, $b$ and $c$ cannot share a neighbor $x_i$ for some $i \geq j + 2$, otherwise $\{x_j, c, x_i, b, x_{s(b)}, x_{s(b)-1}, \ldots, x_{s(b)-l+1}\}$ or $\{x_j, b, x_i, c, x_{s(c)}, \ldots, x_{s(c)-l+1}\}$ induces a banner$_l$ depending on which one is larger among $s(b)$ and $s(c)$. Similarly, $c$ and $b$ cannot share a white neighbor outside $P$. By similar arguments, these properties are also true for the two pairs $c, d$ and $b, d$.

(8) $s(c) \geq j-2$, similarly, $r(c) \leq j+2$. Moreover, if $s(c) = j-2$, then $r(c) = j+1$. Similarly, if $r(c) = j+2$, then $s(c) = j-1$. Indeed, suppose that $s(c) \leq j-4$. Then $c \sim x_{j-2}$, otherwise $\{x_{j-1}, x_{j-2}, x_{j-3}, c, x_{r(c)}, x_{r(c)+1}, \ldots, x_{r(c)+l-1}\}$ induces a banner$_l$ or $\{x_{j-2}, x_{j-1}, c, x_{s(c)}, x_{s(c)-1}, \ldots, x_{s(c)-l+1}, x_{r(c)}, x_{r(c)+1}, x_{r(c)+l-1}\}$ induces an $S_{2,l,l}$ depending on $c \sim x_{j-3}$ or not. But now, $c$ is adjacent to $\{x_{s(c)}, x_{s(c)+1}, x_{j+1}, x_j, x_{j-1}, x_{j-2}\}$, a contradiction to (6). Now, if $s(c) = j-3$, then $c \sim x_{j-2}$ by (2) and $r(c) = j+1$ by (6). Hence, $\{c, x_{j-l-3}, \ldots, x_{j-4}, x_{j-3}, \ldots, x_{j+1}, x_{j+2}, \ldots, x_{j+l+1}\}$ induces an $R_l^3$, a contradiction. Moreover, if $s(c) = j-2$ and $r(c) = j+2$, then $\{c, x_{j-l-2}, \ldots, x_{j-3}, x_{j-2}, \ldots, x_{j+1}, x_{j+2}, \ldots, x_{j+l+2}\}$ induces an $R_l^3$, a contradiction.

(9) $r(b) = j$ or $s(b) = j-1$, similarly, $r(d) = j+1$ or $s(d) = j$. Indeed, if $r(b) \geq j+3$ and $s(b) \leq j-2$, then $\{x_j, x_{j+1}, x_{j+2}, b, x_{s(b)}, x_{s(b)-1}, \ldots, x_{s(b)-l+1}\}$ induces a banner$_l$ or $\{x_{j+1}, x_j, b, x_{s(b)}, x_{s(b)-1}, \ldots, x_{s(b)-l+1}, x_{r(b)}, x_{r(b)+1}, \ldots, x_{r(b)+l-1}\}$ induces an $S_{2,l,l}$ depending on $b \sim x_{j+2}$ or not, a contradiction.

(10) $s(b) \geq j-3$, similarly, $r(d) \geq j+3$. Indeed, suppose that $s(b) \leq j-4$. Then $r(b) = j$, by (9). Now $b$ is not adjacent to $x_{j-2}$ and $x_{j-3}$ at the same time, otherwise either $\{b, x_{j-l-4}, \ldots, x_{j-5}, x_{j-4}, \ldots, x_j, x_{j+1}, \ldots, x_{j+l}\}$ induces an $R_l^3$ or $b$ is adjacent to three pairs of consecutive vertices of $P$, a contradiction to (6). Hence, $b \nsim x_{j-2}$, otherwise $\{x_{j-3}, x_{j-2}, b, x_{s(b)}, x_{s(b)-1}, \ldots, x_{s(b)-l+1}, x_j, x_{j+1}, \ldots, x_{j+l-1}\}$ induces an $S_{2,l,l}$, a contradiction. Suppose that $b \sim x_{j-3}$. Then $c \sim x_{j-2}$, otherwise $\{b, x_{j-3}, x_{j-2}, x_{j-1}, c, x_{r(c)}, x_{r(c)+1}, \ldots, x_{r(c)+l-2}\}$ induces a banner$_l$, a contradiction. Now, $r(c) = j+1$ by (8), $r(d) \geq j+2$ by (1), and $s(d) = j$ by (9). Hence, $\{x_{j-2}, c, x_j, b, x_{s(b),x_{s(b)-1}}, \ldots, x_{s(b)-l+2}, d, x_{r(d)}, x_{r(d)+1}, \ldots, x_{r(d)+l-2}\}$ induces an $S_{2,l,l}$, a contradiction. Thus, $b \nsim x_{j-3}$. Now, $\{x_{j-3}, x_{j-2}, x_{j-1}, b, x_{s(b)}, \ldots, x_{s(b)-l+2}, c, x_{r(c)}, \ldots, x_{r(c)+l-2}\}$ induces an $S_{2,l,l}$, a contradiction.

(11) $r(b) = j$, similarly, $s(d) = j$. Indeed, suppose that $r(b) \geq j+3$. Then by (9), $s(b) = j-1$. Moreover, $s(c) = j-2$, $r(c) = j+1$, $r(d) \geq j+2$, and $s(d) = j$ by (1), (8), and (9). Now, $\{x_{r(b)-1}, b, x_j, c, x_{j-2}, x_{j-3}, \ldots, x_{j-l}, d, x_{r(d)}, x_{r(d)+1}, \ldots, x_{r(d)+l-2}\}$ or $\{x_{r(d)}, d, x_j, c, x_{j-2}, x_{j-3}, \ldots, x_{j-l}, b, x_{r(b)}, x_{r(b)+1}, \ldots, x_{r(b)+l-2}\}$ induces an $S_{2,l,l}$ depending on $r(d) > r(b)$ or $r(b) > r(d)$ (note that by (2) and (7), if $r(b) > r(d)$, then $r(b) > r(d)+1$).

(12) $s(c) = j-1$, similarly, $r(c) = j+1$. Indeed, suppose that $s(c) = j-2$. Then $r(c) = j+1$ by (8), $s(b) = j-1$ by (1), (2), and (7) and $r(d) \geq j+2$ by (1). Among $x_j$ and $x_{j-1}$, there exists only one white vertex. Consider the other white neighbor of $b$, say $\bar{b}$. Then $\{\bar{b}, b, x_j, c, x_{j-2}, x_{j-3}, \ldots, x_{j-l}, d, x_{r(d)}, x_{r(d)+1}, \ldots, x_{r(d)+l-2}\}$ induces an $S_{2,l,l}$, a contradiction.

(13) $x_j$ is black, otherwise $\{\bar{c}, c, x_j, b, x_{s(b)}, \ldots, x_{s(b)-l+2}, d, x_{r(d)}, \ldots, x_{r(d)+l-2}\}$ induces an $S_{2,l,l}$, a contradiction. Now, by the symmetry, we have three remaining cases, which are considered follows.

*Case 3.1.* $b$ is adjacent to $x_{j-2}$ and $x_{j-3}$, $d$ is adjacent to $x_{j+2}$ and $x_{j+3}$. Then $\{x_j, x_{j-l-2}, \ldots, x_{j-3}, b, x_{j-1}, c, x_{j+1}, d, x_{j+3}, \ldots, x_{j+l+2}\}$ induces an $R_l^3$, a contradiction.

*Case 3.2.* $s(b) = j-2$ and $r(d) = j+2$. Then $\{x_j, x_{j-l-1}, \ldots, x_{j-2}, \bar{b}, b, x_{j-1}, c, x_{j+1}, d, \bar{d}, x_{j+2}, \ldots, x_{j+l+1}\}$ induces an $R_l^4$, a contradiction.

*Case 3.3.* $s(b) = j-2$ and $d$ is adjacent to $x_{j+2}$ and $x_{j+3}$. Then $\{x_j, x_{j-l-1}, \ldots, x_{j-2}, \bar{b}, b, x_{j-1}, c, x_{j+1}, d, x_{j+2}, x_{j+3}, \ldots, x_{j+l+1}\}$ induces an $R_l^5$, a contradiction.

Our purpose here is to detect an augmenting extended-chain whose path-part is of length at least $2l + 2$. We first find candidates $(L, R)$ as described above. Note that such candidates can be enumerated in polynomial time. Then perform Steps (a) through (d) for each such pair:

(a) remove all black vertices that have a neighbor in $L$ or in $R$,
(b) remove the vertices of $L$ and $R$ except for $x_l$ and $x_{2p-l}$, and
(c) remove all the vertices that are the center of a claw in the remaining graph,
(d) then in the resulting claw-free graph, determine whether there exists an alternating chain between $x_l$ and $x_{2p-l}$ by the method described in [28, 33].

For each candidate, Steps (a) through (d) can be implemented in time $O(n^4)$. Hence, we have the conclusion of the lemma.

## Appendix 4: Proof of Lemma 7

The proof is consisted of the six following observations.

**Lemma 10** *If $G$ contains no augmenting $P_3$, then an augmenting tree$^1$ (if any) can be found in time $O(n^{17})$.*

*Proof* Refer to Figure 2, tree$^1$ with parameter $r$. If $r = 1$, then tree$^1$ is a $P_3$. Assume that $G$ contains an augmenting graph tree$^1$, for some $r \geq 2$. Therefore, $G$ contains an induced $P_5 = (b_1, a_1, x, a_2, b_2)$, where $b_1, b_2 \in B^1$ and $b_1, b_2$ are non-adjacent to any vertex of $W\{a_1, x, a_2\}$. If $G$ contains no such an initial structure, then it contains no augmenting tree$^1$. If such a structure exists, then we proceed as follows.

Let us denote $A = \{a \in W(x)\backslash\{a_1, a_2\} : a \nsim b_1, b_2\}$ and for $a \in A$, let $K(a)$ denote the set of black neighbors of $a$ in $B_1$ not adjacent to any vertex of $\{x, a_1, a_2, b_1, b_2\}$. Notice that a desired augmenting tree exists only if $K(a) \neq \emptyset$ for every $a \in A$. Finally, let $V' = \bigcup_{a \in A} K(a)$. Since $K(a) \subseteq B^1$ for every $a \in A$, $K(a) \cap K(a') = \emptyset$ for every pair of distinct vertices $a, a' \in A$.

Consider any vertex $a \in A$, we show that $K(a)$ induces a clique for every $a \in A$. Indeed, suppose that $K(a)$ contains two non-adjacent vertices $b_1, b_2$. Then $\{b_1, a, b_2\}$ induces an augmenting $P_3$, a contradiction. It follows that a desired augmenting tree$^1$ exists if and only if $\alpha(G[V']) = |A|$.

We show that $G[V']$ must be $P_5$-free. Indeed, consider an induced $P_4 = (p_1, p_2, p_3, p_4)$ in $G[V']$ and let $a \in A$ be such that $p_1 \in K(a)$. Then none of the vertices $p_3, p_4$ is adjacent to $a$ because $K(a)$ is a clique. Thus, $p_2 \in K(a)$, otherwise $\{b_1, a_1, x, a_2, b_2, a, p_1, p_2, p_3, p_4\}$ induces an $S_{2,2,5}$, a contradiction. Hence, if $G[V']$ induces a $P_4 = (p_1, p_2, p_3, p_4)$, then $p_1$ and $p_2$ have a common white neighbor, while $p_2$ and $p_3$ have no common white neighbor, a contradiction to when consider an induced $P_4 = (p_2, p_3, p_4, p_5)$ in the $P_5 = (p_1, p_2, p_3, p_4, p_5)$.

Since the $P_5$-free graph class is MIS-solvable in time $O(n^{12})$ [22], one can find a simple augmenting tree containing the $P_5$ $(b_1, w_1, b, w_2, b_2)$ in $O(n^{12})$. With an exhaustive search, all candidate $P_5$ of augmenting trees can be found in time $O(n^5)$. For such candidates $P_5$'s, $V'$ can be built in $O(n^3)$. Hence, we have the conclusion of the lemma.

**Lemma 11** *If $G$ contains neither augmenting $P_3$ nor $P_7$, then an augmenting tree$^2$ (if any) can be found in time $O(n^{14})$.*

*Proof* Refer to Figure 2, tree$^2$ with parameter $r$ and $s$. We may restrict ourselves to finding a tree$^2$ with $r, s \geq 2$, since any tree$^2$ with, say $r = 1$, either equals to $P_7$ or contains a redundant subset $U$ of size two such that tree$^2 - U$ is of the form tree$^1$.

As a candidate, consider the subgraph of tree$^2$ (see Figure 2) induced by $\{a_1, a_2, b_1, b_2, c_1, c_2, d_1, d_2, x, y, z\}$ such that $b_1, b_2, d_1, d_2 \in B^1$ and $x, z$ share no common white neighbor other than $y$.

Let us denote $A = (W(x) \cup W(z)) \setminus \{a_1, a_2, c_1, c_2, y\}$. For $a \in A$, let $K(a)$ denote the set of black neighbors of $a$ in $B^1$ not adjacent to any vertex of $\{x, b_1, b_2, d_1, d_2\}$. Note that, by the assumption, every vertex of $A$ is either adjacent to $x$ or $y$. Notice that a desired augmenting tree exists only if $K(a) \neq \emptyset$ for every $a \in A$.

We show that $K(a)$ induces a clique. Indeed, suppose that $K(a)$ contains two non-adjacent vertices $b_1, b_2$. Then $\{b_1, a, b_2\}$ induces an augmenting $P_3$, a contradiction.

Since for every $a \in A$, $K(a) \in B^1$, $K(a) \cap K(a') = \emptyset$ for every pair of distinct vertices $a, a' \in A$.

Finally, let $V' = \bigcup_{a \in A} K(a)$. It follows that a desired augmenting tree$^2$ exists if and only if $\alpha(G[V']) = |A|$.

We now show that $G[V']$ is $P_3$-free. Suppose, to the contrary, that $(p_1, p_2, p_3)$ is an induced $P_3$ in $G[V']$. Let $a \in A$ such that $p_1 \in K(a)$. Since $K(a)$ is a clique, $p_3$ is not adjacent to $a$. Assume that $p_3 \sim a'$. Then since $p_2 \in B^1$, $p_2$ is not adjacent to at least one vertex among $a, a'$. Without loss of generality, assume that $p_2 \nsim a$, and $a$ is adjacent to $x$, but not to $z$. Then $\{d_2, c_2, z, c_1, d_1, y, x, a, p_1, p_2\}$ induces an $S_{2,2,5}$, a contradiction.

Hence, $G[V']$ is a disjoint union of cliques, i.e. a maximum independent set in $G[V']$ can be found in linear time. All candidates of the form tree$^2$ whose $r = s = 2$

can be found by an exhaustive search in time $O(n^{11})$. For such candidates $P_5$'s, $V'$ can be built in $O(n^3)$. Hence, we have the conclusion of the lemma.

**Lemma 12** *If G contains neither augmenting $P_3$ nor $P_5$, then an augmenting tree$^3$ or an augmenting tree$^4$ (if any) can be found in time $O(n^{31})$.*

*Proof* First, note that tree$^4$ is a special case of tree$^3$. We refer to Figure 2, tree$^3$ for indices. Moreover, we may restrict ourselves to finding a tree$^3$ with $s \geq 3$, since any tree$^3$ with, say, $s \leq 2$ is either of the form tree$^1$ or contains a redundant subset $U$ of size four such that tree$^3 - U$ is of the form tree$^1$.

As a candidate, consider the subgraph of tree$^3$ (see Figure 2) induced by $\{d_1, c_1, b_1^1, a_1^1, x, a_1^2, b_1^2, c_2, d_2, a_1^3, b_1^3, c_3, d_3\}$ such that $b_1^1, b_1^2, b_1^3 \in B^2$, $d_1, d_2, d_3 \in B^1$. Let us denote $A = W(x) \backslash \{a_1^1, a_1^2, a_1^3\}$. For $a \in A$, let $K(a)$ denote the set of black neighbors $b$ of $a$ in $B^1 \cup B^2$ and not adjacent to any vertex of $\{x, b_1^1, b_1^2, b_1^3, d_1, d_2, d_3\}$ such that if $b \in B^2$, then $G$ contains a pair of adjacent vertices $c_b$ and $d_b$ such that $c_b \notin W(x)$, $W(b) = \{a, c_b\}$, $d_b \in B^1$, and $d_b$ is not adjacent to any vertex of $\{x, b_1^1, b_1^2, b_1^3, d_1, d_2, d_3, b\}$ (note that $d_b$ may coincide with $d_1, d_2$, or $d_3$). Let $V' = \bigcup_{a \in A} K(a)$. And again, by the existence of a desired augmenting tree$^3$, $K(a)$ is not empty for all $a \in A$. Note that by the assumption, $K(a) \cap K(a') = \emptyset$ for every pair of distinct vertices $a, a' \in A$.

Consider any vertex $a \in A$, we show that $K(a)$ induces a clique. Indeed, suppose that $K(a)$ contains two non-adjacent vertices $b, b'$. By the symmetry, we consider the three following cases.

**Case 1.** $b, b' \in B^1$. Then $\{b, a, b'\}$ induces an augmenting $P_3$, a contradiction.

**Case 2.** $b' \in B^1$ and $b \in B^2$. Then $\{b', a, b, c_b, d_b\}$ induces an augmenting $P_5$, a contradiction.

**Case 3.** $b, b' \in B^2$. Then $c_b \neq c_{b'}$, otherwise $\{b, c_b, b', a, x, a_1^1\}$ induces a banner$_2$, a contradiction. Now, $\{c_{b'}, b', a, b, c_b, x, a_1^i, b_1^i, c_i, d_i\}$ induces an $S_{2,2,5}$, for $c_i$ is among $c_1, c_2, c_3$ different from $c_b, c_{b'}$, a contradiction.

It follows that a desired augmenting tree$^3$ exists if and only if $\alpha(G[V']) = |A|$.

Given $a, a' \in A$ and $b \in K(a) \cap B^2$, $b' \in K(a')$ such that $b \nsim b'$ and if $b' \in B^2$, assume that $d_b \neq d_{b'}$, we show that $b' \nsim d_b$. Indeed, suppose that $b' \sim d_b$. Then $b' \nsim c_b$, otherwise $c_{b'} = c_b$, and hence, $d_{b'} = d_b$, a contradiction. Thus, $\{b_1^1, a_1^1, x, a_1^2, b_1^2, a', b', d_b, c_b, b\}$ induces an $S_{2,2,5}$, a contradiction. Now, if $b' \in B^2$, then $d_b \nsim d_{b'}$, otherwise $\{b_1^1, a_1^1, x, a_1^2, b_1^2, a', b', c_{b'}, d_{b'}, d_b\}$ induces an $S_{2,2,5}$, a contradiction.

Hence, for every pair of non-adjacent vertices $b, b'$ such that $b \in K(a) \cap B^2$, $b' \in K(a')$ for two distinct vertices $a, a' \in A$, $\{b, b', d(b)\}$ is independent. Moreover, if $b' \in B^2$, then $\{b, b', d_b, d_{b'}\}$ is independent.

Now, assume that $B'$ is a maximum independent set of $G[V']$. Let $C' := \{c_b : b \in B' \cap B^2\}$, $D' := \{d_b : b \in B' \cap B^2\}$. Then by above arguments, $B' \cup D'$ is independent. And in the case that $|B'| = |A|$, $H := G[A \cup B' \cup C' \cup D']$ is an augmenting graph of the form tree$^3$ of $G$.

As in Lemma 10, we show that $G[V']$ is $P_5$ free. Indeed, consider an induced $P_4 = (p_1, p_2, p_3, p_4)$ in $G[V']$ and let $a \in A$ such that $p_1 \in K(a)$. Then none of the vertices $p_3, p_4$ is adjacent to $a$ because $K(a)$ is a clique. But now, $p_2 \in K(a)$, otherwise $\{b_1^1, a_1^1, x, a_1^2, b_1^2, a, p_1, p_2, p_3, p_4\}$ induces an $S_{2,2,5}$, a contradiction. Hence, if $G[V']$ induces a $P_4 = (p_1, p_2, p_3, p_4)$, then $p_1$ and $p_2$ have a common white neighbor, while $p_2$ and $p_3$ have no common white neighbor, a contradiction to when consider an induced $P_4 = (p_2, p_3, p_4, p_5)$ in the $P_5 = (p_1, p_2, p_3, p_4, p_5)$.

All candidates can be found by an exhaustive search in time $O(n^{19})$. For such candidates, $V'$ can be built in $O(n^3)$. Again, by the solution for the MIS problem in $P_5$-free graphs [22], we have the conclusion of the lemma.

**Lemma 13** *An augmenting tree*$^5$ *(if any) can be found in time* $O(n^{14})$.

*Proof* Refer to Figure 2, tree$^5$ with parameter $r$ and $s$. We may restrict ourselves to finding a tree$^5$ with $r, s \geq 1$ and $r \geq 2$, since a tree$^5$ with, say, $r = 0$ contains a redundant set $U$ of size four such that tree$^5 - U$ is of the form tree$^1$, and a tree$^5$ with $r = s = 1$ can be found in time $O(n^9)$.

As a candidate, consider the subgraph of tree$^5$ (see Figure 2) induced by $\{a_1, a_2, b_1, b_2, c_1, d_1, u, v, x, y, z\}$ such that $b_1, b_2, v, d_1 \in B^2$ and $x, y$ share no common white neighbor other than $u$. Let us denote $A_x = W(x)\backslash\{a_1, a_2, u\}$ and $A_y = W(y)\backslash\{c_1, u\}$ and for $a \in A := A_x \cup A_y$, let $K(a)$ denote the set of common black neighbors of $a$ and $z$ in $B^2$ not adjacent to any vertex of $\{x, y, b_1, b_2, v, d_1\}$.

Note that by the assumption, every vertex of $A$ is either adjacent to $x$ or $y$. Since $K(a) \subseteq B^2$ for every $a \in A$, $K(a) \cap K(a') = \emptyset$, for every pair of distinct vertices $a, a' \in A$.

Consider a pair of distinct vertices $b, b' \in K(a)$ for some $a \in A$. If $b \nsim b'$, then $\{b, a, b', z, v, u\}$ induces a banner$_2$, a contradiction. Hence, $K(a)$ is a clique for all $a \in A$.

Now, let $V'(x) := \bigcup_{a\in A_x} (K(a))$, $V'(y) := \bigcup_{a\in A_y} (K(a))$, and $V' := V'(x) \cup V'_y$.

Note that, $V'(x) \cap V'(y) = \emptyset$ by the definition. Then a desired augmenting tree$^5$ exists if and only if $K(a) \neq \emptyset$ for every $a \in A$ and $\alpha(G[V']) = |A|$.

As in Lemma 11, we show that $G[V']$ is $P_3$-free. Suppose, to the contrary, that $(p_1, p_2, p_3)$ is an induced $P_3$ in $G[V']$. Let $a \in A$ such that $p_1 \in K(a)$. Since $K(a)$ is a clique, $p_3$ is not adjacent to $a$. Assume that $p_3 \sim a'$. Since $p_2 \in B^2$, $p_2$ is not adjacent to at least one vertex among $a, a'$. Without loss of generality, assume that $p_2 \nsim a$ and $a$ is adjacent to $y$, but not to $x$. Then $\{b_2, a_2, x, b_1, a_1, u, y, a, p_1, p_2\}$ induces an $S_{2,2,5}$, a contradiction. Hence, a maximum independent set can be found in $G[V']$ in linear time.

All candidates can be found by an exhaustive search in time $O(n^{11})$. For such candidates, $V'$ can be build in $O(n^3)$. Hence, we have the conclusion of the lemma.

**Lemma 14** *An augmenting tree*$^6$ *(if any) can be found in time* $O(n^{27})$.

*Proof* Refer to Figure 2, tree$^6$ with parameter $r$ and $s$. We may restrict ourselves to finding a tree$^6$ with $r, s \geq 2$, since a tree$^6$ with, say, $r = 1$, contains a redundant set $U$ of size four such that tree$^6 - U$ is of the form tree$^1$.

As a candidate, consider the subgraph of tree$^6$ (see Figure 2) induced by $\{a_1, a_2, b_1, b_2, c_1, c_2, d_1, d_2, x, y, z\}$ such that $b_1, b_2, c_1, c_2 \in B^2$ and $x, z$ share no common white neighbor.

Let us denote $A_x = W(x)\backslash\{a_1, a_2\}$ and $A_z = W(z)\backslash\{d_1, d_2\}$. For $a \in A := A_x \cup A_z$, let $K(a)$ denote the set of common black neighbors of $a$ and $y$ in $B^2$ and not adjacent to any vertex of $\{x, b_1, b_2, c_1, c_2, z\}$. Note that $A_x \cap A_z = \emptyset$ by the assumption. Since for every $a \in A$, $K(a) \subseteq B^2$, $K(a) \cap K(a') = \emptyset$ for every pair of distinct vertices $a, a' \in A$.

Consider a pair of distinct vertices $b, b' \in K(a)$ for some $a \in A$. If $b \nsim b'$, then $\{b, a, b', y, c_1, d_1\}$ induces a banner$_2$ in the case that $a \in A_x$ (similar for the case $a \in A_z$), a contradiction. Hence, $K(a)$ is a clique for all $a \in A$.

Now, let $V'(x) := \bigcup_{a\in A_x} (K(a))$, $V'(z) := \bigcup_{a\in A_z} (K(a))$, and $V' := V'(x) \cup V'_z$. Note that, $V'(x) \cap V'(z) = \emptyset$. Then a desired augmenting tree$^6$ exists if and only if $K(a) \neq \emptyset$ for every $a \in A$ and $\alpha(G[V']) = |A|$.

As in Lemma 10, we show that $G[V'_x]$ and $G[V'_z]$ are $P_5$-free. Indeed, consider an induced $P_4 = (p_1, p_2, p_3, p_4)$ in $G[V'_x]$ or $G[V'_z]$, let $a \in A$ be such that $p_1 \in K(a)$. Then none of the vertices $p_3, p_4$ is adjacent to $a$ because $K(a)$ is a clique. But now, $p_2 \in K(a)$, otherwise $\{b_1, a_1, x, a_2, b_2, a, p_1, p_2, p_3, p_4\}$ or $\{c_1, d_1, z, d_2, c_2, a, p_1, p_2, p_3, p_4\}$ induces an $S_{2,2,5}$ depending on $a \in A_x$ or $a \in A_z$, a contradiction. Hence, if $G[V'_x]$ or $G[V'_z]$ induces a $P_4 = (p_1, p_2, p_3, p_4)$, then $p_1$ and $p_2$ have a common white neighbor, while $p_2$ and $p_3$ have no common white neighbor, a contradiction to when consider an induced $P_4 = (p_2, p_3, p_4, p_5)$ in the $P_5 = (p_1, p_2, p_3, p_4, p_5)$.

Moreover, assume that there exists a pair of vertices $b, b'$ such that $b \in K(a)$, $b' \in K(a')$ for some $a \in A(x)$, $a' \in A_z$, and $b \sim b'$. Then $\{b_1, a_1, x, a_2, b_2, a, b, b', a', z\}$ induces an $S_{2,2,5}$, a contradiction. Hence, there is no edge connecting a vertex in $G[V'_x]$ and a vertex in $G[V'_z]$. So, $G[V']$ is $P_5$-free.

Note that all candidates can be found by an exhaustive search in time $O(n^{15})$. For such candidates, $V'$ can be build in $O(n^3)$. Hence, by the result of Lokshtanov et al. [22] we have the conclusion of the lemma.

**Lemma 15** *If $G$ contains no augmenting $P_3$, nor $P_5$, nor $P_7$, then an augmenting tree$^7$ (if any) can be found in time $O(n^{19})$.*

*Proof* Refer to Figure 2 for indices. We may restrict ourselves to finding a tree$^7$ with $s \geq 3$, since a tree$^7$ with $s \leq 2$ is of the form tree$^3$ or contains a redundant set $U$ of size at most eight such that tree$^7 - U$ is of the form tree$^3$.

As a candidate, consider the subgraph of tree$^7$ (see Figure 2) induced by $\{x, a_1^1, b_1^1, c_1, d_1, e_1, f_1, a_1^2, b_1^2, c_2, d_2, e_2, f_2, a_1^3, b_1^3, c_3, d_3, e_3, f_3\}$ such that $b_1^1, d_1 \in B^2$ and $f_1 \in B^1$. Let us denote $A = W(x)\backslash\{a_1^1, a_1^2, a_1^3, e_1, e_2, e_3\}$. For $a \in A$, let $K(a)$ denote the set of black neighbors $b$ of $a$ in $B^1 \cup B^2$ not adjacent to any vertex of $\{x, b_1^1, d_1, e_1, f_1, b_1^2, d_2, e_2, f_2, b_1^3, d_3, e_3, f_3\}$ and such that if $b \in B^2$, then $G$ contains either

- two vertices $c_b, d_b$ such that $c_b \notin W(x)$, $W(b) = \{a, c_b\}$, $d_b \in B^1$, and $d_b$ is not adjacent to any vertex of $\{x, b_1^1, b_1^2, b_1^3, d_1, d_2, d_3, f_1, f_2, f_3, b\}$ or
- an induced alternating (black white vertices) $P_4$ $(c_b, d_b, e_b, f_b)$ such that $e_b \in W(x)\backslash\{a_1^1, c_1, a_1^2, c_2, a_1^3, c_3\}$, $c_b \notin W(x)$, $W(b) = \{a, c_b\}$, $W(d_b) = \{c_b, e_b\}$, $W(f_b) = \{e_b\}$, and $d_b, f_b$ are not adjacent to any vertex of $\{x, b_1^1, b_1^2, b_1^3, d_1, d_2, d_3, f_1, f_2, f_3, b\}$.

Let $V' = \bigcup_{a \in A} K(a)$.

By the existence of a desired augmenting tree[7], $K(a)$ is not empty for all $a \in A$. Note that, by assumption, $K(a) \cap K(a') = \emptyset$ for every pair of distinct vertices $a, a' \in A$.

Given a vertex $b \in K(a) \cap B^2$ for some $a \in A$, we show that $d_b \notin K(e_b)$. Indeed, suppose that $d_b \notin K(e_b)$. Since $d_b \in B^2$, $c_b = c_{d_b}$, $d_{d_b} = b$, and $e_{d_b} = a$. Hence, there exists some vertex $b' \in B^1$, such that $f_{d_b} = b'$, i.e. $b' \sim a$ and $b'$ is not adjacent to $b, d_b$. Hence, $b' \nsim f_b$, otherwise $\{c_b, b, a, b', f_b, x, a_1^i, b_1^i, c_i, d_i\}$ induces an $S_{2,2,5}$, for $c_i$ is a vertex among $c_1, c_2, c_3$ different from $c_b$, a contradiction. Now, $\{b', a, b, c_b, d_b, e_b, f_b\}$ induces an augmenting $P_7$, a contradiction.

Suppose that there exist two vertices $b, b'$ such that $b \in K(a) \cap B^2$ and $b' \in K(a') \cap B^2$ for two distinct vertices $a, a' \in A$ and $d_b, d_{b'}$ are different and adjacent to some vertex $a'' \in W(x)\backslash\{a, a', a_1^1, a_1^2, a_1^3\}$ different from $a, a'$. Then $\{c_b, d_b, a'', d_{b'}, c_{b'}, x, a_1^i, b_1^i, c_i, d_i\}$ induces an $S_{2,2,5}$ where $c_i$ is a vertex among $c_1, c_2, c_3$ different from $c_b, c_{b'}$, a contradiction. Hence, for every pair of vertices $b, b'$ such that $b \in K(a) \cap B^2$, $b' \in K(a') \cap B^2$ for two distinct vertices $a, a' \in A$, $e_b \neq e_{b'}$.

Consider any vertex $a \in A$, we show that $K(a)$ induces a clique. Indeed, suppose that $K(a)$ contains two non-adjacent vertices $b, b'$. By the symmetry, we consider the three following cases.

**Case 1.** $b, b' \in B^1$. Then $\{b, a, b'\}$ induces an augmenting $P_3$, a contradiction.

**Case 2.** $b' \in B^1$ and $b \in B^2$. We have the three following subcases.

**2.1.** $d_b \in B^1$. Then $\{b', a, b, c_b, d_b\}$ induces an augmenting $P_5$, a contradiction.

**2.2.** $d_b \in B^2$ and $b' \nsim f_b$. Then $\{b', a, b, c_b, d_b, e_b, f_b\}$ induces an augmenting $P_7$, a contradiction.

**2.3.** $d_b \in B^2$ and $b' \sim f_b$. Then $\{f_b, b', a, b, c_b, x, a_1^i, b_1^i, c_i, d_i\}$ induces an $S_{2,2,5}$, for $c_i$ is a vertex among $c_1, c_2, c_3$ different from $c_b$, a contradiction.

**Case 3.** $b, b' \in B^2$. Then $c_b \neq c_{b'}$, otherwise $\{b, c_b, b', a, x, a_1^1\}$ induces a banner$_2$, a contradiction. Now, $\{c_{b'}, b', a, b, c_b, x, a_1^i, b_1^i, c_i, d_i\}$ induces an $S_{2,2,5}$, for $c_i$ is a vertex among $c_1, c_2, c_3$ different from $c_b, c_{b'}$, a contradiction.

It follows that a desired augmenting tree[7] exists if and only if $\alpha(G[V']) = |A|$.

Given $a, a' \in A$, $b \in K(a) \cap B^2$, and $b' \in K(a')$ such that $b \nsim b'$, if $b' \sim d_b$, then $b' \nsim c_b$, otherwise $c_{b'} = c_b$ and then $d_{b'} = d_b$, a contradiction. Then $\{b_1^1, a_1^1, x, a_1^2, b_1^2, a', b', d_b, c_b, b\}$ induces an $S_{2,2,5}$, a contradiction. Now, if $b' \in B^2$, then $d_b \nsim d_{b'}$, otherwise $\{b_1^i, a_1^i, x, a_1^j, b_1^j, a', b', c_{b'}, d_{b'}, d_b\}$ induces an $S_{2,2,5}$, for $i, j \in \{1, 2, 3\}$ such that $c_b$ is different from $c_i, c_j$, a contradiction. Note that for every $b \in K(a) \cap B^2$ for some $a \in A$, $f_b \in K(e_b)$. Hence, for every pair of non-adjacent vertices $b, b'$ such that $b \in K(a) \cap B^2$, $b' \in K(a')$ for two distinc vertices $a, a' \in A$, $\{b, b', d_b, f_b\}$ is independent. Moreover, if $b' \in B^2$, then $\{b, b', d_b, d_{b'}, f_b, f_{b'}\}$ is independent.

Now, assume that $B'$ is a maximum independent set of $G[V']$. Let $C' := \{c_b : b \in B' \cap B^2\}$, $D' := \{d_b : b \in B' \cap B^2\}$. Then by above arguments, $B' \cup D'$ is independent. And in the case that $|B'| = |A|$, $H := G[A \cup B' \cup C' \cup D']$ is an augmenting graph of the form tree$^7$ of $G$. Hence, a maximum independent set of $G[V']$ in the case that $\alpha(G[V']) = |A|$ gives us an augmenting of the form tree$^7$.

As in Lemma 10, we show that $G[V']$ is $P_5$-free. Indeed, consider an induced $P_4 = (p_1, p_2, p_3, p_4)$ in $G[V']$, and let $a \in A$ be such that $p_1 \in K(a)$. Then none of the vertices $p_3, p_4$ is adjacent to $a$ because $K(a)$ is a clique. But now, $p_2 \in K(a)$, otherwise $\{b_1^1, a_1^1, x, a_1^2, b_1^2, a, p_1, p_2, p_3, p_4\}$ induces an $S_{2,2,5}$, a contradiction. Hence, if $G[V']$ induces a $P_4 = (p_1, p_2, p_3, p_4)$, then $p_1$ and $p_2$ have a common white neighbor, while $p_2$ and $p_3$ have no common white neighbor, a contradiction to when consider an induced $P_4 = (p_2, p_3, p_4, p_5)$ in the $P_5 = (p_1, p_2, p_3, p_4, p_5)$.

All candidates can be found by an exhaustive search in time $O(n^{19})$. For such candidates, $V'$ can be built in $O(n^3)$. By the result of Lokshtanov et al. [22], we have the conclusion of the lemma.

## Appendix 5: Proof of Theorem 5

So, we modify the concept of augmenting vertex [30] as follows.

**Definition 4** Let $S$ be an independent set of a graph $G = (V, F)$ and $v \in V \backslash S$, $s \in N_S(v)$. We say that $v$ is *augmenting* for $S$ associated with $s$ if $G[N(s) \cap H(v, S)]$ contains an independent set $S_{v,s}$ such that $|S_{v,s}| \geq |N_S(v)|$.

Moreover, with an addition assumption that a maximum independent set of $G[N(s) \cap H(v, S)]$ can be found in polynomial time for every $s \in N_S(v)$, we can also choose $s$ such that $\alpha(G[N(s) \cap H(v, S)])$ is maximum.

Refer to Algorithm 4, where $p$ is a constant defined as in Lemma 3, an extended version of Algorithm Alpha in [29], a maximal independent set of $G$ can be found (say by some greedy method) in time $O(n^2)$. One can compute the set $H(v, S)$ in time $O(n^2)$. Note that an augmenting of at most $2m-1$ vertices can be found in time $O(n^{2m+1})$. Moreover, by Lemmas 6, 10, …, 15, an augmenting graph of the forms mentioned in the **while** condition can be found in polynomial time. The **while** loop is repeated at most $n$ time. Hence, we observe the following result, an extension of Theorem 7 in [29].

**Lemma 16** *Given two integers $l$ and $m$, an ($S_{2,2,5}$,banner$_2$,domino,$M_m$, $R_l^3$, $R_l^4$, $R_l^5$)-free graph $G = (V, E)$, a maximal independent set of $G$ $S$, and $v \in V \backslash S$, if one can find a maximum independent set of $G[N(s) \cap H(v, S)]$ for every $s \in N_S(v)$ in polynomial time, then one can find a maximum independent set of $G$ in polynomial time.*

**Algorithm 4** MISAugVer($G$)

**Input:** a ($S_{2,2,5}$,banner$_2$,domino,$M_m$)-free graph $G$
**Output:** $S$, A maximum independent set of $G$.
1: Find an arbitrary maximal independent set $S$ in $G$;
2: **while** There exists an $H$-augmentations to $S$ where $H$ contains at most $2m-1$ vertices, or $H$ is an augmenting $(4, p)$-extended-chain, an augmenting apple, or $H$ is of the form tree$^1$, ..., tree$^7$ or can be reduced to such forms by some redundant set or some reduction set of size at most 32, or $S$ admits an augmenting vertex $v$ associated with some vertex $s$ **do**
3: **if** $S$ admits an $H$-augmentation **then**
4: Apply an augmenting $H$ for $S$;
5: **end if**
6: **if** $S$ admits an augmenting vertex $v$ associated with $s$ **then**
7: $S := (S \backslash N_S(v)) \cup \{v\} \cup S_{v,s}$;
8: **end if**
9: **end while**
10: **return** $S$

Let $G = (V, E)$ be an ($S_{2,2,5}$,banner$_2$,domino,$M_m$, $R_l^3$, $R_l^4$, $R_l^5$, $K^{(h)}$)-free graph with $n$ vertices and $S$ be a maximal independent set of $G$. Assume that one can solve the MIS problem for ($S_{2,2,5}$,banner$_2$,domino,$M_m$, $R_l^3$, $R_l^4$, $R_l^5$, $K$)-free graphs in polynomial time. The goal is to show that one can carry out Step 2 of Algorithm 4 in polynomial time. We use the technique described in [30]. Let us say that a vertex $v \in V$ is a *trivial augmenting vertex* for $S$ if $v$ is augmenting for $S$ and $|N_S(v)| \leq h$. Then one can check if a vertex $v \in V$ is a trivial augmenting vertex for $S$ in time $O(n^{h+1})$, by verifying if $G[H(v, S)]$ contains an independent set $S^*$ of $|N_S(v)|$ vertices. Such $S^*$ is called the independent set associated with the augmenting vertex $v$.

Assume that $G$ admits no trivial augmenting vertex for $S$ and that there exists $v \in V \backslash S$ augmenting for $S$ (in particular, $h < |N_S(v)|$). Thus, $G[H(v, S)]$ contains an independent set $T$ with $|N_S(v)| \leq |T|$. Since $G$ is ($S_{2,2,5}$,banner$_2$,domino,$M_m$)-free together with an additional assumption that $G$ contains no augmenting graph contains at most $2m-1$ vertices, no augmenting graph of the forms tree$^1$, ..., tree$^7$, no augmenting $(4, p)$-extended-chain, no augmenting apple, no augmenting graph that can be reduced to such forms by some redundant set or reduction set, by Lemmas 3 and 4, $H' := (T \cup \{v\}, N_S(v), E(H'))$ is an augmenting bipartite-chain.

Let us write $T = \{t_1, \ldots, t_r\}$ ($r \geq |N_S(v)| \geq h$), with $N_S(t_i) \subset N_S(t_{i+1})$ for any index $i$. Since $G$ admits no trivial augmenting vertex for $S$, one has $|N_S(t_k)| \geq k$ for $k = 1, \ldots, h$. For any $t \in H(v; S)$, let us write $M(t) = \{w \in H(v, S) : N_S(w) \supset N_S(t), |N_S(w)| \geq h\}$. Then $T \subset \{t_1, \ldots, t_h\} \cup (M(t_h) \backslash N(\{t_1, \ldots, t_h\}))$. Note that $M(t_h)$ is $K$-free, otherwise $M(t_h) \cup \{s_1, s_2, \ldots, s_h\} \cup \{v\}$ induces a $K^{(h)}$ for $s_1, \ldots, s_h \in N_S(t_h)$, a contradiction.

Now, since Step 2 of Algorithm 4 considers all the vertices in $V \backslash S$, to check if $S$ admits an augmenting vertex one has not to solve the MIS problem in $H(v, S)$ for every $v \in V \backslash S$. In fact, for every $v \in V \backslash S$, it is sufficient to verify: (i) if $v$ is a trivial augmenting vertex for $S$, and then (ii) if $v$ is augmenting, by assuming that $S$ admit

**Algorithm 5** Procedure Green $(v)$

**Input:** a vertex $v \in V\backslash S$

**Output:** a possible proof that $v$ is augmenting associated with $T = \{t_1, \ldots, t_h\}$ and an independent set $S^*$ associated with $v$.

1: $S^* := \emptyset$; $T := \emptyset$;
2: **if** $|N_S(v)| \leq h$ **then**
3: **if** $H(v; S)$ contains an independent set $Q$ of $|N_S(v)|$ vertices **then**
4: set $S^* := Q$; $\{v$ is (trivially) augmenting for $S\}$;
5: **end if**
6: **else**
7: **for all** independent set $U$ of $h$ vertices of $G[H(v, S)]$, i.e. $U = \{t_1, \ldots, t_h\}$, with $N_S(t_i) \subset N_S(t_{i+1})$, and $|N_S(t_i)| \geq i$ **do**
8: $S' :=$ MISAugVer($G[M(t_h)\backslash N(\{t_1, \ldots, t_h\})]$);
9: **if** $|S' \cup \{t_1, \ldots, t_h\}| > |S^*|$ **then**
10: $S^* := S' \cup \{t_1, \ldots, t_h\}$; $T := \{t_1, \ldots, t_h\}$;
11: **end if**
12: **end for**
13: **end if**
14: **if** $|S^*| \geq |N_S(v)|$ **then**
15: **return** $v$ is augmenting for $S$ associated with $T$ and $S^*$
16: **end if**

no trivial augmenting vertex. That can be formalized by the procedure Algorithm 5 [30], whose input is any vertex $v$ of $V\backslash S$ which can be executed in time $O(n^{h+d+1})$.

Note that, given an augmenting vertex $v$ (for $S$), Procedure Green($v$) could not recognize it as an augmenting vertex: that can happen whenever $H(v, S)$ contains a trivial augmenting vertex. Now, we give the new definition for augmenting vertex $v$ as following.

**Definition 5** Let $S$ be an independent set of a graph $G = (V, E)$, $h$ be an integer, and $v \in V\backslash S, t_1, t_2, \ldots, t_h \in H[v, S]$. We say that $v$ is $h$-augmenting for $S$ associated with $\{t_1, \ldots, t_h\}$, where $N_S(t_i) \subset N_S(t_{i+1})$ for every index $i$, if $G[M(t_h)\backslash N(\{t_1, \ldots, t_h\})]$ contains an independent set $S_{v,t_1,\ldots,t_h}$ such that $|S^*| \geq |N_S(v)|$ where $S^* := S_{v,t_1,\ldots,t_h} \cup \{t_1, t_2, \ldots, t_h\}$. $S^*$ is called the independent set associated with the augmenting vertex $v$.

To summarize, in order to define an efficient method to solve the MIS problem in ($S_{2,2,5}$,banner$_2$,domino,$M_m$, $K^{(h)}$)-free graphs, one can rewrite Step 2 of Algorithm 4 as in Algorithm 6.

**Algorithm 6** New Step 6

1: **for all** $v \in V \backslash S$ **do**
2: Procedure Green($v$);
3: **if** $v$ is augmenting for $S$ associated with $S^*$ **then**
4: $S := (S \backslash N_S(v)) \cup S^*$; **stop**;
5: **end if**
6: **end for**

## References


1. V. E. Alekseev, On the number of maximal stable sets in graphs from hereditary classes, in *Combinatorial-Algebraic Methods in Applied Mathematics* (Gorkiy University, Gorky, 1991), pp. 3–13 (in Russian)
2. V. E. Alekseev, A polynomial algorithm for finding maximum independent sets in fork-free graphs. Discret. Anal. Oper. Res. Ser. **1**, 3–19 (1999) (in Russian)
3. C. Berge, Two theorems in graph theory. Proc. Natl. Acad. Sci. U. S. A. **43**, 842–844 (1957)
4. R. Boliac, V.V. Lozin, An augmenting graph approach to the stable set problem in $P_5$-free graphs. Discret. Appl. Math. **131**(3), 567–575 (2003)
5. J.A. Bondy, U.S.R. Murty, Graph theory, in *Graduate Text in Mathematics*, vol. 244 (Springer, Berlin, 2008)
6. A. Brandstädt, V.B. Le, S. Mahfud, New applications of clique separator decomposition for the maximum weight stable set problem. Theor. Comput. Sci. **370**(1–3), 229–239 (2007)
7. B. Brešar, F. Kardoš, J. Katrenič, G. Semanišin, Minimum $k$-path vertex cover. Discret. Appl. Math. **159**(12), 1189–1195 (2011)
8. K. Cameron, Induced matching. Discret. Appl. Math. **24**(1–3), 97–102 (1989)
9. D.M. Cardoso, M. Kamiński, V.V. Lozin, Maximum $k$-regular induced subgraphs. J. Comb. Optim. **14**(4), 455–463 (2007)
10. A.K. Dabrowski, V.V. Lozin, D. de Werra, V. Zamaraev, Combinatorics and algorithms for augmenting graphs. Graphs Comb. **32**(4), 1339–1352 (2016)
11. J. Edmonds, Paths, trees, and flowers. Can. J. Math. **17**, 449–467 (1965)
12. P. Festa, P.M. Pardalos, M.G.C. Resende, Feedback set problems, in *Handbook of Combinatorial Optimization* (Kluwer Academic Publishers, Dordrecht, 1999), pp. 209–259
13. J.F. Fink, M.S. Jacobson, $n$-Domination $n$-dependence and forbidden subgraphs, in *Graph Theory with Applications to Algorithms and Computer* (Wiley, New York, 1985), pp. 301–311
14. M.U. Gerber, V.V. Lozin, On the stable set problem in special $P_5$-free graphs. Discret. Appl. Math. **125**(2–3), 215–224 (2003)
15. M.U. Gerber, A. Hertz, V.V. Lozin, Stable sets in two subclasses of banner-free graphs. Discret. Appl. Math. **132**(1–3), 121–136 (2003)
16. M.U. Gerber, A. Hertz, D. Schindl, $P_5$-free augmenting graphs and the maximum stable set problem. Discret. Appl. Math. **132**(1–3), 109–119 (2004)
17. A. Hertz, V.V. Lozin, The maximum independent set problem and augmenting graphs, in *Graph Theory and Combinatorial Optimization* (Springer Science and Business Media, Inc., New York, 2005), pp. 69–99
18. A. Hertz, V.V. Lozin, D. Schindl, Finding augmenting chains in extensions of claw-free graphs. Inf. Process. Lett. **86**(3), 311–316 (2003)
19. N.C. Lê, Augmenting approach for some maximum set problems. Discret. Math. **339**(8), 2186–2197 (2016)

20. N.C. Lê, C. Brause, I. Schiermeyer, New sufficient conditions for $\alpha$-redundant vertices. Discret. Math. **338**(10), 1674–1680 (2015)
21. N.C. Lê, C. Brause, I. Schiermeyer, The maximum independent set problem in subclasses of $S_{i,j,k}$. Electron. Notes Discret. Math. **49**, 43–49 (2015)
22. D. Lokshtanov, M. Vatshelle, Y. Villanger, Independent set in $P_5$-free graphs in polynomial time, in *Proceedings of the 25th Annual ACM-SIAM Symposium on Discrete Algorithms* (2014), pp. 570–581
23. V.V. Lozin, M. Milanič, On finding augmenting graphs. Discret. Appl. Math. **156**(13), 2517–2529 (2008)
24. V.V. Lozin, R. Mosca, Independent sets in extensions of $2K_2$-free graphs. Discret. Appl. Math. **146**(1), 74–80 (2005)
25. V.V. Lozin, R. Mosca, Maximum independent sets in subclasses of $P_5$-free graphs. Inf. Process. Lett. **109**(6), 319–324 (2009)
26. V.V. Lozin, R. Mosca, Maximum regular induced subgraphs in $2P_3$-free graphs. Theor. Comput. Sci. **460**(16), 26–33 (2012)
27. M. Milanič, Algorithmic developments and complexity results for finding maximum and exact independent sets in graphs. PhD thesis, Rutgers, The State University of New Jersey, 2007
28. G.J. Minty, On maximal independent sets of vertices in claw-free graphs. J. Comb. Theory Ser. B **28**(3), 284–304 (1980)
29. R. Mosca, Polynomial algorithms for the maximum stable set problem on particular classes of $P_5$-free graphs. Inf. Process. Lett. **61**(3), 137–143 (1997)
30. R. Mosca, Some results on maximum stable sets in certain $P_5$-free graphs. Discret. Appl. Math. **132**(1–3), 175–183 (2004)
31. R. Mosca, Independent sets in ($P_6$,diamond)-free graphs. Discret. Math. Theoretical Comput. Sci. **11**(1), 125–140 (2009)
32. R. Mosca, Stable sets for ($P_6$, $K_{2,3}$)-free graphs. Discuss. Math. Graph Theory **32**(3), 387–401 (2012)
33. N. Sbihi, Algorithme de recherche d'un stable de cardinalite maximum dans un graphe sans etoile. Discret. Math. **29**(1), 53–76 (1980)

# Optimal Patrol on a Graph Against Random and Strategic Attackers

Richard G. McGrath

## 1 Background

Patrol problems are encountered in many real-world situations. Generally speaking, a patrol is the movement of a guard force through a designated area of interest (AOI) for the purpose of observation or security. Patrols are often conducted by authorized and specially trained individuals or groups, and are common in military and law-enforcement settings. The use of patrols, instead of fixed, continuous surveillance, is often necessary because of real-world limitations on time and resources. Patrollers must operate with the intent of maximizing the likelihood of detection of adversaries, infiltration, or attacks. The objective in solving patrol problems is to determine the actions or policies that will maximize this likelihood. In most patrol problems, consideration must be made for the time required for a patroller to travel between specific locations within an AOI, and the time required to conduct an inspection in order to detect illicit activities at a particular location.

There are several military and non-military applications of patrol problems. Military applications include the routing of an unmanned aerial vehicle (UAV) on a surveillance mission or the conduct of ground patrols to interdict the placement of improvised explosive devices (IEDs). Non-military applications include the movement of security guards through museums or art galleries; police forces patrolling streets in a city; security officials protecting airport terminals; and conductors checking passenger tickets on trains in order to detect fare evaders.

This work is motivated by the need to provide for effective security, usually with limited resources, and often against very sophisticated and capable enemies. Not only does the solution to a patrol problem need to be mathematically sound, it also

R. G. McGrath (✉)
United States Naval Academy, Annapolis, MD, USA
e-mail: rmcgrath@usna.edu



B. Goldengorin (ed.), *Optimization Problems in Graph Theory*,
Springer Optimization and Its Applications 139,
https://doi.org/10.1007/978-3-319-94830-0_10

needs to be executable. Additionally, it is often important to ensure that the solution to a patrol problem incorporates sufficient randomization, and thus be unpredictable to potential adversaries.

## 1.1 Problem Description

We consider a problem where multiple locations within an AOI are subject to attack. A patroller (defender) is assigned to the area in order to detect attacks before they can be completed. An attack is considered to be any activity that the patroller wants to interdict or prevent, such as planting or detonating an explosive device, stealing a valuable asset, or breaching a perimeter. The patroller moves between locations and conducts inspections at those locations in order to detect any illicit activity. A specified travel time is required for movements between locations. It then takes the patroller an additional specified amount of time to inspect a new location after he arrives. At the end of the time required to complete an inspection, the patroller can move to any other location in the area.

We explicitly model the patrol problem on a graph, where potential attack locations are represented by vertices. We consider the inclusion of inspection times at each vertex and travel times for the patroller to move along edges between vertices in the graph. We consider this problem in continuous time and structure the patrol model on a complete graph, where the edge length represents the travel time between each pair of vertices.

The time at which an attacker arrives at a location to conduct an attack is random, and occurs according to a Poisson process. When an attacker arrives at a location he begins an attack immediately. The time required to complete an attack is random, with a probability distribution that is known to the attacker and the patroller. The patroller detects any ongoing attacks at a location at the end of his inspection. We consider an attacker to be detected if both the patroller and attacker occupy the same location at the end of the patroller's inspection. The amount of time it takes to complete an attack, as well as the amount of damage that an undetected attack will cause, is specific to each location.

The patroller's objective is to determine a path of locations to visit and inspect that will minimize the long-run cost incurred due to undetected attacks. For instance, where the cost of an attack is the same at all locations, this objective is equivalent to maximizing the probability of detecting an attack.

We consider two patrol models that are closely related:

1. *A single patroller against random attackers*: In the random-attacker case, an attacker will choose a location to attack according to a probability distribution that is known to the patroller. This situation may occur when there is intelligence available regarding potential enemy attack locations. It may be possible from this intelligence to assign a likelihood of attack to specific locations.

2. *A single patroller against strategic attackers*: In the strategic-attacker case, an attacker will actively choose a location to attack in order to inflict the maximum expected damage. Conversely, the patroller seeks to conduct his patrol so as to sustain the least expected damage. This situation may occur with a more capable or better-resourced enemy, who can analyze the expected damage among several attack locations.

In each of these cases, we assume that the attacker cannot observe the real-time location of the patroller. In other words, once an attacker initiates an attack, he will carry on with the attack until either completing the attack or getting detected. An attacker cannot time his attack, nor can he abandon an attack, based on real-time information about the patroller's location.

In the case of a single patroller against random attackers, we present a linear program to determine an optimal patrol policy. This linear program is constructed as a minimum cost-to-time ratio cycle problem on a directed graph. We also present two heuristic methods based on the graph structure that utilize aggregate index values to determine a heuristic patrol policy.

In the case of a single patroller against strategic attackers, we present a linear program to determine an optimal patrol policy. This linear program is a modification of the minimum cost-to-time ratio cycle linear program used for a random attacker that minimizes the largest expected cost among all locations and provides a direct mapping to a mixed strategy. We also present two heuristic methods for this case. The first is a combinatorial method based on the shortest Hamiltonian cycle in the graph. The second is an iterative method based on fictitious play. We also present a linear program that provides a lower bound to an optimal solution, which helps evaluate our heuristic policy when an optimal solution is not available.

## 1.2 Literature Review

A patrol problem can be considered more generally as a type of search problem. Many types of search and patrol problems have been studied in diverse literatures. Early work on search theory focused on two general categories: one-sided search and search games. One-sided search refers to the assumption that a target does not respond to, or is even necessarily aware of, the searcher's actions. In this type of problem, the objective is often to maximize the probability of detection before a deadline, or to minimize the expected time or cost of a search [9].

The two-sided search problem, more commonly referred to as a search game, involves a searcher and a target who knows that he is being pursued. These type of search problems are generally formulated as game-theoretic problems. The information that the target has concerning the searcher will vary anywhere from complete information on the searcher's strategy to a complete lack of information [9]. In these scenarios, a searcher and target can be working in competition, whereby the target wishes to evade detection. Alternatively, a searcher and a target can be working in cooperation, such as a search and rescue scenario, where the objective for both is to minimize the time (or cost) of the search.

Patrol problems are a specific type of search problem. In a patrol problem, a searcher utilizes a patrol strategy to cover an area where an attacker or target may or may not be present [5]. There are several types of game-theoretic patrol problems that relate to our work. An accumulation game is a type of patrol problem where a patroller visits several locations to collect materials hidden by an attacker. If the patroller finds a certain amount of the materials, he wins; otherwise, he loses [4], [14]. An infiltration game is a type of patrol problem where an intruder attempts to penetrate an area without being intercepted by a patroller [2, 7, 10, 21, 23]. An inspection game is a type of patrol problem where the patroller attempts to interdict an attacker during an attack [8, 24]. The infiltration and inspection game categories are most similar to the models that we examine.

There are several examples in search-game literature where the search area is modeled as a graph or network. Kikuta and Ruckle [13] study initial point searches on weighted trees. Kikuta [12] studies search games with traveling costs on a tree. Alpern [3] examines search games on trees with asymmetric travel times.

The works most closely related to this problem are those by McGrath and Lin [16], Lin et al. [15], and Alpern et al. [6]. Alpern et al. examine optimized random patrols where a facility to be patrolled is modeled on a graph with interconnected vertices representing individual locations within the facility. This work focuses on the case of strategic attackers, where an attacker actively chooses a location to attack, and assumes that the time to complete an attack is deterministic and is the same for all locations. Lin et al. examine a patrol problem on a graph with both random and strategic attackers. They use an exact linear program to compute an optimal solution. Since this method quickly becomes computationally intractable as a problem size increases, they introduce index heuristics based on Gittins et al. [11] to determine a patrol policy. They use an aggregate index, where index values are accumulated as a patroller looks ahead into the future, to produce effective patrol policies in a game-theoretic setting. Both of these works use discrete-time models, require the same inspection time at all locations, and prescribe an adjacency structure for their graphs—which puts constraints on how a patroller can move between locations.

## 2 Single Patroller Against Random Attackers

We consider the case of a single patroller against random attackers. In this patrol problem, the patroller's objective is to determine a patrol policy that minimizes the long-run average cost due to undetected attacks. Section 2.1 introduces a patrol model on a graph, where an attacker chooses to attack a specific location based on a probability distribution that is known to the patroller. In Section 2.2, we present a linear program that determines an optimal solution to the patrol problem. Since the linear program quickly becomes computationally intractable as the size of the problem grows, we also present two heuristic methods for determining a solution in Section 2.3. We conduct extensive numerical experiments for several scenarios and present the results in Section 2.4. We make recommendations on how to best utilize the heuristic methods based on the experimental results.

## 2.1 Patrol Model

We consider a problem where multiple heterogeneous locations dispersed throughout an area of interest (AOI) are subject to attack. A patroller (defender) is assigned to patrol the area and inspect locations in order to detect attacks before they can be completed. An attack is considered to be any type of activity that the patroller wants to interdict and prevent, such as planting an explosive device, stealing a valuable asset, or breaching a perimeter. The patroller moves between locations and conducts inspections at those locations in order to detect illicit activity. We consider an attacker to be detected and his attack defeated if both the patroller and attacker occupy the same location at the end of an inspection.

We model this problem as a graph with $n$ vertices, where each vertex represents a location that is subject to attack. We define a set of vertices $N = \{1, \ldots, n\}$ to represent potential attack locations. A random attacker will choose to attack vertex $i$ with probability $p_i \geq 0$, for $i \in N$, and $\sum_{i=1}^{n} p_i = 1$. The time required for an attacker to complete an attack at vertex $i$ is a random variable, which follows a distribution function $F_i(\cdot)$, for $i \in N$, that is known to the attacker and the patroller.

The patroller detects any ongoing attacks at a vertex at the end of an inspection. We assume that there are no false negatives; that is, the attacker will successfully detect all ongoing attacks at a vertex at the end of his inspection. An attack is considered to be unsuccessful if it is detected by the patroller. An attack is successful if it is completed before it is detected.

We assume that an attacker arrives at a location in the AOI to commence an attack according to a Poisson process with rate $\Lambda$. The Poisson process has stationary and independent increments, which implies that attacks are equally likely to occur at any time and that prior attacks do not help the patroller predict future attacks. Attackers arrive at a specific vertex $i$ to begin an attack at a rate of $\lambda_i = p_i \Lambda$, for $i \in N$. These attacker arrivals at specific vertices constitute independent Poisson processes.

In most situations, the attacker arrival rate $\Lambda$ is very small. In the formulation of our problem, the value of $\Lambda$ is inconsequential because we ignore interruptions from attacks. In other words, several attackers can operate simultaneously on the graph, or even at the same vertex, with each acting independently. By minimizing the long-run cost rate, we also minimize the average cost from each attack with $\Lambda$ acting as a scaling constant. Thus, an optimal solution does not depend on the value of $\Lambda$.

It takes a specified amount of time to travel between vertices and conduct inspections. These times are fixed in our problem. The time required for a patroller to travel between vertices is denoted by an $n \times n$ distance matrix $D = [d_{ij}]$, for

$i, j \in N$, where $d_{ij} \geq 0$ for all pairs of vertices $i \neq j$ and $d_{ii} = 0$. The time required for a patroller to complete an inspection at a vertex is denoted by $(v_1, \ldots, v_n)$. From these values, we construct an $n \times n$ transit time matrix denoted by $T = [t_{ij}]$, where $t_{ij} = d_{ij} + v_j$, to indicate the time required for a patroller to travel from vertex $i$ to vertex $j$ and complete an inspection at vertex $j$. The damage inflicted due to an undetected attack at a vertex is denoted by $(c_1, \ldots, c_n)$. An attack inflicts no damage if it is detected before it is completed.

The patroller travels between vertices in the graph and conducts inspections in order to detect attacks. A *patrol policy* consists of a sequence of vertices that the patroller will visit and inspect. We seek to determine an optimal patrol policy that minimizes the long-run cost rate incurred due to undetected attacks.

Fundamentally, the patroller is making a series of sequential decisions under uncertainty in order to determine a patrol policy. Decisions are made at decision epochs, which occur at a specific point in time (in this case at the end of an inspection). At each decision epoch, the patroller observes the state of the system as the amount of time elapsed since he last completed an inspection at each vertex. Based on this information, he chooses an action. The choice of action is which vertex to visit next. The action incurs a cost and causes the system to transition to a new state at a subsequent point in time. The cost incurred is the expected cost due to attacks that will be completed during the time it takes for the patroller to travel to and inspect the next vertex. At the end of the inspection time at the chosen vertex, the system will transition to a new state. At this point, the patroller reaches another decision epoch and the process repeats.

In our problem, we wish to determine an optimal choice of action for the patroller at each decision epoch. The essential elements of this sequential decision model are [18]

1. A set of system states.
2. A set of available actions.
3. A set of state-dependent and action-dependent costs.
4. A set of state-dependent and action-dependent transition times and transition probabilities.

We incorporate all of these elements into a sequential decision model in order to determine an optimal patrol policy.

## 2.2 Optimal Policy

In order to find an optimal solution to our patrol problem, we must determine a patrol policy that minimizes the long-run cost rate. To do so, we define a state space $\Omega$ that consists of all feasible states of the system. The state of the system at any given time can be delineated by

$$\mathbf{s} = (s_1, s_2, \ldots, s_n), \quad (1)$$

where $s_i$ denotes the time elapsed since the patroller last completed an inspection at vertex $i$, for $i \in N$. Based on the assumption that a patroller detects all ongoing attacks at a vertex at the end of an inspection, the state of a vertex returns to 0 immediately upon completion of an inspection. Since we consider this problem in continuous time, the state of a vertex can assume any non-negative value. We write the state space of the system as

$$\Omega = \{(s_1, \ldots, s_n) : s_i \geq 0, \forall i \in N\}. \tag{2}$$

At the end of each inspection, the patroller reaches a decision epoch and will decide to stay at his current vertex to conduct an additional inspection or proceed to another vertex. The action space can be defined as

$$A = \{j : j \in N\}. \tag{3}$$

A deterministic, stationary patrol policy can be specified by a map $\pi$ from the state space to the action space:

$$\pi : \Omega \to A. \tag{4}$$

This patrol policy is deterministic because, for any state of the system, a specific action is prescribed with certainty. It is stationary, or time-homogeneous, because the decision rules associated with a particular patrol policy do not change over time. For any given state of the system, the future of the process is independent of its past. The resulting state depends only on the action chosen by the patroller. If the patroller just inspected vertex $k$ and next wants to inspect vertex $j$, that action will take time $d_{kj} + v_j$; and the system that started in state $\mathbf{s}$ will transition to state

$$\tilde{\mathbf{s}} = (\tilde{s}_1, \tilde{s}_2, \ldots, \tilde{s}_n), \tilde{s}_j = 0; \tilde{s}_i = s_i + d_{kj} + v_j, \forall i \neq j\,. \tag{5}$$

In order to identify the vertex where a patroller has just finished an inspection and is currently located at a decision epoch, we define

$$\omega(\mathbf{s}) = \arg\min_i s_i, \tag{6}$$

since the state of the vertex where an inspection has just been completed will be 0 and the state of all other vertices will be greater than 0.

In our model, the times between decision epochs and state transitions are deterministic. They depend on previous system states and actions only through the current state of the system. We define

$$\tau(\mathbf{s}, j) = d_{\omega(\mathbf{s}),j} + v_j, \tag{7}$$

as the time between decision epochs and the time between state transitions, if the patroller decides to visit vertex $j$, when the system is in state $\mathbf{s}$. At a decision epoch, the patroller will decide his action based only on the current state of the system. For these reasons, our model falls in the category of a semi-Markov decision process (SMDP).

The cost function for this SMDP can be calculated based on the distribution of time required to complete an attack $F_i(\cdot)$ and the cost $c_i$ incurred due to a successful attack at vertex $i$. To illustrate how expected costs are incurred, suppose that the patroller has just finished an inspection at vertex $k$ and the current state of the system is $\mathbf{s}$, where $\omega(\mathbf{s}) = k$. The patroller can then elect to travel to another vertex or remain at vertex $k$ and conduct an additional inspection. There will be an expected cost incurred for each vertex in the graph based on the cost of a successful attack and the number of attacks expected to be completed at that vertex during the transition time between state $\mathbf{s}$ and state $\tilde{\mathbf{s}}$.

To determine the expected number of attacks that are completed at a particular vertex in a time interval, recall from Section 2.1 that the arrival of attackers at a vertex constitutes a Poisson process. Consider an attacker arriving to a vertex at time $y$ after the last inspection was completed, and suppose that the patroller completes his next inspection at that vertex at time $s$. The attacker will complete his attack if the attack time is no greater than $s - y$. Using Poisson sampling (see Proposition 5.3 in Ross [20]), the number of successful attacks at vertex $i$ will follow a Poisson distribution with expected value

$$\lambda_i \int_0^s P(X_i \leq s - y)\, dy = \lambda_i \int_0^s P(X_i \leq t)\, dt, \tag{8}$$

where $X_i$ denotes the time required to complete an attack at vertex $i$, for $i \in N$.

If we know the expected number of attacks that will be completed at vertex $i$ in a time interval, then we can determine the expected cost incurred at vertex $i$ by multiplying (8) by $c_i$. Thus, the expected cost incurred at vertex $i$ when the system is in state $\mathbf{s}$ and the patroller elects to transit to vertex $j$ is

$$C_i(\mathbf{s}, j) = c_i \lambda_i \left( \int_0^{s_i + \tau(\mathbf{s}, j)} P(X_i \leq t)\, dt - \int_0^{s_i} P(X_i \leq t)\, dt \right). \tag{9}$$

The cost at each vertex can be summed across all $n$ vertices in the graph in order to determine the total expected cost when the system starts in state $\mathbf{s}$ and the patroller transits to vertex $j$. The overall cost function for this SMDP is

$$C(\mathbf{s}, j) = \sum_{i=1}^{n} C_i(\mathbf{s}, j). \tag{10}$$

As currently defined, the state space has an infinite number of states; however, in order to be able to compute an optimal policy, we need a finite state space. To do so, we assume that there is an upper limit on the attack time distribution at each vertex. Specifically, let $B_i$ denote the maximum time required to complete an attack at vertex $i$. For the case where $s_i = S \geq B_i$, (9) becomes

$$C_i(\mathbf{s}, j) = c_i\lambda_i \left( \int_S^{S+\tau(\mathbf{s},j)} P(X_i \leq t)\, dt \right)$$

$$= c_i\lambda_i(S + \tau(\mathbf{s}, j) - S) = c_i\lambda_i\tau(\mathbf{s}, j), \tag{11}$$

which remains a constant function over time for any state $s_i \geq B_i$. Therefore, once the state of a vertex has reached the bounded attack time, any additional expected cost will accrue at a constant rate. The bounded attack times allow us to restrict the state of a vertex so that $s_i \leq B_i$, and the state space becomes

$$\Omega = \{(s_1, \ldots, s_n) : 0 \leq s_i \leq B_i, \forall i \in N\}. \tag{12}$$

We consider cases where the attack times at all vertices are bounded. Thus, if the patroller has just inspected vertex $k$ and next wants to inspect vertex $j$, the resulting state at the end of the inspection at vertex $j$ is

$$\tilde{\mathbf{s}} = (\tilde{s}_1, \tilde{s}_2, \ldots, \tilde{s}_n), \tilde{s}_j = 0; \tilde{s}_i = \min\{s_i + d_{kj} + v_j, B_i\}, \forall i \neq j\,. \tag{13}$$

Using (13), we define a transition function to identify the resulting state if the patroller decides to visit vertex $j$ when the system is in state $\mathbf{s}$:

$$\phi(\mathbf{s}, j) = \tilde{\mathbf{s}}. \tag{14}$$

The objective of the patrol problem is to determine a policy for the patroller that minimizes the long-run cost. Recall that the action space in this SMDP is finite because the number of vertices is finite. Therefore, by Theorem 11.3.2 in Puterman [18], there exists a deterministic, stationary optimal policy. Thus, we only need to consider deterministic, stationary policies in our problem. We define

$$\psi_\pi(\mathbf{s}) = \phi(\mathbf{s}, \pi(\mathbf{s})) \tag{15}$$

as the resulting state if the patroller applies policy $\pi$ when in state $\mathbf{s}$. We can define this function because the state transitions are deterministic. From an initial state $\mathbf{s_0}$, policy $\pi$ will produce an indefinite sequence of states, $\{\psi_\pi^\kappa(\mathbf{s_0}), \kappa = 0, 1, 2, \ldots\}$. This sequence must eventually visit some state for a second time since the state space if finite; and since the state transitions are deterministic under the same policy $\pi$, this sequence will then continue to repeat indefinitely. The sequence of vertices that correspond to this repeating cycle of states will constitute a *patrol pattern*.

We define $V_i$ as the long-run expected cost rate at vertex $i$. If we apply the deterministic, stationary policy $\pi$ to any initial state $\mathbf{s_0}$, then the long-run expected cost rate at vertex $i$ is

$$V_i(\pi, \mathbf{s_0}) = \lim_{\xi \to \infty} \frac{\sum_{\kappa=0}^{\xi} C_i(\psi_\pi^\kappa(\mathbf{s_0}), \pi(\psi_\pi^\kappa(\mathbf{s_0}))}{\sum_{\kappa=0}^{\xi} \tau(\psi_\pi^\kappa(\mathbf{s_0}), \pi(\psi_\pi^\kappa(\mathbf{s_0}))}. \tag{16}$$

We seek to determine the minimum long-run cost rate across all vertices, which will give an optimal solution

$$C^{\mathrm{OPT}}(\mathbf{s_0}) = \min_{\pi \in \Pi} \sum_{i=1}^{n} V_i(\pi, \mathbf{s_0}), \tag{17}$$

where $\Pi$ is the set of all feasible deterministic, stationary patrol policies. Dividing (17) by $\Lambda$ will give the minimum average cost incurred for each attack.

We note that $V_i(\pi, \mathbf{s_0})$ depends on $\pi$ and $\mathbf{s_0}$. However, in a connected graph, an optimal cost rate $C^{\mathrm{OPT}}(\mathbf{s_0})$ does not depend on $\mathbf{s_0}$. Since determining an optimal patrol policy is equivalent to finding an optimal patrol pattern, we can develop a policy $\pi$ in a connected graph that will produce any feasible patrol pattern from any starting state $\mathbf{s_0}$. Therefore, when we determine $C^{\mathrm{OPT}}$ in (17), it will be the same for all initial states since we minimize across all feasible patrol policies $\pi \in \Pi$. Thus we can drop the notational dependence of $C^{\mathrm{OPT}}$ on $\mathbf{s_0}$.

### 2.2.1 Linear Program Formulation

One method to solve this SMDP is to construct another graph that uses the state space of the system modeled as a network. To do so, we redefine the problem on a *directed* graph, $G(\mathcal{N}, \mathcal{A})$. Each *node* $k \in \mathcal{N}$ will represent one state of the system, and each *arc* $(k, l) \in \mathcal{A}$ will represent a feasible transition between states. This network will be of order $|\mathcal{N}| = |\Omega|$ and size $|\mathcal{A}| = |\Omega| n$. Each arc is assigned a transit time $t_{kl}$ as determined by the vertex-pair specific distance and inspection times, where $t_{kl} = \tau(k, \omega(l))$; and cost $c_{kl}$ as determined by the cost function (10), where $c_{kl} = C(k, \omega(l))$.

The objective is to find a directed cycle in the network with the smallest ratio of total cost to total transit time. This is a sufficient solution to the problem because any directed cycle in this network will constitute a valid patrol policy, regardless of the length of the cycle. This is an example of a minimum cost-to-time ratio cycle problem, also known as the *tramp steamer problem*, which is described in Sect. 5.7 of Ahuja et al. [1].

To solve this problem, we formulate the following linear program, which we refer to as the random-attacker linear program (RALP):

$$\min_{x} \sum_{(k,l)\in\mathscr{A}} c_{kl} x_{kl} \tag{18a}$$

$$\text{subject to} \sum_{l|(k,l)\in\mathscr{A}} x_{kl} - \sum_{l|(l,k)\in\mathscr{A}} x_{lk} = 0, \forall k \in \mathscr{N} \tag{18b}$$

$$\sum_{(k,l)\in\mathscr{A}} t_{kl} x_{kl} = 1, \tag{18c}$$

$$x_{kl} \geq 0, \forall (k,l) \in \mathscr{A}. \tag{18d}$$

The variable $x_{kl}$ represents the long-run rate at which the patroller uses arc $(k, l)$. The objective function value in (18a) represents the long-run cost rate. The total rate at which the system enters state $k$ must be equal to the total rate that the system exits state $k$, which is ensured by the network balance of flow constraint in (18b). For a single patroller, the rate that he uses arc $(k, l)$ times the amount of time required to transit from node $k$ to node $l$ indicates the fraction of time that he will spend on arc $(k, l)$. The fractions of time must sum to 1, which is ensured by the total-rate constraint in (18c). Finally, the long-run rate at which the patroller uses arc $(k, l)$ cannot be negative, which is ensured by the non-negativity constraint in (18d).

The states on an optimal cycle directly correspond to vertices on the graph, which can be determined by the function $\omega(\mathbf{s})$. Thus, this linear program will produce a specific patrol pattern consisting of a repeating sequence of vertices for the patroller to visit and inspect. This patrol pattern represents an optimal solution to the patrol problem.

The number of decision variables in this linear program is $|\Omega|n$. The size of the constraint matrix is on the order of $|\Omega|$. The value of $|\Omega|$ grows as a function of the number of vertices in the graph, the attack time distributions, and the transit times.

### 2.2.2 Size of State Space

To understand the size of the state space, consider the case where the maximum attack time at all vertices is $B$, the travel time between all vertices is $d$, and the inspection time at all vertices is $v$. Define Z as

$$Z = \left\lceil \frac{B}{d+v} \right\rceil. \tag{19}$$

The number of states in the system for a graph with $n$ vertices and $Z \geq n$ is given by

$$|\Omega| = \sum_{i=0}^{n-1} \binom{n}{1} \binom{n-1}{i} \binom{Z-1}{n-1-i} (n-1-i)!, \tag{20}$$

**Table 1** Examples of state space size

| $n$ | $B$ | $d$ | $v$ | $Z$ | $\|\Omega\|$ |
|---|---|---|---|---|---|
| 5 | 9.8 | 1.0 | 0.2 | 9 | 16,965 |
| 7 | 11.5 | 1.2 | 0.3 | 8 | >260,000 |
| 8 | 15.5 | 0.9 | 0.6 | 11 | >20,000,000 |
| 12 | 18.3 | 0.8 | 0.8 | 12 | >40,000,000,000 |

since for each state of the system there will be exactly one vertex in state 0, as indicated by the first term; $i$ of the remaining $n-1$ vertices at the bounded attack time state $B$, as indicated by the second term; and each of the remaining $n-1-i$ vertices in a distinctive state between $d+v$ and $(d+v)(Z-1)$, as indicated by the third and fourth terms. Some examples of state space size are shown in Table 1. The number of states grows exponentially with the number of vertices, and grows even larger when combined with higher bounded attack times and shorter transit times.

Although we can compute an optimal patrol policy using linear programming, this method quickly becomes computationally intractable as the number of vertices increases and the ratio of the bound of the attack times to transit times increases. Hence, there is a need to develop efficient heuristics.

## 2.3 Heuristic Policies

In this section, we consider solutions based on index heuristic methods. To begin, consider a special case of our problem when $v_i = 1$ and $d_{i,j} = 0$, for $i, j \in N$. This special case coincides with the model presented in Lin et al. [15]. By adding a Lagrange multiplier $w > 0$, they show the optimization problem can be broken into $n$ separate problems, each concerning a single vertex. The Lagrange multiplier $w$ can be interpreted as a service charge incurred for each patrol visit to a vertex. The objective is to decide how frequently to summon a patroller at each vertex in order to minimize the long-run cost rate due to undetected attacks and service charges. For a given state of the system, the solution to this problem can be used to determine an index value for each vertex in the graph. We can develop a heuristic policy where, based only on the current state of the system, the patroller can choose to travel to and inspect the vertex that has the highest index value. We next explain how to extend this method to our patrol model.

### 2.3.1 Single Vertex Problem

We consider the problem at a single vertex where each visit from the patroller incurs a service charge $w > 0$. For a given value of $w$, our objective is to determine a policy that minimizes the total long-run cost rate due to undetected attacks and service charges. Generally speaking, a policy is a mapping from a state to an action. For the single vertex problem, the state of the system $s \geq 0$ is the amount of time since the patroller last completed an inspection at the vertex. The action space for the patroller simplifies to a binary decision: Inspect the vertex at time $s$ or continue to wait.

Although the state space is infinite, the action space is finite for every $s \in \Omega$. Therefore, we only need to consider deterministic, stationary policies [18]. In addition, since each inspection brings the state of the vertex back to 0, any deterministic, stationary policy reduces to the following format: Inspect the vertex once every $s$ time units.

Recall from (8) that the number of successful attacks in the time interval $[0, s)$ between patroller inspections follows a Poisson distribution with expected value

$$\lambda \int_0^s P(X \leq t)\, dt\,. \tag{21}$$

Since each successful attack costs $c$, and a patrol visit costs $w$, the average long-run cost given a policy that inspects the vertex every $s$ time units is

$$f(s) = \frac{c\lambda \int_0^s P(X \leq t)\, dt + w}{s}, \; s > 0. \tag{22}$$

For a given value of $w$, we find $s$ in order to minimize $f(s)$. To minimize $f(s)$, we take the first derivative of $f(s)$, which gives

$$f'(s) = \frac{c\lambda}{s} P(X \leq s) - \frac{c\lambda}{s^2} \int_0^s P(X \leq t)\, dt - \frac{w}{s^2}, \tag{23}$$

and set $f'(s) = 0$ to obtain

$$0 = c\lambda s P(X \leq s) - c\lambda \int_0^s P(X \leq t)\, dt - w. \tag{24}$$

We solve this equation for $w$ as a new function of $s$:

$$W(s) = c\lambda \left( s P(X \leq s) - \int_0^s P(X \leq t)\, dt \right), \tag{25}$$

where $W(s)$ indicates the corresponding service charge such that it is optimal for the vertex to summon patrol visits once every $s$ time units.

Since attack times at each vertex are bounded by a constant $B$, for cases where $s \geq B$ we note that

$$\begin{aligned} W(s) &= c\lambda \left( s - \int_0^s P(X \leq t)\, dt \right) = c\lambda \int_0^s P(X > t)\, dt \\ &= c\lambda \int_0^B P(X > t)\, dt = c\lambda \mathrm{E}[X], \end{aligned} \tag{26}$$

which remains the same for all $s \geq B$.

### 2.3.2 Index Heuristic Time Method

Since $W(s)$ represents the per-visit cost for an optimal policy that visits a vertex in state $s$, we can define an index value for vertex $i$ based on (25) as

$$W_i(s) = c_i\lambda_i \left( sP(X_i \le s) - \int_0^s P(X_i \le t)\,dt \right), \tag{27}$$

if the last inspection at vertex $i$ was completed $s$ time units ago.

A straightforward heuristic method for the patroller at a decision epoch is to compute the index values based on the current state of each vertex and choose to visit the vertex that has the highest index value. This method will produce a feasible patrol pattern; however, it does not account for different travel times between vertices. To solve this problem, we develop methods for the patroller to look further ahead and compute *aggregate* index values before choosing which vertex to visit next. When computing an aggregate index in our continuous-time model, we consider the amount of time that different actions will take. To do so, we select a fixed look-ahead time window $\delta$ and consider all feasible paths and partial paths beginning from the patroller's current vertex $\omega(\mathbf{s})$ that can be completed during time $\delta$. We call this the index heuristic time (IHT) method. A value for $\delta$ is selected based on the structure of the graph and is discussed at the end of this section.

To illustrate the IHT method, we select a look-ahead window $\delta$ and examine an arbitrary patrol sequence over the next $\delta$ time units. For the time window $[0, \delta]$, let $S_i(t), t \in [0, \delta]$ denote the state of vertex $i$ at time $t$. By definition, $S_i(0) = s_i$ and $S_i(t)$ increases over time at slope 1 until the patroller next completes an inspection at vertex $i$, when its value returns to 0. The aggregate index values accumulated at vertex $i$ over the time window $[0, \delta]$ can be written as

$$\int_0^\delta W_i(S_i(t))\,dt, \forall i \in N. \tag{28}$$

For a given patrol sequence, the total index value for all $n$ vertices over the time window $[0, \delta]$ is

$$\sum_{i=1}^{n} \int_0^\delta W_i(S_i(t))\,dt\,. \tag{29}$$

To determine a patrol pattern using the IHT method, we select a starting state of the system $\mathbf{s_0}$ and enumerate all possible paths over the next $\delta$ time units. We compute the total aggregate index value for each of these paths using (29), and choose the path with the highest *aggregate index value per unit time*. The first vertex along that path becomes the vertex that the patroller inspects next. We repeat this process using the new state of the system as the starting state, and continue to repeat the process to form a path of vertices. Recall that since the state space is finite, this sequence must eventually visit some state for a second time. The process terminates when a state repeats and a cycle has been found. The vertices corresponding to the states of the system on this cycle is the patrol pattern that results from using the IHT method.

In order to select a value for $\delta$ in the IHT method, we determine the average transit time $r$ between all vertices in the graph as

$$r = \frac{\sum_{i=1}^{n} \sum_{j=1}^{n} (d_{ij} + v_j)}{n^2}. \tag{30}$$

We then choose a look-ahead time window in terms of multiples of $r$. For example, if we choose $\delta = 3r$ as a look-ahead window, then we are choosing an amount of time that on average will allow the patroller to visit any sequence of three vertices from his current vertex. We can choose a multiple of $r$ more generally, such as $n/2$, which will on average allow the patroller to look ahead over about half the vertices in the graph from his current location. We make recommendations on how to select specific values for $\delta$ based on our numerical experiments. These recommendations are presented in Section 2.4.3.

Although we can choose any state from which to start the IHT method, for consistency in our numerical experiments we identify the vertex that has the maximum value of $W(s)$ when $s \geq B$, as defined in (26). We choose as $\mathbf{s_0}$ the state of the system where this vertex has just completed an inspection and the state of all other vertices is at the bounded attack time. In other words, we determine

$$k = \arg\max_{i \in N} \{c_i \lambda_i \mathrm{E}[X_i]\}, \tag{31}$$

and select as $\mathbf{s_0}$ the state where $s_k = 0$ and $s_j = B_j$, for $j \in N$, $j \neq k$.

### 2.3.3 Index Heuristic Epoch Method

Instead of looking ahead for a fixed time period, as in the IHT method, we consider another heuristic which looks ahead for a fixed number of decision epochs. We call this the index heuristic epoch (IHE) method. To compute an aggregate index using the IHE method, we select a number of decision epochs $\eta$ for the patroller to look ahead. The number $\eta$ can be any positive integer value. For example, if we choose $\eta = 3$ as a look-ahead window, the patroller considers all paths of three vertices from his current vertex, since a decision epoch in our model occurs at the end of each inspection. As with the IHE method, we choose the path with the highest aggregate index value per unit time, and the first vertex along that path is the vertex that the patroller inspects next. We can also choose the look-ahead window more generally, such as $\eta = \lceil \frac{n}{2} \rceil$, which allows the patroller to look ahead over at least half the vertices in the graph.

We choose a starting state $\mathbf{s_0}$ for the IHE method using the same criteria as we did for the IHT method. We enumerate all feasible paths from $\mathbf{s_0}$ that consist of exactly $\eta$ decision epochs and then proceed in the same manner as the IHT method described in Section 2.3.2 to determine a path of vertices based on the highest aggregate index value per unit time, until a patrol pattern has been obtained.

## 2.4 *Numerical Experiments*

To test the IHT and IHE methods, we conduct several numerical experiments. We compare the results obtained from these heuristic methods with an optimal solution. We also report the computation time required. Based on these results, we make conclusions on the efficacy of the heuristic methods, as well as make recommendations for the selection of look-ahead parameters to be used in both the IHT and IHE methods.

As inputs for the problem, we use a probability vector $(p_1, \ldots, p_n)$ indicating the likelihood of an attacker to choose to attack a specific vertex; an attack time distribution parameter matrix; a vector $(c_1, \ldots, c_n)$ of the cost incurred due to a successful attack at each vertex; a distance matrix $D$ of the time it takes for a patroller to travel between each pair of vertices; a vector $(v_1, \ldots, v_n)$ of the time required for a patroller to conduct an inspection at each vertex; and an overall attacker arrival rate $\Lambda$. Recall from Section 2.1 that an optimal solution does not depend on the value of $\Lambda$; therefore, without loss of generality, we set the overall attacker arrival rate to be $\Lambda = 1$ in our numerical experiments. We also set the cost incurred from a successful attack to $c_i = 1$, for $i \in N$, which allows the results to be interpreted as the long-run proportion of attackers that will evade detection.

We consider three general cases of patrol problems. In the first case, which we use as a baseline, the patroller spends about half of the time traveling and half of the time inspecting vertices. For this case, we choose average travel times that are comparable to average inspection times. In the second case, we choose average inspection times that are twice as long as average travel times. In other words, each vertex takes more time to inspect, but the vertices are closer together. In the third case, we choose average travel times that are twice as long as average inspection times. In other words, each vertex takes less time to inspect, but the vertices are farther apart.

All computations are done on a 64-bit Windows 7 desktop computer (Intel Core i7 860@2.8 GHz; 8.0 GB RAM). All linear programs that determine an optimal solution or a lower bound are implemented using GAMS 23.8.2.

### 2.4.1 Generation of Problem Instances

We conduct our numerical experiments on a graph with $n = 5$ vertices, which is a problem size that allows for the computation of an optimal solution. We choose parameters in order to generate and test cases where an optimal detection probability is in the neighborhood of 0.5. This is the case where the development of a good patrol policy can be most helpful.

To generate a random graph of $n$ patrol locations for our experiments, let $(X_i, Y_i)$ denote the Cartesian coordinate of vertex $i$, for $i \in N$, and draw $X_i$ and $Y_i$ from independent uniform distributions over [0, 1]. Letting $d_{ij}$ denote the travel distance between vertices $i$ and $j$, we compute

$$d_{i,j} = \sqrt{(X_i - X_j)^2 + (Y_i - Y_j)^2}, \quad \forall i, j \in N. \tag{32}$$

The expected value of $d_{i,j}$ is $\mathrm{E}[d_{ij}] = 0.5215$ and the variance of $d_{i,j}$ is $\mathrm{Var}(d_{ij}) = 0.0615$.

Based on this average distance and variance, we generate an inspection time at each vertex by drawing from a uniform distribution over $[0.3857, 0.6573]$. This distribution gives an expected inspection time of $\mathrm{E}[v_i] = 0.5215$, which is comparable to the average travel time between vertices. The variance of the inspection times is 0.00615, which is approximately 1/10 of the variance of the vertex distance values. We choose these parameters in order to prevent very small inspection times at vertices, which could lead to excessively large state spaces and prevent the computation of an optimal solution.

For the attack time at each vertex, we use a triangular distribution. A triangular distribution requires three parameters: lower limit (minimum) $a$, upper limit (maximum) $b$, and mode $c$, where $a < b$ and $a \leq c \leq b$. We generate values for $(a, b, c)$ independently from a uniform distribution over $[1.043, 4.172]$. This distribution gives a minimum attack time that is comparable to the average travel time between any two vertices plus the inspection time at the second vertex, which in this case is $0.5215 \times 2 = 1.043$. The expected value of this distribution is comparable to the time required for a patroller to travel and complete inspections over approximately half of the vertices in the graph, which for the case of $n = 5$ is $1.043 \times 5/2 = 2.6075$. From this minimum and expected value, we determine a maximum attack time for use in our experiments as $2 \times 2.6075 - 1.043 = 4.172$. More generally, we can generate attack time distribution parameters from a uniform distribution on $[1.043, 1.043(n - 1)]$ for problems with any number of vertices $n > 2$.

For the likelihood of an attacker to choose a vertex to attack, we create a probability vector $(p_1, \ldots, p_n)$. We spread 0.5 of the total attack probability equally across all $n$ vertices and then randomly assign the remaining 0.5 probability. This ensures that the minimum probability of attack at any vertex is $0.5/n$, which will encourage a patrol policy that visits many or all of the vertices rather than completely excluding one or several vertices simply due to a low probability of attack. To create this vector, we generate $n$ uniform random variables $u_i$ on U[0, 1] and then normalize them so that $p_i = (0.5/n)+(0.5u_i/\sum_{j=1}^{n} u_j)$, for $i \in N$. In our experiments with $n = 5$, this ensures that each vertex has at least a 0.1 probability of selection for attack and no more than a 0.6 probability.

### 2.4.2 Baseline Problems

For our baseline problem, we consider the case where a patroller spends about half of the time traveling and half of the time inspecting vertices. We randomly generate 1000 problem scenarios and determine an optimal solution using the RALP from Section 2.2 and a solution using the heuristic methods from Section 2.3. The RALP on average uses 5920 decision variables and 7105 constraints for a problem size with 1184 states. An optimal solution takes on average 20.68 s to compute. We compare the solution obtained from the heuristic method to an optimal solution. For the look-ahead depth parameter $\delta$ used in the IHT method, we chose an initial value of $\delta = (n/2)r$, with $r$ defined in (30) as the average transit time between vertex pairs in each problem instance. For $n = 5$, this starting value is $\delta = 2.5r$. We also test additional parameter values by increasing and decreasing the look-ahead depth in $0.5r$ increments.

As the IHT method looks further ahead, the computation time increases due to the higher number of paths that must be considered. Performance does not always improve when using deeper looks, and in many cases it may be worse. Two different look-ahead parameter values, $2.5r$ and $3r$ in the IHT method, for example, may return the same patrol pattern or two distinct patrol patterns with different long-run cost rates. If the same problem is solved using multiple look-ahead parameters, we select the best solution that is obtained.

We consider single look-ahead parameter values and also consider sets of multiple look-ahead values in our numerical experiments. For the sets of multiple look-ahead values, we run the selected heuristic method for each individual value and then choose the patrol policy that yields the minimum cost, regardless of which specific look-ahead parameter produced that policy. This method tends to improve overall performance, but with a proportional increase in computation time based on the number and size of the look-ahead parameter values.

Results for the IHT method are shown in Table 2. When using a single look-ahead depth parameter, the best performance, as determined by the smallest excess over optimum for the mean and 90th percentile of problem instances, is obtained with a look-ahead time value of $\delta = 2.5r$. For the hybrid method of using up to three look-ahead parameters and then choosing the best patrol pattern, the best performance using similar criteria is obtained with a look-ahead depth set of $\{2r, 2.5r, 3r\}$.

We repeat the same experiments using the IHE method. For the look-ahead depth parameter $\eta$ used in the IHE method, we chose an initial value of $\eta = \lceil \frac{n}{2} \rceil$. For $n = 5$ this starting value is $\eta = 3$. This indicates that, at each decision epoch, the patroller will consider all possible paths consisting of three decision epochs. We test additional IHE depth parameter values by increasing and decreasing the look-ahead depth in $\eta = 1$ increments.

The IHE method is like the IHT method in that, as it looks further ahead, computation time increases due to the higher number of paths that must be considered. Similarly, the performance does not always improve when using deeper looks. For this reason, we test the IHE method using single look-ahead parameters and also using the hybrid method of comparing the results from multiple look-ahead parameters and selecting the best solution. Results are shown in Table 3. When

**Table 2** Performance of the IHT method on a complete graph with $n = 5$ vertices for 1000 randomly generated problem scenarios with average inspection times comparable to average travel times, using the best solution that was obtained in each problem scenario for the indicated look-ahead depth parameter sets

| | Percent over optimum | | | | |
|---|---|---|---|---|---|
| IHT look-ahead depth ($\delta$) | Mean | 50th | 75th | 90th | Time (s) |
| $2r$ | 3.31 | 0.38 | 4.13 | 8.65 | 2.19 |
| $2.5r$ | 1.22 | 0.00 | 1.60 | 3.60 | 2.47 |
| $3r$ | 1.36 | 0.00 | 1.34 | 5.51 | 3.64 |
| $3.5r$ | 1.88 | 0.00 | 2.03 | 6.52 | 6.75 |
| $4r$ | 3.26 | 1.24 | 5.61 | 7.96 | 18.22 |
| $\{2r, 3r\}$ | 0.55 | 0.00 | 0.23 | 1.56 | 5.83 |
| $\{2.5r, 3r\}$ | 0.62 | 0.00 | 0.49 | 2.15 | 6.11 |
| $\{2r, 2.5r, 3r\}$ | 0.49 | 0.00 | 0.20 | 1.38 | 8.30 |
| $\{2.5r, 3r, 3.5r\}$ | 0.49 | 0.00 | 0.23 | 1.39 | 12.86 |
| $\{3r, 4r\}$ | 1.11 | 0.00 | 1.07 | 4.26 | 21.86 |
| $\{2r, 3r, 4r\}$ | 0.54 | 0.00 | 0.23 | 1.56 | 24.05 |

Mean, 50th, 75th, and 90th percentile performance is indicated as the percentage excess over an optimal solution

**Table 3** Performance of the IHE method on a complete graph with $n = 5$ vertices for 1000 randomly generated problem scenarios with average inspection times comparable to average travel times, using the best solution that was obtained in each problem scenario for the indicated look-ahead depth parameter sets

| | Percent over optimum | | | | |
|---|---|---|---|---|---|
| IHE look-ahead depth ($\eta$) | Mean | 50th | 75th | 90th | Time (s) |
| 2 | 12.72 | 11.25 | 18.48 | 23.33 | 3.22 |
| 3 | 3.09 | 0.67 | 5.33 | 7.60 | 2.76 |
| 4 | 1.62 | 0.24 | 2.41 | 5.61 | 3.78 |
| 5 | 2.81 | 1.14 | 3.90 | 7.98 | 11.25 |
| {2, 3} | 2.87 | 0.28 | 4.32 | 7.36 | 5.98 |
| {3, 4} | 1.04 | 0.00 | 0.95 | 4.32 | 6.54 |
| {2, 3, 4} | 0.97 | 0.00 | 0.92 | 3.85 | 9.76 |
| {4, 5} | 1.30 | 0.00 | 1.49 | 4.36 | 15.03 |
| {3, 4, 5} | 0.89 | 0.00 | 0.63 | 3.68 | 17.79 |
| {2, 3, 4, 5} | 0.89 | 0.00 | 0.63 | 3.68 | 21.01 |

Mean, 50th, 75th, and 90th percentile performance is indicated as the percentage excess over an optimal solution

using a single look-ahead depth parameter, the best performance, as determined by the smallest excess over optimum for the mean and 90th percentile of problem instances, is obtained with a decision epoch look-ahead value of $\eta = 4$. For the hybrid method of running the IHE method with several look-ahead parameters and then choosing the best patrol pattern, the best performance, as determined by a comparison of the excess over optimum and computation time required, is obtained using look-ahead depth sets of {2, 3, 4} and {3, 4, 5}.

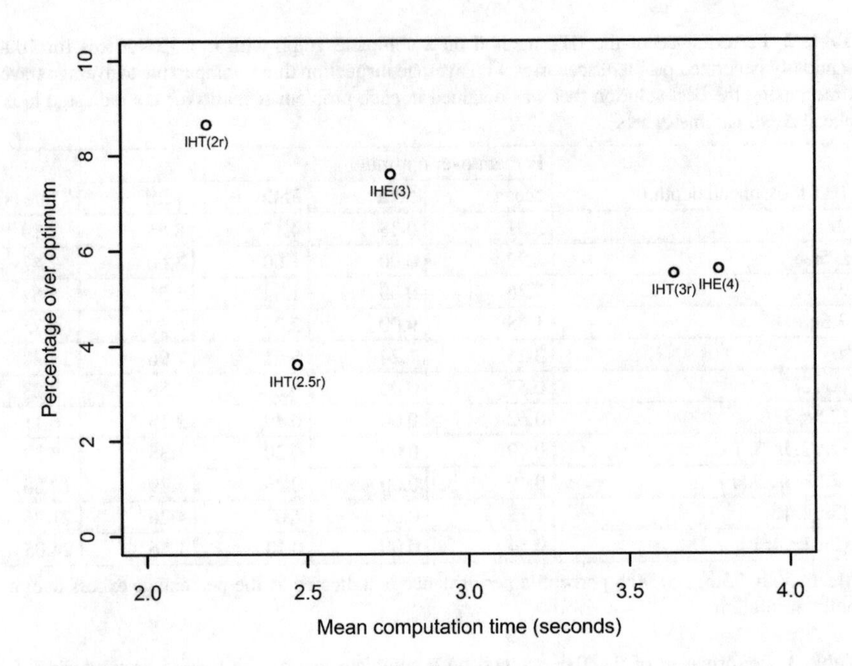


**Fig. 1** IHT and IHE 90th percentile performance with average travel times comparable to average inspection times

Performance of the IHT and IHE methods in the baseline case with a single look-ahead parameter is presented in Figure 1. This figure shows a comparison of performance versus computation time required for different heuristic methods and look-ahead parameters. Although both methods perform well in the experiments, we tend to see better performance using the IHT method in the single look-ahead parameter cases.

In an effort to obtain the best possible results, we also use a hybrid set of look-ahead depth parameters that combine both the IHT and IHE methods. We selected various combinations of parameters based on the results from the individual IHT and IHE experiments. Results are shown in Table 4. Very good performance is obtained with a hybrid IHT look-ahead set of $\{2r, 2.5r, 3r\}$ and the performance improves when incrementally adding IHE look-ahead parameters.

Performance of the combined IHT and IHE methods in the baseline case for different look-ahead depth parameters is presented in Figure 2. This figure shows a comparison of performance versus computation time required for different hybrid combinations of heuristic methods and look-ahead parameters. Both methods again perform well in the experiments, but we tend to see better performance using the IHT method in the hybrid set look-ahead cases, similar to the results from the single look-ahead parameter cases.

**Table 4** Performance of combined IHT and IHE methods on a complete graph with $n = 5$ vertices for 1000 randomly generated problem scenarios with average inspection times comparable to average travel times, using the best solution that was obtained in each problem scenario for the indicated look-ahead depth parameter sets

| IHT($\delta$) and IHE($\eta$) | Percent over optimum | | | | |
|---|---|---|---|---|---|
| look-ahead depth set | Mean | 50th | 75th | 90th | Time (s) |
| {IHT(2.5$r$), IHE(3)} | 0.88 | 0.00 | 0.95 | 3.45 | 5.18 |
| {IHT(2.5$r$), IHE(4)} | 0.61 | 0.00 | 0.49 | 2.12 | 6.19 |
| {IHT(2$r$, 2.5$r$, 3$r$)} | 0.49 | 0.00 | 0.20 | 1.38 | 8.30 |
| {IHE(2, 3, 4)} | 0.97 | 0.00 | 0.92 | 3.85 | 9.67 |
| {IHT(2.5$r$, 3$r$), IHE(3, 4)} | 0.42 | 0.00 | 0.15 | 1.30 | 12.65 |
| {IHT(2$r$, 2.5$r$, 3$r$), IHE(2, 3, 4)} | 0.30 | 0.00 | 0.00 | 0.92 | 17.89 |

Mean, 50th, 75th, and 90th percentile performance is indicated as the percentage excess over an optimal solution

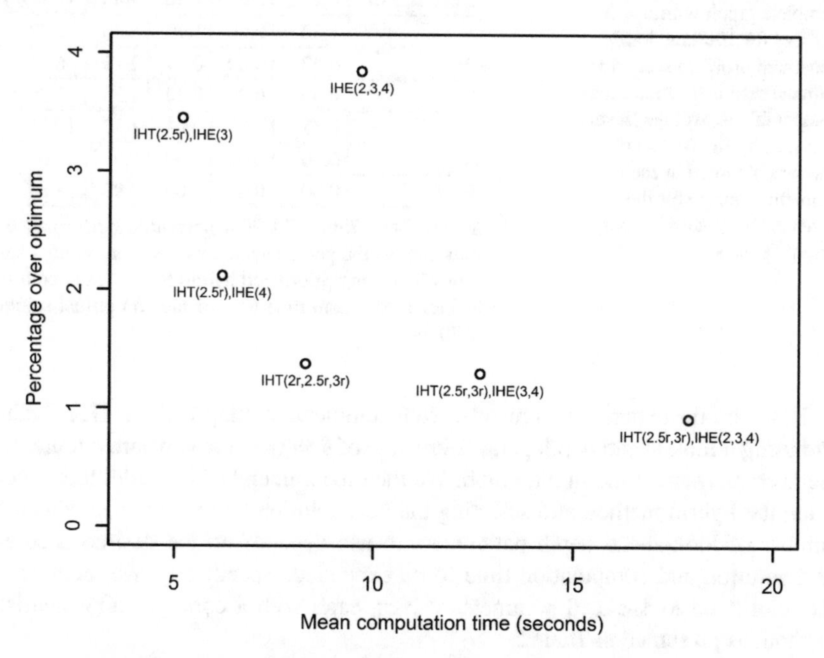


**Fig. 2** IHT and IHE hybrid 90th percentile performance with average travel times comparable to average inspection times

### 2.4.3 Recommendations Based on Numerical Experiments

We see very favorable results using the IHT and IHE methods with many combinations of look-ahead parameters. In general, we have found that looking ahead over about half of the graph structure provides a good balance of performance versus computation time required. We recommend choosing look-ahead depth parameter values as a function of $n$, which represents the number of vertices that are assigned to a patroller.

**Table 5** Prioritized heuristic methods and look-ahead depth parameters

| | Heuristic method and look-ahead depth parameter |
|---|---|
| 1 | $\mathrm{IHT}(\frac{n}{2}r)$ |
| 2 | $\mathrm{IHT}\left(\frac{(n+1)}{2}r\right)$ |
| 3 | $\mathrm{IHT}\left(\frac{(n-1)}{2}r\right)$ |
| 4 | $\mathrm{IHE}(\lceil\frac{n}{2}\rceil)$ |
| 5 | $\mathrm{IHE}(\lceil\frac{n}{2}\rceil+1)$ |
| 6 | $\mathrm{IHE}(\lceil\frac{n}{2}\rceil-1)$ |

**Table 6** Performance of the IHT and IHE methods on a complete graph with $n = 5$ vertices for 1000 randomly generated problem scenarios with average inspection times comparable to average travel times, using the best solution that was obtained in each problem scenario for the indicated look-ahead depth parameter sets

| | Percent over optimum | | | | |
|---|---|---|---|---|---|
| Heuristic set | Mean | 50th | 75th | 90th | Time (s) |
| 1 | 1.22 | 0.00 | 1.60 | 3.60 | 2.47 |
| 2 | 0.62 | 0.00 | 0.49 | 2.15 | 6.11 |
| 3 | 0.49 | 0.00 | 0.20 | 1.38 | 8.30 |
| 4 | 0.37 | 0.00 | 0.01 | 1.29 | 10.96 |
| 5 | 0.30 | 0.00 | 0.00 | 0.92 | 14.67 |
| 6 | 0.30 | 0.00 | 0.00 | 0.92 | 17.89 |

Mean, 50th, 75th, and 90th percentile performance is indicated as the percentage excess over an optimal solution when using prioritized hybrid look-ahead depth sets as indicated. Mean time to compute an optimal solution is 20.68 s

Based on the experimental results, we recommend starting with the IHT method and using a look-ahead depth parameter value of $\delta = (n/2) \times r$, where $r$ represents the average transit time in the graph. We then recommend adding additional looks using the hybrid method and selecting the best solution that is obtained. The total number of look-ahead depth parameters to use depends on the desired accuracy of a solution and computation time to be expended. Specifically, we recommend six prioritized look-ahead parameter values, each with a corresponding heuristic method, as presented in Table 5.

In a problem with $n = 5$, for example, after executing the heuristic method using $\mathrm{IHT}(2.5r)$ we would next use $\mathrm{IHT}(3r)$ and then continue in a similar manner until completing the desired number of looks. The IHE method is introduced at the fourth iteration of the heuristic method in order to complement the results obtained from using the IHT method.

We test the prioritized look-ahead depth parameter set method using the baseline problem case. Results are presented in Table 6. The results indicate a steady improvement in performance, along with a corresponding increase in computation time required, as the number of looks increases. We observe that the heuristic method will return an optimal solution in at least half of the problem instances when

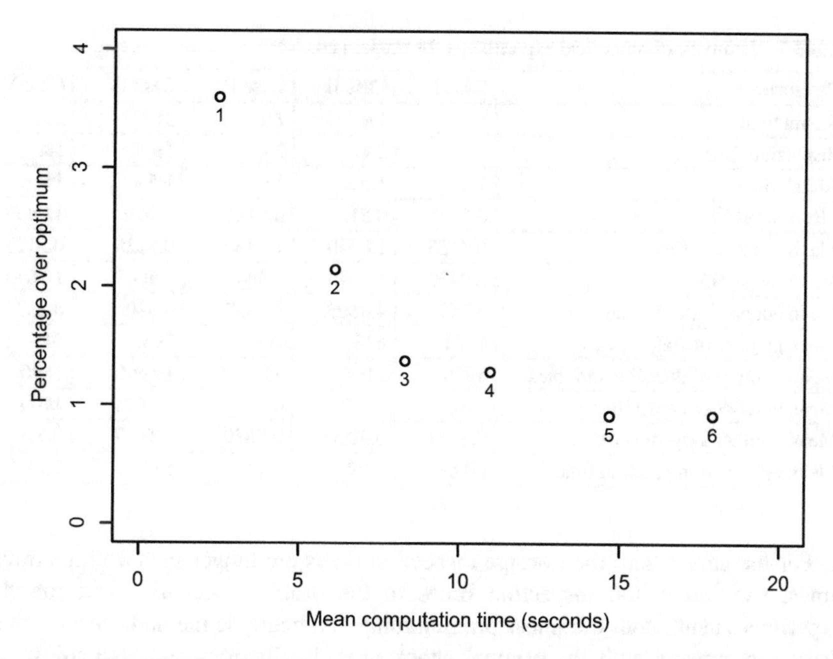


**Fig. 3** Combined IHT and IHE 90th percentile hybrid performance with average travel times comparable to average inspection times, using prioritized heuristic methods and look-ahead depth parameter sets

using a single look-ahead parameter IHT$(2.5r)$. The heuristic method will return a solution that is within 0.01 percent of optimal in at least 75 percent of the problem instances when using the fourth look-ahead set $\{$IHT$(2r, 2.5r, 3r)$, IHE(3)$\}$. Finally, we observe that the heuristic method will return a solution that is within 1 percent of optimal in at least 90 percent of the problem instances when using the fifth look-ahead set, $\{$IHT$(2r, 2.5r, 3r)$, IHE(3,4)$\}$.

These results are also presented in Figure 3 to show the rate of improvement of the prioritized hybrid look-ahead depth sets as computation time increases. We observe the best rate of improvement in performance as a function of computation time required through the third look-ahead depth set $\{$IHT$(2r, 2.5r, 3r)\}$. We test these recommendations further using several additional problem cases.

### 2.4.4 Sensitivity Analysis

In addition to the baseline problems, we consider the case where a patroller needs to spend more time conducting inspections than he does traveling between vertices and the case where the patroller needs to spend more time traveling between vertices than he does conducting inspections. The problem cases considered in the numerical experiments are summarized in Table 7.

**Table 7** Summary of numerical experiments for random attackers

| Parameter | Case I | Case II | Case III | Case IV | Case V |
|---|---|---|---|---|---|
| Travel time | 1× | 1× | 1× | 2× | 2× |
| Inspection time | 1× | 2× | 2× | 1× | 1× |
| Attack time | 1× | 1.5× | 1× | 1.5× | 1× |
| Mean travel time | 0.5125 | 0.5125 | 0.5125 | 1.0430 | 1.0430 |
| Mean inspection time | 0.5125 | 1.0430 | 1.0430 | 0.5125 | 0.5125 |
| Mean transit time | 1.0430 | 1.5645 | 1.5645 | 1.5645 | 1.5645 |
| Mean bounded attack time | 3.2537 | 4.8805 | 3.2537 | 4.8805 | 3.2537 |
| Mean number of states, $\lvert\Omega\rvert$ | 1184 | 633 | 102 | 3938 | 318 |
| Mean number of decision variables | 5920 | 3165 | 510 | 19,690 | 1590 |
| Mean number of constraints | 7105 | 3799 | 613 | 23,674 | 1909 |
| Mean optimal long-run cost | 0.3921 | 0.4200 | 0.5679 | 0.4617 | 0.5198 |
| Mean optimal computation time (s) | 20.68 | 4.99 | 0.11 | 574.85 | 2.11 |

For the case where the average inspection times are longer than average travel times, we double the inspection times in the problem scenarios and run the experiment using both the linear programming and heuristic methods. We conduct these experiments with the original attack time distributions and also adjust the attack distributions as a separate case to maintain an overall probability of detection rate of approximately 0.5. We do this by increasing the attack time distribution parameters at each vertex by a factor of 1.5. The rest of the problem scenario parameters remain the same.

For the case where the average travel times are longer than average inspection times, we double the travel times in the problem scenarios and the run the experiment using both the linear programming and heuristic methods. We use the same original and adjusted attack distributions at each vertex that were used in the cases of increased inspection times as described above. The rest of the problem scenario parameters remain the same. Case I, the baseline case, had the lowest long-run cost on average. Case III generated the smallest number of states and had the highest long-run cost on average. Case IV generated the largest number of states on average.

Results for problem cases II through V using the prioritized look-ahead parameter sets from Section 2.4.3 are presented in Table 8. In each of these problem cases, very favorable results were obtained using the recommended method of incrementally increasing the heuristic method and look-ahead parameter sets. We note that the heuristic performed slightly better in problem cases involving shorter travel times. The average computation time required in each case increases significantly as the average size of the state space grows. We particularly note this for problem Case IV, which had an average state space approximately three times larger than the baseline case, but required computation times that were approximately 25 times greater.

**Table 8** Performance of IHT and IHE methods for problem cases as indicated in Table 7, using prioritized look-ahead depth parameter sets

| Case | Optimal solution time (s) | Heuristic set | Percent over optimum | | | | Heuristic solution time (s) |
|---|---|---|---|---|---|---|---|
| | | | Mean | 50th | 75th | 90th | |
| I | 20.68 | | See Table 6 | | | | |
| II | 4.99 | 1 | 0.69 | 0.00 | 0.81 | 2.25 | 0.84 |
| | | 2 | 0.35 | 0.00 | 0.13 | 1.47 | 1.95 |
| | | 3 | 0.29 | 0.00 | 0.00 | 1.10 | 2.66 |
| | | 4 | 0.26 | 0.00 | 0.00 | 0.84 | 3.45 |
| | | 5 | 0.15 | 0.00 | 0.00 | 0.52 | 4.91 |
| | | 6 | 0.14 | 0.00 | 0.00 | 0.36 | 5.68 |
| III | 0.11 | 1 | 0.99 | 0.00 | 0.94 | 2.99 | 0.09 |
| | | 2 | 0.70 | 0.00 | 0.64 | 2.38 | 0.75 |
| | | 3 | 0.41 | 0.00 | 0.01 | 1.31 | 0.81 |
| | | 4 | 0.41 | 0.00 | 0.01 | 1.31 | 0.87 |
| | | 5 | 0.35 | 0.00 | 0.00 | 1.12 | 1.08 |
| | | 6 | 0.18 | 0.00 | 0.00 | 0.35 | 1.12 |
| IV | 574.85 | 1 | 2.03 | 0.01 | 2.48 | 6.90 | 51.12 |
| | | 2 | 0.61 | 0.00 | 0.50 | 2.09 | 164.94 |
| | | 3 | 0.41 | 0.00 | 0.01 | 1.21 | 203.11 |
| | | 4 | 0.41 | 0.00 | 0.01 | 1.21 | 267.43 |
| | | 5 | 0.39 | 0.00 | 0.00 | 0.82 | 320.75 |
| | | 6 | 0.39 | 0.00 | 0.00 | 0.82 | 403.52 |
| V | 2.11 | 1 | 2.44 | 0.00 | 2.02 | 7.15 | 0.49 |
| | | 2 | 1.06 | 0.00 | 0.67 | 3.97 | 1.96 |
| | | 3 | 0.53 | 0.00 | 0.02 | 1.12 | 2.21 |
| | | 4 | 0.44 | 0.00 | 0.00 | 0.96 | 2.43 |
| | | 5 | 0.41 | 0.00 | 0.00 | 0.86 | 2.97 |
| | | 6 | 0.41 | 0.00 | 0.00 | 0.86 | 3.18 |

Performance is indicated as the percentage excess over an optimal solution

In general, the heuristic returns a solution within 0.01 percent of optimal in at least half of the problem instances using a single look-ahead parameter, IHT$(2.5r)$. The heuristic returns a solution within 0.01 percent of optimal in at least 75 percent of the problem instances using the third look-ahead set, $\{\text{IHT}(2r, 2.5r, 3r)\}$. Finally, we observe that the heuristic returns a solution within 1 percent of optimal in at least 90 percent of the problem instances using the sixth look-ahead set, $\{\text{IHT}(2r, 2.5r, 3r), \text{IHE}(2, 3, 4)\}$. We also note in certain problem cases that this method may require more computation time than what is required to determine an optimal solution using the RALP.

# 3 Single Patroller Against Strategic Attackers

We consider the case of a single patroller against strategic attackers. Section 3.1 introduces a patrol model on a graph, where an attacker will actively choose a location to attack in order to incur the highest cost. In Section 3.2, we present a linear program that determines an optimal solution to the patrol problem. Since the linear program quickly becomes computationally intractable as the size of the problem grows, we also present heuristic methods for determining a solution in Section 3.3. In Section 3.4, we present a method to compute a lower bound for an optimal solution, which allows us to evaluate the heuristic methods when an optimal solution is unavailable. We conduct extensive numerical experiments for several scenarios and present the results in Section 3.5. We make recommendations on how to best utilize the heuristic methods based on the experimental results.

## *3.1 Patrol Model*

We consider a patrol model similar to the random-attacker model presented in Section 2.1, except that in this case, an attacker will actively choose which vertex to attack in order to incur the highest expected cost. In other words, the attacker and the patroller play a simultaneous-move two-person zero-sum game where the attacker is trying to maximize the cost incurred due to a successful attack and the patroller is trying to minimize it. The patroller chooses how to patrol the graph while the attacker chooses which vertex to attack. Except for trivial cases, an optimal strategy for either player in a two-person zero-sum game is often a mixed strategy, which is a probability distribution on the set of a player's pure strategies [17].

To formulate this problem, we modify the model that was used for the random-attacker case in Section 2. Recall from (16) that for a given patrol policy $\pi$, $V_i(\pi)$ is the long-run cost rate at vertex $i$. While the attacker is trying to maximize the expected cost incurred by choice of vertex to attack, the patroller is simultaneously trying to minimize it by choice of patrol policy. The patroller's objective function in this two-person zero-sum game against a strategic attacker is

$$\min_{\pi \in \Pi^{\mathrm{R}}} \max_{i \in N} \frac{V_i(\pi)}{\lambda_i}, \tag{33}$$

where $\Pi^{\mathrm{R}}$ is the set of randomized patrol policies.

## *3.2 Optimal Policy*

It is possible to determine an optimal solution to this problem by formulating and solving a linear program. Recall the linear program from Section 2.2.1 that was used to find an optimal solution for the case of random attackers, where the objective function represented the overall long-run cost rate. In the case of strategic attackers, the objective is to minimize the largest expected cost per attack across each individual vertex, rather than the overall long-run cost rate for the entire graph.

To solve this problem, we again use the directed graph of the state space $G(\mathcal{N}, \mathcal{A})$, where each node $k \in \mathcal{N}$ represents one state of the system and each arc $(k, l) \in \mathcal{A}$ represents a feasible transition between states. Each arc is assigned a transit time $t_{kl}$ as determined by the vertex-pair specific distance and inspection times, where $t_{kl} = \tau(k, \omega(l))$. Each arc is also assigned cost data that represents the expected cost incurred *at each vertex* when the system transitions from state $k$ to state $l$. We write $c_{kl}^{(i)}$ as the expected cost incurred at vertex $i$ for the state pair $(k, l)$, as determined by (9), for $i \in N$.

If $x_{kl}$ represents the long-run fraction of time that arc $(k, l)$ is utilized during the patrol pattern, the long-run cost rate at vertex $i$ is

$$\sum_{(k,l)\in\mathcal{A}} c_{kl}^{(i)} x_{kl}. \tag{34}$$

Dividing this total by the arrival rate of attackers at vertex $i$, we can define the zero-sum game between the patroller and strategic attacker as

$$\min_{x} \max_{i\in N} \sum_{(k,l)\in\mathcal{A}} \frac{c_{kl}^{(i)} x_{kl}}{\lambda_i}. \tag{35}$$

Note that $c_{kl}^{(i)} x_{kl}$ scales proportionately with $\lambda_i$, so the long-run average cost at vertex $i$ does not depend on the value of $\lambda_i$. Hence, we let $\lambda_i = 1$, for all $i \in N$.

To determine an optimal solution for the strategic-attacker problem, we modify the linear program in Section 2.2.1 to minimize the largest long-run average cost per attack among all vertices, which we refer to as the strategic-attacker linear program (SALP):

$$\min_{x} z^{\text{OPT}} \tag{36a}$$

$$\text{subject to} \sum_{(k,l)\in\mathcal{A}} c_{kl}^{(i)} x_{kl} \le z^{\text{OPT}}, \forall i \in N \tag{36b}$$

$$\sum_{l|(k,l)\in\mathcal{A}} x_{kl} - \sum_{l|(l,k)\in\mathcal{A}} x_{lk} = 0, \forall k \in \mathcal{N} \tag{36c}$$

$$\sum_{(k,l)\in\mathscr{A}} t_{kl}x_{kl} = 1, \tag{36d}$$

$$x_{kl} \geq 0, \forall\, (k,l) \in \mathscr{A}. \tag{36e}$$

In an optimal solution, the positive values of $x_{kl}$ indicate the arcs that belong to the cycle with the lowest total cost per unit time. The states on these cycles directly correspond to vertices on the graph, which can be determined by the function $\omega(s)$. Therefore, an optimal mixed strategy patrol policy can be determined. For each state of the system, the patrol policy specifies the probability that the patroller will choose to move to each vertex. We map the solution from the linear program to a patrol policy using

$$p_{kl} = \frac{x_{kl}}{\sum_{l|(k,l)\in\mathscr{A}} x_{kl}}, \text{ for } \sum_{l|(k,l)\in\mathscr{A}} x_{kl} > 0, \tag{37}$$

where $p_{kl}$ is the probability that the patroller will choose to next go to vertex $\omega(l)$ when the system is in state $k$.

As the problem size grows, it quickly becomes computationally intractable to use this method. Therefore, there is a need for efficient heuristic policies.

### 3.3 Heuristic Policies

In this section, we consider heuristics to determine a strategy for the patroller. This method introduces a different kind of randomized strategy, by letting the patroller choose a patrol pattern from a predetermined set and repeat the patrol pattern indefinitely.

For the patrol problems we consider, there are an infinite number of feasible patrol patterns. As it would be impossible to consider an infinite number of patrol patterns, we propose a heuristic method to define a finite set of patrol patterns from which the patroller can select a mixed strategy. If it were possible to consider every feasible patrol pattern, then this method would find an optimal solution. Similarly, if we consider a finite subset of all the feasible patrol patterns, such that all patrol patterns that are part of an optimal solution are elements of that subset, then this method would also find an optimal solution.

We develop strategy reduction techniques that allow us to consider a comprehensive, but reasonable, number of patrol patterns for use in this heuristic method. To do so, we create a finite set $S$ of feasible patrol patterns, ideally with elements that are identical or very similar to the patrol patterns that are part of an optimal solution. In the best case, $S$ would contain all patrol patterns that are part of an optimal solution.

Once we determine a finite set of patrol patterns, $S = \{\xi_1, \xi_2, \ldots, \xi_m\}$, we formulate a different two-person zero-sum game between the attacker and the patroller in a standard matrix form. In this game matrix, row $i$ corresponds to the attacker choosing to attack vertex $i$ and column $j$ corresponds to the patroller choosing patrol pattern $\xi_j$, for $i \in N$ and $j = 1, \ldots, m$. A linear program can then be formulated to solve this two-person zero-sum matrix game [22]. The solution to this game will provide a mixed strategy for both the attacker and the patroller, and the value of the game will be the expected cost due to an undetected attack.

### 3.3.1 Patrol Cost Determination

For any feasible patrol pattern, we can determine the expected cost incurred at each vertex due to an undetected, and therefore successful, attack. We denote the expected cost at vertex $j$ by $\rho_j$. These expected costs are used to populate the game matrix used in the heuristic method. There are three cases to consider when computing the expected cost at a vertex, which are based on the structure of the patrol pattern.

Case one occurs if the patrol pattern never visits vertex $j$. In this case, the expected cost for an attack on vertex $j$ is $c_j$, due to the fact that if the attacker chooses to attack vertex $j$ then the attack will always succeed. Thus,

$$\rho_j = c_j. \tag{38}$$

Case two occurs if the patroller visits vertex $j$ exactly once during a patrol pattern of total time length $\tau$. Recall from Section 2.2 that we can compute the expected number of successful attacks at vertex $j$ when vertex $j$ is inspected once every $\tau$ time units as

$$\lambda_j \int_0^\tau F_j(\tau - t)\, dt = \lambda_j \int_0^\tau F_j(s)\, ds\,. \tag{39}$$

Divide this by the expected number of attackers that will arrive at vertex $j$ during time interval $\tau$, which is $\lambda_j \tau$, to determine the probability of a successful attack:

$$\frac{\lambda_j \int_0^\tau F_j(s)\, ds}{\lambda_j \tau} = \frac{\int_0^\tau F_j(s)\, ds}{\tau}. \tag{40}$$

The expected cost at vertex $j$ will therefore be the cost of a successful attack $c_j$ times the probability of a successful attack:

$$\rho_j = \frac{c_j \int_0^\tau F_j(s)\, ds}{\tau}. \tag{41}$$

Case three occurs if the patroller visits vertex $j$ two or more times during the patrol pattern. In this case, we break the patrol pattern into intervals based on each time the patroller returns to the vertex. If a patroller visits the vertex $m \geq 2$ times during a patrol pattern of total time length $\tau$, we define $t_1$ as the time interval between the $m$-th (final) visit and the first visit to the vertex. The second interval $t_2$

is the time between the first and second visit. The last interval $t_m$ is the time between visit $m-1$ and visit $m$. We compute the expected number of successful attacks at the vertex during each interval and divide that sum by the time to complete a full patrol cycle $\tau$. Thus, the probability of a successful attack at vertex $j$, with $m \geq 2$ visits to vertex $j$, during a patrol pattern of total length $\tau = t_1 + t_2 + \cdots + t_m$ is

$$\frac{\lambda_j \int_0^{t_1} F_j(s)\,ds + \cdots + \lambda_j \int_0^{t_m} F_j(s)\,ds}{\lambda_j \tau}$$

$$= \frac{\int_0^{t_1} F_j(s)\,ds + \cdots + \int_0^{t_m} F_j(s)\,ds}{\tau}, \tag{42}$$

and the expected cost is

$$\rho_j = \frac{c_j \left( \int_0^{t_1} F_j(s)\,ds + \cdots + \int_0^{t_m} F_j(s)\,ds \right)}{\tau}. \tag{43}$$

### 3.3.2 Selection of Patrol Patterns

We consider two groups of patrol patterns to include in $S$. The first group is a combinatorial selection of patrol patterns based on the shortest Hamiltonian cycle in the graph. The second group is determined through an iterative method based on fictitious play.

### 3.3.3 Patrol Patterns Based on Shortest Path

Consider a case where the patroller chooses to use a single patrol pattern, or in other words, he uses a pure strategy. He would likely choose a pattern that visited each vertex at least once, since if he were to never visit a vertex, then an attack at that vertex would always be successful and would incur the full cost. Furthermore, he would likely try to minimize the time between inspections at each vertex.

To minimize the time between inspections at each vertex while visiting each vertex at least once during the patrol pattern, the patroller will follow a shortest Hamiltonian cycle in the graph. This patrol pattern is designated as the first element in the set $S$ and we refer to it as the shortest-path patrol pattern. Finding the shortest-path patrol pattern is an example of solving a *traveling salesman problem*, as described in Sect. 16.5 of Ahuja et al. [1], in which the vertices represent locations that are subject to attack and the weight on each edge is the time required to travel between those locations and complete an inspection at the arrival location.

From (41), the expected cost at vertex $j$ using a shortest-path patrol pattern with total transit time $\tau$ is

$$\rho_j = \frac{c_j \int_0^{\tau} F_j(s)\,ds}{\tau}, \forall\, j \in N. \tag{44}$$

If a patroller were to use this patrol pattern as a pure strategy against strategic attackers, then the long-run cost of this policy is

$$V = \max_{j \in N} \rho_j, \tag{45}$$

since an attacker will employ his own pure strategy of always choosing to attack the vertex that incurs the highest cost.

Since we want to consider the option of a mixed strategy for the patroller, we must add additional patrol patterns to $S$. We start by considering subsets of the shortest-path patrol pattern. Specifically, we consider $n$ additional patrol patterns, which consist of the cycle where one vertex is skipped in the shortest-path patrol pattern and the patroller proceeds to the next vertex in the sequence. These are good patrol patterns to consider because they are consistent with the reasoning of using the shortest-path patrol pattern to minimize time spent on traveling, but they can also account for the heterogeneous qualities of potential attack locations. Due to differences among vertices in attack time distributions $F_i(\cdot)$ or cost incurred due to a successful attack $c_i$, a patroller may want to use a mixed strategy that periodically skips a visit to one or more vertices in order to occasionally direct more resources toward other vertices.

As an example, if the shortest-path patrol pattern in a graph with $n = 5$ vertices is $\{1 - 2 - 3 - 4 - 5 -\}$, then the first subset of patrol patterns is

$$\begin{aligned}
\{2-\ 3-\ 4-\ 5-,\\
1-\ 3-\ 4-\ 5-,\\
1-\ 2-\ 4-\ 5-,\\
1-\ 2-\ 3-\ 5-,\\
1-\ 2-\ 3-\ 4-\}.
\end{aligned}$$

For similar reasons, we also consider all paths of length $n-2$, where two vertices are removed from the shortest-path patrol pattern. In our example, there will be $\binom{5}{3} = 10$ of these patterns to consider:

$$\begin{aligned}
\{3-\ 4-\ 5-,\ 2-\ 4-\ 5-,\\
2-\ 3-\ 5-,\ 2-\ 3-\ 4-,\\
1-\ 4-\ 5-,\ 1-\ 3-\ 5-,\\
1-\ 3-\ 4-,\ 1-\ 2-\ 5-,\\
1-\ 2-\ 4-,\ 1-\ 2-\ 3-\}.
\end{aligned}$$

We continue this process by removing vertices until all subsets of the shortest-path patrol pattern that consist of only one vertex have been considered. For paths of length greater than three, the sequence of vertices can be reordered as required, so that the patroller will be utilizing the shortest Hamiltonian cycle within a particular subgraph of vertices. The total number of patrol patterns considered when using this method is $2^n - 1$. We refer to this set of patterns as the shortest-path (SP) patrol patterns.

In addition to the shortest-path patrol pattern and its subsets, we consider patrol patterns where the patroller chooses one vertex to visit twice during his patrol while visiting each remaining vertex only once. Ideally, we would choose the time for a revisit to a vertex in the patrol pattern such that the time between inspections is as close to even as possible. To determine these patterns, we continue to use the shortest-path patrol pattern as a baseline and insert a revisit to each vertex at all possible points in the pattern, such that the patroller does not complete a revisit to a vertex immediately after completing an inspection at that vertex. Using this method, we will consider an additional $n(n-2)$ patrol patterns. We refer to this set of patrol patterns as the shortest-path with one revisit (SPR1) patrol patterns.

To continue the example from above, for a graph with $n = 5$ vertices and shortest-path patrol pattern $\{1-2-3-4-5-\}$, the SPR1 set would consist of the following additional 15 patrol patterns:

$$\begin{aligned}
\{&1-2-1-3-4-5-,\\
&1-2-3-1-4-5-,\\
&1-2-3-4-1-5-,\\
&1-2-3-2-4-5-,\\
&1-2-3-4-2-5-,\\
&1-2-3-4-5-2-,\\
&1-3-2-3-4-5-,\\
&1-2-3-4-3-5-,\\
&1-2-3-4-5-3-,\\
&1-4-2-3-4-5-,\\
&1-2-4-3-4-5-,\\
&1-2-3-4-5-4-,\\
&1-5-2-3-4-5-,\\
&1-2-5-3-4-5-,\\
&1-2-3-5-4-5-\}.
\end{aligned}$$

Similarly, we can continue this method of generating additional patrol patterns based on the shortest-path patrol pattern by allowing multiple revisits to a vertex. We consider the case of the shortest path with two revisits (SPR2) by starting with the SPR1 patrol patterns and, for each of these patrol patterns, conducting an additional visit to each vertex. We consider paths that revisit all combinations of two vertices, including two revisits to the same vertex, such that there are no immediate revisits to any vertex.

The number of patrol patterns that are generated for a particular number of revisits is based on the number of vertices $n$ in the graph. For the case of two revisits, such as in the SPR2 method, there are an additional $n(n-2)[(n-1)(n-1)+(n-3)]$ patrol patterns to consider. The SPR3 method follows a similar process by conducting revisits to all combinations of three vertices such that there are no immediate revisits to any vertex. The length of the patrol patterns and the size of the sets that are generated in each of these methods are summarized in Table 9.

**Table 9** Shortest path patrol pattern sets

| Path generation method | Length | Number of patterns |
|---|---|---|
| Shortest path (SP) | $\leq n$ | $2^n - 1$ |
| Shortest path with one revisit (SPR1) | $n+1$ | $n^2 - 2n$ |
| Shortest path with two revisits (SPR2) | $n+2$ | $n^4 - 3n^3 + 4n$ |
| Shortest path with three revisits (SPR3) | $n+3$ | $n^6 - 3n^5 - 5n^4 + 19n^3 - 20n$ |

**Table 10** Example numbers of shortest-path patrol patterns

| Pattern set | $n = 5$ | $n = 6$ | $n = 7$ | $n = 10$ | $n = 11$ | $n = 12$ |
|---|---|---|---|---|---|---|
| SP | 31 | 63 | 127 | 1023 | 2047 | 4095 |
| SPR1 | 15 | 24 | 35 | 80 | 99 | 120 |
| SPR2 | 270 | 672 | 1400 | 7040 | 10,692 | 15,600 |
| SPR3 | 5400 | 20,832 | 61,600 | 668,800 | 1,240,272 | 2,168,400 |

A summary of representative patrol pattern sizes for the type of problems that we consider is presented in Table 10. As revisits are increased to four and beyond, there are very large increases in the number of patrol patterns without much further improvement in performance.

### 3.3.4 Patrol Patterns Based on Fictitious Play

We consider an additional group of patrol patterns that are generated using fictitious play as described by Robinson [19]. She shows that an iterative method can be used to generate mixed strategies in a two-person zero-sum game that will converge to an optimal solution. In this iterative method of play, each player arbitrarily chooses a pure strategy in the first round. In subsequent rounds, each player chooses a pure strategy that will produce the best expected value against the mixture of strategies used by the other player in all the previous rounds.

We compute the attacker's mixed strategy $(p_1, \ldots, p_n)$ based on the mixture of strategies used by the patroller in the previous rounds. Based on that probability vector, we can use the IHT and IHE heuristic methods from the random-attacker case presented in Section 2 to generate a new patrol pattern for the patroller. The following algorithm is adapted from Lin et al. [15]:

1. In round 1, each player picks a strategy.
   a. Denote by $\xi^{(d)}$ the patrol pattern used by the patroller in round $d$. Choose $\xi^{(1)}$ to be the shortest-path patrol pattern.
   b. Let the attacker pick the vertex $j$ that has the highest cost in the shortest-path patrol to attack. Use $r_i$, for $i \in N$, to keep track of the number of times vertex $i$ is picked by the attacker. Initialize $r_j = 1$ and $r_i = 0$, for $i \in N, i \neq j$.
2. Repeat the following steps for the predetermined number of rounds, $\nu$. In round $d \geq 2$,

a. Set $p_i = r_i / \sum_{k=1}^{n} r_k$, which represents the attacker's mixed strategy based on his attack history from rounds 1 to $d-1$. Use the random-attacker heuristic method to generate a patrol pattern $\xi^{(d)}$.
b. Find the best vertex for the attacker to attack by assuming the patroller uses patrol pattern $\xi^{(j)}$, $j = 1, \ldots, (m-1)$, each with probability $1/(m-1)$. If attacking vertex $i$ yields the highest expected cost, set $r_i \leftarrow r_i + 1$.

Thus, we can generate two groups of patrol patterns for use in the strategic-attacker heuristic method: the shortest-path patrol pattern and its associated derived patrol patterns, and a set of patrol patterns determined by an iterative method using fictitious play. The heuristic method in the case of fictitious play will have two parameters, the set $L$ of look-ahead depth parameters to be used with the IHT and IHE methods, and the number of iterations of fictitious play, $\nu$.

For a graph with $n$ vertices, we generate $2^n - 1 + n(n-2)$ patrol patterns in the first group when using the SP and SPR1 patrol pattern sets. In the second group we generate up to $|L| \times \nu$ patrol patterns. The actual number of patrol patterns considered in the problem is often much smaller than $[2^n + n^2 - 2n - 1] + [|L| \times \nu]$, since many of the patrol patterns generated during the fictitious-play algorithm will be identical or will produce identical performance.

## 3.4 Lower Bound

When an optimal solution cannot be determined due to the size of a problem, it is valuable to have a way to evaluate a heuristic solution. For this purpose, we provide a method to compute a lower bound for an optimal solution in the strategic-attacker problem. This is a modification of the discrete-time method presented in Lin et al. [15] for our continuous-time problem.

To determine a lower bound for an optimal solution, we formulate a linear program. We define $y_{ir}$ as the rate at which an inspection is completed at vertex $i$, with the last inspection at that vertex having been completed exactly $r$ time units ago.

For example, consider a patrol pattern of total length $\tau = 17$ where inspections are completed at vertex 1 at times $2-5-7-10-14-17$. The times between inspections are $2-3-2-3-4-3$. The inspection rates at vertex 1 using this patrol pattern are $y_{12} = 2/17$, $y_{13} = 3/17$, and $y_{14} = 1/17$. It follows that there is a total inspection rate constraint for any vertex $i$ that is inspected during a patrol pattern:

$$\sum_{r=1}^{\infty} y_{ir} r = 1. \tag{46}$$

If a vertex is not visited at all during a patrol pattern, then the total inspection rate at that vertex will be 0. Therefore, in order to create a total-rate constraint for all vertices and all patrol policies, we use

**Table 11** Example case of time-interval inspections

| $q$ | Interval | Inspections |
|---|---|---|
| 1 | [0, 1.2) | 0 |
| 2 | [1.2, 2.4) | 2 |
| 3 | [2.4, 3.6) | 3 |
| 4 | [3.6, 4.8) | 1 |
| 5 | [4.8, 6.0) | 0 |
| 6 | [6.0, 7.2) | 0 |
| 7 | [7.2, 8.4) | 0 |
| 8 | [8.4, ∞) | 0 |

$$\sum_{r=1}^{\infty} y_{ir} r \leq 1, \forall i \in N. \tag{47}$$

Since we consider this problem in continuous time, we must modify the definition of the inspection rate in order to use it as a variable in a linear program. Recall that the attack time at vertex $i$ is bounded by $B_i$. We divide the time interval $[0, B_i]$ at vertex $i$ into $m$ equal length subintervals. We then define an inspection rate $y_{iq}$, for $q = 1, \ldots, (m-1)$, as the rate at which vertex $i$ is inspected with the previous inspection having been completed at time in $\left[\frac{(q-1)B_i}{m}, \frac{qB_i}{m}\right)$, and $y_{im}$ as the rate at which vertex $i$ is inspected with the previous inspection having been completed at least $(\frac{m-1}{m})B_i$ time units ago.

Again consider the example of a patrol pattern of total length $\tau = 17$ where inspections are completed at vertex 1 at times $2-5-7-10-14-17$. Suppose that $B_1 = 9.6$ and we choose $m = 8$. Table 11 indicates the number of inspections that are completed in each time interval.

Thus, the inspection rates $y_{iq}$ at vertex $i = 1$ for this patrol pattern are $y_{12} = 2/17$, $y_{13} = 3/17$, $y_{14} = 1/17$, and $y_{11} = y_{15} = y_{16} = y_{17} = y_{18} = 0$.

Since the inspection times are broken into $m$ discrete-time intervals, the identity in (47) becomes

$$\sum_{q=1}^{m} y_{iq} \frac{(q-1)B_i}{m} \leq 1, \forall i \in N. \tag{48}$$

We now focus on a single vertex in order to quantify the long-run cost at that vertex. Define $R_i(t)$ as the expected cost that can be avoided for completing an inspection at vertex $i$ if the previous inspection was completed $t$ time units ago. This is equivalent to the expected number of ongoing attacks at vertex $i$ at time $t$ multiplied by $c_i$, so

$$R_i(t) = c_i \lambda_i \int_0^t P(X_i > s)\, ds\,. \tag{49}$$

We also define

$$R_{iq} = R_i\left(\frac{qB_i}{m}\right), q = 1, \ldots, m, \tag{50}$$

as the cost that can be avoided at vertex $i$ for completing an inspection at time $q(B_i/m)$.

Although we do not know the exact value of the expected cost at vertex $i$, we do know that

$$\left(c_i - \frac{1}{\lambda_i}\sum_{q=1}^{m} y_{iq}R_{iq}\right) \leq \text{[expected cost at vertex } i\text{]} \leq \left(c_i - \frac{1}{\lambda_i}\sum_{q=1}^{m} y_{iq}R_{i(q-1)}\right).$$

Therefore, the expected cost incurred at vertex $i$ will be at least

$$c_i - \frac{1}{\lambda_i}\sum_{q=1}^{m} y_{iq}R_{iq}, \forall i \in N, \tag{51}$$

because the expression in (51) will take credit for avoiding cost in the entire interval $\left[0, \frac{qB_i}{m}\right)$ at the constant value represented by $R_i(\frac{qB_i}{m})$ times the inspection rate $y_{iq}$. Thus, the value in (51) represents a lower bound for the expected cost for each attack at vertex $i$.

To formulate a linear program to determine a lower bound for an optimal solution, we also incorporate constraints that account for graph structure. Define $x_{ij}$ as the rate at which a patroller travels from vertex $i$ to vertex $j$ and conducts an inspection at vertex $j$, for $i, j \in N$. Recall that $t_{ij}$ represents the time required for a patroller to travel from vertex $i$ to vertex $j$ and conduct an inspection at vertex $j$. On a graph with a single patroller, the following total-rate constraint applies:

$$\sum_{i,j\in N} x_{ij}t_{ij} = 1. \tag{52}$$

Since the total rate of arrivals to a vertex must equal the total rate of departures from a vertex, we also observe that

$$\sum_{j\in N} x_{ij} = \sum_{j\in N} x_{ji}, \forall i \in N. \tag{53}$$

The variables $x_{ij}$ and $y_{iq}$ are connected through the equation

$$\sum_{q=1}^{m} y_{iq} = \sum_{j\in N} x_{ij}, \forall i \in N, \tag{54}$$

since both sides represent the long-run inspection rate at vertex $i$.

We now formulate a linear program to determine the lower bound for an optimal solution in the single patroller against strategic attackers problem, which we refer to as the lower bound linear program (LBLP):

$$\min_{x,y} z^{\mathrm{LB}} \tag{55a}$$

$$\text{subject to} c_i - \frac{1}{\lambda_i}\sum_{q=1}^{m} y_{iq} R_{iq} \le z^{\mathrm{LB}}, \forall i \in N, \tag{55b}$$

$$\sum_{q=1}^{m} y_{iq} \frac{(q-1)B_i}{m} \le 1, \forall i \in N, \tag{55c}$$

$$\sum_{j\in N} x_{ij} - \sum_{j\in N} x_{ji} = 0, \forall i \in N, \tag{55d}$$

$$\sum_{q=1}^{m} y_{iq} - \sum_{j\in N} x_{ji} = 0, \forall i \in N, \tag{55e}$$

$$\sum_{i,j\in N} x_{ij} t_{ij} = 1, \tag{55f}$$

$$x_{ij} \ge 0, \forall i, j \in N, \tag{55g}$$

$$y_{iq} \ge 0, \forall i \in N; q = 1, \ldots, m. \tag{55h}$$

The decision variables in this problem are $x_{ij}$, the rate that the patroller transits from vertex $i$ to vertex $j$; and $y_{iq}$, the rate that an inspection is completed at vertex $i$ with the time since the last inspection falling in $\left[\frac{(q-1)B_i}{m}, \frac{qB_i}{m}\right)$.

In this linear program, we seek to minimize the maximum expected cost for each attack across all $n$ vertices, which is ensured by constraint (55b). We observe the total inspection rate constraints at each vertex with (55c). We also observe the network balance of flow and total arrival and inspection rate equality constraints in (55d) and (55e). Finally, we observe the total transit rate constraint on a single patroller in (55f), and the non-negativity constraint on patroller transit rates and inspection rates in (55g) and (55h).

While the preceding linear program will produce a valid lower bound, it can be quite loose. We add additional constraints to the linear program in order to tighten the lower bound by limiting the rate of reinspections at a vertex and by considering the transit time that is required between vertices.

To account for the action of a patroller electing to stay at a vertex to conduct an additional inspection, define

$$a_i = \left\lceil \frac{v_i}{(B_i/m)} \right\rceil, \forall i \in N, \tag{56}$$

as the number of subintervals needed for the patroller to inspect vertex $i$ again without leaving vertex $i$; and require that

$$\sum_{q=1}^{a_i} y_{iq} \geq x_{ii}, \forall i \in N, \tag{57}$$

which ensures the total rate of inspections at vertex $i$ in the time interval it takes to conduct an inspection is at least equal to the rate of reinspections at vertex $i$.

We also add constraints to the linear program to account for the patroller's transit rate from vertex $i$ to $j$ and back to vertex $i$, denoted by $u_{iji}$, for $i \neq j$, as follows:

$$u_{iji} \leq x_{ij}, \forall i, j \in N; i \neq j, \tag{58a}$$

$$u_{iji} \leq x_{ji}, \forall i, j \in N; i \neq j, \tag{58b}$$

$$x_{ij} - \sum_{k \neq i} x_{jk} \leq u_{iji}, \forall i, j \in N; i \neq j. \tag{58c}$$

Since the rate that a patroller transits from vertex $i$ to $j$ must be at least equal to the rate that the patroller transits from vertex $i$ to $j$ and back to vertex $i$, we include constraint (58a). The same reasoning applies to constraint (58b). We also observe in (58c) that the rate the patroller transits from vertex $i$ to $j$ and back to vertex $i$ must be at least equal to the rate that he transits from vertex $i$ to $j$, minus the rate he transits from vertex $j$ to any vertex other than $i$.

It also holds that the inspection rate at vertex $i$ must be at least equal to the rate that the patroller transits from vertex $i$ to $j$ and back to vertex $i$. To incorporate this constraint, define

$$g_{iji} = \left\lceil \frac{t_{ij} + t_{ji}}{(B_i/m)} \right\rceil, \forall i, j \in N; i \neq j, \tag{59}$$

and require that

$$\sum_{q=1}^{g_{iji}} y_{iq} \geq x_{ii} + u_{iji}, \forall i, j \in N; i \neq j, \tag{60}$$

where $x_{ii}$ is the rate that the patroller remains at vertex $i$ to conduct an additional inspection and $u_{iji}$ is the rate that the patroller transits from vertex $i$ to $j$ and back to vertex $i$.

We can continue this same idea to account for paths that visit at least two vertices prior to returning to vertex $i$ and define $w_{ijki}$ as the rate at which the patroller transits from vertex $i$ to vertex $j$ to vertex $k$ and returns immediately to vertex $i$. Based on the patroller's transit rate from vertex $i$ to $j$ to $k$ and back to vertex $i$, for $i \neq j, k$, we add the following additional constraints to the linear program:

$$w_{ijki} \leq x_{ij}, \forall i, j, k \in N; i \neq j, k, \tag{61a}$$

$$w_{ijki} \leq x_{jk}, \forall i, j, k \in N; i \neq j, k, \tag{61b}$$

$$w_{ijki} \leq x_{ki}, \forall i, j, k \in N; i \neq j, k, \tag{61c}$$

$$x_{ij} - \sum_{l \neq k} x_{jl} - \sum_{l \neq i} x_{kl} \leq w_{ijki}, \forall i, j, k \in N; i \neq j, k. \tag{61d}$$

Since the rate that a patroller transits from vertex $i$ to $j$ must be at least equal to the rate that the patroller transits from vertex $i$ to $j$ to $k$ and back to vertex $i$, we include constraint (61a). The same reasoning applies to constraints (61c) and (61d). We also observe in (61d) that the rate the patroller transits from vertex $i$ to $j$ to $k$ and back to vertex $i$ must be at least equal to the rate that he transits from vertex $i$ to $j$, minus the rate he transits from vertex $j$ to any vertex other than $k$ and the rate he transits from vertex $k$ to any vertex other than $i$.

It also holds that the inspection rate at vertex $i$ must be at least equal to the rate that the patroller transits from vertex $i$ to $j$ to $k$ and back to vertex $i$. To incorporate this constraint, define

$$h_{ijki} = \left\lceil \frac{t_{ij} + t_{jk} + t_{ki}}{(B_i/m)} \right\rceil, \forall i, j, k \in N; i \neq j, k, \tag{62}$$

and require that

$$\sum_{q=1}^{h_{ijki}} y_{iq} \geq x_{ii} + u_{iji} + w_{ijki}, \forall i, j, k \in N; i \neq j, k, \tag{63}$$

where $x_{ii}$ is the rate that the patroller remains at vertex $i$ to conduct an additional inspection; $u_{iji}$ is the rate that the patroller transits from vertex $i$ to $j$ and back to vertex $i$; and $w_{ijki}$ is the rate that the patroller transits from vertex $i$ to $j$ to $k$ and then back to vertex $i$.

We add constraints (57), (58a), (58b), (58c), (60), (61a), (61b), (61c), (61d), and (63) to the LBLP, which considerably tightens the lower bound. We could continue this same idea to account for paths that visit three or more vertices before returning to a starting vertex; however, for the size of the graphs that we consider, that would involve many more variables with negligible gains in performance. The number of decision variables in this linear program is $n^2 + mn$. The number of constraints is $5n^3 + 5n^2 + (m - 10)n + 1$. For a problem with $n = 5$ and $m = 100$, there are 525 decision variables and 1,201 constraints. In our numerical experiments, it takes on average 0.61 s to compute a lower bound for a problem of this size.

## 3.5 *Numerical Experiments*

To test the shortest-path and fictitious-play (FP) heuristic methods, we conduct several numerical experiments. We compare the results obtained from using the heuristic methods to an optimal solution. We also report the computation time required. Additionally, we compute a lower bound for an optimal solution using the linear program described in Section 3.4. Based on these results, we make conclusions on the efficacy of the heuristics, as well as make recommendations for the best use of the shortest-path and fictitious-play methods.

We test the same five problem cases for strategic attackers that we did for random attackers in Section 2. In each case, we use the same 1000 problem scenarios that were randomly generated for the random-attacker experiments. The attack probability vector is omitted for the strategic-attacker problems, but all other data remain the same. We conduct our baseline experiments on a graph with $n = 5$ vertices.

In our experimental results, an optimal solution that is obtained from using the SALP is indicated by $z^{\text{OPT}}$. The lower bound that is obtained from using the LBLP is indicated by $z^{\text{LB}}$. Solutions obtained from using a heuristic method are indicated by $z^{\text{H}}$, where H indicates the heuristic method that was used.

### 3.5.1 Baseline Problems

For our baseline problem, we consider the case where a patroller spends about half of the time traveling and half of the time inspecting vertices. We determine an optimal solution using the SALP from Section 3.2 and a solution using the heuristic methods from Section 3.3. The SALP on average uses 5920 decision variables and 7110 constraints for a problem size with 1184 states. An optimal solution takes on average 20.68 s to compute. We compare the solution obtained from the heuristic method to an optimal solution. We also determine a lower bound for an optimal solution using the LBLP in Section 3.4, and compare that result to an optimal solution.

Using 1000 problem instances, we test the shortest-path method with the SP, SPR1, SPR2, and SPR3 patrol pattern sets. We also test the FP method with 10, 20, 30, and 50 iterations. Results of the baseline experiments are presented in Table 12. Excellent performance is observed with both the shortest-path SPR2 and SPR3 methods and the FP method with 50 iterations. Each of these methods returns a solution within 1.11 percent of an optimal solution in at least 90 percent of the problem instances. The shortest-path method uses considerably less computation time than the FP method in all cases. A tight lower bound for an optimal solution was also obtained, with an average difference between the lower bound and an optimal solution of 1.20 percent.

We also test combinations of the two-person zero-sum game matrices that are produced from each heuristic method. When the game matrices are combined,

**Table 12** Performance of the shortest-path and fictitious-play heuristic methods on a complete graph with $n = 5$ vertices, based on 1000 randomly generated problem instances with average inspection times that are comparable to average travel times

| Heuristic method | Percent over optimum | | | | |
|---|---|---|---|---|---|
| | Mean | 50th | 75th | 90th | Time (s) |
| Shortest-path (SP) | 1.95 | 1.18 | 2.53 | 4.45 | < 0.01 |
| SP with one revisit (SPR1) | 0.72 | 0.39 | 0.93 | 1.82 | 0.04 |
| SP with two revisits (SPR2) | 0.39 | 0.12 | 0.47 | 1.11 | 0.52 |
| SP with three revisits (SPR3) | 0.28 | 0.05 | 0.28 | 0.80 | 6.15 |
| Fictitious play ($\nu = 10$) | 3.76 | 3.11 | 5.23 | 8.18 | 85.51 |
| Fictitious play ($\nu = 20$) | 1.85 | 1.39 | 2.42 | 4.13 | 167.56 |
| Fictitious play ($\nu = 30$) | 0.79 | 0.45 | 0.90 | 2.11 | 255.45 |
| Fictitious play ($\nu = 50$) | 0.32 | 0.22 | 0.43 | 0.73 | 425.45 |
| Lower bound | −1.20 | −0.29 | −1.17 | −3.35 | 0.61 |

Mean, 50th, 75th, and 90th percentile performance is indicated as the percentage excess over an optimal solution. The lower bound is reported as $(z^{LB} - z^{OPT})/z^{OPT}$ in percentage

**Table 13** Mean performance of the shortest-path and fictitious-play heuristic methods on a complete graph with $n = 5$ vertices, based on 1000 randomly generated problem instances with average inspection times that are comparable to average travel times, reported as the percentage excess over an optimal solution

| FP/SP | Percent over optimum | | | | | Time (s) |
|---|---|---|---|---|---|---|
| | – | SP | SPR1 | SPR2 | SPR3 | |
| – | – | 1.95 | 0.72 | 0.39 | 0.28 | |
| FP 10 | 3.76 | 1.70 | 0.57 | 0.32 | 0.23 | 85.51 |
| FP 20 | 1.85 | 0.99 | 0.36 | 0.19 | 0.16 | 167.56 |
| FP 30 | 0.79 | 0.50 | 0.24 | 0.13 | 0.10 | 255.45 |
| FP 50 | 0.32 | 0.26 | 0.13 | 0.11 | 0.08 | 425.45 |
| Time (s) | | < 0.01 | 0.04 | 0.52 | 6.15 | |

the resulting performance can be no worse than what is obtained with each of the individual methods since additional patrol patterns are being considered. The mean and 90th percentile performance results are presented in Table 13 and Table 14, respectively. We see an improvement in performance when the methods are combined, but it is generally not significant enough to justify the additional computation time required by the FP method. It requires at least 20 iterations of FP combined with the SPR2 set and at least 30 iterations of FP combined with the SPR1 set to improve upon the performance obtained from using the SPR3 patrol pattern set alone.

**Table 14** 90th percentile performance of the shortest-path and fictitious-play heuristic methods on a complete graph with $n = 5$ vertices, based on 1000 randomly generated problem instances with average inspection times that are comparable to average travel times, reported as the percentage excess over an optimal solution

| | Percent over optimum | | | | | |
|---|---|---|---|---|---|---|
| FP/SP | – | SP | SPR1 | SPR2 | SPR3 | Time (s) |
| – | – | 4.45 | 1.82 | 1.11 | 0.80 | |
| FP 10 | 8.18 | 3.98 | 1.75 | 1.04 | 0.72 | 185.51 |
| FP 20 | 4.13 | 2.43 | 1.16 | 0.66 | 0.42 | 167.56 |
| FP 30 | 2.11 | 1.40 | 0.69 | 0.43 | 0.27 | 255.45 |
| FP 50 | 0.73 | 0.67 | 0.43 | 0.28 | 0.17 | 425.45 |
| Time (s) | | <0.01 | 0.04 | 0.52 | 6.15 | |

### 3.5.2 Recommendations Based on Numerical Experiments

We see very favorable results with the SP method. In at least 90 percent of the problem instances, we observe results within 1.11 percent of an optimal solution when using the SPR2 method and within 0.80 percent of an optimal solution when using the SPR3 method. For problems with $n = 5$, the SPR2 method required 0.52 s on average and the SPR3 method required 6.15 s on average to return a solution. The advantage to the SP method is that it provides excellent results for very little computation time.

We can generate additional effective patrol patterns for consideration in determining a randomized patrol policy, and further refine the overall solution, by considering the patterns obtained from multiple iterations of FP. The solution improves as the number of iterations of FP increases, but comes at a cost of significantly increased computation time. In at least 90 percent of problem instances, we see solutions within 2.11 percent of optimal when using 30 iterations of FP and within 0.73 percent of optimal when using 50 iterations of FP. These problem instances required on average 4.25 min and 7 min, respectively, to return a solution. Based on the experimental results, we recommend using the SPR2 method for the strategic-attacker problem.

### 3.5.3 Performance on Smaller and Larger Graphs

In addition to problems with $n = 5$, we test the heuristic methods on smaller and larger size graphs. For graphs with $n = 3, 4$, and 5, we compare the performance of the SPR2 heuristic to an optimal solution. Results are presented in Table 15.

We note that the SPR2 heuristic method works extremely well for graphs smaller than $n = 5$, returning a solution that is within 0.17 percent of optimal in 90 percent of the problem instances with computation times of less than 0.1 s. For graphs with $n = 6, 7, 8$, and 9, we compare the performance of the heuristic to the lower bound. Results are presented in Table 16. We use the lower bound for a comparison because, in our experiments, it is not practical to compute an optimal solution for graphs with $n > 5$ due to computer memory limitations.

**Table 15** Performance of the SPR2 shortest-path heuristic on a complete graph, based on 1000 randomly generated problem instances with average inspection times comparable to average travel times

| Vertices | Percent over optimum | | | | Time (s) | | |
|---|---|---|---|---|---|---|---|
| ($n$) | Mean | 50th | 75th | 90th | $z^{\text{SPR2}}$ | $z^{\text{OPT}}$ | Lower bound |
| 3 | 0.00 | 0.00 | 0.00 | 0.00 | 0.03 | <0.01 | 0.00 |
| 4 | 0.10 | 0.00 | 0.04 | 0.17 | 0.08 | 0.23 | −0.04 |
| 5 | 0.39 | 0.12 | 0.47 | 1.11 | 0.52 | 20.68 | −1.27 |

Mean, 50th, 75th, and 90th percentile performance is indicated as the percentage over the optimum solution. The mean lower bound is reported as $(z^{\text{LB}} - z^{\text{OPT}})/z^{\text{OPT}}$ in percentage

**Table 16** Performance of the SPR2 shortest-path heuristic on a complete graph, based on 1000 randomly generated problem scenarios with average inspection times that are comparable to average travel times

| Vertices | Percent over lower bound | | | | |
|---|---|---|---|---|---|
| ($n$) | Mean | 50th | 75th | 90th | Time (s) |
| 3 | 0.00 | 0.00 | 0.00 | 0.00 | 0.03 |
| 4 | 0.14 | 0.03 | 0.08 | 0.22 | 0.08 |
| 5 | 1.66 | 0.75 | 1.57 | 3.15 | 0.52 |
| 6 | 3.58 | 2.03 | 4.63 | 9.71 | 0.58 |
| 7 | 4.93 | 3.03 | 5.75 | 11.98 | 1.35 |
| 8 | 5.84 | 4.54 | 8.64 | 12.47 | 3.34 |
| 9 | 7.56 | 5.67 | 10.49 | 15.93 | 7.98 |

Mean, 50th, 75th, and 90th percentile performance is indicated as the percentage excess above the lower bound, reported as $(z^{\text{SPR2}} - z^{\text{LB}})/z^{\text{LB}}$ in percentage

We note that the SPR2 shortest-path heuristic method returns results that are within 10 percent of the lower bound in 90 percent of the problem instances for $n = 6$, and within 16 percent of the lower bound in 90 percent of problem instances for $n = 9$. These solutions take on average 0.58 s and 7.98 s, respectively, to compute.

### 3.5.4 Performance on Additional Graph Structures

In addition to problems on a complete graph, we test the SPR2 heuristic method on several additional graph structures. Specifically, we consider line graphs, circle graphs, and random trees. We use the procedures from Section 2.4.1 to generate 1000 random problem instances for problem cases with $n = 4$, 5, 6, and 7 vertices.

To construct a line graph, we randomly assign $n - 1$ edges between $n$ vertices, such that the degree of each vertex is at least one but no more than two. To construct a circle graph, we randomly assign $n$ edges between $n$ vertices, such that the degree of each vertex is exactly two. To construct a random tree, we randomly assign $n - 1$ edges between $n$ vertices, such that the degree of each vertex is at least one and there is at least one vertex of degree greater than two, which excludes line graphs from the random tree category.

We still allow a patroller to travel between any two vertices in order to determine a patrol policy. For these additional graph structures, a patroller may have to travel through one or more interim vertices (without conducting inspections at those vertices) in order to arrive at the destination vertex.

We consider cases where average travel times are comparable to average inspection times. To do this, we scale the travel times between each pair of vertices based on the graph structure. Specifically for any particular graph, we determine the average number of edges between each pair of vertices and divide the travel times by that average value. This produces average total travel times between each pair of vertices that are comparable to average inspection times. We construct a distance matrix $D$ using these scaled travel times. The distance $d_{ij}$ is the total travel time along the shortest path in the graph between each pair of vertices $i$ and $j$, for $i, j \in N$.

Results for these additional graph structures with $n = 4, 5, 6$, and 7 are presented in Table 17. For graphs with $n \leq 5$, we compare the performance of the heuristic to an optimal solution as well as to the lower bound. For graphs with $n \geq 6$, we compare the heuristic to the lower bound, since an optimal solution cannot be determined for problems of this size.

**Table 17** Mean performance of the SPR2 heuristic method on additional graph structures, based on 1000 randomly generated problem scenarios for average inspection times that are comparable to average travel times

| | Vertices | Performance (%) | | Time (s) | |
|---|---|---|---|---|---|
| Graph | ($n$) | $z^{SPR2}/z^{OPT}$ | $z^{SPR2}/z^{LB}$ | $z^{SPR2}$ | $z^{OPT}$ |
| Complete | 4 | 0.10 | 0.12 | 0.08 | 0.23 |
| Complete | 5 | 0.39 | 1.66 | 0.52 | 20.68 |
| Complete | 6 | – | 3.58 | 0.58 | – |
| Complete | 7 | – | 4.93 | 1.35 | – |
| Line | 4 | 0.08 | 0.10 | 0.09 | 0.28 |
| Line | 5 | 0.26 | 0.90 | 0.46 | 35.84 |
| Line | 6 | – | 8.11 | 0.53 | – |
| Line | 7 | – | 11.12 | 1.31 | – |
| Circle | 4 | 0.12 | 0.15 | 0.08 | 0.29 |
| Circle | 5 | 0.50 | 1.18 | 0.50 | 22.25 |
| Circle | 6 | – | 2.32 | 0.54 | – |
| Circle | 7 | – | 3.73 | 1.29 | – |
| Random tree | 4 | 0.05 | 0.14 | 0.09 | 0.23 |
| Random tree | 5 | 0.15 | 0.84 | 0.52 | 28.62 |
| Random tree | 6 | – | 4.79 | 0.55 | – |
| Random tree | 7 | – | 5.99 | 1.35 | – |

Performance is indicated as the mean percentage over optimum for problems where an optimal solution can be determined using the SALP, and the mean percentage over lower bound for all problems

These results indicate that the shortest-path heuristic method can be used very effectively for the strategic-attacker problem on several different graph structures and sizes. For problems with $n = 5$, where an optimal solution can be determined, the SPR2 method returns a solution on average that is within 0.50 percent of optimal. These solutions take approximately 0.5 s to compute, which is 40 times less than the time required to compute an optimal solution. For problems with $n = 7$, where an optimal solution cannot be determined, the heuristic produces on average a result within 3.73 percent of the lower bound on a circle graph, and within 11.12 percent of the lower bound on a line graph. These solutions take less than 1.5 s to compute.

### 3.5.5 Sensitivity Analysis

In addition to the baseline problems, we consider the case where a patroller needs to spend more time conducting inspections than he does traveling between vertices; and the case where the patroller needs to spend more time traveling between vertices than he does conducting inspections. The five specific cases we consider in the numerical experiments are summarized in Table 18. Case III generated the smallest number of states and had the highest long-run cost on average. It also generated the tightest lower bound for an optimal solution. Case IV generated the largest number of states and had the lowest long-run cost on average. It also generated the loosest lower bound for an optimal solution.

The mean performance results for problem cases II through V using both the SP and FP methods are presented in Table 19. The 90th percentile performance results are presented in Table 20. In each of the problem cases, very favorable results are obtained using the SP heuristic method. In at least 90 percent of the problem instances, the SPR2 method returns a solution within 1.51 percent of optimal. These solutions take 0.52 s to compute on average.

**Table 18** Summary of numerical experiments for strategic attackers

| Parameter | Case I | Case II | Case III | Case IV | Case V |
|---|---|---|---|---|---|
| Travel time | 1× | 1× | 1× | 2× | 2× |
| Inspection time | 1× | 2× | 2× | 1× | 1× |
| Attack time | 1× | 1.5× | 1× | 1.5× | 1× |
| Mean number of states, $\|\Omega\|$ | 1,184 | 633 | 102 | 3,938 | 318 |
| Mean number of decision variables | 5,920 | 3,165 | 510 | 19,690 | 1,590 |
| Mean number of constraints | 7,110 | 3,804 | 613 | 23,679 | 1,914 |
| Mean optimal long-run cost | 0.4892 | 0.5085 | 0.6589 | 0.4761 | 0.6224 |
| Mean optimal computation time (s) | 20.68 | 4.99 | 0.11 | 574.85 | 2.11 |
| Lower bound | −1.20 | −0.20 | −0.03 | −4.81 | −0.88 |

The mean lower bound is reported as $(z^{LB} - z^{OPT})/z^{OPT}$ in percentage

**Table 19** Mean performance of the shortest-path and fictitious-play methods, based on 1000 randomly generated problem scenarios for each case

| Case | FP/SP | Percent over optimum (mean) | | | | Time (s) |
|---|---|---|---|---|---|---|
| | | – | SP | SPR2 | SPR3 | |
| II | – | – | 1.26 | 0.21 | 0.14 | 0.52 |
| | FP 10 | 3.32 | 1.23 | 0.18 | 0.12 | 29.71 |
| | FP 20 | 1.20 | 0.67 | 0.13 | 0.10 | 59.52 |
| | FP 30 | 0.60 | 0.44 | 0.10 | 0.07 | 89.92 |
| | FP 50 | 0.30 | 0.27 | 0.08 | 0.04 | 151.99 |
| III | – | – | 0.41 | 0.22 | 0.17 | 0.50 |
| | FP 10 | 1.66 | 0.39 | 0.19 | 0.15 | 2.25 |
| | FP 20 | 0.74 | 0.27 | 0.16 | 0.12 | 4.75 |
| | FP 30 | 0.50 | 0.15 | 0.10 | 0.07 | 7.38 |
| | FP 50 | 0.37 | 0.15 | 0.09 | 0.05 | 12.79 |
| IV | – | – | 2.65 | 0.50 | 0.34 | 0.50 |
| | FP 10 | 4.49 | 2.15 | 0.34 | 0.26 | 717.60 |
| | FP 20 | 2.19 | 1.42 | 0.26 | 0.19 | 1337.60 |
| | FP 30 | 1.08 | 0.80 | 0.12 | 0.09 | 1977.97 |
| V | – | – | 0.90 | 0.53 | 0.44 | 0.47 |
| | FP 10 | 2.96 | 0.74 | 0.45 | 0.38 | 14.30 |
| | FP 20 | 1.37 | 0.51 | 0.31 | 0.26 | 29.78 |
| | FP 30 | 0.83 | 0.60 | 0.22 | 0.17 | 47.16 |
| | FP 50 | 0.51 | 0.17 | 0.16 | 0.11 | 78.87 |

Performance is indicated as the percentage excess over an optimal solution. Shortest-path computation time is indicated for the SPR2 heuristic

# 4 Conclusion

We examine methods to determine effective patrol policies against both random and strategic attackers. We consider two cases: a single patroller against random attackers and a single patroller against strategic attackers.

In the case of a single patroller against random attackers, we determine an optimal solution by modeling the state space of the system as a network and solve a minimum cost-to-time ratio cycle problem using linear programming. The solution represents a patrol policy, which is a repeating pattern of locations for a patroller to visit and inspect that minimizes the long-run cost incurred due to undetected attacks. Although the linear program returns an optimal solution, it quickly becomes computationally intractable for problems of moderate size. We therefore develop and test two aggregate-index heuristic methods, the index heuristic time (IHT) method and the index heuristic epoch (IHE) method. Both of these methods consider the structure of the graph, to include travel and inspection time requirements. The IHT method utilizes a predetermined look-ahead time window for the patroller to

**Table 20** 90th percentile performance of the shortest-path and fictitious-play methods, based on 1000 randomly generated problem scenarios for each case

| Case | FP/SP | Percent over optimum (90th PCTL) | | | | Time (s) |
|---|---|---|---|---|---|---|
| | | – | SP | SPR2 | SPR3 | |
| II | – | – | 3.21 | 0.69 | 0.49 | 0.52 |
| | FP 10 | 5.95 | 2.94 | 0.66 | 0.42 | 29.71 |
| | FP 20 | 2.47 | 1.64 | 0.49 | 0.33 | 59.52 |
| | FP 30 | 1.35 | 1.05 | 0.37 | 0.24 | 89.92 |
| | FP 50 | 0.79 | 0.47 | 0.23 | 0.16 | 151.99 |
| III | – | – | 1.06 | 0.60 | 0.53 | 0.50 |
| | FP 10 | 3.08 | 1.04 | 0.45 | 0.39 | 2.25 |
| | FP 20 | 1.47 | 0.78 | 0.39 | 0.32 | 4.75 |
| | FP 30 | 1.12 | 0.69 | 0.30 | 0.24 | 7.38 |
| | FP 50 | 0.77 | 0.36 | 0.28 | 0.19 | 12.79 |
| IV | – | – | 5.44 | 1.51 | 1.06 | 0.50 |
| | FP 10 | 8.63 | 4.58 | 0.87 | 0.76 | 717.60 |
| | FP 20 | 4.72 | 3.84 | 0.82 | 0.68 | 1337.60 |
| | FP 30 | 2.78 | 2.18 | 0.27 | 0.21 | 1977.97 |
| V | – | – | 1.90 | 1.26 | 1.16 | 0.47 |
| | FP 10 | 5.79 | 1.77 | 1.02 | 0.85 | 14.30 |
| | FP 20 | 3.07 | 1.30 | 0.73 | 0.61 | 29.78 |
| | FP 30 | 1.94 | 1.27 | 0.60 | 0.49 | 47.16 |
| | FP 50 | 1.32 | 0.42 | 0.34 | 0.24 | 78.87 |

Performance is indicated as the percentage excess over an optimal solution. Shortest-path computation time is indicated for the SPR2 heuristic

decide his next action by considering all possible paths and partial paths that can be completed during the time window when starting from his current vertex. For each of these paths, aggregate index values per unit time are computed and the patroller chooses his action based on those index values. He then repeats the process from the next vertex using the same look-ahead time window. This process continues until a patrol pattern is determined. The IHE method works in a similar fashion. However, in this method, a patroller looks ahead a predetermined number of decision epochs, and determines his action by considering all possible paths from the current vertex that consist of the specified number of decision epochs, regardless of the total time those paths will take. We see very favorable results using these methods in numerical experiments. In our baseline experiments, a solution within 1 percent of optimal was returned in at least 90 percent of the problem instances.

In the case of a single patroller against strategic attackers, we determine an optimal solution by modeling the state space of the system as a network and solve a linear program to minimize the largest expected cost per attack among all vertices. The solution consists of a patrol policy, which is a randomized strategy for the patroller that minimizes the long-run expected cost due to an undetected attack. Although the linear program returns an optimal solution, it quickly becomes computationally intractable for problems of moderate size. We therefore develop two heuristic methods, the shortest-path (SP) and fictitious-play (FP) methods. The SP method uses a combinatorial selection of patrol patterns based on the shortest Hamiltonian cycle in the graph. The FP method is an iterative method based on fictitious play. We also present a linear program that determines a lower bound for an optimal solution, so that we can evaluate our heuristics when an optimal solution is not available. We see very favorable results using both methods in numerical experiments; however, the FP method uses considerably more computation time than the SP method. In our baseline experiments, a solution within 1.2 percent of optimal was returned in at least 90 percent of the problem instances.

## References

1. R. Ahuja, T. Magnanti, J. Orlin, *Network Flows: Theory, Algorithms, and Applications* (Prentice Hall, Englewood Cliffs, NJ, 1993)
2. S. Alpern, Infiltration games on arbitrary graphs. J. Math. Anal. Appl. **163**(1), 286–288 (1992)
3. S. Alpern, Search games on trees with asymmetric travel times. SIAM J. Control Optim. **48**(8), 5547–5563 (2010)
4. S. Alpern, R. Fokkink, Accumulation games on graphs. Networks **64**(1), 40–47 (2014)
5. S. Alpern, S. Gal, Searching for an agent who may or may not want to be found. Oper. Res. **50**(2), 311–323 (2002)
6. S. Alpern, A. Morton, K. Papadaki, Patrolling games. Oper. Res. **59**(5), 1246–1257 (2011)
7. J. Auger, An infiltration game on *k* arcs. Nav. Res. Logist. **38**(4), 511–529 (1991)
8. R. Avenhaus, Applications of inspection games. Math. Model. Anal. **9**(3), 179–192 (2004)
9. S. Benkoski, M. Monticino, J. Weisinger, A survey of the search theory literature. Nav. Res. Logist. **38**, 469–464 (1991)
10. A. Garnaev, G. Garnaeva, P. Goutal, On the infiltration game. Int. J. Game Theory **26**(2), 215–221 (1997)
11. J. Gittins, K. Glazebrook, R. Weber, *Multi-armed Bandit Allocation Indices*, 2nd edn. (Wiley, Hoboken, NJ, 2011)
12. K. Kikuta, A search game with traveling cost on a tree. J. Oper. Res. Soc. Jpn. **38**(1), 70–88 (1995)
13. K. Kikuta, W. Ruckle, Initial point search on weighted trees. Nav. Res. Logist. **41**, 821–831 (1994)
14. K. Kikuta, W. Ruckle, Continuous accumulation games on discrete locations. Nav. Res. Logist. **49**(1), 60–77 (2002)
15. K. Lin, M. Atkinson, T. Chung, K. Glazebrook, A graph patrol problem with random attack times. Oper. Res. **61**(3), 694–710 (2013)
16. R. McGrath, K. Lin, Robust patrol strategies against attacks at dispersed heterogeneous locations. Int. J. Oper. Res. **30**(3), 340–358 (2017)
17. G. Owen, *Game Theory*, 3rd edn. (Academic, San Diego, CA, 1995)
18. M. Puterman, *Markov Decision Processes: Discrete Stochastic Dynamic Programming* (Wiley-Interscience, New York, NY, 1994)
19. J. Robinson, An iterative method of solving a game. Ann. Math. **54**(2), 296–301 (1951)
20. S. Ross, *Introduction to Probability Models*, 10th edn. (Academic, San Diego, CA, 2010)

21. W. Ruckle, *Geometric Games and Their Applications* (Pitman, Boston, MA, 1983)
22. A. Washburn, *Two-Person Zero-Sum Games*, 3rd edn. (INFORMS, Linthicum, MD, 2003)
23. A. Washburn, K. Wood, Two-person zero-sum games for network interdiction. Oper. Res. **43**(2), 243–351 (1995)
24. K. Zoroa, P. Zoroa, M. Fernandez-Saez, Weighted search games. Eur. J. Oper. Res. **195**(2), 394–411 (2009)

# Network Design Problem with Cut Constraints

**Firdovsi Sharifov and Hakan Kutucu**

## 1 Introduction

In this paper, we focus on the following minimum cost network design problem under conditions that the number of edges crossing a cut is lower bounded, in the resulting network. We call it as the network design problem with cut constraints (NDPC). Let $G = (V, E)$ be an undirected graph with node set $V$ and edge set $E$, where nodes in $V$ represent a given set of locations and edges of $E$ correspond to potential links. Given a cost $c_e \geq 0$ associated with each edge $e$, some of the edges $e$ in $E$ may be marked as magisterial links whose at least $l_e$ parallel copies must be installed in a resulting network, where $l_e \geq 0$ is a given integer number for each of them. Let $l_e = 0$ for unmarked edges $e$. Let $x_e \geq l_e$ be the number of parallel copies of $e$ which will be installed in a resulting network for each $e \in E$. We want to define a minimum cost spanning subgraph of $G$ in which at least $l_e$ parallel copies of edges $e \in E$ must be installed; moreover, the summation of $x_e$'s for edges crossing the cut determined by any nonempty subset $S \subset V$ is bounded below by the value of a set function $f(S)$ defined on the subsets $S$ of $V$. We assume that there is an oracle returning $f(S)$ for each query $S$. We view the latter conditions as survivability requirements by well-known Menger's theorem in terms of cuts.

Throughout the paper, we denote the family of subsets $\{S; \emptyset \neq S \subset V\}$ by $\mathscr{F}(V)$ and the set of edges with one end node in $S$ and the other end node in $V \setminus S$ by $\delta(S)$. In other words, $\delta(S)$ is the minimal cut of $G$ determined by a nonempty subset $S \subset V$. For all $S \subseteq V$, we use $\overline{S}$ to denote $V \setminus S$.

---

F. Sharifov (✉)
V.M. Glushkov Institute of Cybernetics, Kyiv, Ukraine

H. Kutucu
Karabuk University, Department of Computer Engineering, Karabuk, Turkey
e-mail: hakankutucu@karabuk.edu.tr



B. Goldengorin (ed.), *Optimization Problems in Graph Theory*,
Springer Optimization and Its Applications 139,
https://doi.org/10.1007/978-3-319-94830-0_11

Since $f(S)$ can be interpreted as a prediction of the capacity of the cut determined by any nonempty subset $S \subset V$, we assume that $f(S)$ is a symmetric submodular function [11] and $f(\emptyset) = f(V) = 0$. For such symmetric submodular function, $f(S) \geq 0$, since $f(S) = f(\overline{S})$ and $2f(S) = f(S) + f(\overline{S}) \geq f(V) + f(\emptyset) = 0$. We consider the case $f(S) > 0$ for each $S \in \mathscr{F}(V)$ which provides connectivity of a spanning subgraph induced by edge set $\{e \in E; x_e \geq l_e > 0\}$.

We use Lemma 2 (see Section 3) to formulate NDPC as a linear integer program. The lemma states that it is enough to include the survivability requirement only for minimal cuts in the form of $\delta(S)$ to model of NDPC. This means that NDPC can be formulated as the following integer program:

$$\psi(IP) = \min \sum_{e \in E} c_e x_e, \tag{1}$$

subject to

$$x(\delta(S)) \geq f(S), S \in \mathscr{F}(V), \tag{2}$$

$$x_e \geq l_e \geq 0, \quad e \in E, \tag{3}$$

$$x_e \quad \text{integral}, \quad e \in E. \tag{4}$$

This problem was brought to our attention by Air Transportation Management Department in National Aviation University of Ukraine in the following case [10].

Air Transport System (ATS) as a communication system should operate with account of all the factors that stimulate the growth and efficiency of air transportation. The growth of air traffic and, accordingly, the efficient performance of all types of services in the structure of ATS cause a number of essential reorganization and technical changes in the major subsystems of the ATS. The main purpose of these reconstructions is to improve the service at airports, to expand the route network geographically and to increase flight frequency. In the process of planning these stage changes, it needs to define the optimal flight frequency, the appropriate structure of the route network. Total cost of works related to organizational and technical transformations results from their summing up [2]. Airport capacity in terms of flight frequency can be determined according to the Development Reference Manual, developed by International Air Transport Association (IATA).

The route network is understood as a network, consisting of many airports as its nodes and of nonstop routes as its edges. The total cost $c_i$ of $i$-th airport reconstruction is defined as the sum of costs per unit at all stages of its reconstruction. The vector $l_{ij}$ represents the minimum passenger traffics (flow) between pair of airports $i$ and $j$ with nonstop communication. For any subset $S$ of airports, $f(S)$ is defined as total potentially passenger traffics which may be predictably arrival to each airport in subset $S$ and may be predictably departure from every airport of $S$. Due to IATA, unknown capacity of any airport can be defined as the sum of unknown passenger traffic flows on nonstop flight lines adjacent to this airport. Since $f(S)$ is a symmetric cut function by the definition, the problem can be naturally formulated as NDPC.

The well-known $NP$-hard problems such as $k$-edge connected network design problem [7] and the survivable network design problem (SNDP) with a given connectivity type vector $(r_v; v \in V)$ [7, 9] are special cases of NDPC, when $f(S) = k$ and

$$f(S) = con(S) = \min\{\max\{r_s; s \in S\}, \max\{r_t : t \in \overline{S}\}\},$$

for any $\emptyset \neq S \subset V$ and $f(\emptyset) = f(V) = 0$. By Menger's theorem, we obtain integer formulations for these problems by letting $x_e = 0 \vee 1$ in (4). Note that the following submodular inequality

$$con(L) + con(S) \geq con(L \cup S) + con(L \cap S)$$

may be held strongly, if

$$\max\{r_s; s \in L \cap S\} > \max\{r_t; t \in \overline{L \cap S}\}$$

for nonempty subsets $L, T \subset V$. The latter inequality holds in a few cases if $|r_v - r_w|$ is not ranged over *big* set of integers for any pair of distinct nodes $v, w \in V$. Since $r_v \in \{0, 1, 2\}$ for many cases of SNDP in practice, $con(S)$ can be considered as a modular function. This fact facilitates the work for finding a violated constraint in solving SNDP by the polyhedral approach. We will also approximate $f(S)$ by a modular function to facilitate the same work.

The NDPC may be solved in polynomial time depending on topology of the graph $G$ and the definition of the function $f$. When $G$ is a complete graph and $f(S)$ is a symmetric cut function that is available via only an oracle (see [3, 14]) and all edge costs equal to 0, then NDPC is the problem of finding a spanning subgraph, so that $f(S)$ is the capacity of the cut $\delta(S)$ determined by each $S \in \mathcal{F}(V)$. The algorithm in [14] constructs such subgraphs using the values $f(\{v, w\})$ for all pair of distinct nodes $v, w \in V$ in $O(n^2)$ time. In Section 3, we use techniques [13–15] to reduce the separation problem for inequalities (2) to the minimum cuts problem for modular case of the function $f$.

In addition to the problems mentioned above, an example of the linear relaxation (1)–(3) can be given in the traffic engineering process of the network design [1, 16], where $x_e$ can be interpreted as road strip (edge) $e$. More precisely, consider an example on some transportation network $G = (V, E)$ and let $\overline{x}_e = (\overline{x}_e; e \in E)$ denote the current capacities of its edges. For any nonempty subset $S \subset V$, an estimated capacity of the cut $\delta(S)$ can be computed using the techniques in [17] on statistic data about rates of routing traffics in $G$. The edge capacities $\overline{x}_e$ can be considered as *normal* if $\Delta(f, \overline{x}) = 0$, where

$$\Delta(f, \overline{x}) = \min\{\overline{x}(\delta(S)) - f(S); S \in \mathcal{F}(V)\}.$$

Indeed, if $\Delta(f, \overline{x}) > 0$, then it is clear that $\overline{x}(\delta(S)) > f(S)$ for any $\emptyset \neq S \subset V$, and hence the capacities $\overline{x}_e$ are superfluous. If $\Delta(f, \overline{x}) < 0$, then there exist subsets $S_1, \ldots, S_q \in \mathscr{F}(V)$ $(q \geq 1)$ such that $\overline{x}(\delta(S_k)) < f(S_k)$ for $k = 1, \ldots, q$. Therefore, the capacities of the cuts $\delta(S_k)$ are not enough for normal operation of the network. It needs to extend the capacities of the cuts $\delta(S_k)$ up to $f(S_k)$. So, the problem with objective (1) is subject to the following constraints:

$$x(\delta(S_k)) \geq f(S_k), \quad k = 1, \ldots, q,$$

$$x_e \geq \overline{x}_e \geq 0, \quad e \in E.$$

For a given vector $x = (x_e; e \in E)$, if $\Delta(f, x) \geq 0$, then it satisfies all inequalities (2). Under the trivial inequalities (3), the separation problem for a vector $x = (x_e; e \in E)$ is $\Delta(f, x)$, which is $NP$-hard, in general. The reason is that if $f$ is a cut function; that is, $f(S) = p(\delta(S))$ for some given vector $p$, and $p_e > x_e$, then $\Delta(f, x)$ is the well-known $NP$-hard maximum cut problem on $G$ with edge capacities $p_e - x_e$.

The theory of linear programming [12] states that a linear program with exponential number of inequalities can be solved in polynomial time if and only if the separation problem associated with these inequalities is polynomially solvable. This result has theoretical consequences. For solving network design problems, approximation solutions and polyhedral approaches use the combination of separation algorithms and linear programming solvers. The difficulty with this approach to solve NDPC is that the separation problem $\Delta(f, x)$ is $NP$-hard. In order to use the same techniques in solving NDPC, we first show that when $\delta(S)$ is not a non-minimal cut, that is, after deleting the edges of $\delta(S)$ the graph $G$ splits more than two connected components, then the inequalities (2) for $S$ and $\overline{S}$ can be dropped by Lemma 2 in Section 3. Thus, only minimal cuts can be considered in solving NDPC. For this reason, we consider the collection $\mathscr{F}(T)$ of the fundamental cut sets to be determined by deleting each edge of a spanning tree $T$ of $G$, since each subset $S \in \mathscr{F}(T)$ determines the minimal cut $\delta(S)$. We restrict $\mathscr{F}(V)$ to $\mathscr{F}(T)$ in (2) and consider the two special cases of NDPC, namely, when NDPC includes constraints (2) only for $S \in \mathscr{F}(T)$ (NDPC1), and when NDPC includes the constraints (2) in the form of equations only for each $S \in \mathscr{F}(T)$ (NDPC2). We will call their standard linear programming (LP) relaxations as *fundamental cut sets relaxations* of NDPC and show that the simplex algorithm finds optimal solutions to the LP-relaxation of NDPC2 in at most $m-n+1$ iterations. We illustrate advantages or disadvantages of the LP-relaxations of NDPC1 and NDPC2 in the solving NDPC in a simple example.

We show that the problem defining a spanning tree for obtaining the fundamental cut sets whose LP-relaxation has smaller optimal objective value is $NP$-hard. For Hamiltonian path as spanning tree $T$, the corresponding fundamental cut sets LP-relaxations of NDPC have an integer-valued optimal solution for any integer-valued function $f$. In particular, this shows that the LP-relaxation of the constrained forest problem (CFP) [6] has an integer-valued optimal solution for the proper function with values 1 on the fundamental cut sets determined with respect to some Hamiltonian path. In Section 2, we will describe these in more detail.

In Section 3, we present a strongly polynomial algorithm for finding a solution to the dual problem of the linear relaxation of NDPC based on the fact that the separation problem $\Delta(w, x)$ can be solved by solving at most $n$ minimum cut problems for a modular function $w$ as approximation $f$. In order to get a suitable modular function for approximation to $f$, we use the convex hull of bases of the extended polymatroid associated with the function $f$. The solution to the dual problem is used to define initial spanning tree $T$ in $G$ and for finding an optimal solution $x \in R^E$ to the fundamental cut sets $LP$-relaxations of NDPC. Then, we use the results in [4] and the branch and cut techniques to convert the vector $x$ to the solution of NDPC. In conclusion, we discuss several possible directions for improving the accuracy of approximate solution to NDPC.

## 2 Fundamental Cut Sets LP-Relaxations of NDPC

In this section, we consider fundamental cut sets LP-relaxations of NDPC as special cases of (1)–(3) to obtain a solution to NDPC. Let $T$ be a spanning tree of $G$. We denote the edges of the tree $T$ by $t_1, \ldots, t_{n-1}$. Removing any edge $t_k$ splits $T$ into two subtrees whose node sets are denoted by $V_k$ and $\overline{V}_k = V \setminus V_k$. So, each edge $t_k \in T$ corresponds to the cut $\delta(V_k)$ and vice versa. We set $\mathscr{F}(T) = \{V_1, \ldots, V_{n-1}\}$ and call it as fundamental cut sets of the spanning tree $T$. Consider the following linear programming problem:

$$\min \sum_{e \in E} c_e x_e,$$

subject to

$$x(\delta(S)) \geq f(S), \quad S \in \mathscr{F}(T),$$

$$x_e \geq l_e \geq 0, \quad e \in E.$$

It is convenient to write it in the variables $y_e = x_e - l_e$ as follows:

$$\varphi_1(T) = \min \sum_{e \in E} c_e y_e, \tag{5}$$

subject to

$$y(\delta(V_k)) \geq b(t_k), \quad k = 1, \ldots, n-1, \tag{6}$$

$$y_e \geq 0, \quad e \in E, \tag{7}$$

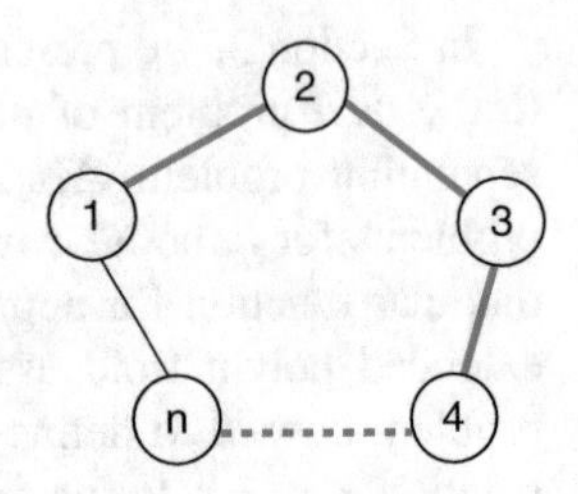


**Fig. 1** Cycle $C_n$, spanning tree $T$, and $n-1$ fundamental cuts

where $b(t_k) = f(V_k) - l(\delta(V_k))$ for all $k = 1, \dots, n-1$. We call this problem *LP-relaxation of NDPC1*. By the similar way, the LP-relaxation of NDPC2 is the following linear program:

$$\varphi_2(T) = \min \sum_{e \in E} c_e y_e, \tag{8}$$

subject to

$$y(\delta(V_k)) = b(t_k), \quad k = 1, \dots, n-1, \tag{9}$$

$$y_e \geq 0, \quad e \in E. \tag{10}$$

Both linear latter programs have $m + n - 1$ constraints and $m$ variables. They will be called *fundamental cut set LP-relaxations of NDPC*. Our NDPC algorithm in the next section solves problems (5)–(7) or (8)–(10) for solving (1)–(3). To show that which of two problems NDPC1 and NDPC would be suitable to solve NPDC, let us consider NDPC1 and NDPC2 on the cycle $C_n$ with $n$ nodes and $n$ edges, shown in Figure 1, where the bold lines (1, 2), (2, 3),..., $(n-1, n)$ are the edges of the spanning tree $T$. Let

$$b(12) = 1, \; b(23) = 2, \dots, b(n-1, n) = n-1$$

and let $c_e = 1$ for all edges $e$ of $C_n$. The vectors

$$y^1 = (y^1_{n-1,n} = 1, y^1_{1,n} = n-2, y^1_e = 0 \text{ for other edges } e)$$

and

$$y^2 = (y^2_{12} = 0, y^2_{23} = 1, \dots, y^2_{n-1n} = n-2, y^2_{1n} = 1)$$

are optimal solutions to the LP-relaxations of NDPC1 and NDPC2, respectively. Clearly,

$$\{v\} = S_v \notin \mathscr{F}(T),$$

for nodes $v = 1, \dots, n-1$ of $C_n$. From $b_t > 0$ for edges of $T$, it follows that $l(\delta(S_v)) < f(S_v)$ for $v = 1, \dots, n-1$ of $C_n$. Hence,

$$x^1(\delta(S_v)) = l(\delta(S_v)) < f(S_v)$$

for the vector $x^1 = y^1 + l$ and

$$x^2(\delta(S_v)) > l(\delta(S_v))$$

for the vector $x^2 = y^2 + l$.

The above examples illustrate that using (8)–(10) in our algorithm is more successful than using (5)–(7). Because we can reduce the number of subproblems obtained in solution of NDPC by the branch and cut techniques using (8)–(10). Based on these observations, using (5)–(7) is reasonable if $f(S) - l(\delta(S))$ is almost a modular function, else using (8)–(10) is reasonable for solving NDPC.

Let $A$ denote the $(n-1) \times m$ constraint matrix of the problems (5)–(7) and (8)–(10). Let the vector $b = (b(t_k); k = 1, \ldots, n-1)$. Since each edge $t_k$ is an edge of the cut $\delta(V_k)$, the matrix $A$ is in the form of $(N, I)$, where $I$ is $(n-1) \times (n-1)$ identity matrix whose columns are indexed by edges $t_k$ of the tree $T$. The rows for $\delta(V_k)$ of matrix $A$ are the 0 or 1 characteristic vectors of the cuts $\delta(V_k)$, for any $t_k$ in $T$ (if $e$ is an edge of the cut $\delta(V_k)$, then this row has 1 in the column $e$, 0 otherwise). A column $h$ in $N$ contains 1 in a row $\delta(V_k)$, if the edge $t_k$ is an edge of the cycle $C(h)$ obtained by adding the edge $h \in E \setminus T$ to the tree $T$. Conversely, if $t_1, \ldots, t_p$ are edges of the cycle $C(h)$, then $h$ is just an edge of the cuts $\delta(V_1), \ldots, \delta(V_p)$. So, we have the following property.

*Property 1* For any edge $h \in E \setminus T$, if $t_1, \ldots, t_p$ are edges of a cycle $C(h)$, then the column $h$ has just 1 in the rows corresponding to cuts $\delta(V_k)$, $k = 1, \ldots, p$. This means that vector columns indexed by $t_1, \ldots, t_p$ and $h$ are linearly dependent.

For any $e \in E$, we write $e \in C(h)$ if $e$ is an edge of the cycle $C(h)$. For edges $e, h \in E \setminus T$, if the cycle $C(e)$ contains only edges $t_k \in C(h)$, then the cycle $C(e)$ is called *a subcycle* of $C(h)$, that is, the edge $e$ is a chord of $C(h)$. Let $a^e$ denote the column $e$ of the matrix $N$ for each edge $e \in E \setminus T$. It follows that $a^h \geq a^e$ for any chord $e$ of the cycle $C(h)$. We also assume that $b(t_k) \geq 0$ for each $V_k \in \mathscr{F}(T)$, since otherwise the problem (8)–(10) is infeasible. Hence, the variables $y_{t_1}, \ldots, y_{t_{n-1}}$ corresponding to the columns of $I$ form basis for (5)–(7) and (8)–(10). The following propositions will be used for solving the problems (5)–(7) and (8)–(10).

**Proposition 1** *For an edge $h \in E \setminus T$, if*

$$\sum_{t_k \in C(h)} c_{t_k} \leq c_h, \tag{11}$$

*then the edge $h$ can be deleted in the graph $G$.*

*Proof* The dual problems of (5)–(7) and (8)–(10)

$$\max \sum_{t_k \in T} b(t_k) u(t_k), \tag{12}$$

$$\sum_{\delta(V_k)\ni e} u(t_k) \le c_e, \quad e \in E \tag{13}$$

include conditions $u(t_k) \le c_{t_k}$ associated with the edges $t_k \in T$. By Property 1,

$$\sum_{\delta(V_k)\ni e} u(t_k) = \sum_{t_k \in C(e)} u(t_k) \le \sum_{t_k \in C(h)} c_{t_k} \le c_h.$$

The constraints (13) for $e = h$ is the sum of those for $e = t_k \in C(h)$. Hence, the edge $h$ can be deleted in $G$. □

**Proposition 2** *Let $C(h)$ be a cycle in $T \cup \{h\}$ and let $e$ be a chord of $C(h)$ for $e, h \in E \setminus T$. If*

$$\sum_{t_k \in C(h)} c_{t_k} - c_h \le \sum_{t_k \in C(e)} c_{t_k} - c_e, \tag{14}$$

*then the edge $h$ can be deleted in the graph $G$.*

*Proof* The dual problem of (5)–(7) can be represented as

$$\min \sum_{t_k \in T} b(t_k) z(t_k), \tag{15}$$

subject to

$$\sum_{\delta(V_k)\ni e} z(t_k) \ge \sum_{t_k \in C(e)} c_{t_k} - c_e, \quad e \in E \setminus T, \tag{16}$$

$$z(t_k) \ge 0, \quad V_k \in \mathscr{F}(T), \tag{17}$$

where

$$z(t_k) = c_{t_k} - u(t_k).$$

The dual problem of (5)–(7) is (15)–(17) with the following additional conditions

$$z(t_k) \le c_{t_k}, \quad V_k \in \mathscr{F}(T).$$

Since the edge $e$ is a chord of $C(h)$, $a^h \ge a^e$, that is,

$$\sum_{\delta(V_k)\ni h} z(t_k) \ge \sum_{\delta(V_k)\ni e} z(t_k).$$

It follows that

$$\sum_{\delta(V_k)\ni h} z(t_k) \ge \sum_{\delta(V_k)\ni e} z(t_k) \ge \sum_{t_k \in C(e)} c_{t_k} - c_e \ge \sum_{t_k \in C(h)} c_{t_k} - c_h,$$

where the last inequality is (14). Therefore, we can delete the constraints (16) for $e = h$. In other words, the edge $h$ can be deleted in the graph $G$. □

By Proposition 2, if (14) holds for one of the chords $e_1, \ldots, e_p \in E \setminus T$ of any cycle $C(h)$, then the constraint (16) for $e = h$ can be deleted. After the examination of (14) for all cycles and for their chords, in the result, we have

$$\sum_{t_k \in C(h)} c_{t_k} - c_h > \sum_{t_k \in C(e_i)} c_{t_k} - c_{e_i} \tag{18}$$

for any cycle $C(h)$ $(h \in E \setminus T)$ and its chords $e_1, \ldots, e_p \in E \setminus T$.

By Propositions 1 and 2, preliminary analysis is necessary in order to find edges $e \in E \setminus T$ for which the conditions (11) and (14) hold. We delete all these edges from the graph $G$. It is clear that this analysis can be carried out in $O(m - n + 1)$ time. Therefore, it can be assumed that

$$\sum_{t_k \in C(h)} c_{t_k} > c_h, \quad \text{for all edges} \quad h \in E \setminus T \tag{19}$$

and the condition (18) holds for any cycle $C(h)$ and its chords $e \in E \setminus T$.

The linear relaxations of NDPC1 and NDPC2 can be solved by one of the well-known simplex algorithms. Despite the fact that, theoretically, the number of pivot steps of the simplex algorithms is exponential, the following theorem states that the number of the classical simplex algorithm with Dantzig's pivot rule is bounded by a linear function of the number of edges for solving the linear program (8)–(10). It seems that to find a polynomial time simplex method and the closely related Hirsch conjecture proof are hard problems, in general.

**Theorem 1** *The simplex algorithm finds an optimal solution to the LP-relaxation (8)–(10) of NDPC2 in at most $m - n + 1$ iterations.*

*Proof* Without loss of generality, we assume $b > 0$, since if $b(t_k) = 0$ for some $t_k$, then $x^0_{t_k} = l_{t_k}$ in an optimal solution and hence the edges of the cut $\delta(V_k)$ can be deleted in $G$ and the problem (8)–(10) can be solved independently for each connected component. Since $b > 0$, we can take

$$y^0_{t_k} = b(t_k) \text{ for all } t_k \in T,$$
$$y^0_e = 0 \text{ for all } \; e \in E \setminus T$$

as the initial basic feasible solution for the problem (8)–(10). Hence, (8)–(10) has an optimal solution by the theory of linear programming. According to (19), the reduced cost is

$$\overline{c}_e = c_e - \sum_{t_k \in C(e)} c_{t_k} < 0,$$

for each $e \in E \setminus T$. Because $y_e$ can be selected as the entering variable for any $e \in E \setminus T$. From (8), it follows that the classical simplex algorithm selects $y_h$ for which

$$\overline{c}_h = \min\{\overline{c}_e; e \in E \setminus T\},$$

for entering into the basic $y^0$ and $y_e^0 = 0$ for any chord $e$ of the cycle $C(h)$. Moreover, $y_{t_q}^0 = 0$, since $y_{t_q} = 0$ is a leaving variable for some edge $t_q \in C(h)$, since columns for edges $t_q \in C(h)$ and $h$ as vectors are linearly dependent by Property 1.

Let the simplex algorithm proceed to a basic solution $y^1 = (y_e^1; e \in E)$ by choosing to bring $y_g$ into basic $\overline{y}$ and removing the variable $y_h$ from the basic $\overline{y}$ for first time. Then, the following two cases are possible for the edges $h$ and $g$ with respect to fundamental cuts.

**Case 1:** In some basis solutions generated before $y^1$, when $y_h$ became a basic variable and $y_{t_q}$ nonbasic variable, there is a chord of $C(h)$, such that $y_t > 0$ for edges $t$ in $C(e_0) = e_0 \cup T$ and $t_q \notin C(e_0)$ and the edge $e_0$ is not a basic edge, i.e., $y_{e_0} = 0$. Since $y_{t_q} = 0$, $C(e_0)$ itself is some cycle and the edge $e_0$ is entering into some basic solution and some edge $t_1 \in C(e_0) \cap T$ is leaving. This can be repeated for some chord $e_1$ of the cycle $C(e_0)$, so that $y_t > 0$ for edges $t$ in $C(e_1) = e_1 \cup T$ and $t_1 \notin C(e_1)$ and the edge $e_1$ is not a basic edge and so on. Assume that this process is carried out for edges $e_i$, where $i = 0, 1, \ldots, p$. By Proposition 1,

$$\sum_{t_k \in C(e_i)} c_{t_k} - c_{e_i} > 0,$$

for edges $e = e_0, e_1, \ldots, e_{p-1}$ and the variables for these edges are in the basic solution generated before $y^1$. When $y_h$ is a leaving basic variable from the basic solution $\overline{y}$ and $y_g$ is an entering basic variable into $\overline{y}$, the existing basic solution can be improved by entering into basic $y_g$ for the chord $g = e_p$ of $C(h)$, i.e., the inequality

$$\sum_{e_i}\{\sum_{t_k \in C(e_i)} c_{t_k} - c_{e_i}\} > \sum_{t_k \in C(h)} c_{t_k} - c_h$$

holds.

Figure 2 displays a piece of $T$ whose edges are bold lines, where $y_t^1$ for bold edges are in the basic solution $y^1$ except $t_q = (6, 7)$. When $y_h = y_{17}$ entered into the basic solution, $y_{t_q}$ became a nonbasic variable. In some solutions generated before $y^1$, for edges $e$ and $t$ on the subcycles $C(3, 6)$ and $C(3, 5)$ of the cycle $C(1, 7)$, the variables $y_e$ and $y_t$ became a basic and a nonbasic variable, respectively, in the following order:

- The $y_{36}$ became a basic variable and $t_{56}$ became nonbasic variable on the subcycle $C(3, 6)$ of $C(1, 7)$;

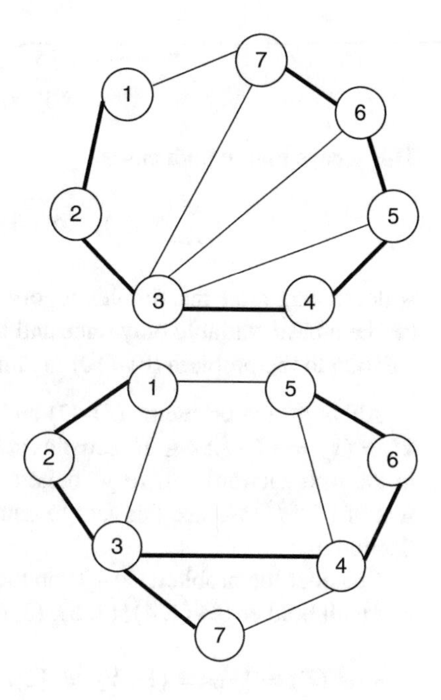


**Fig. 2** $h = (1, 7)$, $t_q = (6, 7)$, $g = (3, 7)$, $e_0 = (3, 6)$, $t_1 = (6, 5)$, $e_1 = (3, 5)$, $t_2 = (4, 5)$

**Fig. 3** $h = (1, 5)$, $t_q = (3, 4)$, $g = (4, 7)$

- The $y_{35}$ became a basic variable and $t_{45}$ became nonbasic variable on the subcycle $C(3, 5)$ of $C(1, 7)$;

After $y_{36}$ and $y_{35}$ became basic variables, $y_g = y_{37}$ is entering variable into the basic solution $y^1$ and $y_h = y_{17}$ is a leaving variable from $y^1$ with respect to the cycle with edges $g = (3, 7)$, $h = (1, 7)$, $t = (1, 2)$, $t = (2, 3)$.

**Case 2:** The edge $t_q$ is on the cycles $C(h)$ and $C(g)$ that are not subcycles of $C(h)$. As in Case 1, in some basic solutions generated before $y^1$, for some edges $t$ in $T \cap C(h)$, the variables $y_t$ and $y_e$ became a nonbasic and a basic variable, respectively, for some subcycles $C(e)$ of $C(h)$. Differently from Case 1, the cycle $C(g)$ is not a subcycle of $C(h)$ and there is a cycle $C$ in $G$ such that the edges $h$, $g$ and some chords of $C(h)$ are on $C$. Figure 3 indicates the cycle $C$ with edges $h$, $(1, 3)$, $(3, 7)$, $g$, $(4, 5)$ for which $y_e^1$ ($e = (1, 3), (4, 5), h$) and $y_t^1 = y_{37}^1$ ($t = (3, 7) \in T$) are basic variables.

Now, from that $y_g$ is an entering basic variable to $\overline{y}$ implies

$$\sum_{\delta(V_k) \ni g} u(t_k) > c_g,$$

for the simplex multipliers $\{u(t_k)\}$ as trial solution to the dual of (8)–(10) which means that the dual constraint associated with the variable $y_g$ is violated. To enter $y_g$ to the basic solution, the simplex algorithm defines new multipliers $\{u(t_k)\}$ satisfying the following equation:

$$\sum_{\delta(V_k)\ni g} u(t_k) = c_g.$$

This means that in both cases,

$$\sum \{u(t_k);\quad \text{for edges } t_k \text{ on } C(h) \text{ and } C(g)\}$$

is decreasing after the simplex algorithm defined the basic $y^1$. Therefore, each $y_h$ can be a basic variable only once and hence the simplex algorithm finds an optimal solution to the problem (8)–(10) in at most $m - n + 1$ iterations. □

Although the problems (5)–(7) and (8)–(10) have $n - 1$ constraints, the vector $x^* = (x_e^* = y_e^* + l_e;\, e \in E)$ satisfies at least $2(n-1)$ constraints in (2) (see Lemma 2 in the next section), where $y^*$ denotes an optimal solution to the fundamental cut sets of NDPC. We use this fact to determine an approximate solution to NDPC in Section 3.

Consider the problem (5)–(7) on the graph $G$ in Figure 4. Let $T$ be the spanning tree with bold edges (1, 4), (2, 5), (3, 6), (4, 5), (5, 6), for which

$$\mathscr{F}(T) = \{V_{14} = \{1\},\, V_{25} = \{2\},\, V_{36} = \{3\},\, V_{45} = \{1, 4\},\, V_{56} = \{3, 6\}\}.$$

The edges crossing with the dashed lines represent edges of the corresponding fundamental cuts whose numbers are shown in rectangles at the end of the dashed lines. Let $b(14) = 5$, $b(25) = 6$, $b(36) = 6$, $b(45) = 7$, and $b(56) = 8$ for the fundamental cut sets in $\mathscr{F}(T)$. The number next to each edge $e$ indicates its cost $c_e$. Since conditions (18) and (19) do not hold at any edge of the cycles $C(12)$, $C(13)$, $C(23)$, $C(46)$, any edge cannot be deleted in $G$. In this example, $y_{14} = 5$, $y_{25} = 6$, $y_{36} = 6$, $y_{45} = 7$, and $y_{56} = 8$ are initial basic variables that correspond to the edges in $T$. In the first iteration, the variable $y_{13}$ is an entering basic variable, $y_{14}$ is

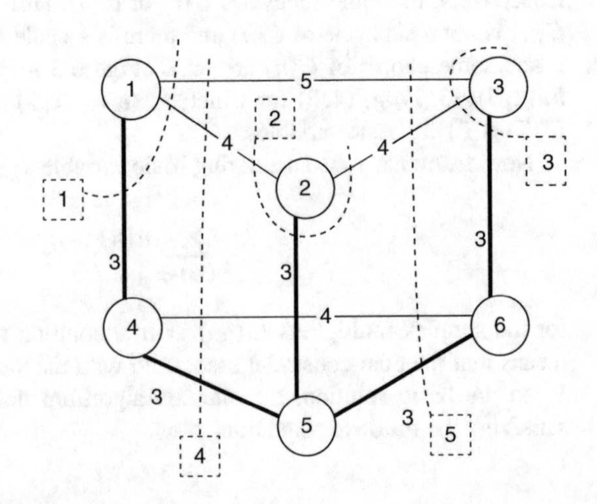


**Fig. 4** Graph $G$, spanning tree $T$, and 5 fundamental cuts

a leaving basic variable. In the next three iterations, $y_{23}$ and $y_{36}$, $y_{46}$ and $y_{56}$, $y_{12}$ and $y_{25}$ become basic and nonbasic variables, respectively. The optimal solution $y^*_{13} = 2.5$, $y^*_{32} = 3.5$, $y^*_{12} = 2.5$, $y^*_{46} = 2$, and $y^*_e = 0$ for the remainder edges.

Now, consider the problem (8)–(10) on the graph in Figure 2, but let $c_e = 1$ for each edge of the graph $G$ and $b(t_k) = 1$ for each edge of the spanning tree $T$ in Figure 2. The simplex algorithm finds the following optimal solution to the problem (8)–(10) in four iterations, where $y^*_{13} = 1/2$, $y^*_{32} = 1/2$, $y^*_{12} = 1/2$, and $y^*_e = 0$ for the remainder edges.

A related problem to NDPC2 is the following optimum communication tree problem (OCTP) [8]. Let a set $V$ of $n$ nodes and a set of requirements $r_{vw}$ for pair of distinct nodes $v$ and $w$ in $V$ be given. The OCST problem is to build a spanning tree connecting these $n$ nodes such that the total cost of the spanning tree is minimum among of all spanning trees. The cost of communication for a pair of nodes $v$ and $w$ is $r_{vw}$ multiplied by the sum of the distances $c_{ij}$ of edges on the unique path connecting $v$ and $w$ in the spanning tree. The cost of a spanning tree is the sum of communications for all pair of nodes in the spanning tree. In [8], it is noted without a proof that the OCST problem is hard to solve, in general.

Let $G = (V, E)$ be a complete graph on $n$ nodes with weight $c_{ij}$ of edges and let $R = (V, E(R))$ denote a graph of given requirements, that is, $(i, j) \in E(R)$ if $r_{ij} > 0$. Let $\delta_R(S)$ denote a cut in the graph $R$ determined by subset $\emptyset \neq S \subset V$. The OCST problem on graph $G$ can be formulated as follows to find a spanning tree $T_*$ of minimum cost:

$$c(T_*) = \sum_{e \in E} c_e x_e \to \min,$$

subject to

$$x_e = r(\delta_R(V_e)),\ V_e \in \mathscr{F}(T_*).$$

For graphs $G$ with cyclomatic number $\nu(G) = m - n + 1$ bounded by some small constant, the spanning tree $T_*$ can be found by enumerating all spanning trees in $G$.

In order to solve the OCST problem in general, one can use the following observations on the problem of OCST. The first observation is that, for a spanning tree $T_0$ of the graph $G$, as a feasible solution to the OCST problem, the vector $y$ with components $y_t = b_t$ for each $t \in T_0$ and $y_e = 0$ for all $e \in E \setminus T_0$ is the initial basic feasible solution to the problem $\varphi_1(T_0)$ with right-hand vector $b = (b_t = r(\delta_R(V_t)); t \in T_0)$, and $y$ satisfies the above constraints when $T_* = T_0$.

Let $\mathscr{T}(T_0)$ denote the set of spanning trees derived from tree $T_0$ by adding any edge $e \in E \setminus T_0$ to $T_0$ and deleting an edge $t \in T_0$ on the unique cycle in $T_0 \cup \{e\}$. The second observation is that if the basic solution $y = (y_e; e \in E)$ is an optimal solution to $\varphi_1(T_0)$, then $c(T_0) \leq c(T)$ for any $T \in \mathscr{T}(T_0)$. Note that heuristics based on Propositions 1 and 2 can be used for defining the spanning tree $T_0$.

It easy to see that the inequality $c(T) \leq \varphi_1(T)$, as cutting plane, allows to eliminate $O(n\nu(G))$ spanning trees from the set of feasible solutions. Taking into account $\nu(G) = O(n^2)$ for a complete graph $G$, the use of inequality $c(T) \leq \varphi_1(T)$ essentially reduces the number of branching iterations for solving the OCST problem on $G$.

Due to the second observation, we get a generalization of the following result. For the case when $r_{ij}$ is any nonnegative number and $c_{ij}$ =1 for all edges $(i, j)$, it was shown that the well-known Gomory and Hu algorithm constructed spanning tree $T(GH)$ of the complete graph with the capacity $r_{ij}$ of edges is an OCST [8]. Indeed, from Propositions 1 and 2, it follows that if nonnegative costs of edges satisfy (11) with respect to $T(GH)$, then the abovementioned initial basic feasible solution for $T(GH) = T_0$ is the optimal solution to the problem $\varphi_1(T(GH))$ and $c(T(GH)) \leq c(T)$ for any spanning tree $T$ in $G$, that is, $T(GH) = T_*$. So, there are a lot of nonnegative edge costs $c_h \neq 1$, that satisfy the inequality (11) for edges $h \in E \setminus T(GH)$.

Consider the problem

$$\varphi_2(T_*) = \min\{\varphi_2(T);\ T \text{ is a spanning tree in } G\}, \tag{20}$$

when $f(S) = r(\delta(S)) + b_0$ for some constant $b_0 > 0$ in NDPC. We show that below the problem (20) is $NP$-hard for arbitrary costs $c_e \geq 0$ of edges $e \in E$. So, the Gomory and Hu algorithm constructed spanning tree may not be optimum for this case of costs. Indeed, let $c_e \geq 0$ be arbitrary costs and let $l_{ij} = r_{ij}$ in NDPC. Then, $b(t_k) = b_0$ for all edges $t_k$ of any spanning tree $T$ in $G$. Let $c_{e_0} = \min\{c_e; e \in E\}$. It is easy to see that

$$c_{e_0} b_0 \leq \varphi_2(T)$$

for any spanning tree $T$ in $G$. Let $\Pi$ be the set of Hamiltonian paths connected the end nodes $v_0$ and $w_0$ of the edge $e_0 = (v_0, w_0)$.

**Lemma 1** *If $T$ is in $\Pi$, then $\varphi_2(T) = c_{e_0} b_0$, otherwise $\varphi_2(T) > c_{e_0} b_0$.*

*Proof* Let $T$ be Hamiltonian path connecting the end nodes of the edge $e_0$ ($T \in \Pi$). Hence, the cycle $C(e_0)$ contains all edges of $T$. Moreover, the vector $y = (y_{e_0} = b_0, y_e = 0,\ e \in E \setminus e_0)$ is a feasible solution to (8)–(10). Since $T \in \Pi$, any edge $e \in E \setminus T$ is a chord of $C(e_0)$. There is a feasible solution $z = (z(t_k) \geq 0, t_k \in T)$ to the dual problem (15)–(17) such that

$$\sum_{\delta(V_k) \ni e_0} z(t_k) = \sum_{t_k \in C(e_0)} c_{t_k} - c_{e_0},$$

by (18). It is easy to see that the complementary slackness conditions hold for the vectors $z$ and $y$. So, $y$ is an optimal solution to (8)–(10) with the objective value $\varphi_2(T) = c_{e_0} b_0$.

Now, let $T \notin \Pi$. By the theory of linear programming, there exists a sufficiently small number $\varepsilon > 0$ such that an optimal solution to (8)–(10) is not changed after setting $c_e = \varepsilon$ for all $c_e = 0$. So, we may assume that $c_e > 0$ for all $e \in E$. From $T \notin \Pi$, it follows that there are at least two cycles $C_1$ and $C_2$ obtained by adding some two edges of $E \setminus T$ to $T$ and there are two edges $e_1 \in C_1$ and $e_2 \in C = C_2 \setminus C_1$ for which $y_{e_1} > 0$ and $y_{e_2} > 0$. So, $e_1 \in E \setminus C$ and

$$\varphi_2(T) = \sum_{e \in E} c_e y_e = \sum_{e \in C} c_e y_e + \sum_{e \in E \setminus C} c_e y_e$$

$$\geq c_{e_2} y_{e_2} + c_{e_0} \sum_{e \in E \setminus C} y_e > c_{e_0} b_0.$$

Thus, $\varphi_2(T) > c_{e_0} b_0$ for any $T \notin \Pi$. □

By Lemma 1, $\varphi_2(T_*) = c_{e_0} b_0$ if and only if $T_*$ is a Hamiltonian path connecting the end nodes $v_0$ and $w_0$ of the edge $e_0 = (v_0, w_0)$. Therefore, the problem of finding $\varphi_2(T_*)$ is equivalent to the $NP$-complete Hamiltonian path problem. It follows that (20) is $NP$-hard in general. Similarly, it can be shown that the problem

$$\varphi_1(T_*) = \min\{\varphi_1(T);\ T \text{ is a spanning tree in } G\}$$

is $NP$-hard, too.

As a conclusion of Section 2, notice the following facts about the polytope defined by (9) and (10). Let $P(T)$ denotes this polytope and let its vertices $v_T$ and $v_{op}$ correspond to initial and an optimal basic solutions. By Theorem 1, it follows that the minimum distance between the vertices $v_T$ and $v_{op}$ is not greater than $m - n + 1$ in $P(T)$. Theorem 1 states that the Hirsch conjecture [5] holds for the facet containing vertices $v_T, v_{op}$, if $m \leq 3(n-1)$ in a graph $G$.

The above examples show that (5)–(7) and (8)–(10) have no integer-valued optimal solutions for $\mathscr{F}(T)$ to be defined with respect to some spanning tree $T$. Note that in (5)–(7) and (8)–(10) it is not required that $f$ is a submodular function. But, when $T$ is any Hamiltonian path, the 1's stay one after another in each column in the matrix $A$ from which it follows that the matrix $A$ is unimodular [12]. This means that the problem (8)–(10) has an integer-valued solution for an integer vector $b$, when $T$ is any Hamiltonian path. The problem (8)–(10) with the vector $b = 1$ is the LP-relaxation of CFP when the proper function $f(S) = 1$ ($f : 2^V \to \{0, 1\}$ [6]) for the fundamental cut sets $S \in \mathscr{F}(T)$. Hence, if the tree $T$ is any Hamiltonian path, then the LP-relaxation of this case of CFP has an integer-valued optimal solution.

## 3 Algorithms for NDPC

In this section, we give an algorithm for solution to NDPC by solving sequence of the fundamental cut sets LP-relaxations of NDPC and the separation problem $\Delta(w, x)$ for a given vector $x$ when $f$ is replaced by a modular function $w$ defined on subsets $S \subseteq V$. First, we need to prove the following useful results to solve NDPC. We say that a node subset $S$ is *minimal cut set* if the graph $G$ has two connected components after deleting the edges in $\delta(S)$, otherwise $S$ is called *non-minimal cut set*.

**Lemma 2** *If $W \subset V$ is a minimal cut set, then the constraints (2) can be deleted for $S = \overline{W}$. If $W$ is a non-minimal cut set, then the constraints (2) can be deleted for $S = W$ and $S = \overline{W}$.*

*Proof* Suppose that for some $K, L \subset V$, the graph $G$ has no edges with one end nodes in $K$ and other end nodes in $L$. Then,

$$x(\delta(F)) = x(\delta(K)) + x(\delta(L)) \geq f(K) + f(L),$$

where $F = K \cup L$. Since

$$f(K) + f(L) \geq f(K \cup L) + f(K \cap L) = f(F) + f(\emptyset) = f(F),$$

the constraints (2) can be deleted for $S = F$ and $S = \overline{F}$. So, if $G$ has no edges with one end node in $K$ and other end nodes in $L$, the constraints (2) for $S = F$ and $S = \overline{F}$ are the sum of those for $S = K$ and $S = L$.

Now, let $W \subset V$ be a minimal cut set. Since $\delta(W) = \delta(\overline{W})$ and $f(W) = f(\overline{W})$, the constraint (2) can be deleted for $S = \overline{W}$. Suppose $W \subset V$ is a non-minimal cut set. We may assume that after deleting the edges of $\delta(W)$ the graph $G$ has three components $G_1$, $G_2$, and $G_3$, since to prove the lemma one of subgraphs $G_1$, $G_2$, and $G_3$ can be disconnected. Either $W = K \cup L$, $\overline{W} = V(G_3)$ or $W = V(G_3)$, $\overline{W} = K \cup L$ for $K = V(G_1)$ and $L = V(G_2)$. Suppose that $W = K \cup L$. Since $W$ is non-minimal cut set, there is not an edge with one end node in $K$ and other end node in $L$. As shown above, the constraints (2) for $S = W$ and $S = \overline{W}$ can be deleted. □

Lemma 2 states that if a vector $x$ satisfies constraints (2) only for minimal cut sets in $\mathscr{F}(V)$, then $x$ is a solution to NDPC. It can be easily shown that an optimal solution to $\Delta(w, x)$ is a minimal cut for a modular function $w$ defined on subsets of $V$. Moreover, the following Lemma 3 also shows some advantages of using a modular function in $\Delta(w, x)$.

An optimal solution to $\Delta(w, x)$ is a minimal cut. So, we want to define appropriate modular function $w$ close to $f$ to treat the hard problem $\Delta(f, x)$. Consider the extended polymatroid

$$EP = \{w = (w_v, v \in V) \in R^V;\ w(S) \leq f(S),\ S \subseteq V\}$$

associated with the function $f$. For each linear ordering of the elements in $V$, the greedy algorithm defines different bases $w^1, \ldots, w^q$ of $EP$, that is, $w^i \in EP$ and $w^i(V) = f(V)$. Let a vector $w$ of dimension $n$ is the convex hull of the bases $w^1, \ldots, w^q$. Then, $w(S) \leq f(S)$ and $w(V) = f(V)$.

Consider the modular function $w(S)$ defined on subsets $S \subseteq V$. When the number $q$ is taken larger and larger, the function $w(S)$ becomes better approximation to the function $f(S)$. However, it makes good sense to take the number $q = O(n^k)$ for some fixed integer $k$, for example, $k = 3, 4$.

**Lemma 3** *For the modular function $w(S)$ defined on subsets of $V$, the $\Delta(w, x)$ can be found by performing at most $n$ minimum cut computations.*

*Proof* First, note that from $w(V) = f(V) = 0$, it follows $w_v > 0$ and $w_v < 0$ for some nodes in $V$. Let

$$V^+ = \{v; w_v > 0, v \in V\}, \ V^- = \{v; w_v < 0, v \in V\}.$$

We introduce a source node $s$ and a sink node $r$ in the network $G = (V, E)$. We connect the source $s$ with each node $v \in V^+$ by edge $(s, v)$ with the capacity $x_{sv} = w_v$ and the sink $r$ with each node $v \in V^-$ by edge $(v, r)$ with the capacity $x_{vr} = -w_v$. All edges $e$ in $E$ have capacity $x_e$. Let $G_{sr}$ denote this network and let $\delta_{sr}(L)$ be a cut in $G_{sr}$ such that $s \in L$ and $r \in V(G_{sr}) \setminus L$. It is clear that $G_{sr}$ does not contain edges $(s, r)$ and paths $s \to v \to r$ for any $v \in V$. Let $L = S \cup \{s\}$ for some $S \subset V$. By the construction of the network $G_{sr}$, if a node $v$ is not adjacent to $s$ or $r$, then $w_v = 0$, that is, $w_v = 0$ for each node $v \notin V^+$ or $v \notin V^-$. Hence,

$$x(\delta_{sr}(L)) = x(\delta(S)) + w(V^+ \setminus S) - w(V^- \cap S).$$

From

$$w(V^+ \setminus S) = w(V^+) - w(V^+ \cap S),$$

it follows

$$\begin{aligned} x(\delta_{sr}(L)) &= x(\delta(S)) + w(V^+) - w(V^+ \cap S) - w(V^- \cap S) \\ &= x(\delta(S)) - w(S) + w(V^+), \end{aligned}$$

where $w(V^+)$ is constant. The problem $\Delta(w, x)$ is reduced to the $s - r$ minimum cut problem

$$\min\{x(\delta_{sr}(L)); \quad L \subset V(G_{sr})\}$$

in the network $G_{sr}$, under conditions that $L \neq \{s\}$ and $L \neq \{s\} \cup V$, that is, $L = S \cup \{s\}$ for some $\emptyset \neq S \subset V$. So, it needs to consider the cases when $L = \{s\}$ or $L = \{s\} \cup V$ is a minimizer for above latter problem. If $L = \{s\}$, then we assign a huge capacity for each edge $(s, v)$ and each time define a minimum cut in $G_{sr}$. If $L = \{s\} \cup V$, then we do the same with respect to edges $(v, r)$. Thus, after solving the $s-r$ minimum cut problem at most $n$ time, we define a minimizer $S_*$ to $\Delta(w, x)$, so that $\emptyset \neq S_* \subset V$. □

To solve the separation problem for a given vector $x \in R^E$ and the constraints (2), besides the above Lemmas, we use the binary branching techniques based on the following result.

Let $a_v = f(V - v)$ for each $v \in V$ and let $\omega(S) = f(S) + a(S)$. The function $\omega(S) = f(S) + a(S)$ is a polymatroid, that is, $\omega(S)$ is submodular, increasing $(\omega(L) \leq \omega(S),$ for $L \subseteq S)$, and normalized $(\omega(\emptyset) = 0)$ [4]. We have

$$\omega(S) + a(\overline{S}) = \omega(S) - a(S) + a(V) = f(S) + a(V).$$

Hence, $x(\delta(S)) \geq f(S)$ if $x(\delta(S)) \geq \omega(S) + a(\overline{S}) - a(V)$. Since $x(\delta(\overline{S})) = x(\delta(S))$, we consider the following problem

$$\Delta(a, x) = \min\{x(\delta(\overline{S})) - a(\overline{S});\ S \subseteq V\}.$$

Similarly to the above reduction of $\Delta(w, x)$, the problem $\Delta(a, x)$ can also be computed by performing at most $n$ minimum cut computations on the network $G_s$ constructed as follows. We add only one node $s$ to $G$ as source, since $a_v \geq 0$ for each $v \in V$, that is, $V^-$ is an empty set and $a(V^- \cap S) = 0$ for each $S \subseteq V$. By the same way, it can be shown that computing $\Delta(a, x)$ is reduced to performing at most $n$ minimum cut computations on the network $G_s$. We need the following lemma on a minimizer $S_*$ of $\Delta(a, x)$.

**Lemma 4** *The following statements hold for the minimizer $S_*$. From*

$$x(\delta(S_*)) - a(S_*) \geq 0, \tag{21}$$

*it follows $x(\delta(S)) \geq f(S)$ for each nonempty $S \subset V$. If*

$$x(\delta(S_*)) - a(S_*) \geq \omega(\overline{S}_*) - a(V), \tag{22}$$

*then $x(\delta(K) \geq f(K)$ for all $K \subseteq \overline{S}_*$ and $x(\delta(L) \geq f(L)$ for all $L \supseteq S_*$. From*

$$x(\delta(S_*)) - a(S_*) \geq \omega(S_*) - a(V), \tag{23}$$

*it follows $x(\delta(S)) \geq f(S)$ for all $S \subseteq S_*$.*

*Proof* Let the inequality (21) hold for the vector $x$ and the minimizer $S_*$. Then, for any nonempty $S \subset V$, we obtain that

$$x(\delta(S)) - a(\overline{S}) = x(\delta(\overline{S})) - a(\overline{S})$$
$$\geq x(\delta(S_*)) - a(S_*) \geq 0 = \omega(V) - a(V) \geq \omega(S) - a(V),$$

since $0 = f(V) = \omega(V) - a(V)$ and $x(\delta(S)) = x(\delta(\overline{S}))$. So, $x(\delta(S)) \geq f(S)$.

Now, let the inequality (22) hold for the vector $x$ and the minimizer $S_*$. Then, for $K \subseteq \overline{S}_*$,

$$x(\delta(K) - a(\overline{K}) = x(\delta(\overline{K}) - a(\overline{K})$$
$$\geq x(\delta(S_*)) - a(S_*)) \geq \omega(\overline{S}_*) - a(V) \geq \omega(K) - a(V).$$

The last inequality follows from that $\omega(\overline{S}_*) \geq \omega(K)$ for all $K \subseteq \overline{S}_*$ ($\omega$ is increasing). So, $x(\delta(K) \geq f(K)$. Now, let $L \supseteq S_*$. Since $L = \overline{K}$ for some $K \subseteq \overline{S}_*$,

$$x(\delta(L)) = x(\delta(\overline{L})) = x(\delta(K)) \geq f(K) = f(\overline{K}) = f(L).$$

So, $x(\delta(L) \geq f(L)$ for any $L \supseteq S_*$.

By the same way, it can be shown that $x(\delta(S) \geq f(S)$ for any $S \subseteq S_*$, when the inequality (23) is true. □

We now turn to the problem (1)–(3). For each nonempty $S \subset V$ and any solution $x = (x_e; e \in E)$ of the problem (1)–(3), we have

$$f(S) - l(\delta(S)) \leq x(\delta(S)) - l(\delta(S)) = z(\delta(S)),$$

where $z = x - l$. It follows that $z$ is a solution to the following problem,

$$\min \sum_{e \in E} c_e z_e, \tag{24}$$

subject to

$$z(\delta(S)) \geq f(S) - l(\delta(S)),\ S \in \mathscr{F}(V), \tag{25}$$
$$z_e \geq 0,\ e \in E. \tag{26}$$

It is easy to see that the problem (1)–(3) can be rewritten as (24)–(26) for $z = x - l$. Consider the dual problem of (24)–(26).

$$\psi(DP) = \max \psi_D(u) = \sum_{S \subset V} (f(S) - l(\delta(S)))u_S, \tag{27}$$

subject to

$$\sum_{S: e \in \delta(S)} u_S \leq c_e,\ e \in E, \tag{28}$$

$$u_S \geq 0,\ \emptyset \neq S \subset V. \tag{29}$$

Now, let $u = (u_S; \emptyset \neq S \subset V)$ be some solution to this dual problem. Then, $\psi_D(u) + \sum_{e \in E} c_e l_e$ is a lower bound for NDPC, since

$$\psi_D(u) + \sum_{e \in E} c_e l_e \leq \psi(DP) + \sum_{e \in E} c_e l_e = \psi(LP) \leq \psi(IP)$$

by the duality theory of linear programming, for the optimal values $\psi(LP)$ and $\psi(IP)$ of the objective functions of the problem (1)–(3) and NDPC. Based on Lemma 3, a solution $u$ to (27)–(29) can be found effectively by the following greedy algorithm.

## 3.1 Defining an Initial Spanning Tree

1. Define $Cut(S(e)) = \max\{w(S) - l(\delta(S)); \quad e \in \delta(S), S \subset V\}$ and a maximizer set $S(e)$, for each edge $e \in E$.
2. Let $Cut(S(e_1)) \geq Cut(S(e_2)) \geq \cdots \geq Cut(S(e_m))$.
3. For each $h = e_1, \ldots, e_m$, do:
   - find $c^*(h) = \min\{c_e;\ h \in \delta(S(e)),\ e \in E\}$;
   - set $u_{S(h)} = c^*(h)$;
   - set $c_e := c_e - c^*(h)$ for each $e \in E$ such that $h \in \delta(S(e))$;

**Theorem 2** *The greedy algorithm that is presented above finds a feasible solution to (27)–(29) in $O(mM)$ time, where $M$ is running time of a minimum cut algorithm.*

*Proof* Let the above greedy algorithm produce a vector

$$u^* = (u^*_S;\ \emptyset \neq S \subset V,\ S \text{ is a minimal set}).$$

From the definition of $u_{S(h)}$ and $c_e$ in the **for** loop at Step 3, it follows that $u^*$ is a feasible solution to (27)–(29). $\Delta(w, x)$ is the problem of finding minimum cut $\delta(S)$ for which $\emptyset \neq S \subset V$ in the network $G_{sr}$ which is constructed in the proof of Lemma 3. This follows that, in $G_{sr}$, a minimum cut $\delta(S) \ni e = (v, w)$ is a minimum one such that either $s, v \in S$ and $w, r \in \overline{S}$ or $s, w \in S$ and $v, r \in \overline{S}$. Hence, a minimum cut $\delta(S) \ni e$ can be defined by solving two minimum cut problems on $G_{sr}$. The first problem is when the node $v$ shrinking to $s$ as the source and the node $w$ shrinking to $r$ as the sink. The second one is when the node $w$ shrinking to $s$ as the source and the node $v$ shrinking to $r$ as the sink. Thus, the above greedy algorithm runs in $O(mM)$ time. □

Now, we can define a spanning tree $T_1$ so that $\mathcal{F}(T_1)$ includes the maximum number of subsets $S$ for which $u^*_S > 0$. Let $x = y^1(T_1) + l$, where $y^1(T_1) = (y^1_e; e \in E)$ is an optimal solution to the problem (8)–(10) when $\mathcal{F}(T) = \mathcal{F}(T_1)$ and $b(t_k) = f(V_k) - l(\delta(V_k))$ for each $V_k \in \mathcal{F}(T_1)$. Since $y^1_e \geq 0$, $x$ satisfies the constraints (3). From (9), it follows that $x$ satisfies (2) also for $\overline{S}$ if $S \in \mathcal{F}(T_1)$. So, it needs to decide whether $x$ satisfies the constraints (2) for each $S \in \mathcal{F}(V)$ and $S \notin \mathcal{F}(T_1)$.

### 3.2 Algorithm

By Lemma 4, the idea of the algorithm is the following. Let a given vector $x = (x_e; e \in E)$ satisfy the constraints (3). First, it needs to define $\Delta(a, x)$, and if the minimizer $S_*$ is a single node subset ($S_* = \{v\}$), then from (22) it follows that the constraints (2) hold for any $S \subset V$. Second, if $S_*$ is not a single set and the inequality (21) does not hold, then the inequality (2) for $S = S_*$ may be violated or held. In this case, it needs to check whether (22) is true for $S_*$.

Suppose that (22) holds. Lemma 4 says that the vector $x$ is a solution of (1)–(3) if it satisfies (2) for cuts $\delta(S)$ determined by subsets $S$ such that either $S \subset S_*$ or $S$ contains some nodes in $S_*$ and in $\overline{S}_*$.

Now, suppose that (22) does not hold for $S_*$, so $x(\delta(S_*)) < f(S_*)$, i.e., the inequality (2) is violated for $S = S_*$. $S_* \notin \mathscr{F}(T_*)$, since $x(\delta(S)) = f(S)$ for all $S \in \mathscr{F}(T_*)$. Thus, after yielding the vector $x$ and the set $S_*$ by above way, the algorithm proceeds its work as follows:

1. Set $k = 1$, $x_e^k = y_e^k + l_e; e \in E$ and $S_k = S_*$.
2. If the inequality (21) holds for $S_k$ and $x^k$, the vector $x^k$ is a solution to (1)–(3).
3. If $S_k$ is a single set ($S_k = \{v\}$) and the inequality (22) holds, then the vector $x^k$ is a solution to (1)–(3).
4. If the inequality (22) does not hold for $S_k$ and $x^k$, call Cut procedure.
5. If the inequality (22) holds for $x^k$ and no single set $S_k$, call Branching procedure to check whether $x^k$ is a solution to (1)–(3). Note that the branching procedure proceeds by letting $x_e = \infty$ or $x_e = x_e^k$ for some fixed variable $x_e$.

**Cut Procedure**

- **Cut1.** Define a spanning tree $T_{k+1}$ so that $S_k \in \mathscr{F}(T_{k+1})$.
- **Cut2.** Set $\mathscr{F}(T) = \mathscr{F}(T_{k+1})$ and $b(t_j) = \max\{0, f(V_j) - x^k(\delta(V_j))$ for $V_j \in \mathscr{F}(T_{k+1})$.
- **Cut3.** Find an optimal solution $y(T_{k+1}) = (y_e^{k+1}; e \in E)$ to (8)–(10).
- **Cut4.** Set $x^{k+1} = x_e^k + y_e^{k+1}$ for $e \in E$.
- **Cut5.** Find minimizer $S_*$ of $\Delta(a, x^{k+1})$ and set $S_{k+1} = S_*$.
- **Cut6.** Set $k := k + 1$ and go to 2.

**Branching Procedure**

- **Branch1.** Set $i = 1$ and $L_1 = S_k, \overline{x}_e = x_e^k$ for all $e \in E$.
- **Branch2.** Set $\overline{x}_e = \infty$ for some edge $e$ of $\delta(L_i)$ and $EDGE[i] = e$ ($EDGE$ is some array of size $m$).
- **Branch3.** Find a minimizer $S_*$ of $\Delta(a, \overline{x})$ and set $L_{i+1} = S_*$.
- **Branch4.** If (22) does not hold for $L_{i+1}$, then call Cut procedure (in this case (2) is violated for $S = L_{i+1}$, since $x_e^k \neq \infty$ for the edges $e$ of the cut $\delta(L_{i+1})$).
- **Branch4.** If $L_{i+1}$ is not single, set $\overline{x}_e = \infty$ for some edge $e$ of $\delta(L_{i+1})$ and $EDGE[i + 1] = e$.

- **Branch5.** If $L_{i+1}$ is single, choose an edge $e = EDGE[q]$ with the minimum position $q$ among edges for which $\overline{x}_e = \infty$ and then set $\overline{x}_e = x_e^k$ (the upward step of the branching procedure in the branch and bound tree).
- **Branch6.** If $\overline{x}_e \neq \infty$ for all $e = EDGE[q]$, issue the vector $x^k$ as a solution of NDPC and stop.
- **Branch7.** Set $i := i + 1$ and go to **Branch3.**

Note that an optimal solution $y^*(T_k)$ to the problem (5)–(7) can be also used in Cut procedure. Let this algorithm define the vector $x^*$ as an approximation solution for (1)–(3). If $x^*$ is not integer solution, we can define integer solution of NDPC by the standard branch and bound algorithm. However, with respect to required accuracy $\varepsilon$ of approximate solution, the integer vector $x_I^*$ obtained by rounding up non-integer components of $x^*$ can be viewed as a solution to NDPC, if the following relative error bound

$$\frac{\psi_I(x_I^*) - \psi_D(u) - \sum_{e \in E} c_e l_e}{\psi_I(x^*)} \leq \varepsilon$$

holds for the values $\psi_I(x^*)$, $\psi_I(x_I^*)$ of (1) on the vectors $x^*$, $x_I^*$ and the value $\psi_D(u)$ of (27) on the vector $u$ defined by the above greedy algorithm.

## 4 An Example for NDPC

In this section, we describe the key technical point of the algorithm and note its flexibility by solving the NDPC on the graph $G = (V, E)$ shown in Figure 5, where the number on an edge is its cost. Let $l_e = 1$ for edges $e$ indicated by the wave lines and $l_g = 0$ for other edges $g$ in the graph $G$. We consider the function

$$f(S) = |S||V \setminus S| \text{ for all } \emptyset \neq S \subset V$$

defined on subsets $S$ of $V$. It can be shown that $f(S)$ is a symmetric submodular function.

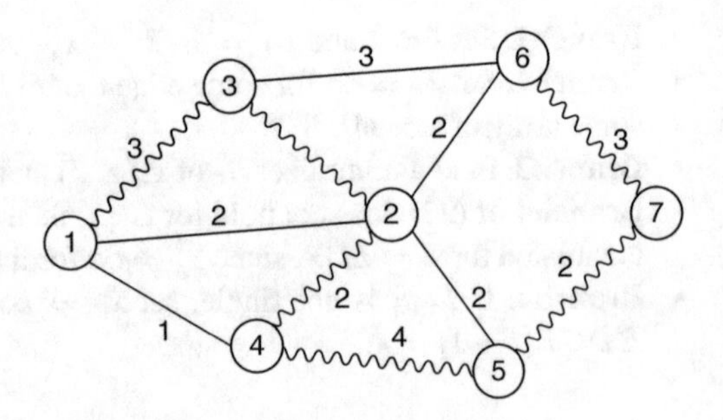

Fig. 5 Graph $G$ as example for NDPC

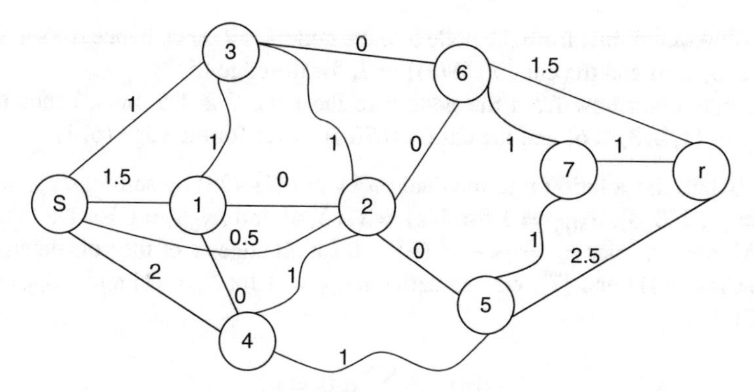


**Fig. 6** Network $G_{sr}$

- $w^1 = (6, 4, 2, 0, -2, -4, -6)$,
- $w^2 = (-6, -4, -2, 0, 2, 4, -6)$,
- $w^3 = (4, 6, 0, 2, -4, -6, -2)$,
- $w^4 = (2, -4, 4, 6, -6, 0 - 2)$.

The vector

$$w = (1.5, 0.5, 1, 2, -2.5, -1.5, -1) = 1/4w^1 + 1/4w^2 + 1/4w^3 + 1/4w^4$$

is the convex hull of the bases $w^1$, $w^2$, $w^3$, and $w^4$. Thus, $V^+ = \{1, 2, 3, 4\}$ and $V^- = \{5, 6, 7\}$ for the base $w$. Clearly, for each edge $e$,

$$Cut(S(e)) = \max\{w(S) - l(\delta(S));\ e = (v, w) \in \delta(S), S \subset V\}$$

can be defined by computing the maximum flow from the node $v$ to the node $w$. To do so, we first construct the network $G_{sr}$ shown in Figure 6, as it was shown in the proof of Lemma 3.

- The maximum flow from the node 1 to the node 2 is 2.5 units, hence $S(e) = \{1, 3, 4\}$ and $Cut(S(e)) = 2, 5$ for the edge $e = (1, 2)$.
- The maximum flow from the node 1 to the node 3 is 2.5 units, hence $S(e) = \{1, 3, 4\}$ and $Cut(S(e)) = 2, 5$ for the edge $e = (1, 3)$.
- The maximum flow from the node 1 to the node 4 is 2 units, hence $S(e) = \{1, 3, 4\}$ and $Cut(S(e)) = 2$ for the edge $e = (1, 4)$.
- For the edges $e = (2, 3)$ and $e = (2, 4)$, the sets $S(e)$ and the cuts $Cut(S(e))$ are the same as for the edges $e = (1, 2)$ and $e = (1, 4)$, respectively.
- For the edges $e = (2, 3)$ and $e = (2, 4)$, the sets $S(e)$ and the cuts $Cut(S(e))$ are the same as for the edges $e = (1, 2)$ and $e = (1, 4)$, respectively.
- For the edges $(v, w) = (3, 6), (2, 6), (2, 5), (4, 5)$, the sets $S(e) = \{1, 2, 3, 4\}$ and the cuts $Cut(S(e)) = 4$, since the maximum flows are 4 from the end node $v$ to the end node $w$ of these edges.

- The maximum flow from the node 5 to the node 7 is 2 units, hence the set $S(e) = \{1, 2, 3, 4, 5\}$ and the cut $Cut(S(e)) = 2$, for the edge (5, 7).
- The maximum flow from the node 6 to the node 7 is 1.5 units, hence the set $S(e) = \{1, 2, 3, 4, 6\}$ and the cut $Cut(S(e)) = 1.5$ for the edge (6, 7).

We obtain the solution $u$ to the dual problem (27)–(29) by setting $u_{S(e)} = 2$ for $S(e) = \{1, 2, 3, 4\}$, $u_{S(e)} = 1$ for $S(e) = \{1, 3, 4\}$ and $u_{S(e)} = 0$ for the other sets $S(e)$. Moreover, since $c_e := c_e - c^*(h) > 0$ for all edges $e$ of the cuts determining by the subsets $\{1\}$ and $\{7\}$, we can define $u_{S(e)} = 1$ for $S = \{1\}$ and $u_{S(e)} = 2$ for $S = \{7\}$. So,

$$\psi_D(u) + \sum_{e \in E} c_e l_e = 67.$$

It is easy to show that only the set $S(e) = \{1, 2, 3, 4\}$ can be separated from remainder nodes in $G$, by some fundamental cut, i.e., there is not a spanning tree whose fundamental cut sets include $S(e) = \{1, 2, 3, 4\}$ and $S(e) = \{2, 3, 4\}$. So, it needs to find a spanning tree whose cut sets contain the subsets $S(e) = \{1, 2, 3, 4\}$. There are a lot of such initial spanning trees. For example, the spanning tree $T_1$ displayed in Figure 7 can be taken as an initial one, since its fundamental cut sets include $S(e) = \{1, 2, 3, 4\}$. For $\mathscr{F}(T_1)$, the problem (8)–(10) has the optimal solution in which

$$y^1_{12} = 5,\ y^1_{25} = 6,\ y^1_{36} = 5,\ y^1_{57} = 3,$$

$y^1_e = 0$ for remainder edges $e$ and the optimal value of (8) is 44.

The algorithm defines a minimizer for $\Delta(a, x)$ on the graph $G$ in Figure 8, where the number $x^1_e = y^1_e + l_e$ on each edge is its capacity.

Since $a_v = 6$ for each node, the capacity of each edge $(s, v)$ is equal to 6. We have $S_* = \{1, 2, 3, 5, 6, 7\}(\overline{S}_* = \{4\})$ and

$$x(\delta(S_*)) - a(S_*) = 2 - 36 = -34,$$

$$\omega(\overline{S}_*) - a(V) = 12 - 42 = -30.$$

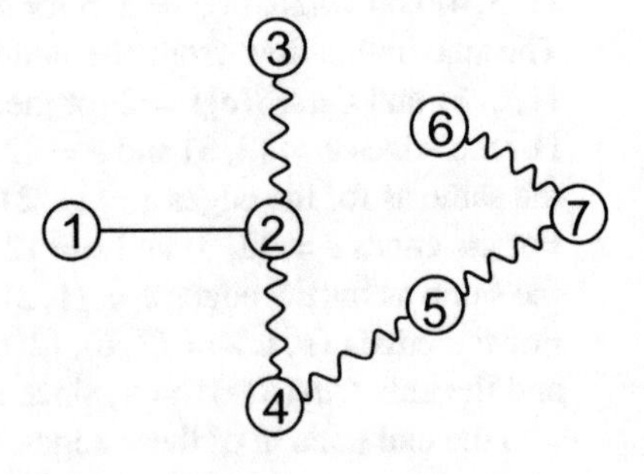


**Fig. 7** Initial spanning tree $T_1$

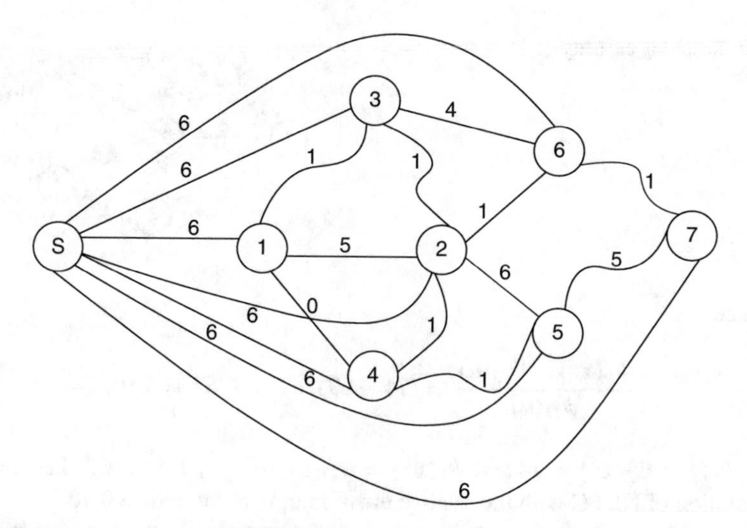


**Fig. 8** Network for finding minimum of $\Delta(a, x)$

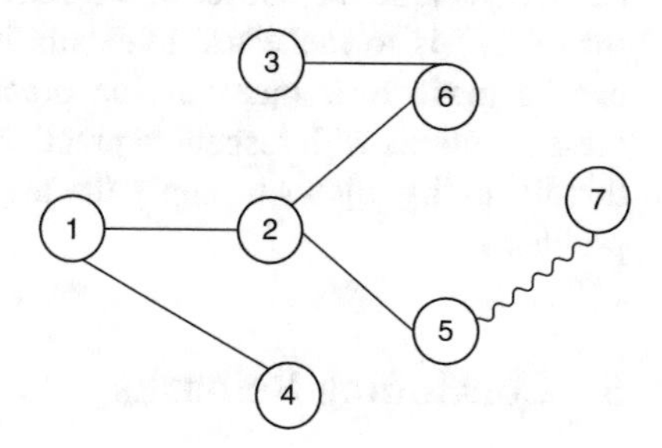


**Fig. 9** Spanning tree $T_2$

So, the constraints (2) are violated for $S = \{1, 2, 3, 5, 6, 7\}$. Cut procedure constructs the spanning tree $T_2$ in Figure 9, such that $\mathscr{F}(T_2)$ contains the subset $S = \{1, 2, 3, 5, 6, 7\}$.

For $\mathscr{F}(T_2)$, the problem (8)–(10) has the optimal solution in which

$$y^2_{14} = 2,\ y^2_{24} = 2,\ y^2_{25} = 2,\ y^2_{26} = 6,$$

$y^2_e = 0$ for remainder edges $e$ and the optimal value of (8) is 22. Again, it needs to find a minimizer to $\Delta(a, x)$ on the graph $G$ in Figure 8, after replacing the capacity $x^1_e$ with $x^2_e = y^1_e + y^2_e + l_e$ for each edge $e$ of the graph $G$. At this time, $S_* = \{1, 2, 3, 5, 6, 7\}$ or $S_* = \{1, 2, 3, 4, 5, 6\}$. So,

$$x(\delta(S_*)) - a(S_*) = 6 - 36 = -30,$$

$$\omega(\overline{S}_*) - a(V) = 12 - 42 = -30.$$

Branch procedure results that the vector $x^2$ is a solution to NDPC. The resulting capacity $x^2_e$ is indicated on each edge $e$ in Figure 10.

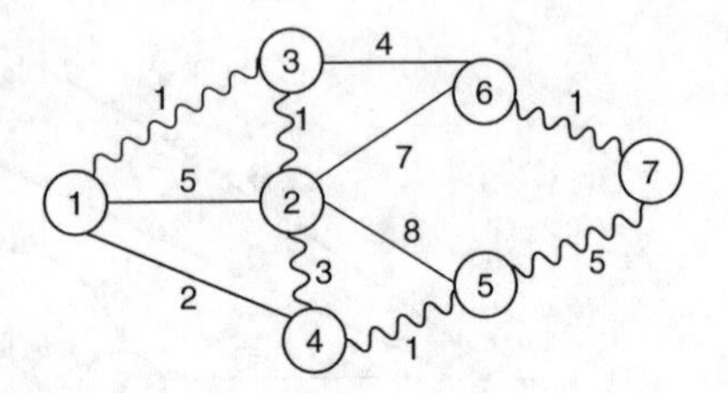


**Fig. 10** Resulting capacity of edges

Since

$$\frac{\psi_I(x^2) - \psi_D(u)}{\psi_D(u)} = (74 - 67)/67 = 7/67 = 0.10,$$

for $\psi_I(x_I^*) = \psi_I(x^2) = 74$ and $\psi_I(x^*) = \psi_D(u) + \sum_{e \in E} c_e l_e = 67$, the vector $x^2$ is a solution of NDPC with the relative error bound not more than 0.10.

Clearly, there are plenty possibles to choose a spanning tree whose fundamental cut sets include the subset $S_k$ in **Cut1**, but a way to handle the problem (8)–(10) for $\mathcal{F}(T_k)$ is to make use of flexibility in choosing spanning tree. When choice can be made over intuitions on practical situation, it is important to formalize these intuitions with respect to practice applications of problem. On the other hand, this flexibility allows to apply the algorithm for solving different network design problems.

## 5 Concluding Remarks

In this paper, we have studied a network design problem with lower bound requirements. When the lower bounds are given with symmetric submodular function, the above proposed Branch-and-Cut algorithm uses a simple linear program or a minimum cut algorithm as a main subroutine to solve NDPC. There are the following interesting questions as future works that we did not touch so far:

1. To prove that the simplex algorithm finds an optimal solution to (5)–(7) in at most $O(m)$ iterations.
2. It easy to see that the cyclomatic number $\nu(G) = m - n + 1 \leq n$ for Halin graphs [9]. Can it be used to show that the problem NDPC on Halin graphs is polynomially solvable?
3. To show that optimal values of flows on edges (solution of $\Delta(a, x^k)$) can be used as initial flow on the edges in solving $\Delta(a, x^{k+1})$ for edge capacities producing by Cut procedure, and the separation problem $\Delta(a, x^{k+1})$ may be faster by this way.
4. To emphasize a class of submodular functions which are approximated by a modular function $W$ such that $W(S) \leq f(S)$ and

$$\max\{f(S) - W(S), \emptyset \neq S \subset V\} < \varepsilon$$

for some given number $\varepsilon$. For example, this inequality holds for $f(S) = W(S) + \lambda|S||V - S|$ when $\lambda < 4\varepsilon/n^2$. Using the function $W$, $\varepsilon$-approximate solution to (1)–(3) can be found as follows: let $\widehat{z} = (\widehat{z_e}; e \in E)$ denote an optimal solution to the problem (24)–(26) when $w(S) = W(S)$. Since $\delta(S) \geq 1$ for any nonempty $S \subset V$, it is not difficult to show that $\widehat{x} = (\widehat{x_e} = \widehat{z_e} + l_e + \varepsilon; e \in E)$ is a feasible solution to the problem (1)–(3). Moreover, since $f(S) \geq W(S)$, the set of all feasible solutions of the problem (24)–(26) contains the set of feasible solutions of (1)–(3) for $w(S) = W(S)$. Hence, an optimal solution $x_*$ to (1)–(3) can be represented as $x_* = z_* + l$, where $z_*$ is a feasible solution of (24)–(26). Since $(c, z_*) \geq (c, \widehat{z})$, then

$$(c, \widehat{x}) = (c, \widehat{z}) + (c, l) + \varepsilon \sum_{e \in E} c_e$$
$$\leq (c, z_*) + (c, l) + \varepsilon \sum_{e \in E} c_e = (c, x_*) + \varepsilon \sum_{e \in E} c_e,$$

that is

$$(c, \widehat{x}) \leq \sum_{e \in E} c_e (x_e^* + \varepsilon).$$

From which it follows that

$$(c, \widehat{x}) \leq \sum_{e \in E} c_e x_e^* + \varepsilon,$$

for a given sufficiently small $\varepsilon$ such that $\varepsilon := \frac{\varepsilon}{\sum_{e \in E} c_e}$.

## References


1. M. Alicherry, R. Bhatia, Y.C. Wan, Designing networks with existing traffic to support fast restoration, in *Approximation, Randomization, and Combinatorial Optimization. Algorithms and Techniques*. Lecture Notes in Computer Science, vol. 3122 (Springer, Berlin, 2004), pp. 1–22
2. D. Bienstock, G. Muratore, Strong inequalities for capacitated survivable network design problems. Math. Program. **89**, 127–147 (2001)
3. W.H. Cunningham, Minimum cuts, modular functions, and matroid polyhedra. Networks **15**, 205–215 (1985)
4. W.H. Cunningham, On submodular function minimization. Combinatorica **5**(3), 185–192 (1985)
5. F. Fritzsche, F.B. Holt, More polytope meeting the conjectured Hirsch bound. Discret. Math. **205**, 77–84 (1999)
6. M.X. Goemans, D.P., Williamson, A general approximation technique for constrained forest problems. SIAM J. Comput. **24**, 296–317 (1995)

7. M. Grotschel, C.L. Monma, M. Stoer, Polyhedral and computational investigations for designing communication networks with high survivability requirements. Oper. Res. **43**, 1012–1024 (1995)
8. T.C. Hu, Optimum communication spanning trees. SIAM J. Comput. **3**(3) ,188–195 (1974)
9. H. Kerivin, A.R. Mahjoub, Separation of partition inequalities for (1, 2) survivable network design problem. Oper. Res. Lett. **30**(4), 265–268 (2002)
10. K.V. Marintseva, F.A. Sharifov, G.N. Yun, A problem of airport capacity definition. Aeronautic **5**, 1–13 (2013)
11. G.L. Nemhauser, L.A. Wolsey, M.L. Fisher, An analysis of the approximation for maximizing submodular set functions. Math. Program. **14**, 265–294 (1978)
12. A. Schrijver, *Combinatorial Optimization: Polyhedra and Efficiency in Algorithms and Combinatorics*, vol. 24 (Springer, Berlin, 2003)
13. F.A. Sharifov, Determination of the minimum cut using the base of an extended polymatroid. Cybern. Syst. Anal. **6**, 856–867 (1997). Translated from Cybernetics and Systems Analysis **6**, 138–152 (1996) (in Russian)
14. F.A. Sharifov, Submodular functions in synthesis of networks. Cybern. Syst. Anal. **37**(4), 603–609 (2001). Translated from Kibernetika i Systemnyi Analis **4**, 166–174 (2001) (in Russian)
15. F.A. Sharifov, *Network Design Problem When Edges of Isomorphic Subgraph Are Deleted, Proceedings* (Evry, Paris, 2003), pp. 521–525
16. F.A. Sharifov, Hulianytskyi L.F., Models and complexity of problems of design and reconstruction of telecommunication and transport systems. Cybern. Syst. Anal. **50**(5), 693–700 (2014) (in Russian)
17. H.D. Sherali, R. Sivanandan, A.G. Hobeika, A linear programming approach for synthesizing origin-destination trip tables from link traffic volumes. Transp. Res. B **28**(3), 213–233 (1994)

# Process Sequencing Problem in Distributed Manufacturing Process Planning

**Dusan Sormaz and Arkopaul Sarkar**

## 1 Introduction

In modern manufacturing, products are generated by transforming raw materials through various processes. Every product starts from design, which captures requirements as geometry, dimensions, and tolerances (called GD&T), which are normally embedded in the design itself. Traditionally, machinists select suitable machine and tools to perform specific operations to transform the raw material into physical replica of the design. With the advent of computer technologies in the manufacturing (defined as computer-integrated manufacturing; CIM), product designers use CAD applications to capture their design intent into digitally interpretable format. On the other hand, the production adopted automation at every phase of product life cycle; computer-aided manufacturing (CAM) applies low-level part programming to generate series of machine instructions and tool path. Computer-aided process planning (CAPP) automatically generates suitable production plan based on the product design, thus acts as a bridge between computer-aided design (CAD) and computer-aided manufacturing (CAM). Modern factories handle numerous product varieties and huge production volume. These factories need sophisticated CAPP systems to select the optimal process plans in terms of cost and resource consumption among many alternative plans for manufacturing a certain product design.

In past research, a number of generative CAPP systems were developed with varying degrees of capabilities; some of which produce only a sequence of feasible processes for manufacturing each feature; others consider machine and tools and

D. Sormaz · A. Sarkar (✉)
Industrial and Systems Engineering, Russ College of Engineering and Technology, Ohio University, Athens, OH, USA
e-mail: sormaz@ohio.edu; sarkara1@ohio.edu





B. Goldengorin (ed.), *Optimization Problems in Graph Theory*,
Springer Optimization and Its Applications 139,
https://doi.org/10.1007/978-3-319-94830-0_12

even specify setup instructions. More advanced process planners evaluate manufacturing cost and resource consumption to enumerate the best possible production plan. Many such past investigations are presented at the end of this chapter. Most of these investigations focus on a particular domain; therefore, constraints considered in those researches vary from one another. Consequently, there is no unified theory, which can serve as a model of process sequencing for any domain. The primary purpose of this chapter to build such a model which considers intricate nuances of process sequencing.

In general, generative process planning is performed in two phases: in the process selection phase, a set of alternative machining processes is selected for each feature of the part design; then, in the process sequencing phase, the sequence at which the features will be machined is determined. This process sequence takes into account a set of constraints imposed by both GD&T requirement and standard manufacturing practice in order to determine the optimal process plan for a given set of machine–tools. One way to represent these constraints is to generate precedence relationships among different features of part design as well as machining processes. A graph data structure is employed to capture these relationships in computer so that various network optimization algorithms can be applied to this network of interlinking alternative process routes to find the optimal process plan.

In this chapter, we will present a formal definition of process sequencing problem, introducing feasibility and validity of a process plan. We will elaborate two types of precedence constraints, strict and relaxed, in the process sequencing problem and their implication in final process plan. Then, we will describe the strategies of representing the solution space of process sequencing problem as a network of alternative process plans. We will also analyze the complexity of such network, and present process clustering method to reduce the size of network. Upon formalizing the properties of such network, we present various network optimization techniques to find the best process plans from the network of alternative process plans. Search algorithms, such as state-space search and A*, will be presented in light of process sequencing problem, along with their parameters, effectiveness, and optimal conditions. A polynomial time transformation to Generalized Traveling Salesman Problem is presented to represent the process sequence problem in the form of a well-known NP-hard problem. In the end, we will describe the test bed and experiments performed on a number of real-life part designs and discuss the efficacy of our approaches by presenting the results of such experiments.

## 2 Problem Statement

As a prerequisite of process planning, a complete and unambiguous CAD design should be available. Before process sequencing can be started, process selection procedure for each feature needs to be performed. The procedure consists of three steps (details of these steps are explained in [24, 25]: process instance selection (based on capability matrix), machine–tool specification, and time and cost estimations. The process selection phase produces sets of alternative processing routes for every feature of the subject part, along with inherent manufacturing constraints.

The goal of process sequencing phase is to find the optimal route for every feature and enumerate the order in which the features will be machined such that the product can be manufactured to its desired quality, by consuming minimum time and resource. In the next section, we formally present the process sequencing problem.

## 2.1 Alternative Machine Routing

Given a mechanical part and a set of manufacturing resources, the process sequencing problem considered here can be described as "the problem of determining the sequence of operations required for producing a part with the objective of minimizing the sum of machine, setup, and tool change costs, while satisfying the precedence constraints among operations" [12]. As we described in the previous section, the precedence among machining operation arises from the precedence among features. Next, we will define the relationship between features and machining process before presenting formal definition of precedence among features.

Let us assume a certain part design $P$ has a set of $p$ features denoted by $F$, where $F = \{F_1, F_2, F_3, \ldots, F_p\}$. Also, at a certain shop floor, there is a set of m machines available, where $M = \{M_1, M_2, M_3, \ldots, M_m\}$. As explained in the earlier subsection, each feature can be produced by alternative process planning or routing. A routing for a feature is defined as a sequence of $v$ machining operations on a subset of $M$ machines, which the feature needs to pass through to be completely manufactured. Although, in reality, many CNC cutting machines can perform different machining process in the same machine (5-axis CNC), it is assumed here that each machine is designated for only one type of machining, such as, drilling machine, milling machine, etc. However, each machine can perform different types of machining operations by using different tools and cutting speeds. A machining operation is defined by the operation performed by certain machine on a workpiece aiming to produce a particular feature without changing tool. It is also assumed that one machine cannot process two machining operations at the same time.

We can represent the relationship between features, corresponding set of routing, and the set of machining operations for individual routing by a Boolean three-dimensional matrix; rows of which represent the features, columns represent possible routing, and the depth represents the possible machining operations. In Figure 1, the three-dimensional matrix is pictorially depicted. Number of rows in this matrix is the total number of features in part $P$ ; number of columns of this matrix is the size of union of all possible routes for every feature in part $P$, i.e., and has depth of length equal to the number of available machining operations, i.e., $m$. The value of cell of the matrix of index $i$, $j$, and $k$, where $i = \{1..p\}$, $j \in$

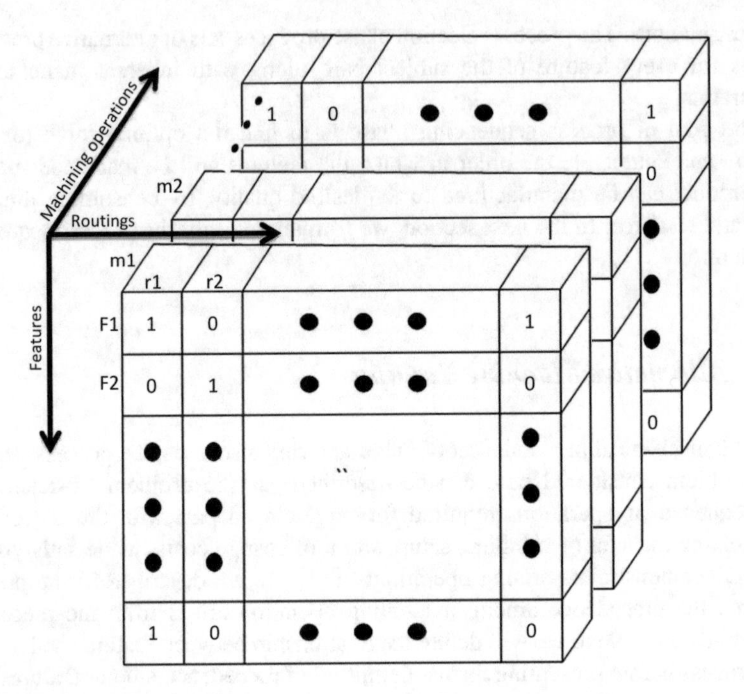


**Fig. 1** Process selection matrix

$\left|\cup_{i\in 1..|p|}\cup_{j\in 1..l_i} r_{ij}\right|$, $k \in \{1..m\}$, is 1 if machining $m_k$ is part of routing $r_j$ , $r_j$ and is a possible routing for feature $F_i$, and 0 otherwise. The $k$-th routing associated with feature $F_i$ is represented as an ordered set of machining operations, denoted by $r_{ik}$, where $r_{ik} = \{m_{ik1}, m_{ik2}, \ldots, m_{ikv_k}\}$, $i = \{1..p\}$, $k \in \{1..l_i\}$, $v_k$ is the number of machining operations needed to produce feature $F_i$ completely if $k$-th routing is selected, and $l_i$ is the number of all alternative routing possible for feature $F_i$. Therefore, $v_k = |r_{ik}|$, and $m_{ikg}$ is $g$-th machining operation to be performed on $i$-th feature if $k$-th route is followed, where $g \in \{1..v_k\}$.

A production plan for part $P$ is constructed by selecting one routing for every feature in the part. A production plan $\Pi$ is defined as a sequence of machining operations, which is the union of all machining operations belonging to the routes chosen for each feature.

$\Pi = \cup_{i\in 1..|F|} r'_i$, where $r'_i$ is the route chosen for feature $F_i$.

The process plan $\Pi$ should be valid. A valid process plan should cover every feature in set F. That means there should be at least one routing available for each feature. The validity conditions are strictly defined below.

For the sake of defining the order of machining operations in sequence $r_{ik}$, we need to define a precedence operator "$\prec$," which is used to state "process $x$ precedes process y" as "$x \prec y$." In terms of scheduling, this means that the completion time of process x should be strictly less than starting time of process $y$. It should be noted that the completion time of process $x$ takes into account the setup and part transfer time along with machining time for that particular process. However for simplicity, we will consider that every machining operation also includes the setup and transfer if necessary. Using the precedence operator, the order of machining operations in routing $r_{ik}$ can be defined as $m_{ikg} \prec m_{ikg+1}$ , $\forall g \in \{1..v_{k-1}\}$.

Due to the fact that every feature can have many different routing and the features in part P can be processed in many different orders after maintaining the precedence constraints mentioned above, it is possible to construe more than one complete production plans for part P. The optimal production plan is the one which minimizes the time and resource consumption. A generic evaluation function $f^w$ is defined which maps a sequence of machining operation to a real value. Many different cost functions, which take into account time, machining time, and setup and transfer time, may be used as an evaluation function. A suitable cost function is introduced in Section 3.2.

At this point, we can define the optimal production plan as $\Pi'$ for which $f^w(\Pi') \leq f^w(\Pi_i)$, where $\Pi_i$ is any other production plan. The order of machining operations in the production plan $\Pi$ should satisfy either of the two precedence constraints, presented below.

The subset of machining operations in the sequence $\Pi$ belonging to the same route should also maintain the same order with other machining belonging to the same parent routing, i.e., if $m_u, m_w \in \Pi$ and $m_u, m_w \in r'_i$ , then $m_u \prec m_w$ when $m_u$ precedes $m_w$ in routing $r'_i$, and $m_w \prec m_u$, if otherwise.

## 2.2 *Feature Precedence*

The features in part $P$ also have precedence constraints on their processing order, as described in the last subsection. These precedence constraints can be generalized as follows. If route $r_{ik}$ is selected for feature $F_i$ and $w$-th route $r_{jw}$ is selected for feature $F_i$ in the final production plan ($\{j \in 1 \cdots p\}$, $w \in \{1 \cdots l_i\}$), then $F_i \prec F_j$ imposes the constraint $m_{ik\cdot} \in r_{ik} \prec m_{jw\cdot} \in r_{jw}$, which means that every operation for the selected route for feature $F_i$ should be finished before any operation for feature $F_j$ can be started. This constraint is called strict precedence constraint. In reality, it may not be required to finish every machining operation of routing $r_{ik}$ before any operation of routing $r_{jw}$ can be started. In some planning strategies, relaxing the strict precedence constraints among features may provide greater flexibility in enumeration of optimum production plan. Such situations are described in Section 3.1 in more detail. The strict precedence can be relaxed by ensuring that for $F_i \prec F_j$, $\exists m_{ik\cdot} \in r_{ik} \prec \forall m_{jw\cdot} \in r_{jw}$, i.e., there is at least some operations of routing $r_{ik}$ need to be finished before any operation of routing $r_{jw}$.

In the case of strict precedence constraint, the process sequencing problem translates to selecting the best routing for each feature among sets of alternative routing available for that feature. Once the final sequence of all routing is found, every routing can be replaced by the corresponding sequence of machining operations. Therefore, three-dimensional matrix shown in Figure 1 section can be reduced to just two-dimensional matrix to represent the available routing for every feature. In this way, strict precedence constraint reduces the complexity of the process planning problem as we do not need to consider the order of the machining operations for individual routing. This simplification is also justified from practical point of view. Every possible routing selected for a feature is a set of machining operations, which are normally done on a single machine in phases, with different tools and speeds, to achieve the desired tolerance specifications. In most of the cases, it is more economical to process the entire routing on a single machine by changing tool before each machining operations in that routing than moving the workpiece repeatedly among different machines because setup and machine transfer time is usually more than tool changing time. This idea is also reflected in the cost function, presented in the Section 2.4.

Considering the above facts, we will only use strict precedence constraint in our further discussions on the application of network optimization algorithms in process sequencing problem. This implies that when any routing is chosen for a feature, every operation in that routing should be finished before next routing can be started for another feature. Following this consideration, we can aggregate the cost of entire routing by a single machining cost. Earlier in Section 2.1, we defined machining operation $m_{ikg}$ as the $g$-th machining operation to be performed on $i$-th feature if $k$-th route is followed. As we considered that the entire routing can be performed on a single machine, we can redefine machining operation as: $m_{ik} = 1$, if $F_i$ can be completely processed by machine $M_j$, and $m_{ik} = 0$, otherwise. Table 1 shows an example of a two-dimensional matrix, containing all possible assignments of machines for every feature of the sample part design shown in Figure 2. Each cell of the table can be considered as a machining operation as defined above.

**Table 1** Machine allocation matrix for part shown in Figure 2

| | Machine | | | | |
|---|---|---|---|---|---|
| Features | M1(Milling) | M2(Drilling) | M3(Milling) | M4(Milling) | M5(Milling) |
| F1(Slab) | 1 | 0 | 1 | 1 | 0 |
| F2(Hole) | 0 | 1 | 1 | 0 | 1 |
| F3(Slab) | 0 | 0 | 1 | 1 | 0 |
| F4(Slot) | 1 | 0 | 0 | 0 | 1 |

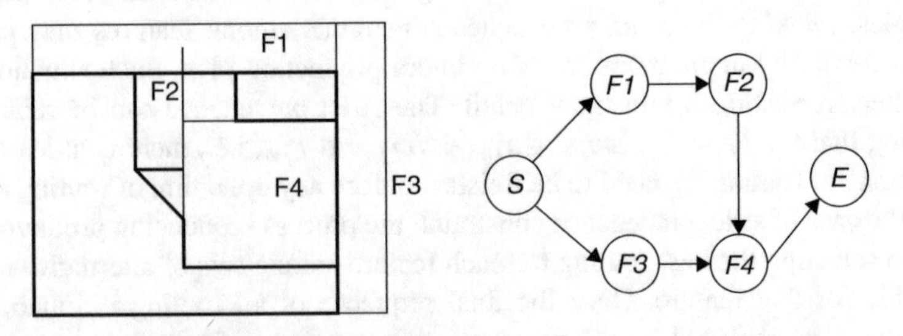


**Fig. 2** Example part design and corresponding FPN

The precedence constraints among features can be captured in the form of a graph. Such graph in which each node represents a feature and each arc represents a precedence relationship is called feature precedence network (FPN). In rigorous term, FPN is defined as a directed graph representing the precedence constraints imposed on the features of a part. We can define FPN as a graph $G =< N, A >$, where $N$ is a set of nodes, each of which corresponds to one feature in the part design. Therefore, $N = \cup_{i\in\{1\cdots p\}} F_i \cup \{S,\ E\}$, where S, E are terminal nodes and $A$ is a set of arcs, each of which corresponds to a precedence relationship between two nodes. Therefore, $A = \{a_{ij} \mid F_i \prec F_j, \forall F_i, F_j \in F\}$.

In other words, FPN is a graph in which the nodes represent the set of primitive features while the edges represent the temporal order between two nodes, with two terminal nodes S and E marking the start and end of the network traversal. It is imperative that an FPN should have at least one node between start and end ($a_{se} \notin A$), assuring that there is at least one feature to be processed and there is no self-connecting arc ($a_{ii} \notin A,\ \forall i \in \{1 \cdots p\}$), indicating that no feature can have precedence relationship with self.

An example of FPN for the part model shown in Figure 2 is represented in the same figure. The part consists of two slabs F1 and F3, a hole F2, and a slot F4. The slabs, F1 and F3, should be machined before hole F2 and slot F4, respectively, in order to reduce machining time for hole and slot and the hole F2 should be drilled before the slot F4 is milled to avoid tool deflection. This information is obtained from part model. The Feature F2 has F1, and F4 has F2 and F3 as its previous attribute and it forms the basis for the feature precedence network generation. The features that qualify to be processed first are F1 and F3. Feature nodes corresponding to F1 and F3 are connected to the start node "S." Thus, the feature node F1 is connected to F2, F2 and F3 connected to F4, and finally F4 is connected to the end node E.

Even though we will only consider strict precedence constraint in this chapter, the precedence requirement among operations is sometimes inevitable for achieving the specified tolerances for a particular design feature. As an exception, designs containing large number of same type of features on the same level of accessibility (Hole set) may benefit from finishing every machining operation with the current tool before changing it for next machining operation. Relaxed precedence constraint may be suitable in this case as the machining operations from different routing can be grouped together. We can actually represent both these precedence with the help of intermediate features by augmenting the initial feature precedence graph with intermediate features. As every machining operation in a route improves the target feature, we can consider the resultant feature, produced after each operation as intermediate feature. The intermediate features can then be inserted into the feature precedence network by maintaining either of strict and relaxed precedence constraints. In this way, the precedence in machining operation can be transferred to precedence of features.

## 2.3 Definition of Process Sequencing Problem

Furthermore, let $T = \mathbb{R}^{m \times m}$ be a machine transfer cost matrix, for which $T_{ij}$ denotes the cost of transporting the in-process workpiece from $i$-th machine to $j$-th machine, and $T_{ii} = 0$, for $\forall i \in \{1 \cdots m\}, \forall j \in \{1 \cdots m\}$.

As explained in the last section, every valid process plan also needs to maintain the precedence relationship among features. Such precedence is represented as two-dimensional matrix $F = \mathbb{Z}^{p \times p}$, for which $F_{ij} = 1$, if $F_i \prec F_j$ for $\forall i \in \{1 \cdots p\}, \forall j \in \{1 \cdots p\}$, and 0 otherwise.

Given matrices $M$, $T$, and $F$, an *optimum production plan* is defined by a valid production plan $\Pi'$ such that $f^w\left(\Pi'\right) \le f^w\left(\Pi_i\right)$, where

$$f^w(\Pi) = \Sigma_{i \in \{1 \cdots p\}} \left( M_{f^c(\Pi_i) f^m(\Pi_i)} + T_{f^m(\Pi_{i+1}) f^m(\Pi_{i+1})} \right) \quad (1)$$

where $\Pi_i$ is any other valid production plan, such that it maintains the following constraints:

1. Every feature needs to be allocated but no feature is allocated twice in the sequence.
2. If $F_{f^c(S_i) f^c(S_j)} = 1$, then $i < j$, for $\forall i, j \in \{1 \cdots p\}$ (there are exactly p allocations in the sequence due to constraint 1), where $f^c(\Pi_i)$ is defined as the feature index of the $i$-th allocation of the process sequence $\Pi$, and $f^m(\Pi_i)$ is defined as the machine index of the $i$-th allocation of the process sequence $\Pi$.

## 2.4 Cost Function

We defined the process sequencing problem in generalized form in the last section. However, in reality that the cost of a realistic production plan depends not only on just machining operations but also on workpiece setup, selection of tool, and tool direction, and most importantly transfer cost of the part from one machine to another. These additional planning dimensions increase the size of the search space, for example, every unit machining operation may choose different tools to achieve comparable results or different setup may be planned as prerequisite operation before a set of machining operations. Moreover, the total cost of a production plan depends on the order of the operations selected because good amount of transfer and tool change cost may be saved by grouping sequence of machining operation together to be processed by the same machine and same tools. Such complexities of finding the optimum production plan are discussed more in Section 5.

However, if we assume that the production plan follows strict precedence constraints and each feature is produced by a routing (sequence of machining operations) without changing machine and setup, then we may express the total cost of the production plan by equation 2, which is adopted from [11]. Notice that setup cost is applied before each routing is started and machine transfer cost is applied in between every routing. Every machining operation for a routing, dedicated to produce one feature, incurs tool change and part handling cost along with machining and tool cost.

Recollecting from the last section, let a production plan $\Pi^p$ be a valid plan for part $P$ with $p$ number of features, thus contains $p$ number of routing, where j-th routing contains $v_j$ machining operations. Then, the cost of $\Pi^p$ is given by:

$$f^w\left(\Pi^P\right) = \Sigma_{j\in\{1..p\}}\left(\frac{C_P^s}{N_P^b} + \frac{C_{j-1,j}^{tr}}{N_P^b} + \Sigma_{k\in\{1\cdots v_j\}} C_j^t\left(t_{jk}^m + t_{jk}^h + \frac{t_{jk}^m}{T_{jk}}\left(t_{jk}^t + \frac{C_{jk}^o}{C_j^t}\right)\right)\right) \quad (2)$$

where,

$C_P^s$ = setup cost for part $P$

$C_{ij}^{tr}$= machine change cost occurs in between $i$-th and $j$-th routing.

$N_p^b$ = batch size for part P

$C_j^t$ = machine operation cost for $j$-th feature (or $j$-th machine as every feature is processed by a single route dedicated to one machine).

$t_{jk}^m$ = machining time of $k$-th operation of $j$-th routing

$t_{jk}^h$= part handling time of $k$-th operation of $j$-th routing

$T_{jk}$= average lifetime of the tool used for $k$-th operation of $j$-th routing

$t_{jk}^t$ = tool change time before $k$-th operation of $j$-th routing

$C_{jk}^o$= cost of the tool used for $k$-th operation of $j$-th routing

## 3 Methodology

Any production plan should satisfy two precedence constraints, precedence in set of machining operations needed to manufacture a single feature and precedence in the set of features in the complete part design. The set of machining operations included in a complete production plan should follow these two precedence constraints. It is necessary that the part features, available machining operations, and their interactions should be translated into a usable form. The graph data structure is extremely suitable in this situation, admitting the fact that precedence relationship between two processes can be represented by a directed arc between two nodes, where each node depicts one of these two processes. In this way, every alternative process plan route can be represented in a single network, which is named as process planning network (PPN). The ultimate goal of process sequencing is to find the optimal route among the network of alternate routes; the PPN acts as a solution space for the process sequencing problem. The graph representation of the solution space enables us to apply different graph optimization and network flow algorithms in order to find the optimal process plan. The following sections describe the methodologies to generate PPN, present some network optimization algorithms, and describe various criterion of these algorithms.

### 3.1 Process Planning Network

The availability of more than one processing sequence from FPN leads to alternate process plans. A process plan may be represented using a resource-based representation, where each node refers to a part feature that is produced using the resource as specified by the node. Such a process plan node represents the machine and the feature that is processed in it.

Using this strategy, a process plan with alternatives may be generated and represented as a network. The PPN may be defined as graph $G = < N, A >$, where $N$ is a set of nodes, each of which corresponds to an ordered pair of a machine and a feature $< M_j,\ F_i >$, signifying that the feature $F_i$ is assigned to machine $M_j$. $N = \bigcup_{i \in 1 \cdots p, j \in 1 \cdots m} < M_j, F_i > \cup \{S,\ E\}$, where S, E are terminal nodes: start, and end, respectively; and $A$ is a set of directed arcs, each of which connects one feature–machine allocation to its next possible feature–machine allocation.

There may be more than one arc originating from one node or terminating at a node. When such branching is available, it signifies that alternative process plans can be constructed from PPN. The costs of subsequent machine allocation are represented on the arcs. The cost includes machining cost at the sink node and setup cost, and transportation cost that might be incurred during the machine transfer. Traversing the network from start to end node through any one of the paths generates a feasible process plan. Thus, a process plan can be defined as a path on the PPN, from start node S to end node E. The total manufacturing cost that will be incurred by use of a particular process plan can be computed by aggregating every cost on the arcs of a path. The path that takes the least total processing time is the process plan that will have least manufacturing cost; that is to say, it is the optimal process plan.

Let us consider the example part and corresponding FPN shown in Figure 2. The part design has four machinable features, including two slabs (F1, F3), a hole (F2), and a slot (F4). It is assumed that the process/machines for the respective features are selected and the FPN is validated before process planning is started. Process plan network generation starts from the node “S,” which marks the start of the planning process. This node may be considered as the raw stock available for the machining. From this node, a search is made to determine the list of features that can be machined next. Table 2 provides a list of parents and their possible children. The set of next possible features are checked against the machine allocation matrix given in Table 1 for generating all possible combinations of feature–machine allocations. Each of this combination is added as a neighbor of node “S.” In this example, it may be noted that features F1 and F3 qualify to be produced first. Nodes with F1 along with its machine candidates and F3 with its machine candidates form the children of the start node. Thus, five nodes are generated at this stage. These nodes

**Table 2** Parent–child representation of FPN shown in Figure 2

| Parent node | Children node |
|---|---|
| S | F1, F3 |
| F1 | F3, F2 |
| F2 | F3, F4 |
| F3 | F1, F2, F4 |
| F4 | E |
| E | |

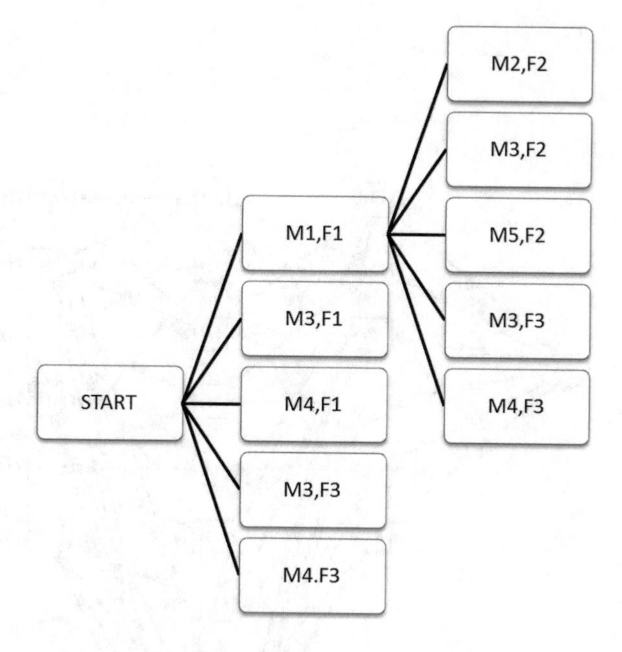


**Fig. 3** A beginning of process network for a hypothetical part shown in Figure 2

are connected to the start node "S," which is the parent node at this stage. The cost of manufacturing each of the features is represented on the arc connecting their respective process plan nodes with their parent node. The portion of the PPN is shown in Figure 3.

## 3.2 Complexity Analysis

The best known difficulty for an unconstrained assignment problem with $n$ tasks and $m$ machine is NP-hard. The number of nodes in the PPN is directly related to the FPN. In worst case, the part may have an exponential number of operation sequences. Considering that all features in set $F$ have equal or no precedence constraints (i.e., features can be machined in any order as shown in FPN in Figure 4), there are at least $F!$ different possible feature sequences. It is also shown by Sormaz et al. that in general for parallel precedence network for $p$ features with $p'$ features in each parallel branch it may be shown that the number of sequences is: $\frac{|F|!}{\Pi_{i\in\{1\cdots b\}} p_i'!}$,

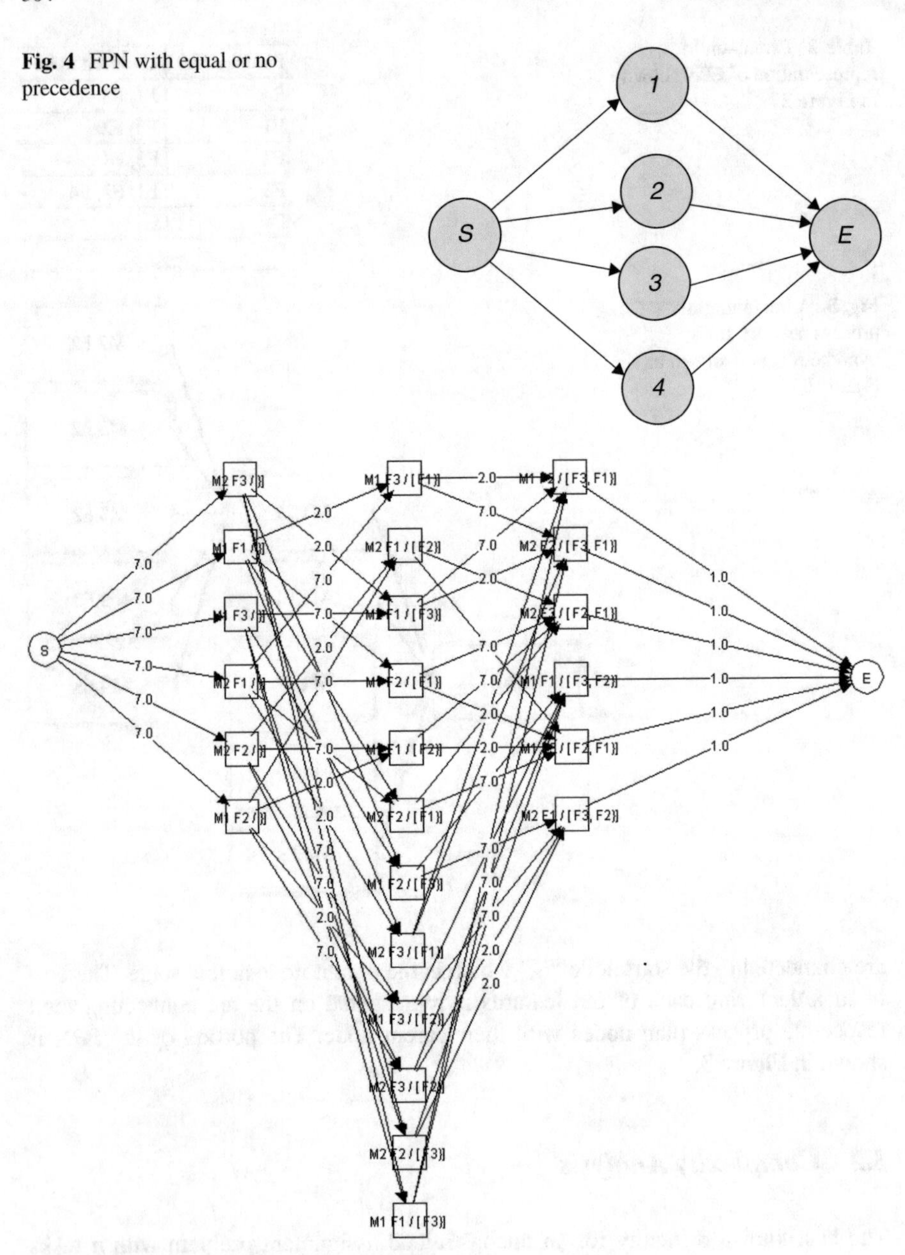


**Fig. 4** FPN with equal or no precedence

**Fig. 5** Complete PPN for a three-feature and two-machine problem

where b is the number of parallel branches in the FPN [27]. The number of process plan nodes generated for an FPN with every node having equal precedence constraints has been shown to be $2\,|F|\,(|F|-1)\,|M|$ [28]. Figure 5 shows a process plan network for three features and two machines, respectively. It can be noted that this PPN has exactly $3 * 2 * 2^2 = 24$ nodes. Clearly, the factor $2\,|F|\,(|F| - 1)\,|M|$ is exponential and thus, the number of process planning node increases exponentially with the number of features.

## 4 Process Clustering

The exponentially expanding solution space of PPN can be reduced by classifying machine–feature nodes in groups that can be performed under the same conditions. In fact, clustering features under the same machining condition has practical benefit as it is always economical to avoid repetitive machining setup, tool change, and part transfer time. These groups can be treated as a single operation in the final process plan. Therefore, we can transform the entire PPN from a network of machine–feature node to a network of process clusters. As the network size can be significantly less in PPN of process cluster than machine–feature nodes [23], process clustering strategy can be applied as a graph reduction algorithm.

Three types of conditions: tool, tool orientation, and machine are mentioned by Šormaz and Khoshnevis [23], which can be used to apply hierarchically grouping features first by the same tool directions, then each tool direction cluster by tools, and then each tool and tool direction group by allocable machines. However, we will consider process clustering only by machines in our reduced version of process sequencing problem, which assumes that every feature will be assigned to a machine and complete the feature before moving to next feature. Further in the discussion, we imply to mean “process clusters by machine” by “process cluster.” Process cluster $PC_j$ is defined below as a set of every feature that can be processed by machine $M_j$:

$$PC_j = \left\{ < M_j,\ F_i > \mid m_{ij} = 1,\ \forall i \in \{1 \cdots p\},\ \forall j \in \{1 \cdots m\} \right\} \tag{3}$$

Once the PPN is reduced by clustering the nodes by common machine, the process sequencing problem is also translated as finding the least cost path from start node to end node of the reduced PPN, traversing through a sequence of process clusters.

In order to derive such path which is a valid process plan, we define a function called $f^c$, which can take a process cluster as input and return the set of features covered by that cluster. If a sequence of $l$ process clusters are selected in a certain process plan $\Pi$, then $\Pi$ is valid if $\cup_{i \in \{1 \cdots l\}} f^c(PC_i) = F$, implying that every feature should be machined, and $f^c(PC_i) \cap f^c(PC_j) = \varnothing, \forall i, j \in \{1 \cdots l\}$, implying that the same feature cannot be part of two clusters in the same process plan.

As an example of the process clustering, let us consider a hypothetical part with seven features and its feature precedence network, shown in Figure 6. In this example, we first consider available features F1, F2, F3, and F4 for machining. For these features, we consider a possible clustering of processes for the same machine

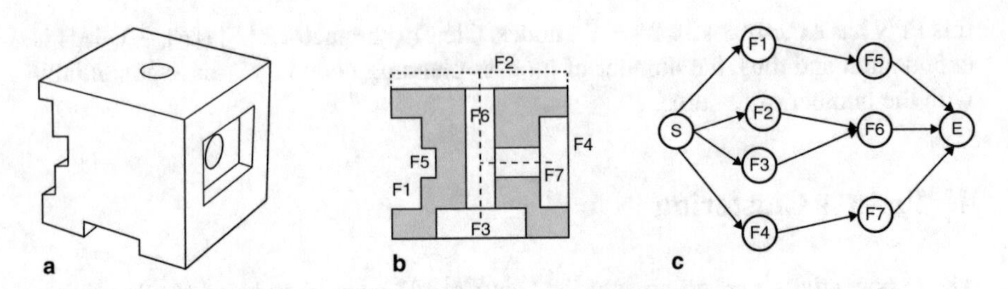


**Fig. 6** (**a**) Netex part design, (**b**) machining features, and (**c**) feature precedence network

**Table 3** Machine allocation matrix for part shown in Figure 5

| | Machines | | | | |
|---|---|---|---|---|---|
| Features | M1 | M2 | M3 | M4 | M5 |
| F1 | 1 | 1 | 1 | 0 | 0 |
| F2 | 1 | 1 | 1 | 1 | 0 |
| F3 | 0 | 1 | 1 | 0 | 0 |
| F4 | 0 | 0 | 1 | 1 | 0 |
| F5 | 0 | 1 | 0 | 1 | 0 |
| F6 | 0 | 0 | 1 | 0 | 1 |
| F7 | 0 | 1 | 0 | 0 | 1 |

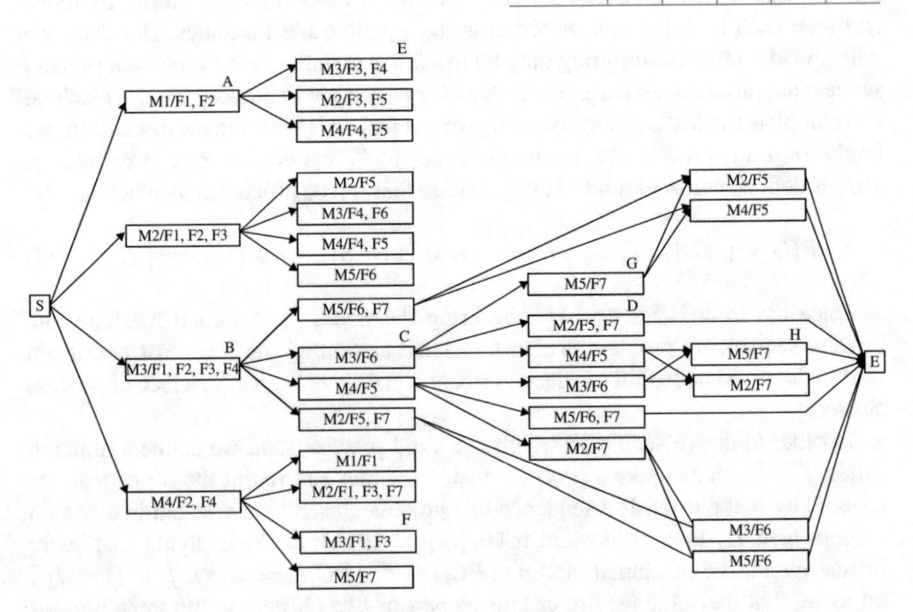


**Fig. 7** Process clustering for the PPN generated for part design shown in Figure 6 [26]

(referring to Table 3). These process clusters are: PC1 (with F1 and F2), PC2 (with F1, F2, and F3), PC3 (with F1, F2, F3, and F4), and P4 (with F2 and F4). The beginning of the PPN generated is shown in Figure 7.

# 5 Process Planning Network Optimization

Process planning network represents alternative process plans; any path from start to finish node is a valid process plan. It is shown in the previous section that a PPN expands exponentially if the network is not clustered. Therefore, a blind search needs to check an exponential number of paths in worst case scenario to find the path which incurs least cost, thus optimal. A number of network optimization algorithms can be employed to improve the search time. We start with state-space search which is an artificial intelligence technique which represents the problem as a state-space graph. Finally, we will show how the process sequencing problem can be represented as Generalized Traveling Salesman Problem.

## *5.1 Space Search*

The state space of the process sequencing problem is defined as a set of 4 elements; a state description, a function to return the set of next possible states based on the current state, a function to find the best possible next state to transition among set of next possible states, and lastly a state-transition function to change the current state of the search with the information of new state. It is imperative that the state description should represent the goal of the search. In this case, we are looking for a complete path from start to end in process plan network. Every state of the state-space search applied in this case should represent any unfinished path so far explored. As the state-space search is applied on a PPN which has already been clustered by machines, any unfinished path can be represented by a set of already sequenced process clusters. Of course, we also need some extra information in the state description which can clearly define at which state the search is currently in and help in finding the next possible states.

## *5.2 State Description*

Any state i in state-space search of PPN is defined by $s_i = (P_i,\ UF_i,\ UNF_i,\ CF_i)$, where $s_i$ is a state from the problem space, $P_i$ is the list of process clusters along a path in the process network, $UF_i$ is a set of used (already machined) features along the path, $UNF_i$ is a set of already discovered features, which are still not processed in the path so far, and $CF_i$ is a set of candidate features available to be processed after state $i$. It is better to recall that each of the process clusters is a group of all features which can be machined in a single machine. The set $UF$ represents all the features which were part of a cluster in set $P$, whereas the set $UNF$ contains the list of unused features, which could have been in one of the clusters selected and in fact they were part of some clusters the state space already explored. They are still not picked up because the cluster they were member of was not selected.

### 5.3 State-Space Operators

The state-space search starts from the stock. In this case, the starting state is $s_0 = (\varnothing, \varnothing, \varnothing, f^n(\{S\}))$, where $S$ is the start node of the FPN, and $f^n$ is a function, which can take a set of nodes belonging to an FPN as input and return the set of next candidate features ($CF_i$), therefore $f^n(F) = \{F_k \mid F_j \prec F_k, \exists k \in 1 \cdots p, \forall F_j \in F\}$.

Each feature in set $CF_i$ may be machined by more than one machine. Therefore, the set of features returned by $f^n$ function can be clustered by grouping them under the same machines. We define the function $f^{pc}$, which can take a set of features and return a set of process clusters$\{PC_{i+1}\}$, by following definition of process cluster given in section process cluster.

Having the state $s_i = (P_i, UF_i, UNF_i, CF_i)$ and new set of process clusters $\{PC_{i+1}\}$, the new state $s_{i+1} = (P_{i+1}, UF_{i+1}, UNF_{i+1}, CF_{i+1})$ is defined with the following functions:

$P_{i+1} = append\left(P_i,\ f^{best}(PC_{i+1})\right)$, where $f^{best}$ function will select the best process cluster to travel next

$$UF_{i+1} = UF_i \cup f^c\left(f^{best}(PC_{i+1})\right)$$
$$UNF_{i+1} = CF_i - f^c\left(f^{best}(PC_{i+1})\right)$$
$$CF_{i+1} = UNF_i \cup f^n\left(f^c\left(f^{best}(PC_{i+1})\right)\right)$$

By applying the state-transition procedure described above, we perform state transitions from the start state $s_0$ to all available states in the state space. As the part has a finite number of features, and the above procedure keeps on discovering more features at every step, it is guaranteed that the procedure will end after all features have been exhausted. When we finish sequencing every feature in part P, the final or goal state is reached; in other words, $CF_i$ is empty.

Every partial and complete sequence of process clusters stored in $C_i$ may be evaluated by a cost function as described in Section 3.2. The cost of each complete and valid process plan found by state-space search can be stored as an upper bound and can be used to prune a particular branch when the cost of the already sequenced clusters is more than upper bound.

### 5.4 Heuristic Search

Intuitively, as the part has a finite number of features, and the above procedure at each step adds more features to the state, it is guaranteed that the procedure will end after all features have been machined in the final or goal state. However, it largely depends on the find-best function, which decides the next best process cluster to visit. Strategies like breadth-first search (BFS) and depth-first search (DFS) exhaustively search the entire network, therefore not applicable in real-life situations, where a part with hundreds or more features will generate a large network and the search will take long time to finish. Best-first search uses a heuristic function to evaluate the promises of every node to reach the goal through an optimal path and then chooses the node with the best promise as a next node to travel. This strategy greatly reduces the search space. Still, certain quality of the heuristic needs to be assured before it can be claimed that best-first search can find an optimal path by using that heuristic.

We will use an evaluation function of form $f(n) = g(n) + h(n)$, where $n$ is any state encountered in the search, $g(n)$ is the total cost manufacturing features covered from start to node $n$, and $h(n)$ is the heuristic estimate of the cost of manufacturing rest of the features. This summation eliminates ties when two candidate nodes have the same heuristic estimates [15]. If the heuristic function h(n) is always less than or equal to the actual cost ($h^*(n)$) for manufacturing rest of the features (in some path from node n to goal), i.e., $h(n) \leq h^*(n)$, then the function $h(n)$ is called the lower bound and the BFS using evaluation function f(n) is called admissible. On the other hand, any search algorithm always terminates in the optimal solution path if such path exists. A*, which is an admissible search algorithm, is used in finding the optimal process plan from the network of alternative process plans.

For computation of the heuristic function, which will estimate cost from the current state to the goal state, it is necessary to consider the cost equations and identify which components of the total cost can be estimated. As we have all the candidate processes generated for each feature, the simplest heuristic function for process sequencing may be defined as a sum of minimal possible cost over a set of candidate processes for unfinished features for a given state. This function may be represented as $h'(s_i) = \Sigma_{j \in F \setminus UF_i} \min_{k \in M | m_{jk}=1} C_{jk}$, where $C_{jk}$ is the cost of machining feature $j$ on machine $k$.

The above heuristic indicates that for a given state we find all remaining features and for each of the features, find a machine with the minimal machining cost. The sum of these costs for remaining features is the minimum cost that has to be added to the current state in order to machine the part completely. It can be observed that this minimum cost can never be more than the actual cost for the reason that any other allocation than the minimum cost machine will cost at least as much as the minimum cost machine. Therefore, we can conclude that $h'(n)$ is admissible. Admissibility of the heuristic is a proof of optimality if tree search is used instead of graph search. However, it is also required to prove that $h'(n)$ is also consistent. It can be easily shown that the function $h'(n)$ is monotonic in nature, i.e., $h'(n_i) - h'(n_j) \leq g(n_j) - g(n_i)$, where $n_j$ is one of the children nodes of $n_i$. The triangular inequality property of monotonicity always guarantees that $h'(n)$ is also consistent [19].

More informed heuristic bound can be found in order to reduce the search space. The purpose of finding such a heuristic function is motivated by the fact that, if the estimation function is tighter or, in other words, closer to the real cost function, the smaller space will be searched and the solution will be found faster. However, such a function may become more complicated to compute, resulting in a slower search process. A polynomial-time heuristic, heuristic similar to the one presented above, is normally suitable.

### 5.5 Transformation to Generalized Traveling Salesman Problem

The process sequencing problem so far discussed can also be represented as a generalized version of Traveling Salesman Problem (TSP), where unlike TSP the tour does not require to be Hamiltonian. We can derive two benefits from such transformation. First, we can prove that process sequencing problem is as hard as a TSP, an NP-complete problem, following several reductions from the Generalized Traveling Salesman Problem (GTSP) to TSP [14, 17]. Second, the transformation from an instance of process sequencing problem to an instance of GTSP can be computed in polynomial time including as many planning variable required. We will present the transformation scheme by considering both machining operation cost and transportation cost for the process sequencing problem.

Given a set $V$ of $n$ vertices, weights $w\ (x \rightarrow y)$ of moving from $x \in V$ to $y \in V$ and a partition $V$ into $m$ nonempty clusters $C_1, C_2, \cdots, C_m$ such that $C_i \cap C_j = \varnothing$ for each $i \neq j$ and $\cup_i C_i = V$, the objective of GTSP is to find a feasible tour $t$ which is shortest in terms of the total length among all possible feasible tours, where a feasible tour is defined as a cycle visiting exactly one vertex from every cluster. More specifically, this particular version of GTSP is called GTSP with nonoverlapping clusters, because no vertex is allowed to be the member of two clusters. In our formulation of process sequencing problem, every machining operation is a unique allocation of feature to the machine. A feasible process plan should have exactly one such operation for every feature. We will consider machine transfer cost along with machine allocation cost per feature in order to demonstrate that such transformation can account for more than one planning dimension. We defined such version of process sequencing problem below. We also transformed the feature precedence constraint to precedence constraint applicable to the GTSP instance, making it an ordered version of GTSP (GTSP-ORD).

Let $M = \mathbb{R}^{p \times m}$ be a feature–machine allocation cost matrix for $p$ features and $m$ machines, where $M_{ij} > 0$, if $i$-th feature can be processed by j-th machine and $M_{ij} = 0$ if $i$-th feature cannot be processed by $j$-th machine for $\forall i \in \{1 \cdots p\}, \forall j \in \{1 \cdots m\}$. This matrix is similar to the machine allocation matrix presented before for sample parts in Tables 1 and 3. However, if feature $i$ is allowed to be allocated to machine $j$, then $M_{ij}$ contains the cost of such assignment. This cost may be calculated aggregating machine operation, tool change, handling, and operator costs. Also, let $T = \mathbb{R}^{m \times m}$ be a machine transfer cost matrix, for which $T_{ij}$ denotes the cost of transporting the workpiece under making from $i$-th machine to $j$-th machine, and $T_{ii} = 0$, for $\forall i \in \{1 \cdots m\}, \forall j \in \{1 \cdots m\}$. Furthermore, let $F = \mathbb{Z}^{p \times p}$ be a feature precedence matrix, for which $F_{ij} = 1$, if $i$-th feature should be processed before $j$-th feature for $\forall i \in \{1 \cdots p\}, \forall j \in \{1 \cdots p\}$. The matrix $T$ can be trivially created from the FPN as $F_{ij} = 1$ when $a_{ij} \in A$ and 0, otherwise.

The process sequence we want to find from the given matrices $M$, $T$, and $F$ will not impose any process clustering as constraint. As machine transfer cost is considered in the problem, it is envisaged that the optimum process sequence $\Pi$ should allocate features on the same machine as consecutively as possible, as machine transfer cost is 0 if the same machine is used for two consecutive operations. However, an optimum process sequence should satisfy the following constraints:

1. Every feature needs to be allocated but no feature is allocated twice in the sequence.
2. If $F_{f^c(S_i)f^c(S_j)} = 1$, then $i < j$, for $\forall i, j \in \{1 \cdots p\}$ (there are exactly p allocations in the sequence due to constraint 1).
3. The total cost of the sequence $cost\,(\Pi) = \Sigma_{i\in\{1\cdots p\}} \left(M_{f^c(\Pi_i)f^m(\Pi_i)} + T_{f^m(\Pi_{i+1})f^m(\Pi_{i+1})}\right)$ is minimum.

$f^c\,(\Pi_i)$ is defined as the feature index of the $i$-th allocation of the process sequence $\Pi$, and $f^m\,(\Pi_i)$ is defined as the machine index of the $i$-th allocation of the process sequence $\Pi$.

Given a process sequencing instance $< M,\ T,\ F >$, we can transform such instance into an instance of GTSP-ORD in the following way:

- Enumerate set $A$ consisting every possible feature allocation and one terminal allocation ($\tau$), which is defined as the beginning and end position of the sequence with no feature allocated to no machine; the machine allocation cost of the terminal allocation is thus 0, and machine transfer cost between terminal allocation and any other allocation is 0. Therefore, the number of nodes in the target GTSP-ORD problem is equal to $|A| = |\{\tau\} \cup \{< i,\ j >|\ M_{ij} > 0,\ \forall i \in \{1 \cdots p\},\ \forall j \in \{1 \cdots m\}\}\,|$.
- The distance from each node to the other node of the GTSP-ORD problem instance is calculated by the following equation and stored in matrix $P = \mathfrak{R}^{|A|\times|A|}$ (row 0 and column 0 is marked as terminal allocation) as follows:

$$P_{i\in A, j\in A} = \begin{cases} M_{f^c(A_j)f^m(A_j)}, & \text{when } i = 0,\ j > 0 \\ M_{f^c(A_j)f^m(A_j)} + T_{f^m(A_i)f^m(A_j)}, & \text{when } i > 0,\ j > 0 \\ 0, & \text{otherwise} \end{cases} \quad (4)$$

- The clusters of nodes are stored in matrix $C = Z^{|A|\times|A|}$, where $C_{ij} = 1$, if allocations $A_i$ and $A_j$ are in the same cluster and $C_{ij} = 0$, if allocations $A_i$ and $A_j$ are in different clusters. Matrix $C$ can be constructed as follows:

$$C_{i\in A, j\in A} = \begin{cases} 1, & \text{when } f^c(A_i) = f^c(A_j), i > 0, j > 0 \\ 0, & \text{otherwise} \end{cases} \quad (5)$$

- The precedence among nodes are stored in Matrix $K = Z^{|A| \times |A|}$ as follows:

$$K_{i \in A, j \in A} = \begin{cases} 1, & \text{when } i = 0, j > 0, F_{tf^c(A_j)} = 1, \forall t \in 1 \cdots p \\ 1, & \text{when } i > 0, j > 0, F_{f^c(A_i) f^c(A_j)} = 1 \\ 0, & \text{otherwise} \end{cases} \tag{6}$$

Given an instance of process sequencing problem, $I_{PSeq} = (M, T, F)$, a corresponding instance of $I_{GTSP-ORD} = (P, C, K)$ can be formed by the four steps mentioned above. The minimum distance tour to be found in $I_{GTSP-ORD}$ is of length $p + 1$ and should start from node $\tau$.

*Claim* The optimum solution of $I_{GTSP-ORD}$ is equal to the optimum solution for $I_{Pseq}$, where $I_{PSeq} \leq_P I_{GTSP-ORD}$ with help of the procedure described above.

*Proof* In order to prove that the optimal solution of $I_{GTSP-ORD}$ is equal to the optimal solution for $I_{PSeq}$, it is enough to show that every solution found in $I_{GTSP-ORD}$ has a corresponding feasible solution for $I_{PSeq}$. This is because if any such tour is minimum of every other tour found in $I_{GTSP-ORD}$, then the corresponding process sequence is also the minimum of every other feasible process sequence. A process sequence is feasible only when no feature is allocated to two different machines and maintains the precedence constraint.

The allocations in any feasible tour $t$ selected for $I_{GTSP-ORD}$ cannot have any feature allocated to two different machines. This is true because any feasible tour for $I_{GTSP-ORD}$ visits exactly one node from each cluster and following Equation (5), every cluster is composed of allocations of only one feature. Therefore, $t$ is also a process sequence in which every feature is allocated to only one machine.

It follows from Equation (6) that if $i < j$, the feature of i-th allocation selected in tour $t$ either precedes feature of j-th allocation or have equal precedence, considering that a feasible tour of GTSP-ORD will maintain the precedence stored in $K$ matrix. Therefore, the corresponding process sequence of $t$ follows the feature precedence constraint defined in $F$.

We present an example of the aforementioned transformation. For this example, the sample part shown in Figure 2 is used. Table 4 contains the machine allocation costs for every feature on five different possible machines. Notice that not all machines are available for every feature. Table 5 contains the machine transfer costs for five machines and we will follow the precedence constraint from the FPN shown in Figure 2.

**Table 4** Machine allocation cost matrix ($M$) for four features and five machines

| | M1 | M2 | M3 | M4 | M5 |
|---|---|---|---|---|---|
| F1 | 1 | 0 | 3 | 15 | 0 |
| F2 | 0 | 5 | 1 | 0 | 1 |
| F3 | 0 | 0 | 2 | 13 | 0 |
| F4 | 10 | 0 | 0 | 0 | 5 |

**Table 5** Machine transfer cost matrix ($T$) for five machines

| | M1 | M2 | M3 | M4 | M5 |
|---|---|---|---|---|---|
| M1 | 0 | 4 | 4 | 9 | 2 |
| M2 | 7 | 0 | 3 | 6 | 6 |
| M3 | 2 | 7 | 0 | 5 | 8 |
| M4 | 5 | 5 | 4 | 0 | 8 |
| M5 | 1 | 1 | 3 | 3 | 0 |

**Table 6** Matrix $P$ storing the distance from one allocation to another

| Allocations (A) | | Af | | F1 | F1 | F1 | F2 | F2 | F2 | F3 | F3 | F4 | F4 |
|---|---|---|---|---|---|---|---|---|---|---|---|---|---|
| | | Am | $\tau$ | M1 | M3 | M4 | M2 | M3 | M5 | M3 | M4 | M1 | M5 |
| Af | Am | Idx | 0 | 1 | 2 | 3 | 4 | 5 | 6 | 7 | 8 | 9 | 10 |
| $\tau$ | 0 | 0 | 1 | 3 | 15 | 5 | 1 | 1 | 2 | 13 | 10 | 5 | |
| F1 | M1 | 1 | 0 | 0 | 0 | 0 | 9 | 5 | 3 | 6 | 22 | 10 | 7 |
| F1 | M3 | 2 | 0 | 0 | 0 | 0 | 12 | 1 | 9 | 2 | 18 | 12 | 13 |
| F1 | M4 | 3 | 0 | 0 | 0 | 0 | 10 | 5 | 9 | 6 | 13 | 15 | 13 |
| F2 | M2 | 4 | 0 | 8 | 6 | 21 | 0 | 0 | 0 | 5 | 19 | 17 | 11 |
| F2 | M3 | 5 | 0 | 3 | 3 | 20 | 0 | 0 | 0 | 2 | 18 | 12 | 13 |
| F2 | M5 | 6 | 0 | 2 | 6 | 18 | 0 | 0 | 0 | 5 | 16 | 11 | 5 |
| F3 | M3 | 7 | 0 | 3 | 3 | 20 | 12 | 1 | 9 | 0 | 0 | 12 | 13 |
| F3 | M4 | 8 | 0 | 6 | 7 | 15 | 10 | 5 | 9 | 0 | 0 | 15 | 13 |
| F4 | M1 | 9 | 0 | 1 | 7 | 24 | 9 | 5 | 3 | 6 | 22 | 0 | 0 |
| F4 | M5 | 10 | 0 | 2 | 6 | 18 | 6 | 4 | 1 | 5 | 16 | 0 | 0 |

**Table 7** Matrix $C$ storing the clusters of allocations

| Allocations (A) | | Af | | F1 | F1 | F1 | F2 | F2 | F2 | F3 | F3 | F4 | F4 |
|---|---|---|---|---|---|---|---|---|---|---|---|---|---|
| | | Am | $\tau$ | M1 | M3 | M4 | M2 | M3 | M5 | M3 | M4 | M1 | M5 |
| Af | Am | Idx | 0 | 1 | 2 | 3 | 4 | 5 | 6 | 7 | 8 | 9 | 10 |
| $\tau$ | | 0 | 1 | 0 | 0 | 0 | 0 | 0 | 0 | 0 | 0 | 0 | 0 |
| F1 | M1 | 1 | 0 | 1 | 1 | 1 | 0 | 0 | 0 | 0 | 0 | 0 | 0 |
| F1 | M3 | 2 | 0 | 1 | 1 | 1 | 0 | 0 | 0 | 0 | 0 | 0 | 0 |
| F1 | M4 | 3 | 0 | 1 | 1 | 1 | 0 | 0 | 0 | 0 | 0 | 0 | 0 |
| F2 | M2 | 4 | 0 | 0 | 0 | 0 | 1 | 1 | 1 | 0 | 0 | 0 | 0 |
| F2 | M3 | 5 | 0 | 0 | 0 | 0 | 1 | 1 | 1 | 0 | 0 | 0 | 0 |
| F2 | M5 | 6 | 0 | 0 | 0 | 0 | 1 | 1 | 1 | 0 | 0 | 0 | 0 |
| F3 | M3 | 7 | 0 | 0 | 0 | 0 | 0 | 0 | 0 | 1 | 1 | 0 | 0 |
| F3 | M4 | 8 | 0 | 0 | 0 | 0 | 0 | 0 | 0 | 1 | 1 | 0 | 0 |
| F4 | M1 | 9 | 0 | 0 | 0 | 0 | 0 | 0 | 0 | 0 | 0 | 1 | 1 |
| F4 | M5 | 10 | 0 | 0 | 0 | 0 | 0 | 0 | 0 | 0 | 0 | 1 | 1 |

After applying the transformation rules given in this section, we derive the following three matrices $P$, $C$, and $K$, which are given in Tables 6, 7, and 8, respectively.

**Table 8** Matrix K storing the precedence relationship among allocations

| Allocations (A) | | Af | | F1 | F1 | F1 | F2 | F2 | F2 | F3 | F3 | F4 | F4 |
|---|---|---|---|---|---|---|---|---|---|---|---|---|---|
| | | Am | $\tau$ | M1 | M3 | M4 | M2 | M3 | M5 | M3 | M4 | M1 | M5 |
| Af | Am | Idx | 0 | 1 | 2 | 3 | 4 | 5 | 6 | 7 | 8 | 9 | 10 |
| $\tau$ | | 0 | 0 | 1 | 1 | 1 | 0 | 0 | 0 | 1 | 1 | 0 | 0 |
| F1 | M1 | 1 | 0 | 0 | 0 | 0 | 1 | 1 | 1 | 0 | 0 | 0 | 0 |
| F1 | M3 | 2 | 0 | 0 | 0 | 0 | 1 | 1 | 1 | 0 | 0 | 0 | 0 |
| F1 | M4 | 3 | 0 | 0 | 0 | 0 | 1 | 1 | 1 | 0 | 0 | 0 | 0 |
| F2 | M2 | 4 | 0 | 0 | 0 | 0 | 0 | 0 | 0 | 0 | 0 | 1 | 1 |
| F2 | M3 | 5 | 0 | 0 | 0 | 0 | 0 | 0 | 0 | 0 | 0 | 1 | 1 |
| F2 | M5 | 6 | 0 | 0 | 0 | 0 | 0 | 0 | 0 | 0 | 0 | 1 | 1 |
| F3 | M3 | 7 | 0 | 0 | 0 | 0 | 0 | 0 | 0 | 0 | 0 | 1 | 1 |
| F3 | M4 | 8 | 0 | 0 | 0 | 0 | 0 | 0 | 0 | 0 | 0 | 1 | 1 |
| F4 | M1 | 9 | 1 | 0 | 0 | 0 | 0 | 0 | 0 | 0 | 0 | 0 | 0 |
| F4 | M5 | 10 | 1 | 0 | 0 | 0 | 0 | 0 | 0 | 0 | 0 | 0 | 0 |

This transformation may enable various combinatorial and graph optimization techniques suitable for GTSP and TSP to be applied in solving process sequencing problem. However, research on GTSP-ORD, as defined here is sparse in the literature. Still, many past investigations on ordered TSP may be extended for solving GTSP-ORD. Even the space search algorithms, presented in this section, can be modified to be suitable for solving $I_{GTSP-ORD}$. Such treatments are beyond the limit of this chapter.

For the example instance of process sequencing problem shown in Figure 2, the ACS-PSEQ algorithm finds minimum cost sequence of allocation as $<F_3, M_3>$, $<F_1, M_1>$, $<F_2, M_5>$, $<F_4, M_5>$ with total cost 13. This is the optimum sequence for this particular instance.

# 6 Experimentation and Results

In this section, we will present some quantitative analysis of process sequencing methods discussed in Section 5. At first, we measured the size of the process planning network as size of the search tree produced by a depth-first search algorithm similar to space search algorithm. The only difference is that we did not use the process-cluster function while generating the next possible states; instead, every available feature machine combination is added in the candidate set $PC_{i,i+1}$. The first goal of our experiment is to show that PPN generated without any process clustering and no or equal precedence among candidate features will generate bigger network than in the cases where process clustering or unequal precedence among features is present. Therefore, the size of the network and the performance of space search can be measured in worst case situation. We generated different machine allocation matrices for a five-feature part and five-machine configuration by varying a parameter called machine allocation percentage (*mp*), which defines the percentage of total number of machines in a shop floor available to each feature. The space search algorithm is written in Java and executed on a computer using Intel Core i7-3630QM processor with 2.4 GHz clock speed and 8 GB physical memory.

Figure 8 shows the plot of PPN network size for different machine allocation percentage. It can be observed that when number of machines available to features increases, the search tree increases in size and thus computation time increases too (see Figure 9). The larger standard error of mean for higher machine allocation matrix is due to the fact that the variation of random assignment piles up when more number of machines are allocated.

We observed in the previous experiment that the size of space search tree increases exponentially. However, not all features are available for processing at the same time due to the presence of precedence among features in a real-life product design. Next, we will evaluate the effect of FPN on space search tree by using two sample parts: (1) Netex part shown in Figure 6, and (2) sample part shown in Figure 10, in Table 9.

It can be observed from Table 9 that inducing precedence among features cuts the size of the space search tree, therefore generates a smaller PPN. This network can further be minimized by employing process clustering described in Section 4. Next experiments show the results of applying our process planning procedure and

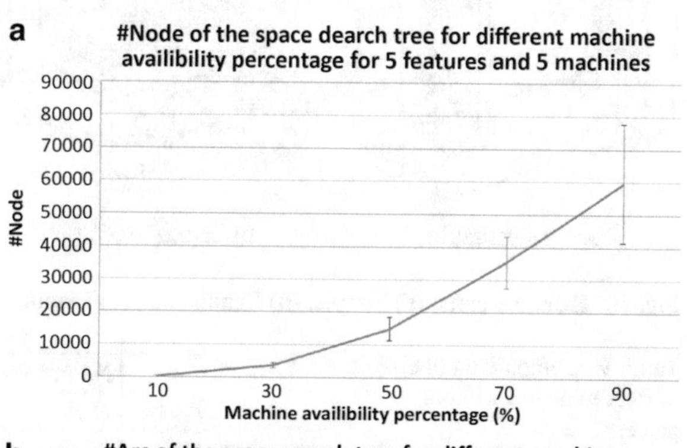


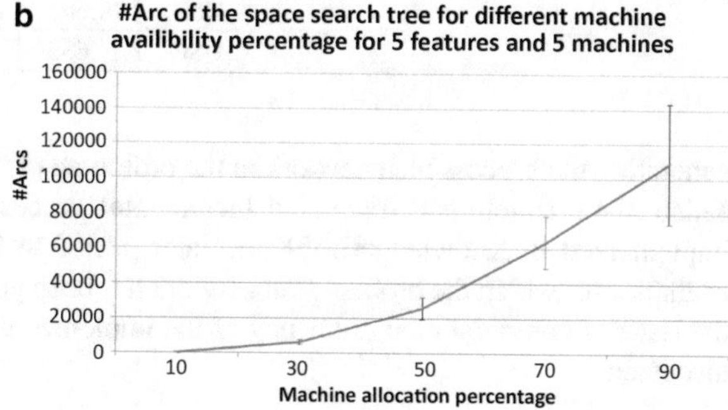


**Fig. 8** Space search tree size in (**a**) number of nodes, (**b**) number of arcs, for different machine allocation percentages in a five-features and five-machines configuration

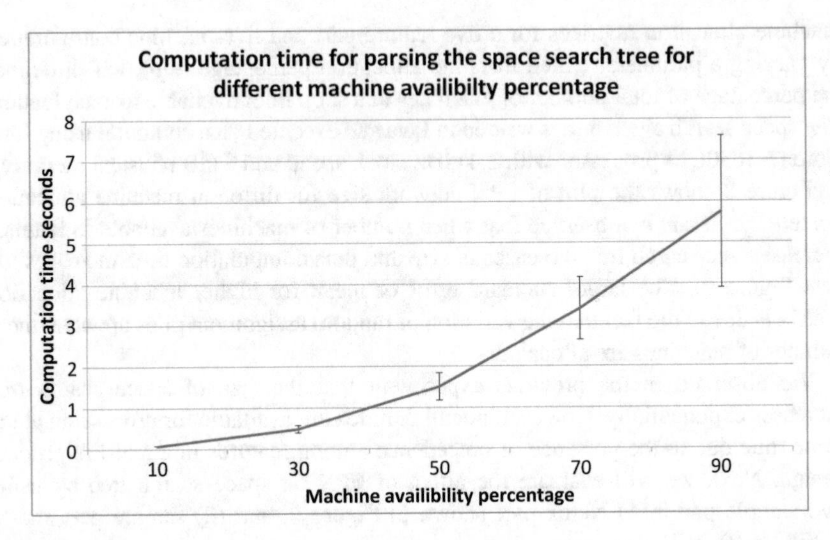


**Fig. 9** Computation time for generating PPNs for different machine allocation percentages in a five features and five machines configuration

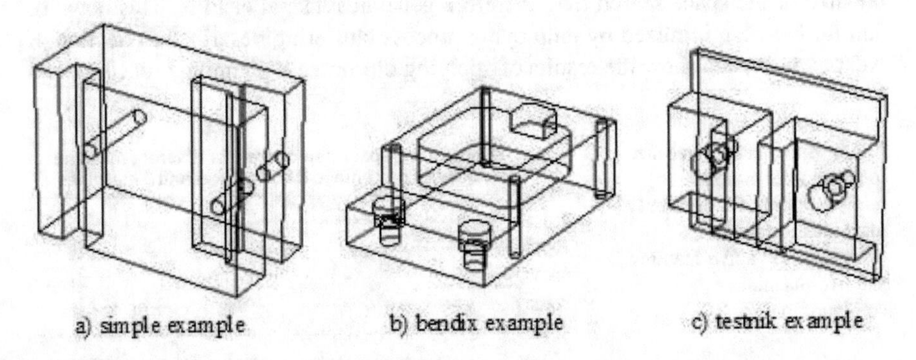


**Fig. 10** Example parts: (**a**) Simple, (**b**) Bendix, and (**c**) Testnik

**Table 9** Comparison of size of PPN with and without precedence

| Part | No precedence | | With precedence | |
|---|---|---|---|---|
| | Node | Arc | Node | Arc |
| Netex | 78,608 | 121,358 | 23,654 | 34,089 |
| USC | 636 | 1034 | 1167 | 2024 |

generating the process plan network in the prototype of the process planning system called 3I-PP (Intelligent Integrated Incremental Process Planner) which has been implemented in KnowledgeCraft©and later ported to LispWorks©. First, several examples for which the process plan network has been generated are provided. Next, the issue of computational efficiency of the implemented space search algorithm is discussed.

**Table 10** Number of features in example part

| Example | Netex | Bendix | Testnik |
|---|---|---|---|
| # of features | 9 | 17 | 13 |

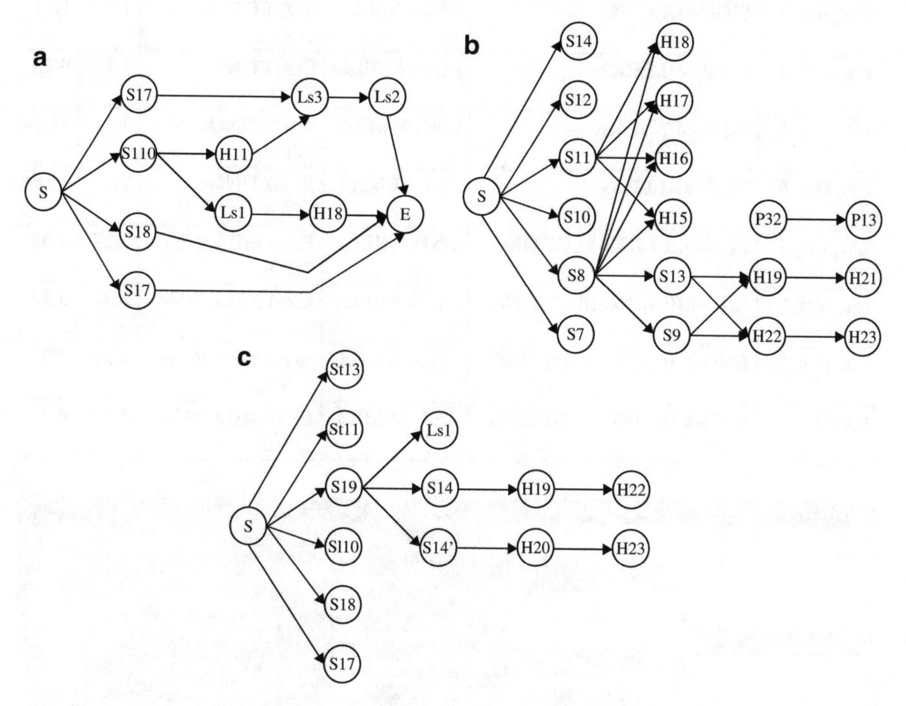


**Fig. 11** The feature precedence network for: (**a**) USC example, (**b**) Bendix example, and (**c**) Testniks example

We used three example parts shown in Figure 10 for generating their individual PPNs, which are also used as problem instances to test the process sequencing algorithm. For all examples, machining features were obtained by running the feature recognition system OOFF (Object-oriented Feature Finder) (Vandenbrande and Requicha 1993). It is interesting to note that one of the examples (shown in c) has two alternative representations. The central slot and the two steps are one representation, and the open pocket is another for the same portion of the removed volume. This example was executed in two separate runs of the system (one for slot-step representation and the other for pocket representation of this volume). We used the FPN with Step representation. The number of features for individual parts is given in Table 10.

Next, the FPN networks are formed for each part by evaluating the possible feature interactions, accessibility, and other geometric constraints. The FPNs for each part are given in Figure 11.

**Table 11** Sample of process candidates

| Features | Process | Machine | Tool | Time | Cost |
|---|---|---|---|---|---|
| Lin_slot_3 | SIDE-MILLING | UNIV-MILL | SLOTTING-TOOL | 7.7 | 10.1 |
| Lin_slot_4 | SIDE-MILLING | PLAIN-MILL | SLOTTING-TOOL | 7.7 | 9.33 |
| Lin_slot_5 | SIDE-MILLING | CNC-V-MILL | SLOTTING-TOOL | 7.7 | 11.64 |
| Lin_slot_6 | SIDE-MILLING | CNC-H-MILL | SLOTTING-TOOL | 7.7 | 11.64 |
| Lin_slot_7 | END-MILLING-SLOTTING | UNIV-MILL | END-MILLING-TOOL | 3.27 | 4.05 |
| Lin_slot_8 | END-MILLING-SLOTTING | VERT-MILL | END-MILLING-TOOL | 3.27 | 3.27 |
| Lin_slot_9 | END-MILLING-SLOTTING | CNC-V-MILL | END-MILLING-TOOL | 3.27 | 4.7 |
| Lin_slot_10 | END-MILLING-SLOTTING | CNC-H-MILL | END-MILLING-TOOL | 3.27 | 4.7 |

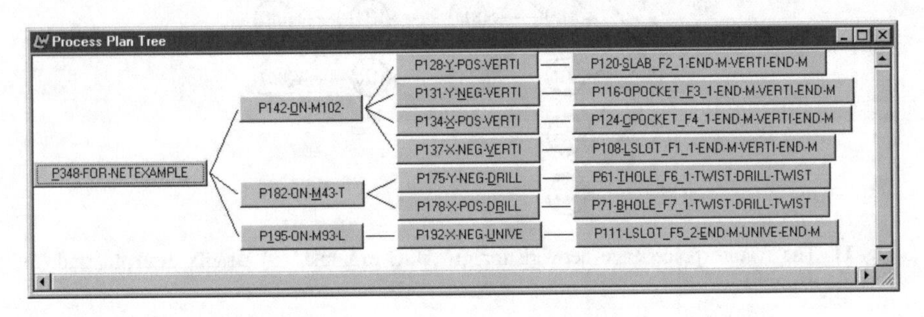


**Fig. 12** PPN for Netex example

For each FPN, the corresponding PPNs are generated with machine clustering already implemented. The number of process candidates is kept the same for every test instance and a small subset of all process candidates are given in Table 11. The optimal process plan for the example in Figure 11c that was generated by applying the space search algorithm is shown in Figure 12. This plan is shown in the form of a process plan tree that shows the process plan levels described in Section 4. For the same example, a process plan network is generated as described in Section 1. The resulting PPN network is the counterpart of the PPN shown in Figure 3. The PPN generated after applying clustering is shown in Figure 13.

The first step in the evaluation of computational complexity of the procedure is the estimation of the problem size for the given input conditions. In our process sequencing problem, size parameters are number of features, number of process candidates per feature (this includes number of processes and number of machines), and the feature precedence network. This number of nodes exists on the highest

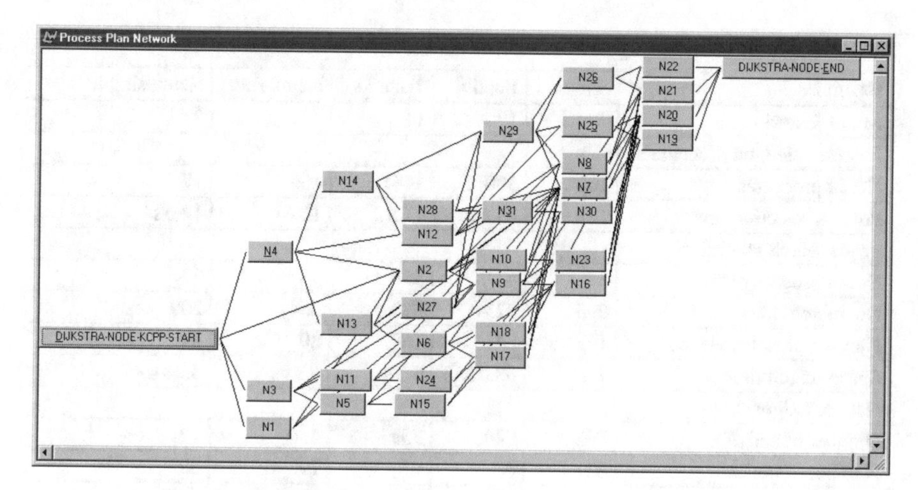


**Fig. 13** Clustered PPN for Netex example

(machine) layer and the number of arcs between them depends on the number of constraints in the FPN. This network explosion seems to lead to the conclusion that the creation of the process plan network will be computationally very expensive. However, the clustering procedure essentially reduces the number of nodes in the process plan network because of the fact that there is a significant overlap of processes for machining features (e.g., end milling may be used for slots, pockets, steps, etc.) and, more importantly, because there is also a significant overlap of machines for machining processes (e.g., a vertical mill is capable of performing end milling, face milling, drilling, and even side milling).

Another reason for a limited network growth is the existence of feature precedence constraints. Due to precedence constraints, the number of features considered at any stage is limited, and combinatorial generation of clusters may be performed only on a small number of features. In a bounding case, where there is a complete precedence constraint between features (i.e., the FPN produces a single sequence in which features have to be machined), clustering is not performed at all and only one feature is considered at each stage of network generation.

The conclusion of the above analysis is that the network growth depends significantly on the particular part and its data (features and process candidates). Consequently, the network procedure has been executed on several realistic examples in order to establish some meaningful relations between the problem size (number of features), the process plan network size (number of nodes), and computational time.

The procedures explained in Section 3 have been compared in the above examples. The results of the comparison (execution times on Intel Pentium III/866 processor) are shown in Table 12. The table shows problem sizes and execution times for both algorithms as applied to five examples. Several observations may be made from this table:

**Table 12** Metrics of process planning conducted on sample parts

| Example | Simple | Bendix | Testniks | Netex-full | Netex-simple |
|---|---|---|---|---|---|
| No. of features | 9 | 17 | 11 | 7 | 7 |
| Process selection procedure | | | | | |
| No. of processes | 102 | 300 | 144 | 100 | 17 |
| Process selection time | 12 s | 43.6 s | 17.5 s | 12.6 s | 12.0 s |
| Space search algorithm | | | | | |
| No. of process plans | 5 | 4 | 55 | 46 | 22 |
| No. of activities | 295 | 723 | 383 | 254 | 209 |
| Number of states visited | 113 | 341 | 168 | 80 | 87 |
| Space search time | 7.2 s | 68.0 s | 6.3 s | 6.3 s | 4.3 s |
| Network algorithm | | | | | |
| Number of activities | 923 | 944 | 759 | 430 | 194 |
| Number of nodes | 63 | 60 | 53 | 29 | 33 |
| Number of arcs | 358 | 361 | 351 | 144 | 85 |
| Network generation | 34.5 s | 56 s | 29.7 s | 10.8 s | 3.8 s |
| Network optimization | 1.8 s | 1.7 s | 1.6 s | 0.5 s | 0.5 s |

1. Problem size is proportional to the number of features, with the number of process candidates in the range of 10–15X number of features (case Net-simple is an exception because it is built with different goal).
2. The state-space search algorithm expands the network into a smaller number of activities due to simultaneous evaluation of current state and avoidance of unpromising paths.
3. Time of the space search algorithm depends also on the actual data in the network.

# 7 Past Researches

During early eighties, generative CAPP system was still an emerging technology, as pointed out by Eversheim and Schneewind in their review, published in 1984 [7]. Ham and Lu first recognized the importance of integrating design and manufacturing as well as techniques of AI in generative process planning research [8]. In a very significant survey published in the following year, Alting and Zhang supported the claims of Ham and Lu as well as pointed machine learning systems as promising candidate for integrated manufacturing [1].

Recent surveys classify various technologies and platforms, used in the last three decades of research, into ten categories: (a) feature based, (b) knowledge based, (c) neural networks, (d) genetic algorithm, (e) fuzzy set theory/logic, (f) Petri nets, (g) agent based, (h) Internet based, (i) STEP-compliant, and (j) emerging/others [29]; [30]. In this chapter, we represent our process sequencing problem assuming the part design as a collection of machining features. Therefore, the process sequencing algorithms presented in this chapter should be part of feature-based process planning. We will narrow our discussions on the past research in feature-based process planning. Still, it should be noted that many feature-based process planning systems heavily borrow techniques from other categories. In fact, past research in process selection phase of generative process planning either used expert systems which automatically select processes based on traditional process selection practices, captured in a knowledge base, or employed different machine learning techniques to train a process selection engine.

There are two primary ways features can be obtained from part design [22]. Feature-based part designs are extremely suitable for feature-based process planning because features can be directly extracted from the part design. For part designs with custom or nonstandard features, a feature recognition procedure needs to be employed to identify the implicit features from the part design. In the past, considerable amount of work has been devoted to transforming the design model represented in CAD format into a feature model for process planning [2]. In the feature model, various structural characteristics of a part design are mapped to one or many standard machining features such as holes, slots, pockets, chamfer, bevel, and many more (STEP AP256 provides an exhaustive list of features for subtractive manufacturing). It has to be noted that the CAD design should also contain the design specifications expressed in terms of tolerances for every feature. A number of studies also investigated the problem of translating the design features, generally having a predefined geometrical structure, into machining feature, which is more suitable for process selection [21].

Earlier known process selection engines are built as expert systems. One of the most frequently referenced rule-based generative process planning systems is GARI [6] which performs planning of the sequence of machining processes for mechanical parts. GARI uses as its input a symbolic representation of the part, features, and relations between features in the form of LISP lists. Most early generative process planning systems used this scheme (LISP list plus rules). HiMapp [3] uses a revised form of a planner called Deviser in its core while RTCAPP [18] is a relatively extended system in comparison to the previous work and includes a manufacturing knowledge base which consists of frame representation of knowledge that connects processes, machines, and tools. SIPS [16] are based on a hierarchical abstraction technique called hierarchical knowledge clustering, where the knowledge is organized in a taxonomic hierarchy of objects. AMPS (Automated Machining Planning System) [4] is a process planning expert system that supports a QTC (Quick Turnaround Cell), an automated manufacturing cell for prismatic parts of one-of-kind type. AMPS uses both frames (in KEE) for declarative knowledge representation and rules for procedural knowledge representation. The combination of object-oriented approach (for feature classification) and rule base (for process selection) was used in XCUT [9].

When the scheduling module receives several alternative process plans for each part, it may implement the scheduling algorithm under a more relaxed set of constraints, hence reaching the better solutions in terms of resource utilization or production cycle minimization. The availability of alternative process plans provides the scheduling module the flexibility to select different process plans for different time windows depending on the status of production resources.

There have been several research efforts devoted to interface between automated process planning systems and CAM (computer-aided manufacturing) and MRP (manufacturing resource planning) software. The latter two activities rely heavily on accurate and updated process plans to carry out manufacturing itself.

The issue of process sequencing and generation of alternative process plans has received early attention in process planning research. Prabhu et al. [20] proposed a method for generation of operations network for rotational parts based on feature precedence constraints. They proposed a tree representation of allowed feature sequences for turning operations. The approach provided directions for further research in the area, but did not address several important issues: processing on more than one machine, processing of prismatic parts (which have more complex precedence relations), and multiple representation of the same process in the tree. Lee et al. presented the precedence among machining operations selected for a part manufacturing in a tree structure and alternative process sequences in a network. Branch and fathoming styled search algorithm is applied to find the best machining sequence from the network of alternative process plans [12]. In a separate study, same authors also presented six different local search strategies, based on Simulated Annealing and Tabu search in order to increase the performance of search [13].

The integration of process plan network into process planning and shop floor control is reported in [5]. Authors describe two-step process planning (offline and online) and hierarchy of manufacturing tasks. The algorithm for conversion of feature graph into task graph is proposed for rotational components. Decomposition of task graph into process hierarchy is based on ISO/STEP representation of process planning.

One method of finding the number of possible paths in a given graph is by finding the Hamiltonian paths. Hamiltonian path in a graph is a path that passes through every vertex only once. Irani et al. [10] adopted Hamiltonian path method to find the number of paths. Authors have described two algorithms for generation of valid Hamiltonian paths from a feature graph. Prabhu et al. remarked that, though the Hamiltonian method enumerates all the possible paths, it is inefficient [20]. This is because it is possible that Hamiltonian path might not be a valid path when precedence constraints are considered. It is important here to note that these reports do not address issue of alternative processes for individual features, do not discuss process network as permanent plan representation, nor completely analyze the fact that more than one machine may be required for manufacturing. These important factors introduce another complexity into process sequencing and require more integrative approach as presented in this paper.

## References


1. L. Alting, H. Zhang, Computer aided process planning: the state-of-the-art survey. Int. J. Prod. Res. **27**(4), 553–585 (1989)
2. B. Babic, N. Nesic, Z. Miljkovic, A review of automated feature recognition with rule-based pattern recognition. Comput. Ind. **59**(4), 321–337 (2008)
3. H.R. Berenji, B. Khoshnevis, Use of artificial intelligence in automated process planning. Comput. Mech. Eng. **5**(2), 47–55 (1986)
4. T.-C. Chang, *Expert Process Planning for Manufacturing* (Addison-Wesley Pub. Co, Reading, MA, 1990)
5. H. Cho, A. Derebail, T. Hale, R.A. Wysk, A formal approach to integrating computer-aided process planning and shop floor control. J. Eng. Ind. **116**(1), 108 (1994)
6. Y. Descotte, J.-C. Latombe, Gari: an expert system for process planning, in *Solid Modeling by Computers* (Springer US, Boston, MA, 1984), pp. 329–346
7. W. Eversheim, J. Schneewind, Computer-aided process planning—State of the art and future development. Robot. Comput. Integr. Manuf. **10**(1–2), 65–70 (1993)
8. I. Ham, S.C.-Y. Lu, Computer-aided process planning: the present and the future. CIRP Ann. Manuf. Technol. **37**(2), 591–601 (1988)
9. K. Hummel, S.L. Brooks, XPS-E revisited: a new architecture and implementation approach for an automated process planning system. CAM-I Publication, DR-88-PP-02 (1988)
10. S.A. Irani, H.-Y. Koo, S. Raman, Feature-based operation sequence generation in CAPP. Int. J. Prod. Res. **33**(1), 17–39 (1995)
11. B. Khoshnevis, J.Y. Park, D. Sormaz, A cost based system for concurrent part and process design. Eng. Econ. **40**(1), 101–124 (1994). https://doi.org/10.1080/00137919408903140
12. D.-H. Lee, D. Kiritsis, P. Xirouchakis, Branch and fathoming algorithms for operation sequencing in process planning. Int. J. Prod. Res. **39**(16), 1649–1669 (2001)
13. D.-H. Lee, D. Kiritsis, P. Xirouchakis, Search heuristics for operation sequencing in process planning. Int. J. Prod. Res. **39**(16), 3771–3788 (2001)
14. Y.-N. Lien, E. Ma, B.W. Wah, Transformation of the generalized traveling salesman problem into the standard traveling salesman problem. Inf. Sci. **74**(1–2), 177–189 (1993)
15. G.F. Luger, *Artificial Intelligence: Structures and Strategies for Complex Problem Solving*, 6th edn. (Addison-Wesley, 2008)
16. D.S. Nau, M. Luce, Knowledge representation and reasoning techniques for process planning: extending SIPS to do tool selection (1987)
17. C.E. Noon, J.C. Bean, An efficient transformation of the generalized traveling salesman problem. Inf. Syst. Oper. Res. **31**(1), 39–44 (1993)
18. J.Y. Park, B. Khoshnevis, A real-time computer-aided process planning system as a support tool for economic product design. J. Manuf. Syst. **12**(2), 181–193 (1993)
19. J. Pearl, *Heuristics: Intelligent Search Strategies for Computer Problem Solving* (Addison-Wesley Longman Publishing Co, Boston, MA, 1984)
20. P. Prabhu, S. Elhence, H. Wang, An operations network generator for computer aided process planning. J. Manuf. Syst. **9**(4), 283–291 (1990)
21. M. Sadaiah, D.R. Yadav, P.V. Mohanram, P. Radhakrishnan, A generative computer-aided process planning system for prismatic components. Int. J. Adv. Manuf. Technol. **20**(10), 709–719 (2002)
22. J.J. Shah, Conceptual development of form features and feature modelers. Res. Eng. Des. **2**(2), 93–108 (1991)
23. D.N. Šormaz, B. Khoshnevis, Process sequencing and process clustering in process planning using state space search. J. Intell. Manuf. **7**(3), 189–200 (1996)
24. D.N. Šormaz, B. Khoshnevis, Process planning knowledge representation using an object-oriented data model. Int. J. Computer Integr. Manuf. **10**(1–4), 92–104 (1997)
25. D.N. Šormaz, B. Khoshnevis, Modeling of manufacturing feature interactions for automated process planning. J. Manuf. Syst. **19**(1), 28–45 (2000)

26. D.N. Sormaz, B. Khoshnevis, Generation of alternative process plans in integrated manufacturing systems. J. Intell. Manuf. **14**(6), 509–526 (2003)
27. D.N. Šormaz, S. Thiruppalli, Relationships between feature precedence complexity and number of alternative processing sequences, in *Proceedings of 10th Industrial Engineering Research Conference*, Dallas, TX (2001)
28. S. Thiruppalli, Incremental generation of alternative process plans for integrated manufacturing. PhD thesis, Ohio university (2002)
29. X. Xu, L. Wang, S.T. Newman, Computer-aided process planning - a critical review of recent developments and future trends. Int. J. Comput. Integr. Manuf. **24**(1), 1–31 (2011)
30. Y. Yusof Kamran Latif, Y.K. Yusof Latif, K Latif, Survey on computer-aided process planning. Int. J. Adv. Manuf. Technol. **75**, 77–89 (2014)

# Sharp Nordhaus–Gaddum-Type Lower Bounds for Proper Connection Numbers of Graphs

**Yuefang Sun**

## 1 Introduction

We refer to book [3] for graph theoretical notation and terminology not described here. An edge-colored graph is said to be *properly colored* if no two adjacent edges share a color. An edge-colored connected graph $G$ is called *properly connected* if between every pair of distinct vertices, there exists a path that is properly colored. The *proper connection number* of a connected graph $G$, denoted by $pc(G)$, is the minimum number of colors needed to color the edges of $G$ to make it properly connected [4]. Clearly, $pc(G) = 1$ if and only if $G = K_n$ and $pc(G) = n - 1$ if and only if $G = K_{1,n-1}$.

The concept of proper connection is not only a natural combinatorial measure but also has applications in communication network [15]: When building a communication network between wireless signal towers, one fundamental requirement is that the network should be connected. If there cannot be a direct connection between two towers $A$ and $B$, for example, if there is a mountain in between, there must be a route through other towers to get from $A$ to $B$. As a wireless transmission passes through a signal tower, to avoid interference, it would help if the incoming signal and the outgoing signal do not share the same frequency. Suppose we assign a vertex to each signal tower, an edge between two vertices if the corresponding signal towers are directly connected by a signal, and assign a color to each edge based on the assigned frequency used for the communication. Then, the number of frequencies needed to assign frequencies to the connections between towers so that there is always a path avoiding interference between each pair of towers is precisely the proper connection number of the corresponding graph.

---

Y. Sun (✉)
Department of Mathematics, Shaoxing University, Zhejiang, People's Republic of China
e-mail: yuefangsun2013@163.com





B. Goldengorin (ed.), *Optimization Problems in Graph Theory*,
Springer Optimization and Its Applications 139,
https://doi.org/10.1007/978-3-319-94830-0_13

Aside from the above application, properly colored paths and cycles appear in a variety of other fields including social sciences [7] and genetics [8–10]. The readers can see a survey [1] dealing with the case where two colors are used on the edges. Recently, there has also been another survey of the area in Chapter 16 of [2].

The definition and study of the proper connection number was inspired by the concept of rainbow connection number which was introduced by Chartrand et al. [5]. A path is called *rainbow* if no two edges in the path share a color. The *rainbow connection number* of a graph $G$, denoted by $rc(G)$, is the minimum number of colors needed such that there is a rainbow path between each pair of vertices in $G$. By replacing "rainbow" with "proper," it is easy to see where the definition of the proper connection number originated. There are more and more researchers investigating the topic of rainbow coloring, the readers can see [17] for a survey and [16] for a monograph on it.

The vertex-coloring version of the proper connection number has been defined and studied independently in [6] and [14]. A vertex-colored graph $G$ is called *proper vertex-connected* if every pair of vertices is connected by a path which has no two consecutive internal vertices of the same color. The *proper vertex-connection number* of $G$, denoted by $pvc(G)$, to be the smallest number of colors needed to make $G$ proper vertex-connected.

For the total-coloring version of the proper connection number, the authors in [6] defined the concept of total proper connection number. Here, we introduce another concept. A total-colored graph $G$ is called *proper total-connected* if for every pair of vertices, there is a path connecting them such that no two adjacent edges share a color and no two consecutive internal vertices share a color. Define the *proper total-connection number* of $G$, denoted by $ptc(G)$, to be the smallest number of colors needed to make $G$ proper total-connected. By definition, we clearly have $ptc(G) \geq \max\{pc(G), pvc(G)\}$.

A Nordhaus–Gaddum-type result is an upper or lower bound on the product or sum of the values of a parameter for a graph $G$ and its complement $\overline{G}$. Nordhaus and Gaddum [18] first established this type of result for the chromatic number of a graph and many analogous results of other graph parameters are obtained since then, such as [11, 12, 19]. In this paper, we will study the Nordhaus–Gaddum-type bounds for the proper connection number $pc(G)$, proper vertex-connection number $pvc(G)$, and proper total-connection number $ptc(G)$ of a graph and get sharp lower bounds for $pc(G) + pc(\overline{G})$ (Theorem 1), $pc(G)pc(\overline{G})$ (Theorem 2), $pvc(G) + pvc(\overline{G})$ (Theorem 4), $pvc(G)pvc(\overline{G})$ (Theorem 5), $ptc(G) + ptc(\overline{G})$ (Theorem 6), and $ptc(G)ptc(\overline{G})$ (Theorem 7), where $G$ is a connected graph of order at least 8.

## 2 Main Results

For a set $S$, we use $|S|$ to denote the size of $S$. Let $c$ be an edge-coloring of $G$, we use $c(e)$ to denote the color of an edge $e$. For a subgraph $H$ of $G$, let $c(H)$ be the set of colors of the edges in $H$.

In [13], Huang, Li, and Wang proved that if $G$ and $\overline{G}$ are both connected, then $4 \leq pc(G) + pc(\overline{G}) \leq n$, and the only graph attaining the upper bound is the tree with maximum degree $\Delta = n - 2$. However, for the lower bound, they did not show that whether it is sharp. Here, we confirm it by giving the following result.

**Theorem 1** *For a connected graph $G$ of order $n \geq 8$ with a connected complement, we have*

$$pc(G) + pc(\overline{G}) \geq 4.$$

*Moreover, the bound is sharp.*

*Proof* We just consider the sharpness of this bound, it suffices to find a connected graph $G$ on $n \geq 8$ vertices such that $pc(G) = pc(\overline{G}) = 2$. We will distinguish four cases.

*Case 1.* We first consider the case that $n = 4k$, where $k \geq 2$.

For the subcase that $k = 2$, let $G$ be a graph with vertex set $\{x\} \cup U \cup V$ and edge set $\{xu_i \mid 1 \leq i \leq 3\} \cup \{u_iv_i \mid 1 \leq i \leq 3\} \cup \{u_1v_2, u_2v_3, u_3v_4\} \cup \{v_iv_j \mid 1 \leq i, j \leq 4\}$, where $U = \{u_i \mid 1 \leq i \leq 3\}$ and $V = \{v_j \mid 1 \leq j \leq 4\}$. We provide an edge-coloring $c$ of $G$ by letting $c(u_1v_1) = c(u_2v_2) = c(u_2v_3) = c(xu_3) = c(u_3v_3) = 1$, $c(e) = 2$ for each other edge. And, we provide an edge-coloring $c^*$ of $\overline{G}$ by letting $c^*(u_1v_3) = c^*(u_2v_4) = c^*(u_2v_1) = c^*(u_3v_2) = c^*(xv_3) = c^*(xv_4) = 1$, $c^*(e) = 2$ for each other edge $e$. Clearly, both $G$ and $\overline{G}$ are properly connected under above colorings and then $pc(G) = pc(\overline{G}) = 2$ in this case.

We now consider the subcase that $k \geq 3$. Let $U = \{u_i \mid 1 \leq i \leq 2k - 1\}$ and $V = \{v_j \mid 1 \leq j \leq 2k\}$. Let $G$ be a graph with vertex set $\{x\} \cup U \cup V$ such that $N(x) = U$, $U$ is an independent set, $G[V]$ is a clique, and for each vertex $u_i$, $u_i$ is adjacent to $v_i, v_{i+1}, \ldots, v_{i+k-1}$, where the subscripts are taken modulo $2k$.

We provide an edge-coloring $c$ of $G$ by letting

$$c(e) = \begin{cases} 1, if\ e = xu_i\ for\ k+1 \leq i \leq 2k-1, \\ 1, if\ e = u_iv_i\ for\ 1 \leq i \leq 2k-1, \\ 1, if\ e = u_kv_{k+1}, \\ 2, otherwise. \end{cases}$$

We then provide an edge-coloring $c^*$ of $\overline{G}$ by letting

$$c^*(e) = \begin{cases} 1, if\ e = v_ju_{j+1}\ for\ 1 \leq j \leq 2k-2, \\ 1, if\ e = u_iu_j\ for\ 1 \leq i, j \leq 2k-1, \\ 1, if\ e = xv_j\ for\ k-1 \leq j \leq 2k-2, \\ 1, if\ e \in \{u_1v_{2k}, u_kv_{k-2}\}, \\ 2, otherwise. \end{cases}$$

Clearly, both $G$ and $\overline{G}$ are properly connected under above colorings and so $pc(G) = pc(\overline{G}) = 2$ in this case.

*Case 2.* We next consider the case that $n = 4k + 1$, where $k \geq 2$. Let $U = \{u_i \mid 1 \leq i \leq 2k\}$ and $V = \{v_j \mid 1 \leq j \leq 2k\}$. Let $G$ be a graph with vertex set $\{x\} \cup U \cup V$ such that $N(x) = U$, $U$ is an independent set, $G[V]$ is a clique, and for each vertex $u_i$, $u_i$ is adjacent to $v_i, v_{i+1}, \ldots, v_{i+k-1}$, where the subscripts are taken modulo $2k$.

We provide an edge-coloring $c$ of $G$ by letting

$$c(e) = \begin{cases} 1, if\ e = xu_i\ for\ k+1 \leq i \leq 2k, \\ 1, if\ e = u_i v_i\ for\ 1 \leq i \leq 2k, \\ 1, if\ e = u_k v_{k+1}, \\ 2, otherwise. \end{cases}$$

It is not hard to show that $G \cong \overline{G}$, furthermore, both $G$ and $\overline{G}$ are properly connected under above colorings. Thus, $pc(G) = pc(\overline{G}) = 2$ in this case.

*Case 3.* We now consider the case that $n = 4k + 2$, where $k \geq 2$.

For the subcase that $k = 2$, let $G$ be a graph with vertex set $\{x\} \cup U \cup V$ and edge set $\{xu_i \mid 1 \leq i \leq 4\} \cup \{u_i v_i \mid 1 \leq i \leq 4\} \cup \{u_1v_2, u_2v_3, u_3v_4, u_4v_5\} \cup \{v_i v_j \mid 1 \leq i, j \leq 5\}$, where $U = \{u_i \mid 1 \leq i \leq 4\}$ and $V = \{v_j \mid 1 \leq j \leq 5\}$. We provide an edge-coloring $c$ of $G$ by letting $c(u_1v_1) = c(u_2v_2) = c(u_2v_3) = c(u_3v_3) = c(u_4v_4) = c(xu_3) = c(xu_4) = 1$, $c(e) = 2$ for each other edge. We then provide an edge-coloring $c^*$ of $\overline{G}$ by letting $c^*(u_1v_4) = c^*(u_2v_5) = c^*(u_2v_1) = c^*(u_3v_2) = c^*(u_4v_3) = c^*(xv_5) = c^*(xv_4) = 1$, $c^*(e) = 2$ for each other edge $e$. Clearly, both $G$ and $\overline{G}$ are properly connected under above colorings and so $pc(G) = pc(\overline{G}) = 2$ in this case.

We now consider the subcase that $k \geq 3$. Let $U = \{u_i \mid 1 \leq i \leq 2k\}$ and $V = \{v_j \mid 1 \leq j \leq 2k+1\}$. Let $G$ be a graph with vertex set $\{x\} \cup U \cup V$ such that $N(x) = U$, $U$ is an independent set, $G[V]$ is a clique, and for each vertex $u_i$, $u_i$ is adjacent to $v_i, v_{i+1}, \ldots, v_{i+k-1}$, where the subscripts are taken modulo $2k + 1$.

We provide an edge-coloring $c$ of $G$ by letting

$$c(e) = \begin{cases} 1, if\ e = xu_i\ for\ k+1 \leq i \leq 2k, \\ 1, if\ e = u_i v_i\ for\ 1 \leq i \leq 2k, \\ 1, if\ e = u_k v_{k+1}, \\ 2, otherwise. \end{cases}$$

We then provide an edge-coloring $c^*$ of $\overline{G}$ by letting

$$c^*(e) = \begin{cases} 1, if\ e = v_j u_{j+1}\ for\ 1 \leq j \leq 2k-1, \\ 1, if\ e = u_i u_j\ for\ 1 \leq i, j \leq 2k, \\ 1, if\ e = xv_j\ for\ k-1 \leq j \leq 2k-1, \\ 1, if\ e \in \{u_k v_{k-2}, u_1 v_{2k+1}\}, \\ 2, otherwise. \end{cases}$$

It is not hard to show that both $G$ and $\overline{G}$ are properly connected under above colorings and then $pc(G) = pc(\overline{G}) = 2$ in this case.

*Case 4.* We finally consider the case that $n = 4k+3$, where $k \geq 2$. Let $U = \{u_i \mid 1 \leq i \leq 2k+1\}$ and $V = \{v_j \mid 1 \leq j \leq 2k+1\}$. Let $G$ be a graph with vertex set $\{x\} \cup U \cup V$ such that $N(x) = U$, $u_{k+1}v_{2k+1} \in E(G)$, $U$ is an independent set, $G[V]$ is a clique, and for each vertex $u_i$, $u_i$ is adjacent to $v_i, v_{i+1}, \ldots, v_{i+k-1}$ where the subscripts are taken modulo $2k+1$.

We provide an edge-coloring $c$ of $G$ by letting

$$c(e) = \begin{cases} 1, \; if \; e = xu_i \; for \; k+1 \leq i \leq 2k+1, \\ 1, \; if \; e = u_iv_i \; for \; 1 \leq i \leq 2k+1, \\ 1, \; if \; e = u_kv_{k+1}, \\ 2, \; otherwise. \end{cases}$$

Then, we provide an edge-coloring $c^*$ of $\overline{G}$ by letting

$$c^*(e) = \begin{cases} 1, \; if \; e = v_ju_{j+1} \; for \; 1 \leq j \leq 2k, \\ 1, \; if \; e = xv_j \; for \; k \leq j \leq 2k, \\ 1, \; if \; e \in \{u_1v_{2k+1}, u_{k+1}v_{k-1}\}, \\ 2, \; otherwise. \end{cases}$$

Clearly, both $G$ and $\overline{G}$ are properly connected under above colorings and then $pc(G) = pc(\overline{G}) = 2$ in this case. □

We know that if both $G$ and $\overline{G}$ are connected, then both of them are noncomplete, so $pc(G)pc(\overline{G}) \geq 4$ by the fact that $pc(G) = 1$ if and only if $G = K_n$. Then, by using the same graphs in Theorem 1, we can get the following result.

**Theorem 2** *For a connected graph $G$ of order $n \geq 8$ with a connected complement, we have*

$$pc(G)pc(\overline{G}) \geq 4.$$

*Moreover, the bound is sharp.*

There are some basic observations about $pvc(G)$ including the following result.

**Proposition 3 ([14])** *If $G$ is a nontrivial connected graph, then*

*(i)* $pvc(G) = 0$ *if and only if $G$ is a complete graph;*
*(ii)* $pvc(G) = 1$ *if and only if* $diam(G) = 2$.

By Proposition 3, we know that $pvc(G) \geq 1$ for any connected noncomplete graph, and furthermore, $pvc(G) + pvc(\overline{G}) \geq 2$ and $pvc(G)pvc(\overline{G}) \geq 1$ if both $G$ and $\overline{G}$ are connected. Especially, we can get the following result.

**Theorem 4** *For a connected graph G of order $n \geq 8$ with a connected complement, we have*

$$pvc(G) + pvc(\overline{G}) \geq 2.$$

*Moreover, the bound is sharp.*

*Proof* According to the above argument, we only need to prove that for $n \geq 8$, there are graphs $G$ and $\overline{G}$ with $n$ vertices such that $diam(G) = diam(\overline{G}) = 2$. We just use the graphs in Theorem 1, it is not hard to check that in each case, the graph $G$ we construct in Theorem 1 satisfies our requirement. □

With a similar argument to that of Theorem 4, the following result also holds.

**Theorem 5** *For a connected graph G of order $n \geq 8$ with a connected complement, we have*

$$pvc(G)pvc(\overline{G}) \geq 1.$$

*Moreover, the bound is sharp.*

We now study the Nordhaus–Gaddum-type lower bounds for the proper total-connected number and get the following result.

**Theorem 6** *For a connected graph G of order $n \geq 8$ with a connected complement, we have*

$$ptc(G) + ptc(\overline{G}) \geq 4.$$

*Moreover, the bound is sharp.*

*Proof* Recall the fact that $ptc(G) \geq \max\{pc(G), pvc(G)\}$ and by Theorems 1 and 4, we have that $ptc(G) + ptc(\overline{G}) \geq 4$.

For the sharpness of this bound, we use the graphs $G$ in Theorem 1 and provide an edge-coloring of $G$ the same to that of Theorem 1. This procedure costs two distinct colors, then we assign any of these two colors to each vertex of $G$. By the argument of Theorem 1, we know that for every pair of vertices in $G$, there is a path such that no two adjacent edges share a color, furthermore, the length of such a path is exactly two, then $G$ is properly total-connected, and so $ptc(G) = 2$. With a similar argument, we can show that $ptc(\overline{G}) = 2$. □

With a similar argument to that of Theorem 6, the following result holds.

**Theorem 7** *For a connected graph G of order $n \geq 8$ with a connected complement, we have*

$$ptc(G)ptc(\overline{G}) \geq 4.$$

*Moreover, the bound is sharp.*

**Acknowledgements** This work was supported by National Natural Science Foundation of China (No. 11401389) and China Scholarship Council (No. 201608330111). The author is very grateful to the referee for helpful comments and suggestions.

## References

1. J. Bang-Jensen, G. Gutin, Alternating cycles and paths in edge-coloured multigraphs: a survey. Discret. Math. **165/166**, 39–60 (1997)
2. J. Bang-Jensen, G. Gutin, *Digraphs Theory, Algorithms and Applications*, 2nd edn. Springer Monographs in Mathematics (Springer, London, 2009)
3. J.A. Bondy, U.S.R. Murty, *Graph Theory*. Graduate Texts in Mathematics, vol. 244 (Springer, Berlin, 2008)
4. V. Borozan, S. Fujita, A. Gerek, C. Magnant, Y. Manoussakis, L. Montero, Z. Tuza, Proper connection of graphs. Discret. Math. **312**(17), 2550–2560 (2012)
5. G. Chartrand, G.L. Johns, K.A. McKeon, P. Zhang, Rainbow connection in graphs. Math. Bohem. **133**(1), 85–98 (2008)
6. E. Chizmar, C. Magnant, P. Salehi Nowbandegani, Note on vertex and total proper connection numbers. AKCE Int. J. Graphs Comb. **13**(2), 103–106 (2016)
7. W.S. Chou, Y. Manoussakis, O. Megalakaki, M. Spyratos, Zs. Tuza, Paths through fixed vertices in edge-colored graphs. Math. Inform. Sci. Hum. **127**, 49–58 (1994)
8. D. Dorninger, On permutations of chromosomes, in *Contributions to General Algebra*, vol. 5 (Verlag Hölder-Pichler-Tempsky/Teubner, Wien/Stuttgart, 1987), pp. 95–103
9. D. Dorninger, Hamiltonian circuits determining the order of chromosomes. Discret. Appl. Math. **50**(2), 159–168 (1994)
10. D. Dorninger, W. Timischl, Geometrical constraints on Bennet's predictions of chromosome order. Heredity **58**, 321–325 (1987)
11. F. Harary, T.W. Haynes, Nordhaus-Gaddum inequalities for domination in graphs. Discret. Math. **155**, 99–100 (1996)
12. F. Harary, R.W. Robinson, The diameter of a graph and its complement. Am. Math. Mon. **92**, 211–212 (1985)
13. F. Huang, X. Li, S. Wang, Proper connection number of complementary graphs. Arxiv:1504.02414
14. H. Jiang, X. Li, Y. Zhang, Y. Zhao, On (strong) proper vertex-connection of graphs. Arxiv:1505.04986
15. X. Li, C. Magnant, Properly colored notions of connectivity – a dynamic survey. Theory Appl. Graphs **0**(1), Art. 2, 1–16 (2015)
16. X. Li, Y. Sun, *Rainbow Connections of Graphs*. Springer Briefs in Mathematics (Springer, New York, 2012)
17. X. Li, Y. Shi, Y. Sun, Rainbow connections of graphs: a survey. Graphs Combin. **29**, 1–38 (2013)
18. E.A. Nordhaus, J.W. Gaddum, On complementary graphs. Am. Math. Mon. **63**, 175–177 (1956)
19. Y. Sun, On rainbow total-coloring of a graph. Discret. Appl. Math. **194**, 171–177 (2015)

Printed in the United States
By Bookmasters